Linear Algebra and Group Theory

V. I. SMIRNOV

Translated, Revised, and Edited by
RICHARD A. SILVERMAN

DOVER PUBLICATIONS, INC.
Mineola, New York

Copyright

Bibliographical Note

This Dover edition, first published in 2011, is an unabridged republication of the work originally published in 1961 by the McGraw-Hill Book Company, Inc., New York.

Library of Congress Cataloging-in-Publication Data

Smirnov, V. I. (Vladimir Ivanovich), 1887–1974.
[Kurs vysshei matematiki. English. Selections]
Linear algebra and group theory / V. I. Smirnov.
p. cm.
Originally published: New York : McGraw-Hill, 1961
Summary: "Derived from an encyclopedic six-volume survey, this accessible text by a prominent Soviet mathematician offers a concrete approach, with an emphasis on applications. Containing material not otherwise available to English-language readers, the three-part treatment covers determinants and systems of equations, matrix theory, and group theory. Problems sets, with hints and answers, conclude each chapter. 1961 edition"—Provided by publisher.
Includes bibliographical references and index.
ISBN-13: 978-0-486-48222-4 (pbk.)
ISBN-10: 0-486-48222-7 (pbk.)
1. Group theory. 2. Algebras, Linear. 3. Matrices. I. Title.

QA174.2.S64 2011
512'.5—dc22

2011006161

Manufactured in the United States by Courier Corporation
48222701
www.doverpublications.com

Preface

This book represents a selection of material from Prof. Vladimir I. Smirnov's encyclopedic six-volume "Course of Higher Mathematics." The present volume, entitled "Linear Algebra and Group Theory," has several unusual features that especially recommend it to the attention of the English-language readership. Unlike many algebra texts, this book does not delve into mathematical byways remote from the applications; instead, the approach adopted is consistently "concrete." Moreover, the book strives for the maximum coverage compatible with its length. Thus, in addition to a detailed treatment of linear algebra, it also contains an excellent introduction to group theory and an extensive discussion of group representations, a topic usually reserved for the specialized treatise. Under the heading of material rarely encountered in first courses on higher algebra, one should also mention Chap. 5 on infinite-dimensional spaces and Chap. 9 on continuous groups. It is apparent that the author's intention was to write an algebra text emphasizing those topics of greatest importance in applied mathematics and theoretical physics. Despite this, there is nothing in the volume that the pure mathematician can afford to ignore.

Because of the great difference between stylistic norms in English and Russian, as well as the absence of grammatical categories in one language that are present in the other, I have felt obliged to apply appropriate "smoothing operations" to ensure the continuity and readability of the translation. In doing so, I have not hesitated to add transitional sentences where I thought they were called for, make theorems out of some propositions not originally labeled as such, or clarify points that I found obscure. Nor have I hesitated to introduce an additional chapter heading not present in the original, renumber the equations in a way that appeared to me more convenient, redraw two of the figures to improve their perspective, and generally make the book conform to what I regard as the needs of its prospective audience. But the two most substantive changes were the following:

1. Chapter 6 of the present volume was originally the Appendix of Russian Vol. 3, Part II, while the Appendix of the present volume was originally Secs. 63 and 93 of Russian Vol. 3, Part II. In making these changes, it was necessary to supply the proof of the Hamilton-Cayley theorem (in the text) and explore the properties of the exponential of a matrix (in the problems).

2. There are no problems in the original, perhaps due to the Russian predilection for the use of special problem collections. It was thought that the addition of copious problems would greatly enhance the value of the English-language edition, as well as permit the incorporation of some important topics not discussed in the text. With this in mind, I have asked Prof. Allen L. Shields of the University of Michigan, Prof. John S. Lomont of the Polytechnic Institute of Brooklyn, and Prof. Jacob T. Schwartz of New York University to assist in preparing and selecting problems for this volume. Their contributions are identified more explicitly in the appropriate places. In addition, we have culled many problems from I. V. Proskuryakov's "A Collection of Problems on Linear Algebra," Moscow, 1957, and from D. K. Faddeyev and I. S. Sominski's "A Collection of Problems on Higher Algebra," Moscow, 1954. Occasional use was also made of problem collections by N. M. Gyunter and R. O. Kuzmin and by V. A. Krechmar. The net result has been to equip this book with over four hundred pertinent problems. Answers to about half of the computational problems and hints for the solution of the less obvious problems involving proofs are given at the end of the book. I have also listed several books for collateral or supplementary reading. There has been no attempt to make this list complete; in fact, it has been confined to books in English.

Finally, two observations should be made: (1) In a preface to the Russian original, Prof. Smirnov thanks D. K. Faddeyev for helping him with the group theory part of the book, in particular for writing Secs. 76, 87, 93, 94, 95 and 96; (2) To make the book self-contained, I have suppressed some references to other volumes of the six-volume course, and I have occasionally replaced others by the phrase "in an earlier volume" or "in a later volume." In every case, the reference was either an allusion to things to come or to elementary material to be found in any good text on advanced calculus, knowledge of which could be presupposed on the part of any reader of this book.

Richard A. Silverman
Translator and Editor

Contents

PART III: GROUP THEORY

PART I

Determinants and Systems of Equations

Chapter 1

Determinants and Their Properties

1. The Concept of a Determinant. We begin this section with a simple algebraic problem, i.e., solving a system of linear equations. Investigation of this problem will lead us to the important concept of a determinant.

We begin by considering the simplest special cases. First we take a system of two equations in two unknowns:

$$\begin{aligned} a_{11}x_1 + a_{12}x_2 &= b_1, \\ a_{21}x_1 + a_{22}x_2 &= b_2. \end{aligned}$$

Here, the coefficients a_{ik} of the unknowns are provided with two indices; the first index shows the equation in which the coefficient occurs, and the second index shows the unknown with which the coefficient is associated. As is well known, the solution of this system has the form

$$x_1 = \frac{b_1a_{22} - a_{12}b_2}{a_{11}a_{22} - a_{12}a_{21}}, \qquad x_2 = \frac{a_{11}b_2 - b_1a_{21}}{a_{11}a_{22} - a_{12}a_{21}}.$$

Next we take three equations in three unknowns:

$$\begin{aligned} a_{11}x_1 + a_{12}x_2 + a_{13}x_3 &= b_1, \\ a_{21}x_1 + a_{22}x_2 + a_{23}x_3 &= b_2, \\ a_{31}x_1 + a_{32}x_2 + a_{33}x_3 &= b_3, \end{aligned}$$

where we use the previous notation for the coefficients. We rewrite the first two equations in the form

$$\begin{aligned} a_{11}x_1 + a_{12}x_2 &= b_1 - a_{13}x_3, \\ a_{21}x_1 + a_{22}x_2 &= b_2 - a_{23}x_3. \end{aligned}$$

Solving these equations with respect to the unknowns x_1 and x_2 by using the previous formulas, we have

$$\begin{aligned} x_1 &= \frac{(b_1 - a_{13}x_3)a_{22} - a_{12}(b_2 - a_{23}x_3)}{a_{11}a_{22} - a_{12}a_{21}}, \\ x_2 &= \frac{a_{11}(b_2 - a_{23}x_3) - (b_1 - a_{13}x_3)a_{21}}{a_{11}a_{22} - a_{12}a_{21}}. \end{aligned}$$

Substituting these expressions into the last equation of the system, we obtain an equation determining the unknown x_3; when this equation is finally solved, we find

$$x_3 = \frac{a_{11}a_{22}b_3 + a_{12}b_2a_{31} + b_1a_{21}a_{32} - a_{11}b_2a_{32} - a_{12}a_{21}b_3 - b_1a_{22}a_{31}}{a_{11}a_{22}a_{33} + a_{12}a_{23}a_{31} + a_{13}a_{21}a_{32} - a_{11}a_{23}a_{32} - a_{12}a_{21}a_{33} - a_{13}a_{22}a_{31}}. \tag{1}$$

We now examine in detail the construction of the formula (1). First of all, we note that its numerator can be obtained from its denominator by substituting the constant terms b_i for the coefficients a_{i3} of the unknown x_3. Thus, the problem is to explain the law of formation of the denominator, which does not contain any terms b_i, but rather is made up exclusively of the coefficients of the system. We write these coefficients in the form of a square array

$$\begin{Vmatrix} a_{11} & a_{12} & a_{13} \\ a_{21} & a_{22} & a_{23} \\ a_{31} & a_{32} & a_{33} \end{Vmatrix}, \tag{2}$$

preserving the order in which they appear in the system itself. This array has three rows and three columns, and the numbers a_{ik} are called its *elements*. The first index gives the row in which the element appears, and the second index gives the column. Writing out the denominator

$$a_{11}a_{22}a_{33} + a_{12}a_{23}a_{31} + a_{13}a_{21}a_{32} - a_{11}a_{23}a_{32} - a_{12}a_{21}a_{33} - a_{13}a_{22}a_{31} \tag{3}$$

of the ratio (1), we see that it consists of six terms. Each term is a product of three elements from the array (2), and in fact each term contains an element from every row and every column. The general form of these products is

$$a_{1p}a_{2q}a_{3r}, \tag{4}$$

where p, q, r are the integers 1, 2, 3 arranged in some definite order, i.e., the first and second indices are both integers (from 1 to 3) and each product (4) contains an element from every row and every column. To obtain all the terms of the expression (3), the second indices p, q, r in the product (4) have to be taken in all possible orders. Clearly, there are six possible *permutations*† of the second indices, namely,

$$1, 2, 3; \quad 2, 3, 1; \quad 3, 1, 2; \quad 1, 3, 2; \quad 2, 1, 3; \quad 3, 2, 1; \tag{5}$$

this gives us all six terms of (3). However, we see that some of the products (4) appear in (3) with a plus sign, while others appear with a minus sign. Thus, it remains only to explain the rule by which the sign is to be chosen. As we see, the products (4) whose second indices are the permutations

$$1, 2, 3; \quad 2, 3, 1; \quad 3, 1, 2 \tag{5a}$$

† See Sec. 2.

appear with a *plus* sign, while the products whose second indices are the permutations

$$1, 3, 2; \qquad 2, 1, 3; \qquad 3, 2, 1 \tag{5b}$$

appear with a *minus* sign.

We now explain how the permutations (5*a*) differ from the permutations (5*b*). We shall call the fact that a larger number comes before a smaller number in a permutation an *inversion*, and we shall calculate the number of inversions in the different permutations (5*a*). In the first of these permutations, there are no inversions at all, i.e., the number of inversions equals zero. Next, we consider the second permutation and compare the size of each number appearing in it with the numbers that follow. We see that there are two inversions here, since the numbers 2 and 3 come before the number 1. Similarly, it can easily be seen that the third of the permutations (5*a*) also contains two inversions. We can summarize this situation by saying that each of the permutations (5*a*) contains an *even* number of inversions. We are now in a position to formulate the following sign rule for the expression (3): The products (4) for which the number of inversions in the permutations formed by the second indices is an *even* number appear in (3) without any change. However, the products (4) for which the permutations formed by the second indices contain an *odd* number of inversions appear in (3) with a *minus* sign. The expression (3) is called the *determinant* (of order 3) corresponding to the array (2). We can now easily generalize these considerations and define a determinant of any order.

Suppose that we are given n^2 numbers arranged in the form of a square array

$$\begin{Vmatrix} a_{11} & a_{12} & \cdots & a_{1n} \\ a_{21} & a_{22} & \cdots & a_{2n} \\ \cdots & \cdots & \cdots & \cdots \\ a_{n1} & a_{n2} & \cdots & a_{nn} \end{Vmatrix} \tag{6}$$

with n rows and n columns. The elements a_{ik} of this array are certain complex numbers, and the indices i and k indicate the row and column in which the number a_{ik} appears. We now take the elements of the array (6) and form all possible products containing one (and only one) element from each row and column. These products have the form

$$a_{1p_1}a_{2p_2} \cdots a_{np_n}, \tag{7}$$

where $p_1, p_2, \ldots, p_n$ is some arrangement of the numbers $1, 2, \ldots, n$. To obtain all possible products of the form (7), we must take all possible permutations of the second indices. As is well known from elementary algebra (see also Sec. 2), the total number of such permutations equals n

factorial, i.e., $1 \cdot 2 \cdots n = n!$ Each permutation has a certain number of inversions as compared with the basic arrangement $1, 2, \ldots, n$. We now ascribe signs to the products (7) by the following rule: We put a plus sign before products whose second indices form a permutation with an even number of inversions, and a minus sign before products whose second indices form a permutation with an odd number of inversions. The sum of all these products, with the appropriate signs, is called the *determinant of order n* corresponding with the array (6). Clearly, this sum contains $n!$ terms.

It is not hard to give this definition in terms of a formula. Let $p_1, p_2, \ldots, p_n$ be a permutation of the numbers $1, 2, \ldots, n$, and denote the number of inversions in this permutation by the symbol

$$[p_1, p_2, \ldots, p_n].$$

Then, the definition just given can be written as the formula

$$\begin{vmatrix} a_{11} & a_{12} & \cdots & a_{1n} \\ a_{21} & a_{22} & \cdots & a_{2n} \\ \cdots & \cdots & \cdots & \cdots \\ a_{n1} & a_{n2} & \cdots & a_{nn} \end{vmatrix} = \sum_{(p_1, p_2, \ldots, p_n)} (-1)^{[p_1, p_2, \ldots, p_n]} a_{1p_1} a_{2p_2} \cdots a_{np_n}, \tag{8}$$

where the summation extends over all possible permutations $p_1, p_2, \ldots, p_n$ of the second indices, and the determinant of an array is indicated by writing the array between vertical lines. (When talking about an array as such and not about its determinant, we put the array between *double* vertical lines.)

We note that in the expression (3), the factors in each product are arranged in such a way that the *first* indices occur in increasing order. Thus, so far, we have been concerned with the permutations formed by the second indices. However, the factors in each product can just as well be arranged in such a way that the *second* indices occur in increasing order. Then (3) becomes

$$a_{11}a_{22}a_{33} + a_{31}a_{12}a_{23} + a_{21}a_{32}a_{13} - a_{11}a_{32}a_{23} - a_{21}a_{12}a_{33} - a_{31}a_{22}a_{13}. \tag{9}$$

Here, the first indices form all possible permutations p, q, r of the integers 1, 2, 3, and it is easily verified that the sign rule for the terms of the expression (9) can be formulated just as before, but with respect to the first indices instead. This leads us to consider not only the sum (8) but also the similar sum

$$\sum_{(p_1, p_2, \ldots, p_n)} (-1)^{[p_1, p_2, \ldots, p_n]} a_{p_1 1} a_{p_2 2} \cdots a_{p_n n}. \tag{10}$$

It is clear that the sum (10) consists of the same products as the sum (8).

Moreover, we shall see below that these products have the same signs as in the sum (8), i.e., the two sums (8) and (10) are the same. (We have just seen that this is the case for $n = 3$.)

Finally, we return to the case $n = 2$. Here, the array has the form

$$\begin{Vmatrix} a_{11} & a_{12} \\ a_{21} & a_{22} \end{Vmatrix},$$

and (8) gives the expression

$$\begin{vmatrix} a_{11} & a_{12} \\ a_{21} & a_{22} \end{vmatrix} = a_{11}a_{22} - a_{12}a_{21} \tag{11}$$

for the second order determinant corresponding to this array.

The above considerations show that if we wish to understand the properties of determinants, we must become more familiar with the properties of permutations. Thus, we now turn to this subject.

2. Permutations. A set of n elements arranged in a definite order is called a *permutation*. There are $n!$ different permutations that can be formed out of n elements. For $n = 2$, this is obvious, since there are only two possible arrangements of two elements. For $n = 3$, the assertion follows immediately from the enumeration (5); here, the elements are the numbers 1, 2, 3, and it is easily verified that (5) gives all possible permutations of these three elements. For arbitrary n, we shall prove our assertion by mathematical induction, i.e., assuming that the assertion is valid for n elements, we shall show that it is also valid for $n + 1$ elements. Thus, we assume that n elements give $n!$ permutations, and we consider any $n + 1$ elements, which we denote by

$$C_1, C_2, \ldots, C_{n+1}.$$

Consider first the permutations whose first element is C_1. To obtain all possible permutations of this type, we must put C_1 in the first position and then write down all possible permutations of the n remaining elements; by assumption, the number of such permutations equals $n!$ In just the same way, the number of permutations whose first element is C_2 also equals $n!$ Thus, the total number of different permutations of the elements $C_1, C_2, \ldots, C_{n+1}$ equals

$$n!(n + 1) = 1 \cdot 2 \cdots n \cdot (n + 1) = (n + 1)!,$$

as was to be proved. (Of course, we can assume that our elements are the positive integers, and we shall henceforth adhere to this convention.)

The operation consisting of interchanging the positions of two elements in a permutation is called a transposition. It is immediately clear that we can obtain any permutation from any other permutation by performing transpositions. For example, take the two permutations

$$1, 3, 4, 2; \qquad 2, 4, 1, 3$$

of four elements. We can go from the first of these permutations to the second by performing the following transpositions

$$1, 3, 4, 2 \rightarrow 2, 3, 4, 1 \rightarrow 2, 4, 3, 1 \rightarrow 2, 4, 1, 3.$$

Here we needed three transpositions to go from the first permutation to the second. If we had used other transpositions, we could have gone from the first permutation to the second differently. In other words, the number of transpositions needed to go from one permutation to another is not uniquely determined, i.e., we can go from one permutation to another by using different numbers of transpositions. Thus, it is important to show that for any two given permutations, the different numbers of transpositions needed to go from one permutation to the other are either all even or all odd; this can be expressed differently by saying that these numbers always have the same *parity*. To see this, we introduce the concept of an inversion, which was already used in the preceding section. Suppose we are given a permutation of the n numbers $1, 2, \ldots, n$. The permutation

$$1, 2, \ldots, n, \tag{12}$$

where the numbers appear in increasing order, will be called the *basic permutation*. By an *inversion*, we mean the fact that two elements of a permutation do not appear in the order in which they appear in the basic permutation (12), or in other words, that a larger number comes before a smaller number. *Permutations in which the number of inversions is an even number will be called permutations of the first class, and those in which the number of inversions is odd will be called permutations of the second class.* The following theorem is basic for our further work:

A transposition changes the number of inversions by an odd number.

PROOF. Take any permutation

$$a, b, \ldots, k, \ldots, p, \ldots, s, \tag{13}$$

and assume that we carry out a transposition of the elements k and p, i.e., that we interchange the positions of these two elements. After such a transposition, the position of the elements k and p with respect to the elements standing to the left of k and to the right of p remains unchanged. The only thing that changes is the position of the elements k and p with respect to the elements of the permutation between k and p and, of course, the position of the elements k and p with respect to each other. We now calculate the total change in the number of inversions. Suppose that in the permutation (13) there are m elements between k and p, and suppose that α of these intermediate elements are in natural order with respect to k and that β form inversions with respect to k, while α_1 are in natural order with respect to p and β_1 form inversions with respect to p. Clearly, we have

$$\alpha + \beta = \alpha_1 + \beta_1 = m. \tag{14}$$

As a result of the transposition, a natural ordering becomes an inversion, and conversely. More precisely, if before the transposition the element k was in natural order with respect to an intermediate element, then after the transposition these two elements form an inversion, and conversely. The same is true for the element p. Thus, the total number of inversions of the elements k and p with respect to the intermediate elements is $\beta + \beta_1$ before the transposition and $\alpha + \alpha_1$ after the transposition, i.e., the change in the number of inversions is

$$\gamma = (\alpha + \alpha_1) - (\beta + \beta_1).$$

Using (14), we can rewrite this number in the form

$$\gamma = (\alpha + \alpha_1) - (m - \alpha + m - \alpha_1) = 2(\alpha + \alpha_1 - m),$$

from which it follows at once that the number γ is even. It remains to consider the relative position of the elements k and p themselves. If they were in natural order before the transposition, then they will form an inversion after the transposition, and conversely, i.e., the change in the number of inversions in this case is equal to 1. Thus, the total change in the number of inversions produced by a transposition is an odd number.

We now derive some consequences of this theorem.

COROLLARY 1. If we write down all $n!$ permutations and transpose two given elements (e.g., the elements 1 and 3) in every one of them, then all permutations of the first class become permutations of the second class, and conversely, and the net effect is to get back the whole set of $n!$ permutations. This implies at once that *the number of permutations of the first class is the same as the number of permutations of the second class.*

COROLLARY 2. Every permutation can be obtained from the basic permutation by transpositions. It immediately follows from the theorem that *the first class consists of permutations which are obtained from the basic permutation by an even number of transpositions, while the second class consists of permutations which are obtained from the basic permutation by an odd number of transpositions.*

COROLLARY 3. *The choice of the basic permutation is completely arbitrary.* Instead of the permutation (12), we could have chosen any other permutation as the basic permutation; of course, in this case, to determine the number of inversions, we must compare any permutation with the basic permutation, i.e., we must start from the order in which the elements appear in the basic permutation. It is not hard to see that if instead of the permutation (12) we choose any permutation of the first class as the basic permutation, then permutations of the first class remain permutations of the first class with respect to the new basic permutation, and permutations of the second class remain permutations of the second class. Conversely, if we choose a permutation of the second class as the basic permutation, then permutations of the second class become permutations of the first class, and permutations of the first class become permutations of the second class.

For example, if in the case of the six permutations of the elements 1, 2, 3, we take 2, 1, 3 as the basic permutation, then the permutations of the first class are

$$2, 1\ 3; \qquad 1, 3, 2; \qquad 3, 2, 1.$$

In the second of these permutations we have two inversions: 1 comes before 2, and 3 before 2, whereas in the basic permutation 2 comes before 1, and 2 before 3. The permutations of the second class are

$$1, 2, 3; \qquad 2, 3, 1; \qquad 3, 1, 2.$$

In the first of these permutations there is one inversion with respect to the basic permutation 2, 1, 3, i.e., 1 comes before 2.

In view of what has just been said, we can formulate the sign rule for the expression (8) as follows: *We write a plus sign in front of a product if the permutation formed by the second indices belongs to the first class, and a minus sign if the permutation belongs to the second class, where* 1, 2, 3 *is taken as the basic permutation.*

We now show one of the basic properties of determinants. Suppose we interchange the first and second columns of the array defining the determinant. Then, instead of the array (6), we have the array

$$\left\| \begin{matrix} a_{12} & a_{11} & a_{13} & \cdots & a_{1n} \\ a_{22} & a_{21} & a_{23} & \cdots & a_{2n} \\ \cdot & \cdot & \cdot & \cdot & \cdot \\ a_{n2} & a_{n1} & a_{n3} & \cdots & a_{nn} \end{matrix} \right\|. \tag{15}$$

Using the definition (8), we can now form the determinant corresponding to the array (15). In this array the columns are numbered in the order 2, 1, 3, . . . , n, and we shall take this as the basic permutation. It is obtained from the previous basic permutation by one transposition and therefore previously belonged to the second class. Thus, with the new choice of basic permutation, the previous permutations of the second class become permutations of the first class, and conversely. Consequently, the determinant corresponding to the array (15) is the sum of the same terms as appear in (8), but since, as just shown, the permutations formed by the second indices have changed classes, the terms in the new sum have opposite signs from the terms in the sum (8). This means that *the value of the determinant changes sign when two columns are interchanged.* We proved this property by interchanging the first and second columns, but just the same kind of proof is valid for the interchange of any two columns. Thus, for example, we have the formula

$$\begin{vmatrix} 1 & 0 & 3 \\ 2 & 7 & 6 \\ 5 & 3 & 0 \end{vmatrix} = - \begin{vmatrix} 1 & 3 & 0 \\ 2 & 6 & 7 \\ 5 & 0 & 3 \end{vmatrix},$$

since the second determinant differs from the first by interchanging the second and third columns.

We now prove another property of determinants. Consider a term of the sum (8):

$$(-1)^{[p_1, p_2, \ldots, p_n]} a_{1p_1} a_{2p_2} \cdots a_{np_n}. \tag{16a}$$

By changing the order of the factors, we can arrange the second indices in natural order, but in doing this, the first indices will form some permutation $q_1, q_2, \ldots, q_n$, and the previous expression takes the form

$$(-1)^{[p_1, p_2, \ldots, p_n]} a_{q_1 1} a_{q_2 2} \cdots a_{q_n n}. \tag{16b}$$

The transition from (16*a*) to (16*b*) can be accomplished by a series of transpositions of the factors. Every such transposition is simultaneously a transposition of both the first and second indices. If the number of transpositions necessary to go from (16*a*) to (16*b*) is even, then it follows that the permutation $p_1, p_2, \ldots, p_n$ belongs to the first class, since it becomes the basic permutation $1, 2, \ldots, n$ after an even number of transpositions and therefore can obviously be obtained from the basic permutation by an even number of transpositions. However, in this case the permutation $q_1, q_2, \ldots, q_n$ also belongs to the first class, since it is obtained at the same time from the basic permutation by the same even number of transpositions. In just the same way, if $p_1, p_2, \ldots, p_n$ belongs to the second class, then $q_1, q_2, \ldots, q_n$ also belongs to the second class. This implies that

$$(-1)^{[p_1, p_2, \ldots, p_n]} = (-1)^{[q_1, q_2, \ldots, q_n]},$$

and consequently we can write

$$(-1)^{[p_1, p_2, \ldots, p_n]} a_{1p_1} a_{2p_2} \cdots a_{np_n} = (-1)^{[q_1, q_2, \ldots, q_n]} a_{q_1 1} a_{q_2 2} \cdots a_{q_n n}.$$

Thus, comparing corresponding terms in the sums (8) and (10), we see that the two sums are identical. The rows in the sum (10) play the same role as the columns in the sum (8). It follows at once from these considerations that *the value of a determinant does not change if we change all the rows in its array to columns and all the columns to rows.* Thus, for example, the following two third order determinants are equal:

$$\begin{vmatrix} 2 & 3 & 5 \\ 7 & 0 & 1 \\ 2 & 1 & 6 \end{vmatrix} = \begin{vmatrix} 2 & 7 & 2 \\ 3 & 0 & 1 \\ 5 & 1 & 6 \end{vmatrix}.$$

3. Basic Properties of Determinants

PROPERTY 1. First of all we restate the property just proved: *The value of a determinant does not change when its rows and columns are interchanged.* Thus, everything which will be proved below for columns will be valid for rows also, and conversely.

PROPERTY 2. In the preceding section we saw that interchanging two columns changes only the sign of a determinant. The same is true of rows: *Interchanging two rows (columns) of a determinant changes only its sign.*

PROPERTY 3. If a determinant has two identical rows and we interchange them, then on the one hand we change nothing, while on the other hand, according to what has been proved, we change the sign of the determinant. Thus, if we denote the value of the determinant by Δ, we have $\Delta = -\Delta$ or $\Delta = 0$. *Therefore, a determinant with two identical rows (columns) equals zero.*

PROPERTY 4. *By a linear homogeneous function of the variables $x_1, x_2, \ldots, x_n$, we mean a polynomial of the first degree in these variables without a constant term, i.e., an expression of the form*

$$\varphi(x_1, x_2, \ldots, x_n) = a_1x_1 + a_2x_2 + \cdots + a_nx_n,$$

where the coefficients a_i do not depend on the x_i. Such a function clearly has the two properties:

$$\varphi(kx_1, kx_2, \ldots, kx_n) = k\varphi(x_1, x_2, \ldots, x_n);$$
$$\varphi(x_1 + y_1, x_2 + y_2, \ldots, x_n + y_n) = \varphi(x_1, x_2, \ldots, x_n) + \varphi(y_1, y_2, \ldots, y_n).$$

The latter property also holds for any number of terms. Referring to (8), we see that each term of the sum in the right-hand side contains as a factor one and only one element from every row. It follows from this that *a determinant is a linear homogeneous function of the elements of any row (or of any column).* Therefore, *if all the elements of a row (or column) contain a common factor, this factor can be brought out in front of the determinant sign.*

As we have seen above, the value of the determinant corresponding to the array (6) is often denoted by

$$\begin{vmatrix} a_{11} & a_{12} & \cdots & a_{1n} \\ a_{21} & a_{22} & \cdots & a_{2n} \\ \cdots & \cdots & \cdots & \cdots \\ a_{n1} & a_{n2} & \cdots & a_{nn} \end{vmatrix},$$

or more briefly, by

$$|a_{ik}| \qquad (i, k = 1, 2, \ldots, n).$$

Then, for example, a special case of the property just proved is

$$\begin{vmatrix} ka_{11} & ka_{12} & ka_{13} \\ a_{21} & a_{22} & a_{23} \\ a_{31} & a_{32} & a_{33} \end{vmatrix} = k \begin{vmatrix} a_{11} & a_{12} & a_{13} \\ a_{21} & a_{22} & a_{23} \\ a_{31} & a_{32} & a_{33} \end{vmatrix}.$$

The second of the indicated properties of linear homogeneous functions leads to the following property of determinants: *If the elements of any row (column) are sums of the same number of terms, then the determinant is equal to the sum of the determinants in which the elements of the row (column) in question are replaced by the separate terms.* Thus, for example, we have

$$\begin{vmatrix} a & b & c + c' \\ d & e & f + f' \\ g & h & i + i' \end{vmatrix} = \begin{vmatrix} a & b & c \\ d & e & f \\ g & h & i \end{vmatrix} + \begin{vmatrix} a & b & c' \\ d & e & f' \\ g & h & i' \end{vmatrix}.$$

We note still another consequence of this linearity and homogeneity: *If all the elements of a row (column) are equal to zero, then the determinant is equal to zero.*

PROPERTY 5. If the ith row and kth column of the array (6), which intersect at the element a_{ik}, are deleted, then there remain $n - 1$ rows and $n - 1$ columns. The corresponding determinant of order $n - 1$ is called the minor of the original determinant of order n belonging to the element a_{ik}. We denote it by Δ_{ik} and form the product

$$A_{ik} = (-1)^{i+k}\Delta_{ik}, \tag{17}$$

which we call the *cofactor* (or *algebraic complement*) of the element a_{ik}. We now show that these cofactors are the coefficients of the linear homogeneous function discussed in connection with the last property, i.e., we show that the formula

$$\Delta = A_{i1}a_{i1} + A_{i2}a_{i2} + \cdots + A_{in}a_{in} \qquad (i = 1, 2, \ldots, n) \tag{18}$$

holds for the ith row and that the formula

$$\Delta = A_{1k}a_{1k} + A_{2k}a_{2k} + \cdots + A_{nk}a_{nk} \qquad (k = 1, 2, \ldots, n) \tag{19}$$

holds for the kth column, where Δ is the value of the determinant. In other words, we have to show that if we group together all the terms in the sum (8) which contain a given element a_{ik}, then the coefficient of this element will be its cofactor A_{ik}, defined by (17). We begin by denoting this coefficient by B_{ik} and noting that the coefficient is a sum of products of $n - 1$ elements; however, these products do not contain elements of the ith row and kth column.

We consider first the case $i = k = 1$ and write down the terms of the sum (8) which contain the element a_{11}:

$$a_{11} \sum_{(p_2, \ldots, p_n)} (-1)^{[1, p_2, \ldots, p_n]} a_{2p_2} \cdots a_{np_n}.$$

Here the summation must range over all possible permutations $p_2, p_3, \ldots, p_n$ of the numbers $2, 3, \ldots, n$. In the complete permutation $1, p_2, \ldots, p_n$, the element 1 is in natural order with respect to all the following elements; therefore, for the number of inversions, we have

$$[1, p_2, \ldots, p_n] = [p_2, \ldots, p_n],$$

where in both cases we take for the basic permutation the one in which the numbers appear in increasing order. Thus, we have the following expression for the coefficient of a_{11}:

$$B_{11} = \sum_{(p_2, p_3, \ldots, p_n)} (-1)^{[p_2, \ldots, p_n]} a_{2p_2} \cdots a_{np_n}.$$

This sum is by definition a determinant, but a determinant which differs from the original one by the absence of the first row and the first column. From this it is clear that

$$B_{11} = \Delta_{11} = (-1)^{1+1}\Delta_{11} = A_{11},$$

i.e., our assertion is proved for $i = k = 1$.

We now turn to the case of arbitrary i and k. We interchange the ith row step by step with the rows above it until it appears at the location of the first row. To do this, we have to make $i - 1$ row interchanges. In just the same way, by making $k - 1$ successive interchanges we move the kth column over to the location of the first column. After these interchanges, the element a_{ik} appears in the upper left-hand corner instead of the element a_{11}. The ith row and the kth column appear first, while the order of the remaining rows and columns is not changed. The result obtained above shows that after these interchanges the coefficient of a_{ik} will be equal to Δ_{ik}. However, we needed to perform $(i - 1) + (k - 1)$ interchanges of rows and columns, and each such interchange multiplies the determinant by a factor -1, so that in all we multiply by the factor

$$(-1)^{(i-1)+(k-1)} = (-1)^{i+k}.$$

Consequently, the final expression for the coefficient B_{ik} is

$$B_{ik} = \frac{\Delta_{ik}}{(-1)^{i+k}} = (-1)^{i+k}\Delta_{ik} = A_{ik},$$

as was to be proved. Thus we have established (18) and (19).

If in the determinant Δ we replace the elements of the ith row in turn by the numbers $c_1, c_2, \ldots, c_n$ without changing the remaining rows, then the factors A_{is} in (18) do not change, and the value of the new determinant is

$$\Delta' = A_{i1}c_1 + A_{i2}c_2 + \cdots + A_{in}c_n. \tag{20}$$

In particular, if we choose the numbers $c_1, c_2, \ldots, c_n$ to be equal to the elements $a_{j1}, a_{j2}, \ldots, a_{jn}$ of another row with index j different from i, then the determinant Δ' has identical ith and jth rows and its value equals zero, i.e., $\Delta' = 0$, so that

$$A_{i1}a_{j1} + A_{i2}a_{j2} + \cdots + A_{in}a_{jn} = 0 \qquad (i \neq j). \tag{21a}$$

In just the same way, we have for columns

$$A_{1k}a_{1l} + A_{2k}a_{2l} + \cdots + A_{nk}a_{nl} = 0 \qquad (k \neq l). \tag{21b}$$

Equations (19) and (21) lead us to the following property of determinants, which will be important later:

If we multiply the elements of a row (column) by their cofactors and add these products, we obtain the value of the determinant. Moreover, if the elements of a row (column) are multiplied by the cofactors corresponding to the elements of another row (column) and if these products are added, then the sum equals zero.

PROPERTY 6. Suppose we add the elements of the second row multiplied by a factor p to the elements of the first row of a determinant Δ. Then the elements of the first row become

$$a_{1s} + pa_{2s} \qquad (s = 1, 2, \ldots, n),$$

and by Property 4, the new determinant is the sum of two determinants, the original determinant and a second determinant, whose first row consists of the elements

$$pa_{2s} \qquad (s = 1, 2, \ldots, n)$$

and whose remaining rows are identical with those of Δ. Factoring p out of the first row, we obtain identical first and second rows, and therefore the second determinant vanishes. Thus, *if to the elements of a row (column) we add the corresponding elements of another row (column) which have first been multiplied by the same factor, the value of the determinant does not change.*

We now introduce some notation which will be used below. Suppose as before that we are given the square array of numbers (6), and let l be a positive integer which is not greater than n. We introduce the following notation for the determinant of order l made up of the rows of the array (6) with indices $p_1, p_2, \ldots, p_l$ and of the columns with indices $q_1, q_2, \ldots, q_l$:

$$A\begin{pmatrix} p_1, & p_2, & \ldots, & p_l \\ q_1, & q_2, & \ldots, & q_l \end{pmatrix} = \begin{vmatrix} a_{p_1q_1} & a_{p_1q_2} & \cdots & a_{p_1q_l} \\ a_{p_2q_1} & a_{p_2q_2} & \cdots & a_{p_2q_l} \\ \cdot & \cdot & \cdot & \cdot \\ a_{p_lq_1} & a_{p_lq_2} & \cdots & a_{p_lq_l} \end{vmatrix}. \tag{22}$$

By the determinant of the first order corresponding to a number a, we shall mean the number a itself, i.e., $A\begin{pmatrix} p \\ q \end{pmatrix} = a_{pq}$. The sequences of positive integers $p_1, p_2, \ldots, p_l$ and $q_1, q_2, \ldots, q_l$ do not have to be arranged in the order of increasing p_s and q_s. If the numbers do appear in increasing order in both of the sequences, then the determinant (22) is called a *minor of order l* of the determinant (8). The determinant (22) is obtained from (8) by deleting $n - l$ rows and $n - l$ columns. Let the numbers of these deleted rows and columns in increasing order be $r_1, r_2, \ldots, r_{n-l}$ and $s_1, s_2, \ldots, s_{n-l}$. The minor

$$A\begin{pmatrix} r_1, & r_2, & \ldots, & r_{n-l} \\ s_1, & s_2, & \ldots, & s_{n-l} \end{pmatrix}$$

is called the *complementary minor* of the minor (22), and the expression

$$(-1)^{p_1+p_2+\cdots+p_l+q_1+q_2+\cdots+q_l} A\begin{pmatrix} r_1, & r_2, & \ldots, & r_{n-l} \\ s_1, & s_2, & \ldots, & s_{n-l} \end{pmatrix} \tag{22a}$$

is called the *cofactor* (or *algebraic complement*) of the minor (22). For a single element a_{ik}, this definition of cofactor agrees with the previous definition (17). We denote the cofactor (22a) by

$$A'\begin{pmatrix} p_1, & p_2, & \ldots, & p_l \\ q_1, & q_2, & \ldots, & q_l \end{pmatrix}.$$

It is completely specified by giving the determinant (22), i.e., by giving the sequences of row indices $p_1, p_2, \ldots, p_l$ and column indices $q_1, q_2, \ldots, q_l$.

Consider a fixed set of row indices $p_1, p_2, \ldots, p_l$. The determinant Δ is clearly a homogeneous polynomial of degree l in the elements of these rows, and it can be shown that Δ can be expressed by the formula (*Laplace's theorem*)

$$\Delta = \sum_{q_1 < q_2 < \cdots < q_l} A\begin{pmatrix} p_1, & p_2, & \ldots, & p_l \\ q_1, & q_2, & \ldots, & q_l \end{pmatrix} A'\begin{pmatrix} p_1, & p_2, & \ldots, & p_l \\ q_1, & q_2, & \ldots, & q_l \end{pmatrix}, \tag{23}$$

where the summation ranges over all subsequences $q_1 < q_2 < \cdots < q_l$ of the sequence $1, 2, \ldots, n$. The number of terms in the sum (23) equals the number of combinations of n elements taken l at a time, i.e.,

$$\binom{n}{l} = \frac{n(n-1)\cdots(n-l+1)}{l!},$$

since the order of the numbers q_s plays no role. (In forming the sum (23) the numbers q_s are taken only in increasing order.) For $l = 1$, we have

$$A\begin{pmatrix} p_1 \\ q_1 \end{pmatrix} = a_{p_1 q_1},$$

and (23) reduces to (18) for $i = p_1$. It is easy to construct a formula similar to (23) which corresponds to an expansion of Δ with respect to the elements of arbitrarily selected columns. We shall have no further use for (23) and therefore do not give its proof.

4. Calculation of Determinants. The calculation of a second order determinant is very easy. According to (11), it is sufficient to write down the array

$$\begin{Vmatrix} a_{11} & a_{12} \\ a_{21} & a_{22} \end{Vmatrix}$$

and take with a plus sign the product of the elements on the diagonal drawn with the solid line and with a minus sign the product of the elements on the diagonal drawn with the broken line.

Consider now the third order determinant given explicitly by (3). It is not hard to see that it can be formed in the following way: Write down the array representing the determinant, and write under it the first and second rows again. We then have an array containing six diagonals, with three elements in each:

$$\begin{Vmatrix} a_{11} & a_{12} & a_{13} \\ a_{21} & a_{22} & a_{23} \\ a_{31} & a_{32} & a_{33} \\ a_{11} & a_{12} & a_{13} \\ a_{21} & a_{22} & a_{23} \end{Vmatrix}.$$

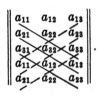

We take with a plus sign the products of the elements lying on the diagonals drawn with a solid line, and with a minus sign the products of the elements lying on the diagonals drawn with a broken line. Then the sum of these six products gives the determinant (*Sarrus' rule*).

This rule cannot be generalized to higher order determinants; then, we must proceed differently in order to shorten the calculations. For example, it is often useful to use Property 6 of determinants (given in the preceding section). We illustrate this by an example. Consider the fourth order determinant

$$\Delta = \begin{vmatrix} 3 & 5 & 1 & 0 \\ 2 & 1 & 4 & 5 \\ 1 & 7 & 4 & 2 \\ -3 & 5 & 1 & 1 \end{vmatrix}.$$

We multiply the third row by -2 and add it to the second row; then we multiply the same row by 3 and add it to the fourth row and subtract it from the first row. In this way, by the property mentioned, we arrive at a determinant which equals the one just written but which has three zeros in the first column:

$$\Delta = \begin{vmatrix} 0 & -16 & -11 & -6 \\ 0 & -13 & -4 & 1 \\ 1 & 7 & 4 & 2 \\ 0 & 26 & 13 & 7 \end{vmatrix}.$$

Then, using (19) to expand with respect to elements of the first column, we obtain

$$\Delta = \begin{vmatrix} -16 & -11 & -6 \\ -13 & -4 & 1 \\ 26 & 13 & 7 \end{vmatrix}.$$

We now multiply the third column by 4 and add it to the second column; then we multiply the third column by 13 and add it to the first column. In this way, we obtain

$$\Delta = \begin{vmatrix} -94 & -35 & -6 \\ 0 & 0 & 1 \\ 117 & 41 & 7 \end{vmatrix} = -\begin{vmatrix} -94 & -35 \\ 117 & 41 \end{vmatrix} = 94 \cdot 41 - 35 \cdot 117$$
$$= -241.$$

5. Examples

1. Suppose that we are required to calculate the volume V of the parallelepiped whose three sides, drawn from one vertex, are the vectors

A, **B**, **C**. As is well known, V can be expressed as the scalar product of the vector **A** with the vector product **B** × **C**:

$$V = \mathbf{A} \cdot (\mathbf{B} \times \mathbf{C}). \tag{24}$$

Here the volume is obtained with a plus sign if the vectors **A**, **B**, **C** have the same orientation as the coordinate axes, and with a minus sign if the orientation is different. The components of the vector product are equal to

$$B_yC_z - B_zC_y, \qquad B_zC_x - B_xC_z, \qquad B_xC_y - B_yC_x,$$

so that the scalar product appearing in (24) is

$$A_x(B_yC_z - B_zC_y) + A_y(B_zC_x - B_xC_z) + A_z(B_xC_y - B_yC_x).$$

It is not hard to see that this sum is the third order determinant

$$V = \begin{vmatrix} A_x & B_x & C_x \\ A_y & B_y & C_y \\ A_z & B_z & C_z \end{vmatrix}. \tag{25}$$

If the determinant is zero, it means that the volume is zero, or in other words, that the three vectors are coplanar, i.e., lie in one plane. If we interchange two rows (columns) in the determinant, e.g., the first and the second, then the order of the vectors **A**, **B**, **C** is changed to the order **B**, **A**, **C**; if in the first order the vectors had the same orientation as the coordinate axes, they must now have the opposite orientation, and conversely. Correspondingly, the value of the determinant changes its sign. Similarly, if we consider two vectors in the xy-plane, with coordinates (A_x, A_y) and (B_x, B_y), then the area of the parallelogram formed by the two vectors is equal to the second order determinant

$$\begin{vmatrix} A_x & B_x \\ A_y & B_y \end{vmatrix}.$$

Consider now the triangle whose vertices are the points

$$(x_1, y_1), (x_2, y_2), (x_3, y_3).$$

In terms of the vectors

$$\mathbf{A} = (x_2 - x_1, y_2 - y_1), \qquad \mathbf{B} = (x_3 - x_1, y_3 - y_1),$$

the area of the triangle can be expressed in the form

$$P = \frac{1}{2}\begin{vmatrix} A_x & B_x \\ A_y & B_y \end{vmatrix} = \frac{1}{2}\begin{vmatrix} x_2 - x_1 & x_3 - x_1 \\ y_2 - y_1 & y_3 - y_1 \end{vmatrix}.$$

It is not hard to see that this second order determinant can be replaced by a third order determinant, i.e., the formula can be written in the form

$$P = \frac{1}{2}\begin{vmatrix} x_1 & x_2 & x_3 \\ y_1 & y_2 & y_3 \\ 1 & 1 & 1 \end{vmatrix}.$$

The vanishing of this determinant gives the condition for the points (x_1, y_1), (x_2, y_2), (x_3, y_3) to lie on one line. In other words, the equation of the straight line going through two given points (x_1, y_1) and (x_2, y_2) can be written in the form

$$\begin{vmatrix} x & x_1 & x_2 \\ y & y_1 & y_2 \\ 1 & 1 & 1 \end{vmatrix} = 0.$$

2. Using determinants, we can easily write the equations of various geometrical objects. For example, suppose we are looking for the equation of the circle going through three given points (x_1, y_1), (x_2, y_2) and (x_3, y_3). It is easy to see that by using a fourth order determinant, we can write this equation as follows:

$$\begin{vmatrix} x^2 + y^2 & x_1^2 + y_1^2 & x_2^2 + y_2^2 & x_3^2 + y_3^2 \\ x & x_1 & x_2 & x_3 \\ y & y_1 & y_2 & y_3 \\ 1 & 1 & 1 & 1 \end{vmatrix} = 0. \tag{26}$$

In fact, expanding with respect to elements of the first column, we find that this equation is an equation of the second degree, in which the coefficients of x^2 and y^2 are identical and the xy term is absent, i.e., (26) corresponds to a circle. Then, if we put $x = x_k$ and $y = y_k$ ($k = 1, 2, 3$) in this equation, the first column of the determinant will be the same as one of the other columns and the equation will be satisfied, i.e., the circle actually goes through the three given points. We observe that if the three given points lie on a straight line, then the coefficient of $x^2 + y^2$ in (26) turns out to be zero, and the equation corresponds to a straight line and not a circle.

In just the same way, in the three-dimensional space with axes OX, OY, OZ, we can write the equation of the plane passing through three given points (x_1, y_1, z_1), (x_2, y_2, z_2), (x_3, y_3, z_3) by using a fourth order determinant:

$$\begin{vmatrix} x & x_1 & x_2 & x_3 \\ y & y_1 & y_2 & y_3 \\ z & z_1 & z_2 & z_3 \\ 1 & 1 & 1 & 1 \end{vmatrix} = 0. \tag{27}$$

If the three given points lie on a straight line, then (27) reduces to the identity $0 = 0$.

3. Consider the determinant D_n of order n, each of whose rows consists of the powers of some number, from the $(n-1)$th power down to the zeroth power:

$$D_n = \begin{vmatrix} x_1^{n-1} & x_1^{n-2} & \cdots & x_1 & 1 \\ x_2^{n-1} & x_2^{n-2} & \cdots & x_2 & 1 \\ \cdot & \cdot & \cdot & \cdot & \cdot \\ x_n^{n-1} & x_n^{n-2} & \cdots & x_n & 1 \end{vmatrix}. \tag{28}$$

For $n = 1$ and $n = 2$, we have

$$D_1 = 1,\ D_2 = x_1 - x_2.$$

To expand the determinant D_3, we replace the number x_1 by the unknown x, obtaining a determinant of the form

$$D_3(x) = \begin{vmatrix} x^2 & x & 1 \\ x_2^2 & x_2 & 1 \\ x_3^2 & x_3 & 1 \end{vmatrix}.$$

Expanding with respect to elements of the first row, we see that $D_3(x)$ is a polynomial of the second degree in x. If we substitute $x = x_2$ or $x = x_3$ in the determinant, then the first row becomes the same as the second or the third, and the determinant vanishes. Thus, the trinomial $D_3(x)$ has the roots x_2 and x_3 and can be written in the form

$$D_3(x) = A_3(x - x_2)(x - x_3),$$

where A_3 is the coefficient of x^2, i.e., the cofactor of the element x^2 which stands in the upper left-hand corner of $D_3(x)$. Since

$$A_3 = \begin{vmatrix} x_2 & 1 \\ x_3 & 1 \end{vmatrix},$$

i.e., A_3 is the determinant of type D_2 made up of the numbers x_2 and x_3, we finally have

$$D_3(x) = (x_2 - x_3)(x - x_2)(x - x_3).$$

Substituting $x = x_1$, we obtain an expression for D_3 in the form of the product of three factors:

$$D_3 = \begin{aligned} &(x_1 - x_2)(x_1 - x_3) \\ &\qquad \times (x_2 - x_3). \end{aligned}$$

In just the same way, using the expression for D_3, we can express D_4 as a product of six factors

$$D_4 = \begin{aligned} &(x_1 - x_2)(x_1 - x_3)(x_1 - x_4) \\ &\qquad \times (x_2 - x_3)(x_2 - x_4) \\ &\qquad\qquad \times (x_3 - x_4), \end{aligned}$$

and for any n, we obtain the following expression for D_n:

$$D_n = \begin{array}{r} (x_1 - x_2)(x_1 - x_3) \cdots (x_1 - x_n) \\ \times (x_2 - x_3) \cdots (x_2 - x_n) \\ \cdots\cdots\cdots \\ \times (x_{n-1} - x_n). \end{array} \tag{29}$$

This expression bears an interesting relation to the basic definition of a determinant. Any determinant of order n can be written in the form

$$\begin{vmatrix} x_{1n} & x_{1,n-1} & \cdots & x_{11} \\ x_{2n} & x_{2,n-1} & \cdots & x_{21} \\ \cdots & \cdots & \cdots & \cdots \\ x_{nn} & x_{n,n-1} & \cdots & x_{n1} \end{vmatrix}. \tag{30}$$

Suppose that in a purely formal way we replace each element x_{ik} by x_i^{k-1}. After we have made such a substitution, the determinant (30) clearly becomes the *Vandermonde determinant* (28). This leads at once to the following rule for forming the sum giving the determinant (30): Multiply out the terms in (29), and in each of the products so obtained, replace x_k^{s-1} by x_{ks}. If a product does not contain some power of the number x_k, then supply the factor x_k^0, which after substitution becomes x_{k1}. We note that this rule can be adopted as the definition of a determinant.

4. Consider the following expression, with which we shall be concerned in later work:

$$\Delta(x) = \begin{vmatrix} a_{11} + x & a_{12} & a_{13} & \cdots & a_{1n} \\ a_{21} & a_{22} + x & a_{23} & \cdots & a_{2n} \\ a_{31} & a_{32} & a_{33} + x & \cdots & a_{3n} \\ \cdots & \cdots & \cdots & \cdots & \cdots \\ a_{n1} & a_{n2} & a_{n3} & \cdots & a_{nn} + x \end{vmatrix}. \tag{31}$$

We expand $\Delta(x)$ in powers of x by first rewriting it in the following form:

$$\Delta(x) = \begin{vmatrix} a_{11} + x & a_{12} + 0 & a_{13} + 0 & \cdots & a_{1n} + 0 \\ a_{21} + 0 & a_{22} + x & a_{23} + 0 & \cdots & a_{2n} + 0 \\ a_{31} + 0 & a_{32} + 0 & a_{33} + x & \cdots & a_{3n} + 0 \\ \cdots & \cdots & \cdots & \cdots & \cdots \\ a_{n1} + 0 & a_{n2} + 0 & a_{n3} + 0 & \cdots & a_{nn} + x \end{vmatrix}. \tag{32}$$

Each of the columns of this determinant is the sum of two terms, so that by repeatedly applying Property 4 of determinants we can represent $\Delta(x)$ as the sum of 2^n determinants, whose columns no longer contain sums. If we delete the second terms in all the columns of the expression

(32), we obtain a determinant which does not contain x and which is just the constant term in the expansion of $\Delta(x)$:

$$\Delta = \begin{vmatrix} a_{11} & a_{12} & \cdots & a_{1n} \\ a_{21} & a_{22} & \cdots & a_{2n} \\ \cdot & \cdot & \cdot & \cdot \\ a_{n1} & a_{n2} & \cdots & a_{nn} \end{vmatrix}. \tag{33}$$

On the other hand, if we delete the first terms in all the columns, we obtain

$$\begin{vmatrix} x & 0 & 0 & \cdots & 0 \\ 0 & x & 0 & \cdots & 0 \\ 0 & 0 & x & \cdots & 0 \\ \cdot & \cdot & \cdot & \cdot & \cdot \\ 0 & 0 & 0 & \cdots & x \end{vmatrix} = x^n,$$

which is the leading term of the polynomial $\Delta(x)$.

We now consider the intermediate terms of the polynomial. Suppose we keep the second terms in the columns with indices $p_1, p_2, \ldots, p_s$ and the first terms in the remaining columns. Then every column with index p_k ($k = 1, 2, \ldots, s$) will consist entirely of zeros except for one element equal to x, which stands on the principal diagonal, i.e., at the intersection of a row and a column with the same indices. Successively expanding the determinant so obtained with respect to the elements of the columns $p_1, p_2, \ldots, p_s$, we obtain the factor x^s, while having to delete the rows with the indices $p_1, p_2, \ldots, p_s$ and the columns with the same indices. After every such deletion, the cofactor of the element involved will be just equal to the corresponding minor, since the index of the deleted row is the same as that of the deleted column. Thus, for any choice of the column indices p_k ($k = 1, 2, \ldots, s$) our determinant will contain x^s with a coefficient equal to the determinant of order $n - s$ obtained from the original determinant (33) by deleting the rows and columns containing the elements $a_{p_1p_1}, a_{p_2p_2}, \ldots, a_{p_sp_s}$ of the principal diagonal. We denote this determinant of order $n - s$ by the symbol

$$\Delta_{p_1p_2\cdots p_s}^{p_1p_2\cdots p_s}.$$

It is called a principal minor of the determinant Δ of order $n - s$. Making various choices of the indices $p_1, p_2, \ldots, p_s$, we finally find that the complete coefficient of x^s in the expression $\Delta(x)$ is the sum of all possible principal minors of order $n - s$, i.e.,

$$\Delta(x) = x^n + S_1x^{n-1} + S_2x^{n-2} + \cdots + S_{n-1}x + S_n,$$

where S_k is the sum of all the principal minors of order k of the determinant Δ; in particular, S_n equals Δ. We can write an explicit expression for the coefficient S_k:

$$S_k = \sum_{p_1<p_2<\cdots<p_{n-k}}^{(1,2,\ldots,n)} \Delta_{p_1p_2\cdots p_{n-k}}^{p_1p_2\cdots p_{n-k}} = \sum_{q_1<q_2<\cdots<q_k} \begin{vmatrix} a_{q_1q_1} & a_{q_1q_2} & \cdots & a_{q_1q_k} \\ a_{q_2q_1} & a_{q_2q_2} & \cdots & a_{q_2q_k} \\ \cdots & \cdots & \cdots & \cdots \\ a_{q_kq_1} & a_{q_kq_2} & \cdots & a_{q_kq_k} \end{vmatrix}. \tag{34}$$

Here the summation extends over all possible combinations $q_1, q_2, \ldots, q_k$ of k numbers taken from the sequence $1, 2, \ldots, n$ and arranged in increasing order.

If in (34) we were simply to sum with respect to each index q_j over all values from 1 to n, then the integers in the permutation $q_1, q_2, \ldots, q_k$ would appear not only in increasing order, but also in all other orders. More precisely, if we summed with respect to all the q_j from 1 to n, then each sequence in natural order would be replaced by $k!$ permutations. Now we note that if we interchange any two numbers q_i and q_j, the value of the determinant in (34) does not change. For example, interchanging q_1 and q_2 interchanges the first and second rows and the first and second columns of the determinant, which has no effect on the value of the determinant. Thus, it follows that if in (34) we simply sum over each of the numbers q_j from 1 to n, each term of the sum (34) is repeated $k!$ times. Therefore, we can write the expression for the coefficient S_k in the form

$$S_k = \frac{1}{k!} \sum_{q_1=1}^{n} \sum_{q_2=1}^{n} \cdots \sum_{q_k=1}^{n} A\begin{pmatrix} q_1, q_2, \ldots, q_k \\ q_1, q_2, \ldots, q_k \end{pmatrix}. \tag{35}$$

6. The Multiplication Theorem for Determinants. In this section we derive the formula for the product of two determinants of the same order. Suppose we have two determinants

$$\Delta = |a_{ik}| \tag{36a}$$

and

$$\Delta_1 = |b_{ik}| \tag{36b}$$

of order n. We form the new determinant whose elements are given by the formula

$$c_{ik} = \sum_{s=1}^{n} a_{is}b_{sk} \qquad (i, k = 1, 2, \ldots, n) \tag{37}$$

and show that this determinant equals the product of the determinants

(36*a*) and (36*b*). We begin with the case $n = 2$. Using (37) and expanding the determinant by using Property 4 of Sec. 3, we obtain

$$\begin{vmatrix} c_{11} & c_{12} \\ c_{21} & c_{22} \end{vmatrix} = \begin{vmatrix} a_{11}b_{11} + a_{12}b_{21} & a_{11}b_{12} + a_{12}b_{22} \\ a_{21}b_{11} + a_{22}b_{21} & a_{21}b_{12} + a_{22}b_{22} \end{vmatrix}$$

$$= \begin{vmatrix} a_{11}b_{11} & a_{11}b_{12} \\ a_{21}b_{11} & a_{21}b_{12} \end{vmatrix} + \begin{vmatrix} a_{11}b_{11} & a_{12}b_{22} \\ a_{21}b_{11} & a_{22}b_{22} \end{vmatrix} + \begin{vmatrix} a_{12}b_{21} & a_{11}b_{12} \\ a_{22}b_{21} & a_{21}b_{12} \end{vmatrix} + \begin{vmatrix} a_{12}b_{21} & a_{12}b_{22} \\ a_{22}b_{21} & a_{22}b_{22} \end{vmatrix}.$$

Removing common factors from elements of the same column, we find that the first and fourth terms are determinants with identical columns, which therefore vanish. Then, interchanging columns in one of the determinants, we have

$$\begin{vmatrix} c_{11} & c_{12} \\ c_{21} & c_{22} \end{vmatrix} = b_{11}b_{22}\begin{vmatrix} a_{11} & a_{12} \\ a_{21} & a_{22} \end{vmatrix} + b_{12}b_{21}\begin{vmatrix} a_{12} & a_{11} \\ a_{22} & a_{21} \end{vmatrix}$$

$$= b_{11}b_{22}\begin{vmatrix} a_{11} & a_{12} \\ a_{21} & a_{22} \end{vmatrix} - b_{12}b_{21}\begin{vmatrix} a_{11} & a_{12} \\ a_{21} & a_{22} \end{vmatrix}$$

$$= \begin{vmatrix} a_{11} & a_{12} \\ a_{21} & a_{22} \end{vmatrix}(b_{11}b_{22} - b_{12}b_{21}) = \begin{vmatrix} a_{11} & a_{12} \\ a_{21} & a_{22} \end{vmatrix}\begin{vmatrix} b_{11} & b_{12} \\ b_{21} & b_{22} \end{vmatrix},$$

as required.

In the general case of order n, we find after applying Property 4 of Sec. 3 that

$$|c_{ik}| = \sum_{s_1, s_2, \ldots, s_n} \begin{vmatrix} a_{1s_1}b_{s_11} & a_{1s_2}b_{s_22} & \cdots & a_{1s_n}b_{s_nn} \\ a_{2s_1}b_{s_11} & a_{2s_2}b_{s_22} & \cdots & a_{2s_n}b_{s_nn} \\ \cdots & \cdots & \cdots & \cdots \\ a_{ns_1}b_{s_11} & a_{ns_2}b_{s_22} & \cdots & a_{ns_n}b_{s_nn} \end{vmatrix}, \tag{38}$$

where the summation variables $s_1, s_2, \ldots, s_n$ range independently over the integers $1, 2, \ldots, n$. The terms of this sum can be written in the form

$$b_{s_11}b_{s_22} \cdots b_{s_nn} \begin{vmatrix} a_{1s_1} & a_{1s_2} & \cdots & a_{1s_n} \\ a_{2s_1} & a_{2s_2} & \cdots & a_{2s_n} \\ \cdots & \cdots & \cdots & \cdots \\ {}_{ns_1} & a_{ns_2} & \cdots & a_{ns_n} \end{vmatrix}. \tag{39}$$

If some of the numbers $s_1, s_2, \ldots, s_n$ are the same, then the determinant in (39) has identical columns and the corresponding term vanishes. Thus, we need only consider terms such that no two of the numbers s_k are the same, i.e., such that the sequence $s_1, s_2, \ldots, s_n$ is some permutation of the numbers $1, 2, \ldots, n$. Multiplying (39) twice by

$(-1)^{[s_1, s_2, \ldots, s_n]}$, which clearly does not change its value, we can rewrite the expression as a product of two factors:

$$(-1)^{[s_1, s_2, \ldots, s_n]} \begin{vmatrix} a_{1s_1} & a_{1s_2} & \cdots & a_{1s_n} \\ a_{2s_1} & a_{2s_2} & \cdots & a_{2s_n} \\ \cdots & \cdots & \cdots & \cdots \\ a_{ns_1} & a_{ns_2} & \cdots & a_{ns_n} \end{vmatrix} \cdot (-1)^{[s_1, s_2, \ldots, s_n]} b_{s_1 1} b_{s_2 2} \cdots b_{s_n n}.$$

In the first factor, we can apply successive transpositions to convert the permutation $s_1, s_2, \ldots, s_n$ to the form $1, 2, \ldots, n$. Each transposition changes both the sign of the quantity $(-1)^{[s_1, s_2, \ldots, s_n]}$ and the sign of the determinant, so that the entire first factor remains unchanged. Thus, (39) can be rewritten in the form

$$\begin{vmatrix} a_{11} & a_{12} & \cdots & a_{1n} \\ a_{21} & a_{22} & \cdots & a_{2n} \\ \cdots & \cdots & \cdots & \cdots \\ a_{n1} & a_{n2} & \cdots & a_{nn} \end{vmatrix} \cdot (-1)^{[s_1, s_2, \ldots, s_n]} b_{s_1 1} b_{s_2 2} \cdots b_{s_n n}.$$

Consequently, returning to the sum (38), we obtain

$$|c_{ik}| = \Delta \sum_{(s_1, s_2, \ldots, s_n)} (-1)^{[s_1, s_2, \ldots, s_n]} b_{s_1 1} b_{s_2 2} \cdots b_{s_n n},$$

where the summation extends over all permutations $s_1, s_2, \ldots, s_n$ of the numbers $1, 2, \ldots, n$. The last sum is just the determinant Δ_1, i.e., $|c_{ik}| = \Delta\Delta_1$, as was to be proved.

The formula (37) for c_{ik} means the following: The elements of the ith row of the determinant Δ are multiplied by the corresponding elements of the kth column of the second determinant, and these products are added. We know that we can interchange rows and columns of a determinant without changing its value. Consequently, the above rule for multiplying rows by columns can be replaced by three other rules, the "row by row" rule, the "column by column" rule, and the "column by row" rule. Thus, we finally state our theorem as follows:

Given two determinants $|a_{ik}|$ and $|b_{ik}|$ of order n, form the new determinant $|c_{ik}|$ whose elements are calculated by one of the following formulas $(i, k = 1, 2, \ldots, n)$:

$$c_{ik} = \sum_{s=1}^{n} a_{is} b_{sk}, \tag{40a}$$

$$c_{ik} = \sum_{s=1}^{n} a_{is} b_{ks}, \tag{40b}$$

$$c_{ik} = \sum_{s=1}^{n} a_{si} b_{sk}, \tag{40c}$$

$$c_{ik} = \sum_{s=1}^{n} a_{si} b_{ks}. \tag{40d}$$

Then the determinant $|c_{ik}|$ is equal to the product of the determinants $|a_{ik}|$ and $|b_{ik}|$.

EXAMPLE. Together with the original determinant $\Delta = |a_{ik}|$, consider the determinant $|A_{ik}|$ formed from the cofactors of its elements. According to the theorem just proved, we can express the product $|a_{ik}|\,|A_{ik}|$ as a determinant by multiplying rows by rows, obtaining a determinant with the following elements:

$$c_{ik} = \sum_{s=1}^{n} a_{is} A_{ks}.$$

By Property 5 of the determinant (Sec. 3), we have

$$c_{ik} = 0 \text{ for } i \neq k,\ c_{ii} = \Delta,$$

i.e.,

$$|a_{ik}|\,|A_{ik}| = \begin{vmatrix} \Delta & 0 & 0 & \cdots & 0 \\ 0 & \Delta & 0 & \cdots & 0 \\ 0 & 0 & \Delta & \cdots & 0 \\ \cdot & \cdot & \cdot & \cdots & \cdot \\ 0 & 0 & 0 & \cdots & \Delta \end{vmatrix},$$

or, as is easily seen,

$$|a_{ik}|\,|A_{ik}| = \Delta^n, \text{ i.e., } \Delta|A_{ik}| = \Delta^n.$$

Assuming that $\Delta \neq 0$ and dividing by Δ, we find that

$$|A_{ik}| = \Delta^{n-1}. \tag{41}$$

If the elements $a_{ik} = a_{ik}^{(0)}$ are such that the determinant Δ vanishes, then we can find values of a_{ik} arbitrarily close to $a_{ik}^{(0)}$ for which the value of Δ is different from zero. For these values of a_{ik} we have (41), and therefore, if we pass to the limit $a_{ik} \to a_{ik}^{(0)}$, (41) is also valid for $a_{ik} = a_{ik}^{(0)}$, i.e., (41) is also valid for $\Delta = 0$. If we express Δ and A_{ik} in terms of the elements a_{ik}, then (41) yields an identity in the a_{ik}.

7. Rectangular Matrices. Subsequently we shall encounter arrays of numbers in which the number of rows may not be the same as the number of columns. Consider such a more general array:

$$\begin{Vmatrix} a_{11} & a_{12} & \cdots & a_{1n} \\ a_{21} & a_{22} & \cdots & a_{2n} \\ \cdot & \cdot & \cdots & \cdot \\ a_{m1} & a_{m2} & \cdots & a_{mn} \end{Vmatrix}. \tag{42}$$

It contains m rows and n columns, where the numbers m and n may be either different or equal. By deleting rows and columns from this array until the number of rows is the same as the number of columns, we can form a determinant from the rows and columns which remain. Such determinants are called *minors* of the array (42). The highest order which they can have is clearly equal to the smaller of the two numbers m and n, and the smallest order they can have is 1, i.e., the minors of order 1 are the elements of the array (42) themselves.

Suppose that all the minors of (42) of order l are zero. Then, it is not hard to see that all the minors of order $l + 1$ are also zero. In fact, every minor of order $l + 1$ can be represented as the sum of the products of the elements of any of its rows with the cofactors of these elements. However, except possibly for sign, the cofactors coincide with certain minors of order l and are therefore all zero. Once all the minors of order $l + 1$ are zero, then in just the same way, all the minors of order $l + 2$ are zero, and so on. Therefore, if all the minors of (42) of a given order vanish, all the minors of higher orders also vanish.

We now introduce the concept of the *rank* of the array (42), or, as one says, of the *matrix* (42); this concept is important later. *By the rank of the matrix* (42) *is meant the order of its largest nonvanishing minor. Thus, if the rank of the matrix is k, then at least one of its minors of order k is nonvanishing, while all its minors of order $k + 1$ vanish.*

Suppose that in addition to the matrix (42), we have a matrix

$$\begin{Vmatrix} b_{11} & b_{12} & \cdots & b_{1m} \\ b_{21} & b_{22} & \cdots & b_{2m} \\ \cdot & \cdot & \cdots & \cdot \\ b_{n1} & b_{n2} & \cdots & b_{nm} \end{Vmatrix} \tag{43}$$

containing n rows and m columns.† We form the m^2 numbers

$$c_{ik} = \sum_{s=1}^{n} a_{is} b_{sk} \qquad (i, k = 1, 2, \ldots, m). \tag{44}$$

The square matrix consisting of the numbers c_{ik} is usually called the *product* of the rectangular matrices (42) and (43).

We now prove a theorem which is a generalization of the multiplication theorem for determinants.

THEOREM. *If $m \leq n$, then*

$$|c_{ik}| = \sum_{r_1 < r_2 < \cdots < r_m} A\begin{pmatrix} 1, & 2, & \ldots, & m \\ r_1, & r_2, & \ldots, & r_m \end{pmatrix} B\begin{pmatrix} r_1, & r_2, & \ldots, & r_m \\ 1, & 2, & \ldots, & m \end{pmatrix}, \tag{45}$$

where the summation extends over all values r_k of the numbers $1, 2, \ldots, n$

† Briefly, an $n \times m$ *matrix.*

which satisfy the indicated inequality. If $m > n$, the determinant $|c_{ik}|$ vanishes.

The meaning of the symbols

$$A\begin{pmatrix}1, & 2, & \ldots, & m\\ r_1, & r_2, & \ldots, & r_m\end{pmatrix}, \qquad B\begin{pmatrix}r_1, & r_2, & \ldots, & r_m\\ 1, & 2, & \ldots, & m\end{pmatrix},$$

was explained in Sec. 3. For example, the second symbol denotes the determinant consisting of the elements of the matrix (43) which belong to the rows $r_1, r_2, \ldots, r_m$ and the columns $1, 2, \ldots, m$. For $m = n$, the sum appearing in (45) contains only one term corresponding to $r_1 = 1, r_2 = 2, \ldots, r_m = m$, and (45) reduces to the multiplication theorem for determinants.

First we consider the case $m < n$. The proof of (45) is similar to the proof of the multiplication theorem for determinants. As in that proof, we have

$$|c_{ik}| = \sum_{s_1, s_2, \ldots, s_m} A\begin{pmatrix}1, & 2, & \ldots, & m\\ s_1, & s_2, & \ldots, & s_m\end{pmatrix} b_{s_1 1} b_{s_2 2} \cdots b_{s_m m}, \tag{46}$$

where each of the numbers s_k can take the values $1, 2, \ldots, n$ and where we can discard terms in which some of the numbers s_k are equal, since these terms vanish. By choosing any definite sequence of numbers $r_1 < r_2 < \cdots < r_m$ from the numbers $1, 2, \ldots, n$ and extracting from the sum (46) those terms for which the numbers $s_1, s_2, \ldots, s_m$ (without regard for order) are the same as the numbers $r_1, r_2, \ldots, r_m$, we obtain the following sum:

$$\sum_{t_1, t_2, \ldots, t_m} A\begin{pmatrix}1, & 2, & \ldots, & m\\ t_1, & t_2, & \ldots, & t_m\end{pmatrix} b_{t_1 1} b_{t_2 2} \cdots b_{t_m m}, \tag{47}$$

which is part of the sum (46). Here the summation extends over all possible permutations $t_1, t_2, \ldots, t_m$ of the numbers $r_1, r_2, \ldots, r_m$. By multiplying each term of the sum (47) twice by $(-1)^{[t_1, t_2, \ldots, t_m]}$, it can be shown in just the same way as in Sec. 6 that this sum equals

$$A\begin{pmatrix}1, & 2, & \ldots, & m\\ r_1, & r_2, & \ldots, & r_m\end{pmatrix} B\begin{pmatrix}r_1, & r_2, & \ldots, & r_m\\ 1, & 2, & \ldots, & m\end{pmatrix}.$$

To obtain the entire sum (46), we must now sum this product with respect to all $r_1 < r_2 < \cdots < r_m$; the result is just (45).

Finally, suppose that $m > n$. Then, we can add $m - n$ columns consisting entirely of zeros to the matrix (42) and $m - n$ rows consisting entirely of zeros to the matrix (43). If now we calculate the c_{ik} not by

(44) but by the formula

$$c_{ik} = \sum_{s=1}^{m} a_{is}b_{sk} \qquad (i, k = 1, 2, \ldots, m), \tag{48}$$

we obtain the previous values of the c_{ik}, since the additional terms in the right-hand side of (48) vanish. On the other hand, after the indicated changes, the matrices (42) and (43) become square matrices with vanishing determinants, and therefore by the multiplication theorem for determinants, the determinant $|c_{ik}|$ also vanishes. This completes the proof of the theorem.

REMARK. If two rectangular matrices both have m rows and n columns, then multiplying rows by rows, i.e.,

$$c_{ik} = \sum_{s=1}^{n} a_{is}b_{ks} \qquad (i, k = 1, 2, \ldots, m),$$

we obtain a determinant $|c_{ik}|$ which vanishes for $m > n$ and which is expressed by the formula

$$|c_{ik}| = \sum_{r_1<r_2<\cdots<r_m} A\begin{pmatrix}1, & 2, & \ldots, & m\\ r_1, & r_2, & \ldots, & r_m\end{pmatrix} B\begin{pmatrix}1, & 2, & \ldots, & m\\ r_1, & r_2, & \ldots, & r_m\end{pmatrix}$$

for $m \leq n$.

COROLLARY. Let there be given two square matrices of order n, consisting of the elements a_{ik} and b_{ik}, and let the numbers c_{ik} be determined by (44). We now seek an expression for the minor

$$C\begin{pmatrix}p_1, & p_2, & \ldots, & p_l\\ q_1, & q_2, & \ldots, & q_l\end{pmatrix}$$

of the determinant $|c_{ik}|$ in terms of minors of the determinants $|a_{ik}|$ and $|b_{ik}|$. It is not hard to see that the square matrix corresponding to the minor

$$C\begin{pmatrix}p_1, & p_2, & \ldots, & p_l\\ q_1, & q_2, & \ldots, & q_l\end{pmatrix}$$

is the product of the following two rectangular matrices:

$$\begin{Vmatrix} a_{p_1 1} & a_{p_1 2} & \cdots & a_{p_1 n}\\ a_{p_2 1} & a_{p_2 2} & \cdots & a_{p_2 n}\\ \cdots & \cdots & \cdots & \cdots\\ a_{p_l 1} & a_{p_l 2} & \cdots & a_{p_l n}\end{Vmatrix}, \qquad \begin{Vmatrix} b_{1q_1} & b_{1q_2} & \cdots & b_{1q_l}\\ b_{2q_1} & b_{2q_2} & \cdots & b_{2q_l}\\ \cdots & \cdots & \cdots & \cdots\\ b_{nq_1} & b_{nq_2} & \cdots & b_{nq_l}\end{Vmatrix}.$$

Applying the theorem just proved, we obtain the desired expression:

$$C\begin{pmatrix} p_1, p_2, \ldots, p_l \\ q_1, q_2, \ldots, q_l \end{pmatrix} = \sum_{r_1<r_2<\cdots<r_l} A\begin{pmatrix} p_1, p_2, \ldots, p_l \\ r_1, r_2, \ldots, r_l \end{pmatrix} B\begin{pmatrix} r_1, r_2, \ldots, r_l \\ q_1, q_2, \ldots, q_l \end{pmatrix}, \quad (49)$$

where the r_k take integral values from 1 to n. Now let R_A, R_B, and R_C be the ranks of the matrices $\|a_{ik}\|$, $\|b_{ik}\|$, and $\|c_{ik}\|$, respectively. If $R_A < n$, say, and we take any $l > R_A$ in (49), then by the definition of R_A, all the minors

$$A\begin{pmatrix} p_1, p_2, \ldots, p_l \\ r_1, r_2, \ldots, r_l \end{pmatrix}$$

vanish, and therefore all the minors

$$C\begin{pmatrix} p_1, p_2, \ldots, p_l \\ q_1, q_2, \ldots, q_l \end{pmatrix}$$

also vanish. It follows that $R_C < l$, i.e., $R_C \leq R_A$. If the rank of the matrix $\|a_{ik}\|$ equals n, then clearly $R_C \leq R_A$, since $R_C \leq n$. In just the same way $R_C \leq R_B$. We shall show later that $R_C = R_A$ if the determinant $|b_{ik}| \neq 0$ and that $R_C = R_B$ if $|a_{ik}| \neq 0$.

PROBLEMS†

1. Evaluate the following second order determinants:

(a) $\begin{vmatrix} 5 & 2 \\ 7 & 3 \end{vmatrix}$; (b) $\begin{vmatrix} 1 & 2 \\ 3 & 4 \end{vmatrix}$; (c) $\begin{vmatrix} 3 & 2 \\ 8 & 5 \end{vmatrix}$; (d) $\begin{vmatrix} 6 & 9 \\ 8 & 12 \end{vmatrix}$; (e) $\begin{vmatrix} a^2 & ab \\ ab & b^2 \end{vmatrix}$;

(f) $\begin{vmatrix} n+1 & n \\ n & n-1 \end{vmatrix}$; (g) $\begin{vmatrix} a+b & a-b \\ a-b & a+b \end{vmatrix}$; (h) $\begin{vmatrix} a^2+ab+b^2 & a^2-ab+b^2 \\ a+b & a-b \end{vmatrix}$;

(i) $\begin{vmatrix} \cos\alpha & -\sin\alpha \\ \sin\alpha & \cos\alpha \end{vmatrix}$; (j) $\begin{vmatrix} \sin\alpha & \cos\alpha \\ \sin\beta & \cos\beta \end{vmatrix}$; (k) $\begin{vmatrix} \cos\alpha & -\sin\alpha \\ \sin\beta & \cos\beta \end{vmatrix}$;

(l) $\begin{vmatrix} \sin\alpha+\sin\beta & \cos\beta+\cos\alpha \\ \cos\beta-\cos\alpha & \sin\alpha-\sin\beta \end{vmatrix}$; (m) $\begin{vmatrix} 2\sin\alpha\cos\alpha & 2\sin^2\alpha-1 \\ 2\cos^2\alpha-1 & 2\sin\alpha\cos\alpha \end{vmatrix}$;

(n) $\begin{vmatrix} 1 & \log_b a \\ \log_a b & 1 \end{vmatrix}$; (o) $\begin{vmatrix} a & c+id \\ c-id & b \end{vmatrix}$; (p) $\begin{vmatrix} a+ib & b \\ 2a & a-ib \end{vmatrix}$;

† All equation numbers (unless otherwise specified) refer to the corresponding equations of Chap. 1. For hints and answers, see p. 431. These problems are by A. L. Shields and the translator, drawing largely from the sources mentioned in the Preface, especially Proskuryakov's book.

(q) $\begin{vmatrix} \cos\alpha + i\sin\alpha & 1 \\ 1 & \cos\alpha - i\sin\alpha \end{vmatrix}$; (r) $\begin{vmatrix} a+ib & c+id \\ -c+id & a-ib \end{vmatrix}$.

In the last four examples $i = \sqrt{-1}$.

2. Using determinants, solve the following systems of equations:

(a) $2x + 5y = 1,\ 3x + 7y = 2$; (b) $2x - 3y = 4,\ 4x - 5y = 10$; (c) $5x - 7y = 1,\ x - 2y = 0$;

(d) $4x + 7y + 13 = 0,\ 5x + 8y + 14 = 0$; (e) $x\cos\alpha - y\sin\alpha = \cos\beta,\ x\sin\alpha + y\cos\alpha = \sin\beta$;

(f) $x\tan\alpha + y = \sin(\alpha+\beta),\ x - y\tan\alpha = \cos(\alpha+\beta)$ $\left(\alpha \neq \frac{\pi}{2} + k\pi, \text{ where } k \text{ is an integer}\right)$.

3. Show that the roots of the equation

$$\begin{vmatrix} a - x & b \\ b & c - x \end{vmatrix} = 0$$

are real, if a, b, and c are real.

4. Show that the rational function

$$\frac{ax+b}{cx+d},$$

where at least one of the numbers c and d is nonvanishing, is independent of x if and only if

$$\begin{vmatrix} a & b \\ c & d \end{vmatrix} = 0.$$

5. Evaluate the following third order determinants:

(a) $\begin{vmatrix} 2 & 1 & 3 \\ 5 & 3 & 2 \\ 1 & 4 & 3 \end{vmatrix}$; (b) $\begin{vmatrix} 3 & 2 & 1 \\ 2 & 5 & 3 \\ 3 & 4 & 2 \end{vmatrix}$; (c) $\begin{vmatrix} 4 & -3 & 5 \\ 3 & -2 & 8 \\ 1 & -7 & -5 \end{vmatrix}$; (d) $\begin{vmatrix} 3 & 2 & -4 \\ 4 & 1 & -2 \\ 5 & 2 & -3 \end{vmatrix}$;

(e) $\begin{vmatrix} 3 & 4 & -5 \\ 8 & 7 & -2 \\ 2 & -1 & 8 \end{vmatrix}$; (f) $\begin{vmatrix} 4 & 2 & -1 \\ 5 & 3 & -2 \\ 3 & 2 & -1 \end{vmatrix}$; (g) $\begin{vmatrix} 1 & 1 & 1 \\ 1 & 2 & 3 \\ 1 & 3 & 6 \end{vmatrix}$; (h) $\begin{vmatrix} 0 & 1 & 1 \\ 1 & 0 & 1 \\ 1 & 1 & 0 \end{vmatrix}$;

(i) $\begin{vmatrix} 5 & 6 & 3 \\ 0 & 1 & 0 \\ 7 & 4 & 5 \end{vmatrix}$; (j) $\begin{vmatrix} 2 & 0 & 3 \\ 7 & 1 & 6 \\ 6 & 0 & 5 \end{vmatrix}$; (k) $\begin{vmatrix} 1 & 5 & 25 \\ 1 & 7 & 49 \\ 1 & 8 & 64 \end{vmatrix}$; (l) $\begin{vmatrix} 1 & 1 & 1 \\ 4 & 5 & 9 \\ 16 & 25 & 81 \end{vmatrix}$;

(m) $\begin{vmatrix} 1 & 2 & 3 \\ 4 & 5 & 6 \\ 7 & 8 & 9 \end{vmatrix}$; (n) $\begin{vmatrix} a & b & c \\ b & c & a \\ c & a & b \end{vmatrix}$; (o) $\begin{vmatrix} a & b & c \\ c & a & b \\ b & c & a \end{vmatrix}$; (p) $\begin{vmatrix} 0 & a & 0 \\ b & c & d \\ 0 & e & 0 \end{vmatrix}$;

(q) $\begin{vmatrix} a & x & x \\ x & b & x \\ x & x & c \end{vmatrix}$; (r) $\begin{vmatrix} a+b & x & x \\ x & b+x & x \\ x & x & c+x \end{vmatrix}$; (s) $\begin{vmatrix} \alpha^2+1 & \alpha\beta & \alpha\gamma \\ \alpha\beta & \beta^2+1 & \beta\gamma \\ \alpha\gamma & \beta\gamma & \gamma^2+1 \end{vmatrix}$;

(t) $\begin{vmatrix} \cos\alpha & \sin\alpha\cos\beta & \sin\alpha\sin\beta \\ -\sin\alpha & \cos\alpha\cos\beta & \cos\alpha\sin\beta \\ 0 & -\sin\beta & \cos\beta \end{vmatrix}$; (u) $\begin{vmatrix} \sin\alpha & \cos\alpha & 1 \\ \sin\beta & \cos\beta & 1 \\ \sin\gamma & \cos\gamma & 1 \end{vmatrix}$;

(v) $\begin{vmatrix} 1 & 0 & 1+i \\ 0 & 1 & i \\ 1-i & -i & 1 \end{vmatrix}$; (w) $\begin{vmatrix} x & a+ib & c+id \\ a-ib & y & e+if \\ c-id & e-if & z \end{vmatrix}$;

(x) $\begin{vmatrix} x & y & x+y \\ y & x+y & x \\ x+y & x & y \end{vmatrix}$; (y) $\begin{vmatrix} (b+c)^2 & a^2 & a^2 \\ b^2 & (c+a)^2 & b^2 \\ c^2 & c^2 & (a+b)^2 \end{vmatrix}$;

(z) $\begin{vmatrix} 1 & 1 & 1 \\ x & y & z \\ x^2 & y^2 & z^2 \end{vmatrix}$.

6. Using determinants, solve the following systems of equations:

(a) $2x + 3y + 5z = 10,\ 3x + 7y + 4z = 3,\ x + 2y + 2z = 3;$ (b) $5x - 6y + 4z = 3,\ 3x - 3y + 2z = 2,\ 4x - 5y + 2z = 1;$ (c) $4x - 3y + 2z + 4 = 0,\ 6x - 2y + 3z + 1 = 0,\ 5x - 3y + 2z + 3 = 0;$
(d) $2ax - 3by + cz = 0,\ 3ax - 6by + 5cz = 2abc,\ 5ax - 4by + 2cz = 3abc.$

7. Solve the system

$$\begin{aligned} x + y + z &= a, \\ x + \epsilon y + \epsilon z &= b, \\ x + \epsilon^2 y + \epsilon^2 z &= c, \end{aligned}$$

where

$$\epsilon = -\frac{1}{2} \pm i\frac{\sqrt{3}}{2} \qquad (\epsilon^3 = 1).$$

8. Solve the system

$$\begin{aligned} (b+c)(y+z) - ax &= b - c, \\ (c+a)(x+z) - by &= c - a, \\ (a+b)(x+y) - cz &= a - b, \end{aligned}$$

where $a + b + c \neq 0$.

9. Solve the system

$$\begin{aligned} x + ay + a^2z + a^3 &= 0, \\ x + by + b^2z + b^3 &= 0, \\ x + cy + c^2z + c^3 &= 0, \end{aligned}$$

where no two of the numbers a, b, c are equal.

10. Find the largest value that a third order determinant can have if all of its elements are equal to -1 or $+1$.

11. Find the number of inversions in the following permutations:

(a) 2, 3, 5, 4, 1; (b) 6, 3, 1, 2, 5, 4; (c) 1, 9, 6, 3, 2, 5, 4, 7, 8;
(d) 7, 5, 6, 4, 1, 3, 2; (e) 1, 3, 5, 7, . . . , $2n - 1$, 2, 4, 6, 8, . . . , $2n$;
(f) 2, 4, 6, . . . , $2n$, 1, 3, 5, . . . , $2n - 1$.

12. Using just the definition of a determinant (cf. (8)), evaluate the following determinants:

(a) $\begin{vmatrix} a_{11} & 0 & 0 & \cdots & 0 \\ a_{21} & a_{22} & 0 & \cdots & 0 \\ a_{31} & a_{32} & a_{33} & \cdots & 0 \\ \cdots & \cdots & \cdots & \cdots & \cdots \\ a_{n1} & a_{n2} & a_{n3} & \cdots & a_{nn} \end{vmatrix}$; (b) $\begin{vmatrix} a_{11} & a_{12} & a_{13} & a_{14} & a_{15} \\ a_{21} & a_{22} & a_{23} & a_{24} & a_{25} \\ a_{31} & a_{32} & 0 & 0 & 0 \\ a_{41} & a_{42} & 0 & 0 & 0 \\ a_{51} & a_{52} & 0 & 0 & 0 \end{vmatrix}$.

13. Which of the following combinations actually occur in the expansion of the determinant (8), and with what sign ($n = 5$)?

(a) $a_{43}a_{21}a_{35}a_{12}a_{54}$; (b) $a_{21}a_{32}a_{43}a_{11}a_{54}$; (c) $a_{12}a_{23}a_{31}a_{45}a_{54}$.

14. Choose j and k so that the following terms occur with the plus sign in the expansion of the determinant (8) ($n = 5$):

(a) $a_{12}a_{2j}a_{3k}a_{45}a_{51}$; (b) $a_{23}a_{4j}a_{5k}a_{12}a_{31}$; (c) $a_{ij}a_{12}a_{23}a_{34}a_{45}$.

15. Which permutation of the sequence $1, 2, \ldots, n$ has the greatest number of inversions, and how many inversions does it contain?

16. Show that one can go from any permutation $a_1, a_2, \ldots, a_n$ of the numbers $1, 2, \ldots, n$ to any other permutation $b_1, b_2, \ldots, b_n$ by using no more than $n - 1$ transpositions. Give an example of a permutation which cannot be brought into the form $1, 2, \ldots, n$ by using fewer than $n - 1$ transpositions.

17. Show that one can go from any permutation $a_1, a_2, \ldots, a_n$ to any other permutation $b_1, b_2, \ldots, b_n$ by using no more than $\frac{1}{2}n(n - 1)$ transpositions of *neighboring* elements.

18. Suppose that the number of inversions in the permutation $a_1, a_2, \ldots, a_n$ is k. How many inversions are there in the permutation $a_n, a_{n-1}, \ldots, a_2, a_1$?

19. What is the total number of inversions in all $n!$ permutations of n elements?

20. Show that one can go from any permutation of the numbers $1, 2, \ldots, n$ containing k inversions to the basic permutation $1, 2, \ldots, n$ by making k (and no fewer) transpositions of *neighboring* elements.

21. How does a determinant change if each of its elements is replaced by the element symmetric to the given element with respect to the "center" of the determinant?

22. How does a determinant change if each of its elements is replaced by the element symmetric to the given element with respect to the diagonal opposite to the principal diagonal (in (8), the diagonal from a_{n1} to a_{nn})?

23. Let $\Delta = |a_{ik}|$ be a determinant of order n, whose elements are in general complex. Show that

(a) If $a_{ik} = -a_{ki}$ for all i, j and if n is odd, then $\Delta = 0$;
(b) If $a_{ik} = \bar{a}_{ki}$ for all i, j, then Δ is a real number. (The overbar denotes the complex conjugate.)

24. How does a determinant of order n change if

(a) Each of its elements is replaced by its negative;
(b) Each of its elements a_{ik} is multiplied by c^{i-k}, where $c \neq 0$?

25. Solve the equation

$$\begin{vmatrix} 1 & 1 & 1 & \cdots & 1 \\ 1 & 1-x & 1 & \cdots & 1 \\ 1 & 1 & 2-x & \cdots & 1 \\ \cdots & \cdots & \cdots & \cdots & \cdots \\ 1 & 1 & 1 & \cdots & n-x \end{vmatrix} = 0.$$

26. Evaluate the determinant

$$\begin{vmatrix} \alpha^2 & (\alpha+1)^2 & (\alpha+2)^2 & (\alpha+3)^2 \\ \beta^2 & (\beta+1)^2 & (\beta+2)^2 & (\beta+3)^2 \\ \gamma^2 & (\gamma+1)^2 & (\gamma+2)^2 & (\gamma+3)^2 \\ \delta^2 & (\delta+1)^2 & (\delta+2)^2 & (\delta+3)^2 \end{vmatrix}.$$

27. Prove that

$$\begin{vmatrix} b+c & c+a & a+b \\ b_1+c_1 & c_1+a_1 & a_1+b_1 \\ b_2+c_2 & c_2+a_2 & a_2+b_2 \end{vmatrix} = 2\begin{vmatrix} a & b & c \\ a_1 & b_1 & c_1 \\ a_2 & b_2 & c_2 \end{vmatrix}.$$

28. Evaluate the following determinants:

(a) $\begin{vmatrix} a & 3 & 0 & 5 \\ 0 & b & 0 & 2 \\ 1 & 2 & c & 3 \\ 0 & 0 & 0 & d \end{vmatrix}$; (b) $\begin{vmatrix} 1 & 0 & 2 & a \\ 2 & 0 & b & 0 \\ 3 & c & 4 & 5 \\ d & 0 & 0 & 0 \end{vmatrix}$; (c) $\begin{vmatrix} a & 1 & 0 & 0 \\ -1 & b & 1 & 0 \\ 0 & -1 & c & 1 \\ 0 & 0 & -1 & d \end{vmatrix}$;

(d) $\begin{vmatrix} x & a & b & 0 & c \\ 0 & y & 0 & 0 & d \\ 0 & e & z & 0 & f \\ g & h & k & u & m \\ 0 & 0 & 0 & 0 & v \end{vmatrix}$.

29. Show that the determinant

$$D_n = \begin{vmatrix} a_1 & 1 & 0 & \cdots & 0 & 0 \\ -1 & a_2 & 1 & \cdots & 0 & 0 \\ 0 & -1 & a_3 & \cdots & 0 & 0 \\ \cdots & \cdots & \cdots & \cdots & \cdots & \cdots \\ 0 & 0 & 0 & \cdots & a_{n-1} & 1 \\ 0 & 0 & 0 & \cdots & -1 & a_n \end{vmatrix}$$

satisfies the recursion formula

$$D_n = a_nD_{n-1} + D_{n-2}.$$

30. Show that

$$\begin{vmatrix} a_1 & \lambda & \lambda & \cdots & \lambda \\ \lambda & a_2 & \lambda & \cdots & \lambda \\ \cdots & \cdots & \cdots & \cdots & \cdots \\ \lambda & \lambda & \lambda & \cdots & a_n \end{vmatrix} = \varphi(\lambda) - \lambda\frac{d\varphi}{d\lambda},$$

where $\varphi(\lambda) = (a_1 - \lambda)(a_2 - \lambda) \cdots (a_n - \lambda)$.

31. How many minors of order k has a determinant of order n?

32. Evaluate the following determinants by using Laplace's theorem:

(a) $\begin{vmatrix} 5 & 1 & 2 & 7 \\ 3 & 0 & 0 & 2 \\ 1 & 3 & 4 & 5 \\ 2 & 0 & 0 & 3 \end{vmatrix}$; (b) $\begin{vmatrix} 1 & 1 & 3 & 4 \\ 2 & 0 & 0 & 8 \\ 3 & 0 & 0 & 2 \\ 4 & 4 & 7 & 5 \end{vmatrix}$; (c) $\begin{vmatrix} 0 & 5 & 2 & 0 \\ 8 & 3 & 5 & 4 \\ 7 & 2 & 4 & 1 \\ 0 & 4 & 1 & 0 \end{vmatrix}$;

(d) $\begin{vmatrix} 1 & 2 & 0 & 0 & 0 & 0 \\ 3 & 4 & 0 & 0 & 0 & 0 \\ 7 & 6 & 5 & 4 & 0 & 0 \\ 2 & 3 & 4 & 5 & 0 & 0 \\ 5 & 1 & 2 & 6 & 7 & 3 \\ 2 & 7 & 5 & 3 & 4 & 1 \end{vmatrix}$.

33. Show that

$$\begin{vmatrix} a_{11} & 0 & a_{12} & 0 & \cdots & a_{1n} & 0 \\ 0 & b_{11} & 0 & b_{12} & \cdots & 0 & b_{1n} \\ a_{21} & 0 & a_{22} & 0 & \cdots & a_{2n} & 0 \\ 0 & b_{21} & 0 & b_{22} & \cdots & 0 & b_{2n} \\ \cdots & \cdots & \cdots & \cdots & \cdots & \cdots & \cdots \\ a_{n1} & 0 & a_{n2} & 0 & \cdots & a_{nn} & 0 \\ 0 & b_{n1} & 0 & b_{n2} & \cdots & 0 & b_{nn} \end{vmatrix}$$

$$= \begin{vmatrix} a_{11} & a_{12} & \cdots & a_{1n} \\ a_{21} & a_{22} & \cdots & a_{2n} \\ \cdots & \cdots & \cdots & \cdots \\ a_{n1} & a_{n2} & \cdots & a_{nn} \end{vmatrix} \begin{vmatrix} b_{11} & b_{12} & \cdots & b_{1n} \\ b_{21} & b_{22} & \cdots & b_{2n} \\ \cdots & \cdots & \cdots & \cdots \\ b_{n1} & b_{n2} & \cdots & b_{nn} \end{vmatrix}.$$

34. Show that

$$\begin{vmatrix} a_1 + b_1 & a_1 + b_2 & \cdots & a_1 + b_n \\ a_2 + b_1 & a_2 + b_2 & \cdots & a_2 + b_n \\ \cdots & \cdots & \cdots & \cdots \\ a_n + b_1 & a_n + b_2 & \cdots & a_n + b_n \end{vmatrix} = 0 \qquad \text{for } n > 2.$$

35. Show that

$$\begin{vmatrix} 1+x & 1 & 1 & 1 \\ 1 & 1-x & 1 & 1 \\ 1 & 1 & 1+z & 1 \\ 1 & 1 & 1 & 1-z \end{vmatrix} = x^2z^2.$$

36. Show that

$$\begin{vmatrix} 0 & x & y & z \\ x & 0 & z & y \\ y & z & 0 & x \\ z & y & x & 0 \end{vmatrix} = (x+y+z)(x-y-z)(x-y+z)(x+y-z).$$

37. Evaluate the following determinant of order n:

$$\begin{vmatrix} 1 & 1 & 1 & \cdots & 1 \\ 1 & 0 & 1 & \cdots & 1 \\ 1 & 1 & 0 & \cdots & 1 \\ \cdots & \cdots & \cdots & \cdots & \cdots \\ 1 & 1 & 1 & \cdots & 0 \end{vmatrix}.$$

38. Evaluate the following determinant:

$$\begin{vmatrix} a & b & c & 1 \\ b & c & a & 1 \\ c & a & b & 1 \\ 1 & 1 & 1 & 1 \end{vmatrix}$$

when $a + b + c = 0$ and when $a + b + c = 3$.

39. Show that

$$\begin{vmatrix} 1 & 1 & 1 & 1 & \cdots & 1 \\ 1 & 2 & 2 & 2 & \cdots & 2 \\ 1 & 2 & 3 & 3 & \cdots & 3 \\ \cdots & \cdots & \cdots & \cdots & \cdots & \cdots \\ 1 & 2 & 3 & 4 & \cdots & n \end{vmatrix} = 1$$

by reducing the determinant to triangular form.

40. Let a, b, c be the sides of a triangle, and let A be the angle opposite side a. Show that

$$\begin{vmatrix} a^2 & b \sin A & c \sin A \\ b \sin A & 1 & \cos A \\ c \sin A & \cos A & 1 \end{vmatrix} = 0.$$

41. Prove that

$$\begin{vmatrix} a_{11} + x & a_{12} + x & \cdots & a_{1n} + x \\ a_{21} + x & a_{22} + x & \cdots & a_{2n} + x \\ \cdots & \cdots & \cdots & \cdots \\ a_{n1} + x & a_{n2} + x & \cdots & a_{nn} + x \end{vmatrix} = \begin{vmatrix} a_{11} & a_{12} & \cdots & a_{1n} \\ a_{21} & a_{22} & \cdots & a_{2n} \\ \cdots & \cdots & \cdots & \cdots \\ a_{n1} & a_{n2} & \cdots & a_{nn} \end{vmatrix} + x \sum_{i=1}^{n} \sum_{k=1}^{n} A_{ik},$$

where A_{ik} is the cofactor of the element a_{ik}.

42. The numbers 204, 527, and 255 are divisible by 17. Prove that the determinant

$$\begin{vmatrix} 2 & 0 & 4 \\ 5 & 2 & 7 \\ 2 & 5 & 5 \end{vmatrix}$$

is also divisible by 17.

43. Show that

$$D_n = \begin{vmatrix} 1 + a_1 & 1 & 1 & \cdots & 1 \\ 1 & 1 + a_2 & 1 & \cdots & 1 \\ 1 & 1 & 1 + a_3 & \cdots & 1 \\ \cdots & \cdots & \cdots & \cdots & \cdots \\ 1 & 1 & 1 & \cdots & 1 + a_n \end{vmatrix}$$

$$= a_1 a_2 \cdots a_n \left(1 + \frac{1}{a_1} + \frac{1}{a_2} + \cdots + \frac{1}{a_n}\right).$$

(It is assumed that none of the a_i vanishes.)

44. Show that the determinant

$$\begin{vmatrix} f_1(a_1) & f_1(a_2) & \cdots & f_1(a_n) \\ f_2(a_1) & f_2(a_2) & \cdots & f_2(a_n) \\ \cdots & \cdots & \cdots & \cdots \\ f_n(a_1) & f_n(a_2) & \cdots & f_n(a_n) \end{vmatrix}$$

vanishes if $f_1(x), f_2(x), \ldots, f_n(x)$ are polynomials in x, each of which is of degree no greater than $n - 2$, and the numbers $a_1, a_2, \ldots, a_n$ are arbitrary.

45. Use determinants to

(a) Calculate the area of the triangle with vertices at the points $(1, -1)$, $(2, 1)$, $(-1, -1)$;
(b) Find the equation of the circle through the points $(2, 1)$, $(5, 1)$, $(5, 5)$;
(c) Find the equation of the plane through the points $(2, 0, 0)$, $(0, -3, 0)$, $(0, 0, 4)$;
(d) Write the condition for four points (x_i, y_i, z_i), $i = 1, 2, 3, 4$, to lie in a plane;
(e) Calculate the volume of the tetrahedron with vertices at $(1, 1, 1)$, $(-1, 1, 1)$, $(1, -1, 1)$, $(1, 1, -1)$.

46. Evaluate the determinant

$$\begin{vmatrix} 1 & x_1 & x_1^2 & \cdots & x_1^{n-2} & x_1^n \\ 1 & x_2 & x_2^2 & \cdots & x_2^{n-2} & x_2^n \\ \cdot & \cdot & \cdot & \cdots & \cdot & \cdot \\ 1 & x_n & x_n^2 & \cdots & x_n^{n-2} & x_n^n \end{vmatrix}.$$

47. Consider the determinant $d(x) = |a_{ik}(x)|$ whose elements are functions of x. Show that the derivative $d'(x)$ is the sum of the n determinants obtained from $d(x)$ by differentiating each row (or column) in turn.

48. Use the result of the preceding problem to show that

$$d(x) = \begin{vmatrix} a_1 + x & a_2 & \cdots & a_n \\ a_1 & a_2 + x & \cdots & a_n \\ \cdot & \cdot & \cdots & \cdot \\ a_1 & a_2 & \cdots & a_n + x \end{vmatrix} = x^{n-1}(x + a_1 + \cdots + a_n).$$

49. Multiply the determinants

$$\begin{vmatrix} 1 & 2 & 3 \\ 3 & 4 & 2 \\ 4 & 5 & 4 \end{vmatrix} \quad \text{and} \quad \begin{vmatrix} 2 & -3 & 1 \\ 1 & -4 & 3 \\ 1 & -5 & 2 \end{vmatrix}$$

in all four possible ways, i.e., multiply rows by rows, rows by columns, columns by rows, and columns by columns (cf. Sec. 6).

50. Evaluate the determinant

$$\begin{vmatrix} a & b & c & d \\ -b & a & d & -c \\ -c & -d & a & b \\ -d & c & -b & a \end{vmatrix}$$

by squaring it.

51. Evaluate the following determinants by expressing them as products:

(a) $$\begin{vmatrix} 1 + x_1y_1 & 1 + x_1y_2 & \cdots & 1 + x_1y_n \\ 1 + x_2y_1 & 1 + x_2y_2 & \cdots & 1 + x_2y_n \\ \cdot & \cdot & \cdots & \cdot \\ 1 + x_ny_1 & 1 + x_ny_2 & \cdots & 1 + x_ny_n \end{vmatrix};$$

(b) $$\begin{vmatrix} \cos(\alpha_1 - \beta_1) & \cos(\alpha_1 - \beta_2) & \cdots & \cos(\alpha_1 - \beta_n) \\ \cos(\alpha_2 - \beta_1) & \cos(\alpha_2 - \beta_2) & \cdots & \cos(\alpha_2 - \beta_n) \\ \cdot & \cdot & \cdots & \cdot \\ \cos(\alpha_n - \beta_1) & \cos(\alpha_n - \beta_2) & \cdots & \cos(\alpha_n - \beta_n) \end{vmatrix};$$

(c) $$\begin{vmatrix} \sin 2\alpha_1 & \sin(\alpha_1+\alpha_2) & \cdots & \sin(\alpha_1+\alpha_n) \\ \sin(\alpha_2+\alpha_1) & \sin 2\alpha_2 & \cdots & \sin(\alpha_2+\alpha_n) \\ \cdots & \cdots & \cdots & \cdots \\ \sin(\alpha_n+\alpha_1) & \sin(\alpha_n+\alpha_2) & \cdots & \sin 2\alpha_n \end{vmatrix}.$$

52. Show that the *circulant*

$$\begin{vmatrix} a_0 & a_1 & a_2 & \cdots & a_{n-1} \\ a_{n-1} & a_0 & a_1 & \cdots & a_{n-2} \\ \cdots & \cdots & \cdots & \cdots & \cdots \\ a_1 & a_2 & a_3 & \cdots & a_0 \end{vmatrix}$$

is equal to

$$\prod_{k=0}^{n-1} (a_0 + a_1\epsilon_k + a_2\epsilon_k^2 + \cdots + a_{n-1}\epsilon_k^{n-1}),$$

where $$\epsilon_k = \cos\frac{2k\pi}{n} + i\sin\frac{2k\pi}{n} \qquad (k = 0, 1, \ldots, n-1)$$

are the nth roots of unity.

53. Apply the preceding problem to evaluate the following determinant of order n:

$$\begin{vmatrix} x & a & a & \cdots & a \\ a & x & a & \cdots & a \\ \cdots & \cdots & \cdots & \cdots & \cdots \\ a & a & a & \cdots & x \end{vmatrix}.$$

Also evaluate this determinant by using Prob. 30.

54. Show that if

$$D(x) = \begin{vmatrix} a_{11}-x & a_{12} & \cdots & a_{1n} \\ a_{21} & a_{22}-x & \cdots & a_{2n} \\ \cdots & \cdots & \cdots & \cdots \\ a_{n1} & a_{n2} & \cdots & a_{nn}-x \end{vmatrix},$$

then the product $D(x)D(-x)$ can be represented in the form

$$\begin{vmatrix} A_{11}-x^2 & A_{12} & \cdots & A_{1n} \\ A_{21} & A_{22}-x^2 & \cdots & A_{2n} \\ \cdots & \cdots & \cdots & \cdots \\ A_{n1} & A_{n2} & \cdots & A_{nn}-x^2 \end{vmatrix},$$

where the A_{ij} are independent of x. Express the A_{ij} in terms of the a_{ij}.

55. Show that the determinant

$$\begin{vmatrix} 1 & \cos\varphi_3 & \cos\varphi_2 \\ \cos\varphi_3 & 1 & \cos\varphi_1 \\ \cos\varphi_2 & \cos\varphi_1 & 1 \end{vmatrix}$$

vanishes, if $\varphi_1 + \varphi_2 + \varphi_3 = 0$.

56. Evaluate the determinant formed from the cofactors of each of the following determinants, both by direct computation and by using formula (41) of Sec. 6:

(a) $\begin{vmatrix} a & b & 0 \\ c & d & 0 \\ 0 & 0 & 1 \end{vmatrix}$; (b) $\begin{vmatrix} 0 & 1 & 1 \\ 1 & 0 & 1 \\ 1 & 1 & 0 \end{vmatrix}$; (c) $\begin{vmatrix} 0 & 0 & 0 & 1 \\ 1 & 1 & 1 & 2 \\ 2 & 1 & 0 & 3 \\ 4 & 2 & 0 & 4 \end{vmatrix}$.

57. Let $\Delta = |a_{ik}|$ be a determinant of order n, and let $|A_{ik}|$ be the determinant formed from the cofactors of Δ. Let M be a minor of order m of Δ, let C be the cofactor of M (defined by (22*a*)), and let M' be the minor of $|A_{ik}|$ corresponding to the minor M, i.e., formed of the cofactors of the elements of Δ appearing in M. Prove that

$$M' = \Delta^{m-1}C.$$

Comment: If the cofactor of the whole determinant is regarded as being 1, then this is a generalization of formula (41) of Sec. 6.

58. Let $\Delta = |a_{ik}|$ be a determinant of order n, let Δ_0 be the minor of order $n - 2$ obtained by deleting rows i, j and columns k, l ($i < j$, $k < l$) from Δ, and let A_{ik} denote the cofactor of a_{ik}. Prove that

$$\begin{vmatrix} A_{ik} & A_{il} \\ A_{jk} & A_{jl} \end{vmatrix} = (-1)^{i+j+k+l}\Delta_0\,\Delta.$$

59. Use the theorem expressed by equation (45) of Sec. 7 to prove *Cauchy's formula*

$$\begin{aligned}(a_1c_1 + \cdots + a_mc_m)(b_1d_1 + \cdots + b_md_m) \\ - (a_1d_1 + \cdots + a_md_m)(b_1c_1 + \cdots + b_mc_m) \\ = \sum_{1\le i<k\le n} (a_ib_k - a_kb_i)(c_id_k - c_kd_i),\end{aligned}$$

where a_i, b_i, c_i, d_i ($i = 1, 2, \ldots, m$) are arbitrary real numbers. Then, use this result to prove the *Cauchy-Schwarz inequality*

$$\left(\sum_{k=1}^{m} a_kb_k\right)^2 \le \sum_{k=1}^{m} a_k^2 \sum_{k=1}^{m} b_k^2$$

(cf. equation (126) of Sec. 29). Show that the equality holds if and only if one set of numbers is a multiple of the other set.

60. By a method similar to that of the preceding problem, prove the complex form of the Cauchy-Schwarz inequality, i.e.,

$$\left|\sum_{k=1}^{m} a_k\bar{b}_k\right|^2 \le \sum_{k=1}^{m} |a_k|^2 \sum_{k=1}^{m} |b_k|^2,$$

where a_i, b_i ($i = 1, 2, \ldots, m$) are now arbitrary complex numbers (cf. equation (126*a*) of Sec. 29). Again show that equality holds if and only if one set of numbers is a multiple of the other set.

61. Multiply the following pairs of matrices, and compute the determinant and the rank of the resulting matrices:

$$\text{(a)} \begin{Vmatrix} 1 & 2 \\ 3 & 4 \\ 5 & 6 \end{Vmatrix}, \begin{Vmatrix} 1 & 0 & 3 \\ 0 & 2 & 0 \end{Vmatrix}; \quad \text{(b)} \begin{Vmatrix} 1 \\ 2 \\ 3 \end{Vmatrix}, \|4 \;\; 2 \;\; 1\|; \quad \text{(c)} \begin{Vmatrix} 1 & 3 & 5 \\ 2 & 4 & 6 \end{Vmatrix}, \begin{Vmatrix} 1 & 0 \\ 0 & 2 \\ 3 & 0 \end{Vmatrix}.$$

62. Find the rank of the following matrices:

(a) $\begin{Vmatrix} 6 & 3 \\ 2 & 1 \end{Vmatrix}$; (b) $\begin{Vmatrix} 1 & 1 & 1 \\ 1 & 1 & 1 \\ 1 & 1 & 1 \end{Vmatrix}$; (c) $\begin{Vmatrix} 0 & -1 & -1 \\ 1 & 0 & -1 \\ 1 & 1 & 0 \end{Vmatrix}$;

(d) $\begin{Vmatrix} 1 & -1 & 1 & -1 \\ -1 & 1 & 1 & 1 \\ 1 & 1 & -1 & 1 \\ 1 & 1 & 1 & -1 \end{Vmatrix}$.

63. Show how the rank of the following matrices depends on λ:

(a) $\begin{Vmatrix} 3 & 1 & 1 & 4 \\ \lambda & 4 & 10 & 1 \\ 1 & 7 & 17 & 3 \\ 2 & 2 & 4 & 3 \end{Vmatrix}$; (b) $\begin{Vmatrix} 1 & \lambda & -1 & 2 \\ 2 & -1 & \lambda & 5 \\ 1 & 10 & -6 & 1 \end{Vmatrix}$.

64. Show that the n points $(x_1, y_1), (x_2, y_2), \ldots, (x_n, y_n)$ in the plane all lie on one line if and only if the rank of the matrix

$$\begin{Vmatrix} 1 & 1 & \cdots & 1 \\ x_1 & x_2 & \cdots & x_n \\ y_1 & y_2 & \cdots & y_n \end{Vmatrix}$$

is less than 3.

Chapter 2

Solution of Systems of Linear Equations

8. Cramer's Rule. Now that we have introduced the concept of a determinant and explained its basic properties, we consider next the application of this concept to the solution of systems of linear equations. First we consider the important case where the number of equations is the same as the number of unknowns. We can write such a system, containing n equations and n unknowns, in the form

$$\begin{aligned} a_{11}x_1 + a_{12}x_2 + \cdots + a_{1n}x_n &= b_1, \\ a_{21}x_1 + a_{22}x_2 + \cdots + a_{2n}x_n &= b_2, \\ \cdots\cdots\cdots\cdots\cdots\cdots&\cdots\cdots \\ a_{n1}x_1 + a_{n2}x_2 + \cdots + a_{nn}x_n &= b_n, \end{aligned} \tag{1}$$

where the designation of the coefficients is the same as that introduced in Sec. 1 for the case of three equations in three unknowns. We shall make one assumption, namely, that the determinant of the system (the determinant corresponding to the coefficient matrix $\|a_{ik}\|$ of the system) is different from zero, i.e.,

$$\Delta = |a_{ik}| \neq 0. \tag{2}$$

We multiply both sides of the equations of the system (1) by the cofactors of the elements of the kth column of the determinant, i.e., we multiply both sides of the first equation of the system by A_{1k}, the second by A_{2k}, and so on. Then we add together the equations so obtained. As a result we obtain an equation whose right-hand side is the sum

$$A_{1k}b_1 + A_{2k}b_2 + \cdots + A_{nk}b_n.$$

In the left-hand side, the coefficient of the unknown x_l is given by the sum

$$A_{1k}a_{1l} + A_{2k}a_{2l} + \cdots + A_{nk}a_{nl} \qquad (l = 1, 2, \ldots, n),$$

which equals zero for $l \neq k$ and equals the determinant Δ for $l = k$. Thus, we arrive at an equation of the form

$$\Delta \cdot x_k = A_{1k}b_1 + A_{2k}b_2 + \cdots + A_{nk}b_n.$$

Carrying out this procedure for every index k, we find that the new system of equations

$$\Delta \cdot x_k = A_{1k}b_1 + A_{2k}b_2 + \cdots + A_{nk}b_n \qquad (k = 1, 2, \ldots, n) \quad (3)$$

is implied by the system (1). Conversely, it is not hard to see that the system (1) is implied by the system (3). In fact, multiply both sides of (3) by a_{lk} and then sum over all k from 1 to n. Again using Property 5 of the determinant (Sec. 3), we easily arrive at the equation

$$\Delta \cdot (a_{l1}x_1 + a_{l2}x_2 + \cdots + a_{ln}x_n) = \Delta \cdot b_l, \quad (4)$$

which, when divided by the nonzero factor Δ, yields the lth equation of the system (1). Of course, this can be done for any l.

Thus, the systems (1) and (3) are equivalent, and we can solve the system (3) instead of the system (1). The latter system can be solved at once and has a unique solution, which is given by the formula

$$x_k = \frac{A_{1k}b_1 + A_{2k}b_2 + \cdots + A_{nk}b_n}{\Delta} \qquad (k = 1, 2, \ldots, n). \quad (5)$$

We note that in view of what was said in Sec. 3, the numerator of this expression is the determinant, which is obtained from the determinant Δ by replacing the elements of the kth column (i.e., the coefficients a_{ik} of x_k) by the constant terms b_i. Thus we have the following theorem:

Cramer's rule: If the determinant Δ of the system (1) *is different from zero, then the system has a unique solution given by* (5). *According to this formula, each of the unknowns is given as the quotient of two determinants. The denominator is the determinant of the system, while the numerator is the determinant obtained from the one in the denominator when the coefficients of the unknown being solved for are replaced by the corresponding constant terms of the system.*

In the case of a large number of equations, the use of Cramer's rule is not convenient. There exist approximate practical methods for solving systems of many equations in many unknowns, which we shall not go into here.

9. The General Case. We now consider the general case of m inhomogeneous linear equations in n unknowns:

$$\begin{aligned}
X_1 &= a_{11}x_1 + a_{12}x_2 + \cdots + a_{1k}x_k + a_{1,k+1}x_{k+1} + \cdots \\
&\qquad + a_{1n}x_n = b_1, \\
X_2 &= a_{21}x_1 + a_{22}x_2 + \cdots + a_{2k}x_k + a_{2,k+1}x_{k+1} + \cdots \\
&\qquad + a_{2n}x_n = b_2, \\
&\cdots\cdots\cdots\cdots\cdots\cdots\cdots\cdots\cdots\cdots\cdots\cdots \\
X_k &= a_{k1}x_1 + a_{k2}x_2 + \cdots + a_{kk}x_k + a_{k,k+1}x_{k+1} + \cdots \\
&\qquad + a_{kn}x_n = b_k, \\
X_{k+1} &= a_{k+1,1}x_1 + a_{k+1,2}x_2 + \cdots + a_{k+1,k}x_k + a_{k+1,k+1}x_{k+1} + \cdots \\
&\qquad + a_{k+1,n}x_n = b_{k+1}, \\
&\cdots\cdots\cdots\cdots\cdots\cdots\cdots\cdots\cdots\cdots\cdots\cdots \\
X_m &= a_{m1}x_1 + a_{m2}x_2 + \cdots + a_{mk}x_k + a_{m,k+1}x_{k+1} + \cdots \\
&\qquad + a_{mn}x_n = b_m.
\end{aligned} \tag{6}$$

To simplify subsequent considerations, we have denoted the entire left-hand side of the sth equation by X_s. The coefficients a_{ik} of this system form a rectangular matrix (array) $\|b_{ik}\|$ of rank k, say. By interchanging rows and columns, i.e., by renumbering the equations and the unknowns, we can cause a nonzero determinant of order k to appear in the upper left-hand corner of $\|a_{ik}\|$. We shall call this determinant the *principal determinant* of the system; it has the form

$$\Delta = \begin{vmatrix} a_{11} & a_{12} & \cdots & a_{1k} \\ a_{21} & a_{22} & \cdots & a_{2k} \\ \cdots & \cdots & \cdots & \cdots \\ a_{k1} & a_{k2} & \cdots & a_{kk} \end{vmatrix}. \tag{7}$$

We now form $m - k$ determinants of order $k + 1$, called the *characteristic determinants of the system; they are obtained from the principal determinant by adding to it another row, consisting of coefficients of an equation with a higher index than k, and another column, consisting of the constant terms.* More precisely, we define the characteristic determinants by the following formula:

$$\Delta_{k+s} = \begin{vmatrix} a_{11} & a_{12} & \cdots & a_{1k} & b_1 \\ a_{21} & a_{22} & \cdots & a_{2k} & b_2 \\ \cdots & \cdots & \cdots & \cdots & \cdots \\ a_{k1} & a_{k2} & \cdots & a_{kk} & b_k \\ a_{k+s,1} & a_{k+s,2} & \cdots & a_{k+s,k} & b_{k+s} \end{vmatrix} \qquad (s = 1, 2, \ldots, m - k). \tag{8}$$

(If $k = m$, i.e., if the rank equals the number of equations, then there are no characteristic determinants.) In addition to the characteristic determinants, we consider other determinants which are obtained from the characteristic determinants by replacing the constant terms in the last column by the left-hand sides of the system of equations, i.e.,

$$\begin{vmatrix} a_{11} & a_{12} & \cdots & a_{1k} & X_1 \\ a_{21} & a_{22} & \cdots & a_{2k} & X_2 \\ \cdot & \cdot & \cdot & \cdot & \cdot \\ a_{k1} & a_{k2} & \cdots & a_{kk} & X_k \\ a_{k+s,1} & a_{k+s,2} & \cdots & a_{k+s,k} & X_{k+s} \end{vmatrix} \qquad (s = 1, 2, \ldots, m - k). \quad (9)$$

These determinants contain the x_j as well as the given coefficients a_{ik}. However, it is easily seen that the determinants (9) vanish identically. In fact, since

$$X_i = a_{i1}x_1 + a_{i2}x_2 + \cdots + a_{in}x_n,$$

the elements of the last column of the determinant (9) consist of n terms. Therefore, by Property 4 of Sec. 3, each determinant (9) can be represented as a sum of terms of the following form:

$$\begin{vmatrix} a_{11} & a_{12} & \cdots & a_{1k} & a_{1j} \\ a_{21} & a_{22} & \cdots & a_{2k} & a_{2j} \\ \cdot & \cdot & \cdot & \cdot & \cdot \\ a_{k1} & a_{k2} & \cdots & a_{kk} & a_{kj} \\ a_{k+s,1} & a_{k+s,2} & \cdots & a_{k+s,k} & a_{k+s,j} \end{vmatrix} x_j.$$

It is easy to see that the determinant which is the coefficient of x_j vanishes. In fact, if $j \leq k$, the last column of this determinant is the same as one of the preceding columns. If, however, $j > k$, the determinant is a minor of order $k + 1$ of the coefficient matrix $\|a_{ik}\|$, and therefore vanishes, since by hypothesis, the rank of this matrix equals k. Subtracting the determinants (9), which are identically zero, from the characteristic determinants and using Property 4 of determinants (Sec. 3), we can represent the characteristic determinants in the following form:

$$\Delta_{k+s} = \begin{vmatrix} a_{11} & a_{12} & \cdots & a_{1k} & b_1 - X_1 \\ a_{21} & a_{22} & \cdots & a_{2k} & b_2 - X_2 \\ \cdot & \cdot & \cdot & \cdot & \cdot \\ a_{k1} & a_{k2} & \cdots & a_{kk} & b_k - X_k \\ a_{k+s,1} & a_{k+s,2} & \cdots & a_{k+s,k} & b_{k+s} - X_{k+s} \end{vmatrix}$$

$$(s = 1, 2, \ldots, m - k). \quad (10)$$

Here the dependence of the determinants on the x_j is only apparent.

We now assume that our system (6) has a solution

$$x_1 = x_1^{(0)}, x_2 = x_2^{(0)}, \ldots, x_n = x_n^{(0)}.$$

Substituting $x_j = x_j^{(0)}$ in the last column of the determinant (10), we obtain a last column consisting only of zeros, i.e., all the characteristic determinants must vanish. Thus we have the following theorem:

THEOREM 1. *A necessary condition for the system* (6) *to have at least one solution is that all the characteristic determinants* (8) *vanish.*

We now prove the sufficiency of this condition and then give a method for finding all the solutions of the system. Thus, assume that all the characteristic determinants vanish. Take them in the form (10) and expand them with respect to elements of the last column. It is not hard to see that the cofactor of the element $b_{k+s} - X_{k+s}$ equals the principal determinant Δ, which is different from zero, and we can write the condition that all the characteristic determinants vanish in the form

$$a_1^{(k+s)}(b_1 - X_1) + a_2^{(k+s)}(b_2 - X_2) + \cdots + a_k^{(k+s)}(b_k - X_k) + \Delta \cdot (b_{k+s} - X_{k+s}) = 0 \qquad (s = 1, 2, \ldots, m - k), \tag{11}$$

where the $a_p^{(q)}$ are numerical coefficients which do not concern us at all. Next we assume that we have a solution of the first k equations of the system, and we imagine that this solution is substituted for the x_j in the identity (11). Then all the differences

$$b_1 - X_1,\ b_2 - X_2,\ \ldots,\ b_k - X_k$$

vanish, and after the substitution we obtain

$$\Delta \cdot (b_{k+s} - X_{k+s}) = 0$$

or

$$b_{k+s} - X_{k+s} = 0 \qquad (s = 1, 2, \ldots, m - k),$$

since $\Delta \neq 0$. Thus, it turns out that if all the characteristic determinants vanish, then every solution of the first k equations of the system satisfies all the subsequent equations as well. In this case, all that remains is to solve the first k equations, which we do by first transposing to the right-hand side all the unknowns with indices greater than k, so that the equations take the form

$$\begin{aligned} a_{11}x_1 + a_{12}x_2 + \cdots + a_{1k}x_k &= b_1 - a_{1,k+1}x_{k+1} - \cdots - a_{1n}x_n, \\ a_{21}x_1 + a_{22}x_2 + \cdots + a_{2k}x_k &= b_2 - a_{2,k+1}x_{k+1} - \cdots - a_{2n}x_n, \\ \cdots\cdots\cdots & \cdots\cdots\cdots \\ a_{k1}x_1 + a_{k2}x_2 + \cdots + a_{kk}x_k &= b_k - a_{k,k+1}x_{k+1} - \cdots - a_{kn}x_n. \end{aligned} \tag{12}$$

We regard these equations as a system for determining $x_1, x_2, \ldots, x_k$. Since, by hypothesis, the determinant of the system is different from zero, we can obtain a unique solution by using Cramer's rule. However, we observe that the constant terms of our system contain the parameters $x_{k+1}, \ldots, x_n$, which can be assigned arbitrary values. It is an immediate consequence of Cramer's rule that the solution of the system (12) has the form

$$x_j = \alpha_j + \beta_{k+1}^{(j)}x_{k+1} + \cdots + \beta_n^{(j)}x_n \quad (j = 1, 2, \ldots, k), \tag{13}$$

where the α_s and $\beta_p^{(q)}$ are certain numerical coefficients, and $x_{k+1}, \ldots, x_n$

are arbitrary. It follows from the foregoing that these formulas give the most general solution of the system (6) under the assumption that all the characteristic determinants are equal to zero. Thus, we have proved the following theorem:

THEOREM 2. *If all the characteristic determinants of the system* (6) *vanish, then it is sufficient to take only those equations of the system which contain the principal determinant and solve them with respect to the unknowns whose coefficients form the principal determinant. The solution can be obtained by Cramer's rule and gives expressions for k unknowns (where k is the rank of the matrix of the coefficients) in the form of linear functions* (13) *of the remaining n-k unknowns, whose values are completely arbitrary. In this way, we obtain all the solutions of the system* (6).

Comparing Theorems 1 and 2, we arrive at the following conclusion:

THEOREM 3. *A necessary and sufficient condition for the existence of solutions of the system* (6) *is the vanishing of all the characteristic determinants of the system.*

We observe that if $k = n$, i.e., if the rank is equal to the number of unknowns, then the right-hand side of (13) does not contain any of the x_j, and all the unknowns from x_1 to x_n can be determined completely. Thus we have the following theorem:

THEOREM 4. *A necessary and sufficient condition for a system to have a unique solution is that all its characteristic determinants vanish and that the rank of the matrix of its coefficients be equal to the number of unknowns.*

We note that all the preceding considerations are obviously valid in the case where the number of equations is equal to the number of unknowns, i.e., $m = n$.

EXAMPLE. Consider the system of four equations in three unknowns:

$$\begin{aligned} x - 3y - 2z &= -1, \\ 2x + y - 4z &= 3, \\ x + 4y - 2z &= 4, \\ 5x + 6y - 10z &= 10. \end{aligned}$$

We write down the matrix of its coefficients:

$$\begin{Vmatrix} 1 & -3 & -2 \\ 2 & 1 & -4 \\ 1 & 4 & -2 \\ 5 & 6 & -10 \end{Vmatrix}.$$

It is easily verified that all the third order determinants contained in this matrix are equal to zero and that the second order determinant appearing in the upper left-hand corner is different from zero. Thus, this determinant can be taken as the principal determinant, and the rank of the system equals 2. We now form the characteristic determinants. In the present example, there are two of them:

$$\Delta_3 = \begin{vmatrix} 1 & -3 & -1 \\ 2 & 1 & 3 \\ 1 & 4 & 4 \end{vmatrix} = 0, \qquad \Delta_4 = \begin{vmatrix} 1 & -3 & -1 \\ 2 & 1 & 3 \\ 5 & 6 & 10 \end{vmatrix} = 0.$$

They are both zero, and hence the given solution is compatible. Thus, it is sufficient to solve the first two equations with respect to x and y, after transposing the terms in z to the right-hand side:

$$\begin{aligned} x - 3y &= 2z - 1, \\ 2x + y &= 4z + 3. \end{aligned}$$

The solution has the form

$$x = \frac{\begin{vmatrix} 2z - 1 & -3 \\ 4z + 3 & 1 \end{vmatrix}}{\begin{vmatrix} 1 & -3 \\ 2 & 1 \end{vmatrix}} = 2z + 8/7, \qquad y = \frac{\begin{vmatrix} 1 & 2z - 1 \\ 2 & 4z + 3 \end{vmatrix}}{\begin{vmatrix} 1 & -3 \\ 2 & 1 \end{vmatrix}} = 5/7,$$

where z is arbitrary.

10. Homogeneous Systems. A system is called homogeneous if all its constant terms b_i are equal to zero. If a homogeneous system has characteristic determinants, then their last columns consist of zeros only, so that they all vanish. It is clear that every homogeneous system has the solution

$$x_1 = x_2 = \cdots = x_n = 0,$$

which will henceforth be called the *trivial* (or *null*) *solution.* The basic problem for a homogeneous system is the following: Does it have solutions other than the trivial solution, and if so, what is the set of solutions like? We consider first the case where the number of equations equals the number of unknowns. Then the system has the form

$$\begin{aligned} a_{11}x_1 + a_{12}x_2 + \cdots + a_{1n}x_n &= 0, \\ a_{21}x_1 + a_{22}x_2 + \cdots + a_{2n}x_n &= 0, \\ \cdots\cdots\cdots\cdots\cdots\cdots\cdots\cdots \\ a_{n1}x_1 + a_{n2}x_2 + \cdots + a_{nn}x_n &= 0. \end{aligned} \tag{14}$$

If the determinant of this solution does not vanish, then by Cramer's rule, the system has a unique solution, namely, the trivial solution However, if the determinant does vanish, then the rank k of the coefficient matrix will be less than the number of unknowns n, and therefore the values of n-k unknowns will be completely arbitrary and we shall have an infinite set of solutions different from the trivial solution. Thus we arrive at the following basic theorem:

THEOREM 1. *A necessary and sufficient condition for the system* (14) *to have a solution different from the trivial solution is that its determinant be equal to zero.*

We now compare the results which we have obtained for the inhomogeneous system (1) and the homogeneous system (14). If the determinant of the system is different from zero, then the inhomogeneous system (1) has a unique solution and the homogeneous system has only the trivial solution. However, if the determinant of the system equals zero, then the homogeneous system (14) has solutions different from zero, while the inhomogeneous system (1) has in general no solutions at all, since in order for it to have a solution, its constant terms must be such as to make all the characteristic determinants vanish. This parallelism of results will play an important role later. In physics, we encounter homogeneous systems in the study of free vibrations and inhomogeneous systems in the study of forced vibrations. The vanishing of the determinant indicates the presence of free vibrations for a homogeneous system and the appearance of the phenomenon of resonance for an inhomogeneous system.

We now investigate in more detail the solutions of the system (14), in the case where its determinant vanishes. If k is the rank of the coefficient matrix, then obviously $k < n$. According to the theorem proved in the preceding section, we can take the k equations containing the principal determinant and solve them for the corresponding k unknowns. We can assume without loss of generality that these unknowns are $x_1, x_2, \ldots,$ x_k. Then the solution can be obtained in the form

$$x_j = \beta_{k+1}^{(j)} x_{k+1} + \cdots + \beta_n^{(j)} x_n \qquad (j = 1, 2, \ldots, k), \tag{15}$$

where the $\beta_p^{(q)}$ are certain numerical coefficients, and $x_{k+1}, \ldots, x_n$ can take arbitrary values.

We now note a general property of the solutions of the system (14). This property is a direct consequence of the linearity and homogeneity of the system, and may be called the *superposition principle for solutions.* If we have several solutions,

$$x_s = x_s^{(1)},\ x_s = x_s^{(2)},\ \ldots,\ x_s = x_s^{(l)} \qquad (s = 1, 2, \ldots, n) \tag{16}$$

of the system, then, multiplying them by arbitrary constants and adding, we obtain another solution,

$$x_s = C_1 x_s^{(1)} + C_2 x_s^{(2)} + \cdots + C_l x_s^{(l)} \qquad (s = 1, 2, \ldots, n),$$

of the system. Just as in the case of linear differential equations, we call the solutions (16) *linearly independent* if there exists no set of constants C_i with at least one C_i different from zero such that the equation

$$\sum_{i=1}^{l} C_i x_s^{(i)} = 0$$

holds for every s. It is not hard to construct a *complete system of solutions*, i.e., $n - k$ linearly independent solutions which when multiplied by arbitrary constants and added yield all the solutions. In fact, referring to (15), which gives the *general* solution of the system (14), we construct solutions in the following way: In the first solution we set $x_{k+1} = 1$ and all the other x_{k+s} equal to zero, in the second solution we set $x_{k+2} = 1$ and all the other x_{k+s} equal to zero, and so forth, until finally in the $(n - k)$th solution, we set $x_n = 1$ and all the other x_{k+s} equal to zero. It is not hard to see that the solutions so constructed are linearly independent, since each solution has an unknown equal to 1 which takes the value 0 in the other solutions. We denote these solutions as follows:

$$x_s = x_s^{(k+1)},\ x_s = x_s^{(k+2)},\ \ldots,\ x_s = x_s^{(m)} \qquad (s = 1, 2, \ldots, n).$$

We now take any solution of the system (14). It can be obtained from the formulas (15) for certain special values of the x_{k+s}:

$$x_{k+1} = \gamma_{k+1},\ x_{k+2} = \gamma_{k+2},\ \ldots,\ x_n = \gamma_n.$$

It is immediately clear that this solution is a linear combination of the solutions just constructed:

$$x_s = \gamma_{k+1} x_s^{(k+1)} + \gamma_{k+2} x_s^{(k+2)} + \cdots + \gamma_n x_s^{(n)} \qquad (s = 1, 2, \ldots, n).$$

We shall return later (Sec. 14) to the study of solutions of the homogeneous system (14), and we shall show that regardless of how they are chosen, the total number of linearly independent solutions is $n - k$.

We now consider the general case of m homogeneous equations in n unknowns. If $m < n$, then the rank k, which cannot exceed m, is also less than n, and therefore $n - k$ unknowns are arbitrary, i.e., *if the number of homogeneous equations is less than the number of unknowns, then the system has solutions different from zero.* In general, $k \leq n$, and for $k = n$, the system has only the trivial solution.

11. Linear Forms. The study of systems of linear forms is closely related to the problem of solving systems of first degree equations. By a *linear form* in the variables $x_1, x_2, \ldots, x_n$, we mean a linear homogeneous function of these variables. Suppose there are m such linear forms:

$$y_s = a_{s1}x_1 + a_{s2}x_2 + \cdots + a_{sn}x_n \qquad (s = 1, 2, \ldots, m). \quad (17)$$

These forms are called *linearly dependent* if there exist constants α_1, α_2, . . . , α_m, where at least one of the α_i is different from zero, such that the relation

$$\alpha_1 y_1 + \alpha_2 y_2 + \cdots + \alpha_m y_m = 0$$

holds identically in the variables $x_1, x_2, \ldots, x_n$. If there are no such constants, the forms (17) are called *linearly independent.* Then the coefficients of all the variables x_l must be set equal to zero in the identity just written, so that this identity is equivalent to the following system of n equations:

$$\begin{aligned} \alpha_1 a_{11} + \alpha_2 a_{21} + \cdots + \alpha_m a_{m1} &= 0, \\ \alpha_1 a_{12} + \alpha_2 a_{22} + \cdots + \alpha_m a_{m2} &= 0, \\ \cdots\cdots\cdots\cdots\cdots\cdots\cdots\cdots \\ \alpha_1 a_{1n} + \alpha_2 a_{2n} + \cdots + \alpha_m a_{mn} &= 0. \end{aligned}$$

Thus, *the forms y_s are linearly independent if and only if this system of homogeneous equations in $\alpha_1, \alpha_2, \ldots, \alpha_m$ has only the trivial solution.*

The results obtained previously lead to a number of consequences concerning the linear dependence of forms. If $m > n$, the homogeneous system just written actually has nontrivial solutions, and the forms are linearly dependent. A necessary and sufficient condition for the forms to be linearly independent is that k, the rank of the matrix of the coefficients a_{pq}, be equal to m, the number of forms. If $m = n$, i.e., *if the number of forms equals the number of unknowns, then a necessary and sufficient condition for the linear independence of the forms is that the determinant of the matrix be nonvanishing.* In this case, we say that we have a *complete system of linearly independent forms.* If $m \leq n$ and the forms (17) are linearly independent (i.e., if $k = m$), then, given any set of values of y_s, the system of equations (17) can be solved with respect to the variables x_l whose coefficients form a nonvanishing determinant of order k, i.e., *a set of linearly independent forms can take any set of values y_s.* In the case $k = m = n$, if any set of values of y_s is specified, then all the values of x_l are uniquely determined.

Suppose now that $k < m$. By appropriately numbering the forms y_s and the variables x_l, we can assume that there is a nonvanishing determinant of order k in the upper left-hand corner of the matrix $\|a_{pq}\|$. Then the first k forms $y_1, y_2, \ldots, y_k$ are linearly independent, while each of the other forms y_{k+t} can be expressed as a linear combination of the first k forms. To see this, we note that in the case of the first k forms, the rank of the coefficient matrix, which is just k, is equal to the number of forms, so that the forms are linearly independent. However, if we take $k + 1$ forms $y_1, \ldots, y_k, y_{k+t}$, then the rank of the matrix formed by their coefficients, which is also k, is less than the number of forms, so that the forms are linearly dependent, i.e., there exist constants β_s, not all zero, such that

$$\beta_1 y_1 + \cdots + \beta_k y_k + \beta_{k+t} y_{k+t} = 0.$$

Here the coefficient β_{k+t} must be different from zero, since otherwise the forms $y_1, \ldots, y_k$ would be linearly dependent. Thus, we obtain a linear expression for y_{k+t} in terms of the first k forms:

$$y_{k+t} = -\frac{\beta_1}{\beta_{k+t}} y_1 - \cdots - \frac{\beta_k}{\beta_{k+t}} y_k.$$

We call the number k the *rank of the system of forms* (17). On the one hand, this number is equal to the rank of the matrix of coefficients; on the other hand, it is equal to the largest number of linearly independent forms in the system (17).

Suppose that we have k linearly independent forms $y_1, y_2, \ldots, y_k$, where $k < n$. We can assume that the determinant of order k appearing in the upper left-hand corner of the matrix $\|a_{pq}\|$ is different from zero. It is not hard to see that we can enlarge this system of k forms to make a complete system of n linearly independent forms. In fact, to do this it is sufficient to set

$$y_{k+1} = x_{k+1}, \ldots, y_n = x_n.$$

The determinant of the resulting n forms is

$$\begin{vmatrix} a_{11} & a_{12} & \cdots & a_{1k} & a_{1,k+1} & \cdots & a_{1n} \\ a_{21} & a_{22} & \cdots & a_{2k} & a_{2,k+1} & \cdots & a_{2n} \\ \cdots & \cdots & \cdots & \cdots & \cdots & \cdots & \cdots \\ a_{k1} & a_{k2} & \cdots & a_{kk} & a_{k,k+1} & \cdots & a_{kn} \\ 0 & 0 & \cdots & 0 & 1 & \cdots & 0 \\ 0 & 0 & \cdots & 0 & 0 & \cdots & 0 \\ \cdots & \cdots & \cdots & \cdots & \cdots & \cdots & \cdots \\ 0 & 0 & \cdots & 0 & 0 & \cdots & 1 \end{vmatrix}.$$

Expanding this determinant first with respect to the elements of the last row, then with respect to the elements of the next to the last row, etc., we see that it equals the determinant of order k which appears in the upper left-hand corner and is therefore nonvanishing. Hence, the forms $y_1, y_2, \ldots, y_n$ are indeed linearly independent, and we see that *any system of linearly independent forms can be enlarged to make a complete system of linearly independent forms.*

12. n-Dimensional Vector Space. We now give a geometrical formulation of the results obtained above, which will be of great use later. Thus, we introduce the concept of a *vector in an n-dimensional space.* By such a vector, we mean a set of n numbers (in general complex) arranged in a definite order, i.e., *every vector* $\mathbf{x}$ *is a sequence of* n *complex numbers called the components of* $\mathbf{x}$ *and written* $\mathbf{x} = (x_1, x_2, \ldots, x_n)$. *The set of all such vectors forms the* n*-dimensional vector space* R_n.

Two vectors are said to be equal if and only if all their components are equal, i.e., given two vectors

$$\mathbf{u} = (u_1, u_2, \ldots, u_n), \ \mathbf{v} = (v_1, v_2, \ldots, v_n),$$

the vector equation $\mathbf{u} = \mathbf{v}$ is equivalent to the n scalar equations $u_1 = v_1$, $u_2 = v_2, \ldots, u_n = v_n$. We now define the operations of multiplication of vectors by numbers and addition of vectors. Multiplication of a vector by a number is defined as multiplication of all the components of the vector by the number, i.e., if the vector $\mathbf{x}$ has components $x_1, x_2, \ldots, x_n$, then the vector $k\mathbf{x}$ has the components $kx_1, kx_2, \ldots, kx_n$. Similarly, addition of vectors is defined as addition of their components, i.e., given two vectors $\mathbf{x} = (x_1, x_2, \ldots, x_n)$ and $\mathbf{y} = (y_1, y_2, \ldots, y_n)$, then the sum $\mathbf{x} + \mathbf{y}$ has the components $x_1 + y_1, x_2 + y_2, \ldots, x_n + y_n$. By the *zero vector*, we mean the vector $(0, 0, \ldots, 0)$ which has all its components equal to zero; we denote this vector temporarily by $\boldsymbol{\theta}$. Obviously, we have $\boldsymbol{\theta} = 0 \cdot \mathbf{x}$ and $x + \boldsymbol{\theta} = x$, where $\mathbf{x}$ is any vector. Subtraction of vectors is defined as follows: The vector $\mathbf{x} - \mathbf{y}$ has the components $x_1 - y_1, x_2 - y_2, \ldots, x_n - y_n$. It is obvious that $\mathbf{x} - \mathbf{x} = \boldsymbol{\theta}$ and $\mathbf{x} - \mathbf{y} = \mathbf{x} + (-1)\mathbf{y}$, i.e., subtracting the vector $\mathbf{y}$ is equivalent to adding the vector $\mathbf{y}$ multiplied by -1.

In what follows, we shall often have to deal with vector equations. Every such equation is equivalent to n scalar equations, which express the fact that the corresponding components of the two sides of the equation are equal. We shall make no further use of the symbol $\boldsymbol{\theta}$, but we must remember that if zero stands on one side of a vector equation, this zero is understood to be the zero vector. The usual properties of addition and multiplication follow at once from the definitions given above. Thus we have

$$\mathbf{x} + \mathbf{y} = \mathbf{y} + \mathbf{x}; \quad \mathbf{x} + (\mathbf{y} + \mathbf{z}) = (\mathbf{x} + \mathbf{y}) + \mathbf{z};$$
$$(k_1 + k_2)\mathbf{x} = k_1\mathbf{x} + k_2\mathbf{x}; \quad k(\mathbf{x} + \mathbf{y}) = k\mathbf{x} + k\mathbf{y}; \quad k_1(k_2\mathbf{x}) = (k_1k_2)\mathbf{x}.$$

Thus, we can interchange or group terms in a sum of vectors consisting of any number of terms. The relations $\mathbf{x} = \mathbf{z} - \mathbf{y}$ and $\mathbf{y} = \mathbf{z} - \mathbf{x}$ follow from $\mathbf{x} + \mathbf{y} = \mathbf{z}$, and conversely $\mathbf{x} = \mathbf{y} + \mathbf{z}$ follows from $\mathbf{x} - \mathbf{y} = \mathbf{z}$.

We now introduce the concept of linear dependence and linear independence of vectors. *The vectors*

$$\mathbf{x}^{(1)}, \mathbf{x}^{(2)}, \ldots, \mathbf{x}^{(l)} \tag{18}$$

are said to be linearly dependent if there exist constants $C_1, C_2, \ldots, C_l$, *not all equal to zero, such that*

$$C_1\mathbf{x}^{(1)} + C_2\mathbf{x}^{(2)} + \cdots + C_l\mathbf{x}^{(l)} = 0. \tag{19}$$

If no such constants exist, then the vectors (18) *are said to be linearly independent.* If we denote the components of the vector $\mathbf{x}^{(j)}$ by $x_1^{(j)}, x_2^{(j)}, \ldots, x_n^{(j)}$, then clearly (19) is equivalent to the following system of n equations in the unknowns $C_1, C_2, \ldots, C_l$:

$$\begin{aligned} x_1^{(1)}C_1 + x_1^{(2)}C_2 + \cdots + x_1^{(l)}C_l &= 0, \\ x_2^{(1)}C_1 + x_2^{(2)}C_2 + \cdots + x_2^{(l)}C_l &= 0, \\ \cdots\cdots\cdots\cdots\cdots\cdots\cdots\cdots \\ x_n^{(1)}C_1 + x_n^{(2)}C_2 + \cdots + x_n^{(l)}C_l &= 0. \end{aligned} \tag{20}$$

We now give a geometrical formulation of our previous results concerning homogeneous systems; this will lead to a variety of interesting consequences. First of all, suppose that $l > n$, i.e., that the number of vectors is greater than the number of dimensions of the space. Then the number of equations in the homogeneous system (20) is less than the number of unknowns, and as we know, the system must have nontrivial solutions, i.e., our vectors must be linearly dependent. In other words, *the number of linearly independent vectors is at most equal to the number of dimensions of the space.* Next consider the case $l = n$. Then the system (20) has the same number of equations and unknowns and will have a nontrivial solution if and only if its determinant vanishes. Thus, if we have n vectors in an n-dimensional space, and if we form a determinant from the n^2 components of the vectors, e.g., by putting the components of each vector in a definite column and making the number of the row the same as the number of the component, then a necessary and sufficient condition for the vectors to be linearly independent is that this determinant be different from zero. The value of the determinant is analogous to the volume of a parallelepiped in three-dimensional space.

In any determinant $|b_{ik}|$ of order n, we can regard the elements $b_{1k}, b_{2k}, \ldots, b_{nk}$ of every column as the components of a vector $\mathbf{b}^{(k)}$; then the value of the determinant will be a function of the n vectors $\mathbf{b}^{(1)}, \mathbf{b}^{(2)}, \ldots, \mathbf{b}^{(n)}$. The vanishing of the determinant implies the linear dependence of the vectors, and conversely. To express the determinant as a function of the vectors $\mathbf{b}^{(k)}$, we write

$$|b_{ik}| = \Delta(\mathbf{b}^{(1)}, \mathbf{b}^{(2)}, \ldots, \mathbf{b}^{(n)}).$$

Recalling that a determinant changes its sign when two of its columns are interchanged, we can assert that the function Δ changes its sign when two of its arguments are interchanged. Such a function is called *antisymmetric.* For example, it is easily seen that the Vandermonde determinant D_n, which we considered in Sec. 5, is also an antisymmetric function of its arguments $x_1, x_2, \ldots, x_n$.

We now return to our study of the system (20) and ask when the vectors $\mathbf{x}^{(1)}, \mathbf{x}^{(2)}, \ldots, \mathbf{x}^{(l)}$ are linearly independent, assuming that $l \leq n$. We denote by k the rank of the matrix formed by the components $x_p^{(q)}$. If $k = l$, then, as we have seen, the system has only the trivial solution, so that the vectors are linearly independent. However, if $k < l$, the system must have a nontrivial solution, i.e., *a necessary and sufficient condition for a set of vectors to be linearly independent is that the rank of the matrix formed from the components of the vectors be equal to the number of vectors in the set.* We now assume that $k < l$, so that the vectors $\mathbf{x}^{(1)}, \mathbf{x}^{(2)}, \ldots, \mathbf{x}^{(l)}$ are linearly dependent. We choose k of these vectors (this can possibly be done in several ways) whose components, regarded as columns of numbers, contain a nonvanishing determinant of order k. According to what has just been proved, these vectors are linearly independent. It is easy to see that each of the remaining vectors can be expressed as a linear combination of these k vectors. In fact, let $\mathbf{x}^{(1)}, \mathbf{x}^{(2)}, \ldots, \mathbf{x}^{(k)}$ be the linearly independent vectors. These vectors, taken together with any other vector $\mathbf{x}^{(k+s)}$, form a set of $k + 1$ linearly dependent vectors, since the rank k of the matrix formed from their coefficients is smaller than the number of vectors. Hence, there exist constants C_i $(i = 1, 2, \ldots, k, k + s)$ which are not all zero, such that

$$C_1\mathbf{x}^{(1)} + C_2\mathbf{x}^{(2)} + \cdots + C_k\mathbf{x}^{(k)} + C_{k+s}\mathbf{x}^{(k+s)} = 0.$$

Moreover, C_{k+s} is certainly not zero, since otherwise the vectors $\mathbf{x}^{(1)}$, $\mathbf{x}^{(2)}, \ldots, \mathbf{x}^{(k)}$ would be linearly dependent. Thus, it follows that

$$\mathbf{x}^{(k+s)} = -\frac{C_1}{C_{k+s}}\mathbf{x}^{(1)} - \frac{C_2}{C_{k+s}}\mathbf{x}^{(2)} - \cdots - \frac{C_k}{C_{k+s}}\mathbf{x}^{(k)},$$

i.e., $\mathbf{x}^{(k+s)}$ is a linear combination of $\mathbf{x}^{(1)}, \mathbf{x}^{(2)}, \ldots, \mathbf{x}^{(k)}$.

Now let $\mathbf{x}^{(1)}, \mathbf{x}^{(2)}, \ldots, \mathbf{x}^{(n)}$ be any n linearly independent vectors, where n is the dimension of the vector space. As an example of such vectors, we can take the vectors

$$(1, 0, 0, \ldots, 0), (0, 1, 0, \ldots, 0), \ldots, (0, 0, 0, \ldots, 1). \quad (21)$$

If we take any other vector $\mathbf{x}$ whatsoever, then as we have just seen, the $n + 1$ vectors $\mathbf{x}^{(1)}, \mathbf{x}^{(2)}, \ldots, \mathbf{x}^{(n)}, \mathbf{x}$ are linearly dependent; thus

$$C_1\mathbf{x}^{(1)} + C_2\mathbf{x}^{(2)} + \cdots + C_n\mathbf{x}^{(n)} + C\mathbf{x} = 0,$$

where the constant C is certainly different from zero, since otherwise the vectors $\mathbf{x}^{(1)}, \mathbf{x}^{(2)}, \ldots, \mathbf{x}^{(n)}$ would be linearly dependent. It follows from all this that *any vector* $\mathbf{x}$ *can be expressed in terms of* n *linearly independent vectors:*

$$\mathbf{x} = \alpha_1\mathbf{x}^{(1)} + \alpha_2\mathbf{x}^{(2)} + \cdots + \alpha_n\mathbf{x}^{(n)} \qquad \left(\alpha_s = -\frac{C_s}{C}\right). \tag{22}$$

Moreover, it is not hard to see that there exists only one representation of $\mathbf{x}$ in terms of $\mathbf{x}^{(1)}, \mathbf{x}^{(2)}, \ldots, \mathbf{x}^{(n)}$. In fact, if besides the representation just given, there were another representation

$$\mathbf{x} = \beta_1\mathbf{x}^{(1)} + \beta_2\mathbf{x}^{(2)} + \cdots + \beta_n\mathbf{x}^{(n)},$$

where the coefficients β_s are different from the corresponding coefficients α_s, then subtracting the two representations term by term, we would obtain

$$(\alpha_1 - \beta_1)\mathbf{x}^{(1)} + (\alpha_2 - \beta_2)\mathbf{x}^{(2)} + \cdots + (\alpha_n - \beta_n)\mathbf{x}^{(n)} = 0,$$

so that, contrary to hypothesis, the vectors $\mathbf{x}^{(1)}, \mathbf{x}^{(2)}, \ldots, \mathbf{x}^{(n)}$ would be linearly dependent. If we take the vectors $\mathbf{x}^{(1)}, \mathbf{x}^{(2)}, \ldots, \mathbf{x}^{(n)}$ to be the vectors (21), then the numbers α_s in (22) will obviously be just the components of the vector $\mathbf{x} = (x_1, x_2, \ldots, x_n)$. More generally, the numbers α_s are called the *components of* $\mathbf{x}$ *with respect to the basis vectors* $\mathbf{x}^{(1)}, \mathbf{x}^{(2)}, \ldots, \mathbf{x}^{(s)}$. By choosing the α_s to be all possible complex numbers, we obtain all the vectors of our n-dimensional space.

Suppose now that we have k linearly independent vectors

$$\mathbf{x}^{(1)}, \mathbf{x}^{(2)}, \ldots, \mathbf{x}^{(k)}, \tag{23}$$

where $k < n$. We say that the set of all vectors given by the formula

$$\mathbf{y} = C_1\mathbf{x}^{(1)} + C_2\mathbf{x}^{(2)} + \cdots + C_k\mathbf{x}^{(k)}, \tag{23a}$$

where the C_s are arbitrary constants, forms a *subspace* L_k of dimension k, which is *spanned* by the vectors (23). As before, we can show that every vector in L_k has a unique representation in terms of $\mathbf{x}^{(1)}, \mathbf{x}^{(2)}, \ldots, \mathbf{x}^{(k)}$. If a vector $\mathbf{z}$ belongs to L_k, i.e., if $\mathbf{z}$ can be expressed as a linear combination of the vectors $\mathbf{x}^{(1)}, \mathbf{x}^{(2)}, \ldots, \mathbf{x}^{(k)}$, then obviously the vector $c\mathbf{z}$, where c is an arbitrary constant, can also be expressed as a linear combination of $\mathbf{x}^{(1)}, \mathbf{x}^{(2)}, \ldots, \mathbf{x}^{(k)}$ and hence belongs to L_k. Similarly, if $\mathbf{z}^{(1)}$ and $\mathbf{z}^{(2)}$ belong to L_k, their sum $\mathbf{z}^{(1)} + \mathbf{z}^{(2)}$ also belongs to L_k. More generally, *if the vectors* $\mathbf{z}^{(1)}, \mathbf{z}^{(2)}, \ldots, \mathbf{z}^{(p)}$ *all belong to* L_k, *then any linear combination* $\gamma_1\mathbf{z}^{(1)} + \gamma_2\mathbf{z}^{(2)} + \cdots + \gamma_p\mathbf{z}^{(p)}$ also belongs to L_k.

We now take m arbitrary vectors belonging to L_k:

$$\mathbf{y}^{(s)} = C_1^{(s)}\mathbf{x}^{(1)} + C_2^{(s)}\mathbf{x}^{(2)} + \cdots + C_k^{(s)}\mathbf{x}^{(k)} \qquad (s = 1, 2, \ldots, m). \tag{24}$$

Because of the linear independence of the vectors (23), a relation of the form

$$\alpha_1\mathbf{y}^{(1)} + \alpha_2\mathbf{y}^{(2)} + \cdots + \alpha_m\mathbf{y}^{(m)} = 0$$

is equivalent to a system

$$\alpha_1 C_q^{(1)} + \alpha_2 C_q^{(2)} + \cdots + \alpha_m C_q^{(m)} = 0 \qquad (q = 1, 2, \ldots, k)$$

of k homogeneous equations in $\alpha_1, \alpha_2, \ldots, \alpha_k$. If this system has nontrivial solutions, then the vectors (24) are linearly dependent. In particular, if $m > k$, there actually are solutions other than the trivial solution, i.e., every set of more than k vectors of the subspace spanned by the vectors (23) is a set of linearly dependent vectors. From this it follows immediately that the subspace spanned by the linearly independent vectors (23) cannot be spanned by any set of $l < k$ linearly independent vectors $\mathbf{z}^{(1)}, \mathbf{z}^{(2)}, \ldots, \mathbf{z}^{(l)}$. In fact, as just noted, this implies, on the one hand, that the subspace has no more than l linearly independent vectors and, on the other hand, that the linearly independent vectors (23), whose number k is greater than l, belong to the subspace, which is contradictory. Moreover, *any k linearly independent vectors $\mathbf{u}^{(1)}, \mathbf{u}^{(2)}, \ldots, \mathbf{u}^{(k)}$ belonging to the subspace L_k span L_k (in the sense described above)*. To see this, we note first that by the definition of a subspace, every linear combination

$$C_1\mathbf{u}^{(1)} + C_2\mathbf{u}^{(2)} + \cdots + C_k\mathbf{u}^{(k)}$$

belongs to L_k. Now suppose we take any vector $\mathbf{y}$ belonging to L_k. The $k + 1$ vectors $\mathbf{u}^{(1)}, \mathbf{u}^{(2)}, \ldots, \mathbf{u}^{(k)}, \mathbf{y}$ belong to L_k and must therefore be linearly independent, i.e.,

$$\beta_1\mathbf{u}^{(1)} + \beta_2\mathbf{u}^{(2)} + \cdots + \beta_k\mathbf{u}^{(k)} + \gamma\mathbf{y} = 0.$$

Since $\mathbf{u}^{(1)}, \mathbf{u}^{(2)}, \ldots, \mathbf{u}^{(k)}$ are linearly independent, the coefficient γ must be different from zero. Thus, every vector $\mathbf{y}$ in L_k can be expressed as a linear combination of $\mathbf{u}^{(1)}, \mathbf{u}^{(2)}, \ldots, \mathbf{u}^{(k)}$, so that these vectors actually span L_k. Finally, if $m = k$ in (24) and if the determinant of the coefficients $C_p^{(q)}$ is nonvanishing, then the vectors $\mathbf{y}^{(1)}, \mathbf{y}^{(2)}, \ldots, \mathbf{y}^{(k)}$ are linearly independent vectors of L_k. More generally, it is easily seen that the number of linearly independent vectors given by (24) equals the rank of the matrix formed by the coefficients $C_p^{(q)}$.

We have seen above that (*a*) if the vector $\mathbf{z}$ belongs to a subspace L, then for any constant c, the vector $c\mathbf{z}$ also belongs to L, and (*b*) if $\mathbf{z}^{(1)}$ and $\mathbf{z}^{(2)}$ belong to L, then $\mathbf{z}^{(1)} + \mathbf{z}^{(2)}$ also belongs to L. This suggests the following new definition of a subspace: *A subspace is a set of vectors L such that (a) if* $\mathbf{z}$ *belongs to L, then* $c\mathbf{z}$ *also belongs to L, and (b) if* $\mathbf{z}^{(1)}$ *and* $\mathbf{z}^{(2)}$ *belong to L, then* $\mathbf{z}^{(1)} + \mathbf{z}^{(2)}$ *also belongs to L.* From this it follows immediately that every linear combination of vectors of L also belongs to L. We have just seen that the previous definition of a subspace implies the properties postulated in the new definition. Conversely, we shall now show that the new definition implies the previous definition. This means that the two definitions are equivalent.

Take any vector $\mathbf{x}^{(1)}$ belonging to L. By the definition of the subspace L, the vectors $C_1\mathbf{x}^{(1)}$, where C_1 is arbitrary, also belong to L. If L consists only of these vectors, then L is the same as a space L_1 as previously defined; otherwise, there is a vector $\mathbf{x}^{(2)}$ in L which is linearly independent of $\mathbf{x}^{(1)}$. Then the vectors $C_1\mathbf{x}^{(1)} + C_2\mathbf{x}^{(2)}$, where C_1 and C_2 are arbitrary, belong to L. If L consists only of these vectors, then L is the same as a space L_2 as previously defined; otherwise, there is a vector $\mathbf{x}^{(3)}$ in L such that $\mathbf{x}^{(1)}$, $\mathbf{x}^{(2)}$, $\mathbf{x}^{(3)}$ are linearly independent. Continuing in this way, we finally find a finite number of linearly independent vectors whose linear combinations exhaust L, since there can be no more than n linearly independent vectors. The total number k of these vectors $\mathbf{x}^{(s)}$ gives the dimension of the space L. (If it turns out that $k = n$, then L is just the whole n-dimensional space.) Thus, the two definitions of a subspace are equivalent.

We note the following fact concerning the structure of subspaces. Suppose that the vectors $\mathbf{x}^{(1)}, \mathbf{x}^{(2)}, \ldots, \mathbf{x}^{(k)}$ are linearly *dependent.* Then, just as before, the vectors (23*a*) define a subspace L. Suppose that the first l vectors $\mathbf{x}^{(1)}, \mathbf{x}^{(2)}, \ldots, \mathbf{x}^{(l)}$ are linearly independent and that each of the other vectors $\mathbf{x}^{(l+1)}, \mathbf{x}^{(l+2)}, \ldots, \mathbf{x}^{(k)}$ can be expressed as a linear combination of the first l vectors. Then the set of vectors defined by (23*a*) obviously coincides with the set of vectors defined by the formula

$$\mathbf{y} = C_1\mathbf{x}^{(1)} + C_2\mathbf{x}^{(2)} + \cdots + C_l\mathbf{x}^{(l)},$$

i.e., the subspace L spanned by the linearly dependent vectors $\mathbf{x}^{(1)}, \mathbf{x}^{(2)}, \ldots, \mathbf{x}^{(k)}$ has the dimension l $(l < k)$.

Finally, consider real three-dimensional space ($n = 3$), and let all the vectors be drawn from a certain fixed point O (the origin). Then, the subspace L_1 is a straight line passing through O, and the subspace L_2 is a plane passing through O.

13. The Scalar Product. First we adopt the following notation. If α is a complex number, the symbol $\bar{\alpha}$ denotes the complex conjugate of α, and the symbol $|\alpha|$ denotes the modulus (absolute value) of α, so that $\alpha\bar{\alpha} = |\alpha|^2$. If α is real, we have $\bar{\alpha} = \alpha$ and $|\alpha|^2 = \alpha^2$. We now introduce a new concept which will play an important role subsequently.

DEFINITION. *By the scalar product of two vectors* $\mathbf{x} = (x_1, x_2, \ldots, x_n)$ *and* $\mathbf{y} = (y_1, y_2, \ldots, y_n)$, *we mean the sum*

$$\sum_{s=1}^{n} x_s \bar{y}_s.$$

We shall denote the scalar product by the symbol $(\mathbf{x}, \mathbf{y})$. Then we have

$$(\mathbf{x}, \mathbf{y}) = \sum_{s=1}^{n} x_s \bar{y}_s, \qquad (\mathbf{y}, \mathbf{x}) = \sum_{s=1}^{n} y_s \bar{x}_s,$$

from which it follows that

$$(\mathbf{y}, \mathbf{x}) = \overline{(\mathbf{x}, \mathbf{y})}.$$

Two vectors are said to be orthogonal or perpendicular (to each other) if their scalar product equals zero. Since the complex conjugate of zero is also zero, the order of the vectors in the scalar product plays no role in the condition for orthogonality. It is easily seen that the zero vector $(0, 0, \ldots, 0)$ is orthogonal to any vector $\mathbf{x}$.

The following properties of the scalar product are immediate consequences of the definition:

$$(\alpha\mathbf{x}, \mathbf{y}) = \alpha(\mathbf{x}, \mathbf{y}); \qquad (\mathbf{x}, \alpha\mathbf{y}) = \bar{\alpha}(\mathbf{x}, \mathbf{y}),$$

where α is a numerical factor. Moreover, we have

$$(\mathbf{x} + \mathbf{y}, \mathbf{z}) = (\mathbf{x}, \mathbf{z}) + (\mathbf{y}, \mathbf{z}); \qquad (\mathbf{x}, \mathbf{y} + \mathbf{z}) = (\mathbf{x}, \mathbf{y}) + (\mathbf{x}, \mathbf{z}),$$

and this *distributive law* holds for any number of summands. One of its consequences is the formula

$$(\mathbf{x} + \mathbf{y}, \mathbf{u} + \mathbf{v}) = (\mathbf{x}, \mathbf{u}) + (\mathbf{x}, \mathbf{v}) + (\mathbf{y}, \mathbf{u}) + (\mathbf{y}, \mathbf{v}).$$

If we form the scalar product of a vector $\mathbf{x} = (x_1, x_2, \ldots, x_n)$ with itself, we obtain

$$(\mathbf{x}, \mathbf{x}) = \sum_{s=1}^{n} x_s \bar{x}_s = \sum_{s=1}^{n} |x_s|^2,$$

i.e., a real number which is positive for any vector $\mathbf{x}$ except the zero vector $(0, 0, \ldots, 0)$ and which is zero for the zero vector. *The (non-negative) square root of the real number* $(\mathbf{x}, \mathbf{x})$ *is called the norm or the length of the vector* $\mathbf{x}$. Denoting the norm by the symbol $\|\mathbf{x}\|$, we can write

$$\|\mathbf{x}\|^2 = (\mathbf{x}, \mathbf{x}) = \sum_{s=1}^{n} |x_s|^2; \qquad \|\mathbf{x}\| = \sqrt{(\mathbf{x}, \mathbf{x})} = \sqrt{\sum_{s=1}^{n} |x_s|^2}.$$

The equation $\|\mathbf{x}\| = 0$ implies that $\mathbf{x}$ is the zero vector.

Suppose that we have three orthogonal vectors $\mathbf{x}$, $\mathbf{y}$ and $\mathbf{z}$, i.e.,

$$(\mathbf{x}, \mathbf{y}) = 0, (\mathbf{x}, \mathbf{z}) = 0, (\mathbf{y}, \mathbf{z}) = 0.$$

It follows from these relations and the distributive law for the scalar product that

$$(\mathbf{x} + \mathbf{y} + \mathbf{z}, \mathbf{x} + \mathbf{y} + \mathbf{z}) = (\mathbf{x}, \mathbf{x}) + (\mathbf{y}, \mathbf{y}) + (\mathbf{z}, \mathbf{z})$$

or

$$\|\mathbf{x} + \mathbf{y} + \mathbf{z}\|^2 = \|\mathbf{x}\|^2 + \|\mathbf{y}\|^2 + \|\mathbf{z}\|^2.$$

This formula is called the *Pythagorean theorem*. It is valid for any number of summands, provided that they are (pairwise) orthogonal vectors. *We now show that if the vectors* $\mathbf{x}^{(1)}, \mathbf{x}^{(2)}, \ldots, \mathbf{x}^{(l)}$ *are nonzero and orthogonal, then they are linearly independent.* To prove this, we write

$$\sum_{s=1}^{l} C_s \mathbf{x}^{(s)} = 0$$

and show that all the numbers C_s must equal zero. Taking the scalar product of both sides of this equation with $\mathbf{x}^{(k)}$, where k is one of the numbers $1, 2, \ldots, l$, we obtain

$$\sum_{s=1}^{l} C_s(\mathbf{x}^{(s)}, \mathbf{x}^{(k)}) = 0.$$

Since the vectors $\mathbf{x}^{(s)}$ are orthogonal in pairs, we have $(\mathbf{x}^{(s)}, \mathbf{x}^{(k)}) = 0$ for $s \neq k$, so that the equation just written reduces to $C_k(\mathbf{x}^{(k)}, \mathbf{x}^{(k)}) = 0$. Thus we have $C_k\|\mathbf{x}^{(k)}\|^2 = 0$, and since $\|\mathbf{x}^{(k)}\|^2 > 0$, this implies that $C_k = 0$ for any k, QED.

14. Geometrical Interpretation of Homogeneous Systems. Consider the homogeneous system

$$\begin{aligned} a_{11}x_1 + a_{12}x_2 + \cdots + a_{1n}x_n &= 0, \\ a_{21}x_1 + a_{22}x_2 + \cdots + a_{2n}x_n &= 0, \\ \cdots\cdots\cdots\cdots\cdots\cdots\cdots\cdots \\ a_{n1}x_1 + a_{n2}x_2 + \cdots + a_{nn}x_n &= 0, \end{aligned} \tag{25}$$

and introduce the vectors

$$\mathbf{a}^{(1)} = (\bar{a}_{11}, \bar{a}_{12}, \ldots, \bar{a}_{1n}), \ldots, \mathbf{a}^{(n)} = (\bar{a}_{n1}, \bar{a}_{n2}, \ldots, \bar{a}_{nn}). \tag{26}$$

Then the system (25) can be written in the following compact form:

$$(\mathbf{x}, \mathbf{a}^{(1)}) = 0, (\mathbf{x}, \mathbf{a}^{(2)}) = 0, \ldots, (\mathbf{x}, \mathbf{a}^{(n)}) = 0. \tag{27}$$

Thus, the problem of solving (25) has been reduced to the problem of finding a vector $\mathbf{x}$ which is orthogonal to all the vectors $\mathbf{a}^{(i)}$. If the determinant $|a_{ik}|$ is nonvanishing, then obviously the determinant $|\bar{a}_{ik}|$, which is its complex conjugate, is also nonvanishing. In this case, the vectors $\mathbf{a}^{(j)}$ are linearly independent and the system (27) has only the trivial solution, since no vector except the zero vector can be orthogonal to each of n linearly independent vectors (in an n-dimensional space).

We now consider the other case, where the determinant of the system (25) vanishes. Suppose that the rank of the system is k. Then, since the determinants appearing in the matrix $\|\bar{a}_{ik}\|$ are the complex conjugates of those appearing in the matrix $\|a_{ik}\|$, the rank of the matrix $\|\bar{a}_{ik}\|$ is also k. Hence, by what has been proved previously, k of the vectors $\mathbf{a}^{(j)}$ are linearly independent, and the rest are linear combinations of these k vectors. Without loss of generality, we can assume that these linearly independent vectors are

$$\mathbf{a}^{(1)}, \mathbf{a}^{(2)}, \ldots, \mathbf{a}^{(k)} \tag{28}$$

and that the rest can be expressed in the form

$$\mathbf{a}^{(k+s)} = \beta_1^{(k+s)}\mathbf{a}^{(1)} + \beta_2^{(k+s)}\mathbf{a}^{(2)} + \cdots + \beta_k^{(k+s)}\mathbf{a}^{(k)} \qquad (s = 1, 2, \ldots, n-k),$$

where the $\beta_p^{(q)}$ are certain numerical coefficients. From this it follows immediately that if $\mathbf{x}$ is orthogonal to the vectors (28), $\mathbf{x}$ is thereby also orthogonal to all the vectors $\mathbf{a}^{(j)}$. In fact, we have

$$(\mathbf{x}, \mathbf{a}^{(k+s)}) = \sum_{i=1}^{k} \beta_i^{(k+s)}(\mathbf{x}, \mathbf{a}^{(i)}) = 0 \qquad (s = 1, 2, \ldots, n-k),$$

since each of the separate terms is zero by hypothesis. Thus, it is sufficient to solve the first k equations of the system. Assuming, as we always do, that a nonvanishing determinant of order k appears in the upper left-hand corner of the matrix $\|a_{ik}\|$, and using the method given in Sec. 12, we obtain $n - k$ linearly independent solutions for $\mathbf{x}$; we denote these solutions by $\mathbf{x}^{(1)}, \mathbf{x}^{(2)}, \ldots, \mathbf{x}^{(n-k)}$. Moreover, any other solution can be represented as a linear combination of these $n - k$ vectors. The situation can be summarized by saying that the vectors given by the formula

$$\mathbf{y} = C_1\mathbf{a}^{(1)} + C_2\mathbf{a}^{(2)} + \cdots + C_k\mathbf{a}^{(k)},$$

where the C_i are arbitrary constants, form a subspace L_k of dimension k, which is a subspace of the whole n-dimensional space; similarly, the vectors $\mathbf{x}^{(1)}, \mathbf{x}^{(2)}, \ldots, \mathbf{x}^{(n-k)}$ span a subspace M_{n-k} of dimension $n - k$. The subspace M_{n-k} is the *orthogonal complement* of the subspace L_k in the sense that every vector in M_{n-k} is orthogonal to every vector in L_k, and conversely. The subspace M_{n-k} consists of the vectors which satisfy the system (27), i.e., which are orthogonal to $\mathbf{a}^{(1)}, \mathbf{a}^{(2)}, \ldots, \mathbf{a}^{(k)}$.

It is not hard to see that the n vectors $\mathbf{a}^{(1)}, \ldots, \mathbf{a}^{(k)}, \mathbf{x}^{(1)}, \ldots, \mathbf{x}^{(n-k)}$ are linearly independent. The proof is as follows: Suppose we have a relation

$$(c_1\mathbf{a}^{(1)} + \cdots + c_k\mathbf{a}^{(k)}) + (d_1\mathbf{x}^{(1)} + \cdots + d_{n-k}\mathbf{x}^{(n-k)}) = 0 \quad (29)$$

between these vectors. The first term in parentheses is a vector $\mathbf{a}$ in L_k, while the second term is a vector $\mathbf{x}$ in M_{n-k}; moreover, we have $\mathbf{a} + \mathbf{x} = 0$ or $\mathbf{a} = -\mathbf{x}$. But the vectors $\mathbf{a}$ and $\mathbf{x}$ are orthogonal. It follows that the vector $\mathbf{a}$ is orthogonal to itself, i.e., $(\mathbf{a}, \mathbf{a}) = 0$ or $\|\mathbf{a}\| = 0$. This implies that the vector $\mathbf{a}$ is the zero vector, and the same is true for the vector $\mathbf{x}$. Thus we have

$$c_1\mathbf{a}^{(1)} + \cdots + c_k\mathbf{a}^{(k)} = 0, \; d_1\mathbf{x}^{(1)} + \cdots + d_{n-k}\mathbf{x}^{(n-k)} = 0.$$

But, by hypothesis, the vectors $\mathbf{a}^{(1)}, \ldots, \mathbf{a}^{(k)}$ are linearly independent, and hence all the constants c_s must vanish; the same is true for the coefficients d_s. It follows that all the coefficients in (29) must vanish, i.e., that the vectors $\mathbf{a}^{(1)}, \ldots, \mathbf{a}^{(k)}, \mathbf{x}^{(1)}, \ldots, \mathbf{x}^{(n-k)}$ are linearly independent, QED. Thus, every vector $\mathbf{x}$ can be uniquely represented in the form

$$x = (\gamma_1\mathbf{a}^{(1)} + \cdots + \gamma_k\mathbf{a}^{(k)}) + (\delta_1\mathbf{x}^{(1)} + \cdots + \delta_{n-k}\mathbf{x}^{(n-k)}),$$

where the first term in parentheses is a vector belonging to L_k, and the second term is a vector belonging to M_{n-k}. As already noted, M_{n-k} consists of all possible solutions of the system (27); since the dimension of M_{n-k} is $n - k$, any complete system of linearly independent solutions consists of $n - k$ solutions. Thus, finally, our study of homogeneous systems yields the following important result: *Given a subspace L_k of dimension k ($k < n$), the vectors orthogonal to L_k form a subspace M_{n-k} of dimension $n - k$, and every vector $\mathbf{x}$ in R_n can be represented in the form of a sum $\mathbf{x} = \mathbf{y} + \mathbf{z}$, where $\mathbf{y}$ belongs to L_k and $\mathbf{z}$ belongs to M_{n-k}.*

To show that this representation is *unique*, suppose that in addition to the representation $\mathbf{x} = \mathbf{y} + \mathbf{z}$, we have another representation $\mathbf{x} = \mathbf{u} + \mathbf{v}$, where $\mathbf{u}$ belongs to L_k and $\mathbf{v}$ to M_{n-k}. We have to prove that $\mathbf{u} = \mathbf{y}$ and $\mathbf{v} = \mathbf{z}$. Since $\mathbf{y} + \mathbf{z} = \mathbf{u} + \mathbf{v}$, we have $\mathbf{y} - \mathbf{u} = \mathbf{v} - \mathbf{z}$. The difference $\mathbf{y} - \mathbf{u}$ belongs to L_k, while the difference $\mathbf{v} - \mathbf{z}$ belongs to M_{n-k}. It follows that $\mathbf{y} - \mathbf{u}$ is orthogonal to itself, i.e.,

$$(\mathbf{y} - \mathbf{u}, \mathbf{y} - \mathbf{u}) = 0$$

or $\|\mathbf{y} - \mathbf{u}\| = 0$, whence $\mathbf{y} - \mathbf{u} = 0$ and $\mathbf{y} = \mathbf{u}$. Then $\mathbf{y} - \mathbf{u} = \mathbf{v} - \mathbf{z}$ implies that $\mathbf{v} = \mathbf{z}$, and the proof is complete.

The vector $\mathbf{y}$ appearing in the representation $\mathbf{x} = \mathbf{y} + \mathbf{z}$ is called the *projection of* $\mathbf{x}$ *onto the subspace* L_k. In this representation, $\mathbf{y}$ and $\mathbf{z}$ are orthogonal and the Pythagorean theorem gives $\|\mathbf{x}\|^2 = \|\mathbf{y}\|^2 + \|\mathbf{z}\|^2$, which implies that $\|\mathbf{y}\| \leq \|\mathbf{x}\|$, with the equality occurring if and only if $\mathbf{z}$ is the zero vector, in which case $\mathbf{x}$ belongs to L_k and $\mathbf{y} = \mathbf{x}$. Similarly, $\|\mathbf{z}\| \leq \|\mathbf{x}\|$, with equality occurring if and only if $\mathbf{x}$ is orthogonal to L_k, in which case $\mathbf{z} = \mathbf{x}$. The subspaces L_k and M_{n-k} are called *orthogonal complements*. If $k = n$, then L_n is the whole space R_n and M_0 reduces to just the zero vector.

As an example, we consider once more the case of real three-dimensional space. Let $k = 2$, so that $n - k = 3 - 2 = 1$. The subspace L_2 is a plane P passing through the origin O, while M_1 is a straight line K passing through O and perpendicular to P. Every vector can be uniquely represented as the sum of two vectors, one of which lies in the plane P and the other on the line K.

In this section we have given a geometrical interpretation of the solution of a homogeneous system in the case where the number of equations equals the number of unknowns. One can do just the same sort of thing in the general case, where the number of vectors $\mathbf{a}^{(s)}$ is not necessarily equal to n. A similar remark applies to the material in the next section.

15. Inhomogeneous Systems. Consider the inhomogeneous system

$$\begin{aligned} a_{11}x_1 + a_{12}x_2 + \cdots + a_{1n}x_n &= b_1, \\ a_{21}x_1 + a_{22}x_2 + \cdots + a_{2n}x_n &= b_2, \\ \cdots\cdots\cdots\cdots\cdots\cdots\cdots\cdots \\ a_{n1}x_1 + a_{n2}x_2 + \cdots + a_{nn}x_n &= b_n. \end{aligned} \tag{30}$$

Solving this system can be interpreted as finding a vector

$$\mathbf{x} = (x_1, x_2, \ldots, x_n)$$

which satisfies the relations

$$(\mathbf{x}, \mathbf{a}^{(1)}) = b_1, \ (\mathbf{x}, \mathbf{a}^{(2)}) = b_2, \ \ldots, \ (\mathbf{x}, \mathbf{a}^{(n)}) = b_n, \tag{31}$$

where the vectors $\mathbf{a}^{(j)}$ are given by (26). If the determinant of the system (30) is nonvanishing, then Cramer's rule gives a unique solution. Suppose now that the determinant vanishes and that the rank of the coefficient matrix $\|a_{ij}\|$ is k. As always, we assume that a nonvanishing determinant of order k appears in the upper left-hand corner of $\|a_{ij}\|$. Next, in addition to the system (30), we consider the system of homogeneous equations whose coefficients are obtained from those of the original system (30) by interchanging rows and columns and changing all numbers to their complex conjugates. This new system is

$$\begin{aligned} \bar{a}_{11}y_1 + \bar{a}_{21}y_2 + \cdots + \bar{a}_{n1}y_n &= 0, \\ \bar{a}_{12}y_1 + \bar{a}_{22}y_2 + \cdots + \bar{a}_{n2}y_n &= 0, \\ \cdots\cdots\cdots\cdots\cdots \\ \bar{a}_{1n}y_1 + \bar{a}_{2n}y_2 + \cdots + \bar{a}_{nn}y_n &= 0. \end{aligned} \tag{32}$$

Just like $\|a_{ij}\|$, the coefficient matrix of (32) has rank k, and the determinant of order k appearing in the upper left-hand corner is nonvanishing. This homogeneous system is said to be the *adjoint* of the system (30). As we have seen above, the general solution of (32) is a linear combination of $n - k$ solutions (vectors), which can be obtained in the following way: Set all the y_{k+s} $(1 \le s \le n - k)$ equal to 0 except one, which is set equal to 1; then solve the first k equations with respect to the unknowns $y_1, y_2, \ldots, y_k$ by Cramer's rule. Thus, for example, setting $y_{k+1} = 1$, we obtain the system

$$\begin{aligned} \bar{a}_{11}y_1 + \bar{a}_{21}y_2 + \cdots + \bar{a}_{k1}y_k &= -\bar{a}_{k+1,1}, \\ \bar{a}_{12}y_1 + \bar{a}_{22}y_2 + \cdots + \bar{a}_{k2}y_k &= -\bar{a}_{k+1,2}, \\ \cdots\cdots\cdots\cdots\cdots \\ \bar{a}_{1k}y_1 + \bar{a}_{2k}y_2 + \cdots + \bar{a}_{kk}y_k &= -\bar{a}_{k+1,k}. \end{aligned}$$

Solving this system and taking complex conjugates, we find that

$$\bar{y}_m = -\frac{\Delta'_m}{\Delta'} \qquad (m = 1, 2, \ldots, k),$$

$$\bar{y}_{k+1} = 1,\ \bar{y}_{k+2} = \bar{y}_{k+3} = \cdots = \bar{y}_n = 0.$$

Here

$$\Delta' = \begin{vmatrix} a_{11} & a_{21} & \cdots & a_{k1} \\ a_{12} & a_{22} & \cdots & a_{k2} \\ \cdots & \cdots & \cdots & \cdots \\ a_{1k} & a_{2k} & \cdots & a_{kk} \end{vmatrix} \neq 0$$

and Δ'_m is obtained from Δ' by replacing the elements of the mth column by $a_{k+1,1}, \ldots, a_{k+1,k}$.

We now write the condition for the vector $\mathbf{b} = (b_1, b_2, \ldots, b_n)$ to be orthogonal to the vector $\mathbf{y}$ just obtained as the solution of (32). This condition is

$$(\mathbf{b}, \mathbf{y}) = -\sum_{m=1}^{k} \frac{\Delta'_m}{\Delta'} b_m + b_{k+1} = 0$$

or

$$-\sum_{m=1}^{k} \Delta'_m b_m + \Delta' b_{k+1} = 0. \tag{33}$$

Next we change columns to rows in the determinant Δ'_m and move the

mth row over to the position of the last row by performing $k - m$ row interchanges. The result is

$$-\Delta_m = \begin{vmatrix} a_{11} & a_{12} & \cdots & a_{1k} \\ \cdot & \cdot & \cdot & \cdot \\ a_{m-1,1} & a_{m-1,2} & \cdots & a_{m-1,k} \\ a_{m+1,1} & a_{m+1,2} & \cdots & a_{m+1,k} \\ \cdot & \cdot & \cdot & \cdot \\ a_{k1} & a_{k2} & \cdots & a_{kk} \\ a_{k+1,1} & a_{k+1,2} & \cdots & a_{k+1,k} \end{vmatrix} \cdot (-1)^{k+1+m},$$

which is just the cofactor of the element b_m in the characteristic determinant

$$\Delta_{k+1} = \begin{vmatrix} a_{11} & a_{12} & \cdots & a_{1k} & b_1 \\ \cdot & \cdot & \cdot & \cdot & \cdot \\ a_{k1} & a_{k2} & \cdots & a_{kk} & b_k \\ a_{k+1,1} & a_{k+1,2} & \cdots & a_{k+1,k} & b_{k+1} \end{vmatrix}.$$

Thus, the condition (33) is equivalent to the vanishing of this characteristic determinant. In the same way, setting $y_{k+s} = 1$, we find quite generally that the condition (33) is equivalent to $\Delta_{k+s} = 0$. Thus, we arrive at the following result: *If the determinant of the system* (30) *vanishes, then a necessary and sufficient condition for the system to have a solution is that the vector* $\mathbf{b} = (b_1, b_2, \ldots, b_n)$ *be orthogonal to all the vectors obtained as solutions of the homogeneous adjoint system* (32). The general solution of the system (30) will be the sum of a particular solution of (30) and the general solution of the corresponding homogeneous system obtained from (30) by replacing all the b_j by zeros. This general solution of the homogeneous system will contain $n - k$ arbitrary constants.

We now mention another geometrical interpretation of the basic theorem on the solution of inhomogeneous systems; this interpretation will be important later. Suppose we have n linear forms:

$$\begin{aligned} y_1 &= a_{11}x_1 + a_{12}x_2 + \cdots + a_{1n}x_n, \\ y_2 &= a_{21}x_1 + a_{22}x_2 + \cdots + a_{2n}x_n, \\ &\cdots\cdots\cdots\cdots\cdots\cdots \\ y_n &= a_{n1}x_1 + a_{n2}x_2 + \cdots + a_{nn}x_n. \end{aligned} \tag{34}$$

Let the x_s take arbitrary complex values, and let $\mathbf{y}$ be the vector with components $y_1, y_2, \ldots, y_n$. Then, if the determinant $|a_{ij}|$ is nonvanishing, definite values of x_k can be found for any given values y_k. Thus, in this case, the vector $\mathbf{y}$ ranges over the whole n-dimensional space when the x_k are given arbitrary values.

Suppose now that the matrix $\|a_{ij}\|$ has rank $r < n$. Without loss of generality, we can assume that the determinant of order r appearing in the upper left-hand corner of this matrix is nonvanishing. Then the basic theorem on the solution of inhomogeneous systems leads to the following conclusion: The set of values $y_1, y_2, \ldots, y_n$ obtained from the formulas (34) has the property that r values $y_1, y_2, \ldots, y_r$ can be chosen arbitrarily. However, when these r values have been chosen, the remaining values $y_{r+1}, y_{r+2}, \ldots, y_n$ are then completely determined; in fact, they are obtained from the condition that the characteristic determinants vanish. Thus, in geometrical language, if the rank of the matrix $\|a_{ij}\|$ is r, the formulas (34) give values of $\mathbf{y} = (y_1, y_2, \ldots, y_n)$ which range over an r-dimensional subspace, i.e., the subspace spanned by the vectors obtained by setting in turn each of the y_s $(s = 1, 2, \ldots, r)$ equal to 1 and the others equal to 0.

In the case just considered, the number of linear forms equals the number of variables x_s. More generally, we have

$$\begin{aligned} y_1 &= a_{11}x_1 + a_{12}x_2 + \cdots + a_{1n}x_n, \\ y_2 &= a_{21}x_1 + a_{22}x_2 + \cdots + a_{2n}x_n, \\ &\cdots\cdots\cdots\cdots\cdots\cdots\cdots\cdots \\ y_m &= a_{m1}x_1 + a_{m2}x_2 + \cdots + a_{mn}x_n; \end{aligned}$$

when the x_s take arbitrary values, the y_s range over an r-dimensional subspace of m-dimensional space, where again r is the rank of the matrix $\|a_{ij}\|$. The proof is identical with the one just given.

16. The Gram Determinant. Hadamard's Inequality. Suppose that we have m vectors

$$\mathbf{x}^{(s)} = (x_1^{(s)}, x_2^{(s)}, \ldots, x_n^{(s)}) \qquad (s = 1, 2, \ldots, m).$$

Using the scalar products $(\mathbf{x}^{(i)}, \mathbf{x}^{(j)})$, we form the determinant

$$G(\mathbf{x}^{(1)}, \mathbf{x}^{(2)}, \ldots, \mathbf{x}^{(m)}) \equiv \begin{vmatrix} (\mathbf{x}^{(1)}, \mathbf{x}^{(1)}) & (\mathbf{x}^{(1)}, \mathbf{x}^{(2)}) & \cdots & (\mathbf{x}^{(1)}, \mathbf{x}^{(m)}) \\ (\mathbf{x}^{(2)}, \mathbf{x}^{(1)}) & (\mathbf{x}^{(2)}, \mathbf{x}^{(2)}) & \cdots & (\mathbf{x}^{(2)}, \mathbf{x}^{(m)}) \\ \cdots & \cdots & \cdots & \cdots \\ (\mathbf{x}^{(m)}, \mathbf{x}^{(1)}) & (\mathbf{x}^{(m)}, \mathbf{x}^{(2)}) & \cdots & (\mathbf{x}^{(m)}, \mathbf{x}^{(m)}) \end{vmatrix}. \tag{35}$$

This determinant is called the *Gram determinant of the vectors* $\mathbf{x}^{(1)}, \mathbf{x}^{(2)}, \ldots, \mathbf{x}^{(m)}$. The general term of the Gram determinant has the form

$$(\mathbf{x}^{(i)}, \mathbf{x}^{(k)}) = \sum_{s=1}^{n} x_s^{(i)}\overline{x_s^{(k)}}.$$

We now consider separately the cases $m = n$, $m < n$, and $m > n$. For $m = n$, the determinant (35) equals the product of the determinants

$$\begin{vmatrix} x_1^{(1)} & x_2^{(1)} & \cdots & x_n^{(1)} \\ x_1^{(2)} & x_2^{(2)} & \cdots & x_n^{(2)} \\ \cdots & \cdots & \cdots & \cdots \\ x_1^{(n)} & x_2^{(n)} & \cdots & x_n^{(n)} \end{vmatrix} \cdot \begin{vmatrix} \overline{x_1^{(1)}} & \overline{x_1^{(2)}} & \cdots & \overline{x_1^{(n)}} \\ \overline{x_2^{(1)}} & \overline{x_2^{(2)}} & \cdots & \overline{x_2^{(n)}} \\ \cdots & \cdots & \cdots & \cdots \\ \overline{x_n^{(1)}} & \overline{x_n^{(2)}} & \cdots & \overline{x_n^{(n)}} \end{vmatrix},$$

where we use the "row by column" rule of multiplication. Since the value of a determinant remains the same when its rows and columns are interchanged, we see that the second factor is the complex conjugate of the first factor. Therefore, for $m = n$, the Gram determinant (35) equals the square of the absolute value of the determinant $|x_k^{(i)}|$ formed from the components $x_k^{(i)}$ of the vectors $\mathbf{x}^{(1)}, \mathbf{x}^{(2)}, \ldots, \mathbf{x}^{(n)}$. Thus the determinant (35) is positive if these vectors are linearly independent and equals zero if they are linearly dependent (Sec. 12). For $m \neq n$, we have the two rectangular matrices

$$\begin{Vmatrix} x_1^{(1)} & x_2^{(1)} & \cdots & x_n^{(1)} \\ x_1^{(2)} & x_2^{(2)} & \cdots & x_n^{(2)} \\ \cdots & \cdots & \cdots & \cdots \\ x_1^{(m)} & x_2^{(m)} & \cdots & x_n^{(m)} \end{Vmatrix} \tag{36a}$$

and

$$\begin{Vmatrix} \overline{x_1^{(1)}} & \overline{x_1^{(2)}} & \cdots & \overline{x_1^{(m)}} \\ \overline{x_2^{(1)}} & \overline{x_2^{(2)}} & \cdots & \overline{x_2^{(m)}} \\ \cdots & \cdots & \cdots & \cdots \\ \overline{x_n^{(1)}} & \overline{x_n^{(2)}} & \cdots & \overline{x_n^{(m)}} \end{Vmatrix}, \tag{36b}$$

and the matrix corresponding to the determinant (35) is the product of these two matrices (Sec. 7). By the theorem proved in Sec. 7, the determinant (35) equals zero for $m > n$. But in this case the vectors $\mathbf{x}^{(1)}, \mathbf{x}^{(2)}, \ldots, \mathbf{x}^{(m)}$ are linearly dependent (Sec. 12). For $m < n$, by the theorem just cited, we have

$$G(\mathbf{x}^{(1)}, \mathbf{x}^{(2)}, \ldots, \mathbf{x}^{(m)}) = \sum_{r_1 < r_2 < \cdots < r_m} X\begin{pmatrix} 1, & 2, & \ldots, & m \\ r_1, & r_2, & \ldots, & r_m \end{pmatrix} Y\begin{pmatrix} r_1, & r_2, & \ldots, & r_m \\ 1, & 2, & \ldots, & m \end{pmatrix},$$

where

$$X\begin{pmatrix} 1, & 2, & \ldots, & m \\ r_1, & r_2, & \ldots, & r_m \end{pmatrix}, \qquad Y\begin{pmatrix} r_1, & r_2, & \ldots, & r_m \\ 1, & 2, & \ldots, & m \end{pmatrix} \tag{36c}$$

denote minors of the matrices (36a) and (36b), respectively. By a previous argument, the quantities (36c) are complex conjugates, and the last formula can be rewritten in the form

$$G(\mathbf{x}^{(1)}, \mathbf{x}^{(2)}, \ldots, \mathbf{x}^{(m)}) = \sum_{r_1 < r_2 < \cdots < r_m} \left| X\begin{pmatrix} 1, & 2, & \ldots, & m \\ r_1, & r_2, & \ldots, & r_m \end{pmatrix} \right|^2. \tag{37}$$

If the vectors $\mathbf{x}^{(1)}, \mathbf{x}^{(2)}, \ldots, \mathbf{x}^{(m)}$ are linearly independent, then the rank of the matrix (36a) equals m (Sec. 12), and at least one of the nonnegative terms appearing in the right-hand side of (37) is positive. However, if these vectors are linearly dependent, then the rank of the matrix (36a) is less than m, and all the determinants of order m appearing in this matrix vanish. In this case, it follows from (37) that $G(\mathbf{x}^{(1)}, \mathbf{x}^{(2)}, \ldots, \mathbf{x}^{(m)}) = 0$. Thus, consideration of all three cases $m = n$, $m > n$, and $m < n$ leads us to the following general theorem:

THEOREM. *The Gram determinant* $G(\mathbf{x}^{(1)}, \mathbf{x}^{(2)}, \ldots, \mathbf{x}^{(m)})$ *is positive if the vectors* $\mathbf{x}^{(1)}, \mathbf{x}^{(2)}, \ldots, \mathbf{x}^{(m)}$ *are linearly independent and vanishes if they are linearly dependent.*

We now prove another formula involving Gram determinants. Let $\mathbf{x}$ be any vector in R_n, and let $\mathbf{x}$ satisfy the equation $\mathbf{x} = \mathbf{y} + \mathbf{z}$, where $\mathbf{y}$ belongs to the subspace spanned by the vectors $\mathbf{x}^{(1)}, \mathbf{x}^{(2)}, \ldots, \mathbf{x}^{(m)}$ and $\mathbf{z}$ is orthogonal to this subspace. The formula which we wish to prove is

$$G(\mathbf{x}^{(1)}, \mathbf{x}^{(2)}, \ldots, \mathbf{x}^{(m)}, \mathbf{x}) = \|\mathbf{z}\|^2 G(\mathbf{x}^{(1)}, \mathbf{x}^{(2)}, \ldots, \mathbf{x}^{(m)}). \tag{38}$$

Since $\mathbf{z}$ is orthogonal to all the vectors $\mathbf{x}^{(s)}$, we have the relations

$$(\mathbf{x}^{(s)}, \mathbf{x}) = (\mathbf{x}^{(s)}, \mathbf{y}), \qquad (\mathbf{x}, \mathbf{x}^{(s)}) = (\mathbf{y}, \mathbf{x}^{(s)}).$$

Using these relations and the formula $(\mathbf{x}, \mathbf{x}) = (\mathbf{y}, \mathbf{y}) + (\mathbf{z}, \mathbf{z})$ (Sec. 13), we can write

$$G(\mathbf{x}^{(1)}, \mathbf{x}^{(2)}, \ldots, \mathbf{x}^{(m)}, \mathbf{x})$$
$$= \begin{vmatrix} (\mathbf{x}^{(1)}, \mathbf{x}^{(1)}) & (\mathbf{x}^{(1)}, \mathbf{x}^{(2)}) & \cdots & (\mathbf{x}^{(1)}, \mathbf{x}^{(m)}) & (\mathbf{x}^{(1)}, \mathbf{y}) \\ (\mathbf{x}^{(2)}, \mathbf{x}^{(1)}) & (\mathbf{x}^{(2)}, \mathbf{x}^{(2)}) & \cdots & (\mathbf{x}^{(2)}, \mathbf{x}^{(m)}) & (\mathbf{x}^{(2)}, \mathbf{y}) \\ \cdots & \cdots & \cdots & \cdots & \cdots \\ (\mathbf{x}^{(m)}, \mathbf{x}^{(1)}) & (\mathbf{x}^{(m)}, \mathbf{x}^{(2)}) & \cdots & (\mathbf{x}^{(m)}, \mathbf{x}^{(m)}) & (\mathbf{x}^{(m)}, \mathbf{y}) \\ (\mathbf{y}, \mathbf{x}^{(1)}) & (\mathbf{y}, \mathbf{x}^{(2)}) & \cdots & (\mathbf{y}, \mathbf{x}^{(m)}) & (\mathbf{y}, \mathbf{y}) + (\mathbf{z}, \mathbf{z}) \end{vmatrix}.$$

Next we write the elements of the last row in the form

$$(\mathbf{y}, \mathbf{x}^{(1)}) + 0, (\mathbf{y}, \mathbf{x}^{(2)}) + 0, \ldots, (\mathbf{y}, \mathbf{x}^{(m)}) + 0, (\mathbf{y}, \mathbf{y}) + (\mathbf{z}, \mathbf{z}),$$

and then represent the determinant as a sum of two determinants by using Property 4 of Sec. 3. The result is

$$G(\mathbf{x}^{(1)}, \mathbf{x}^{(2)}, \ldots, \mathbf{x}^{(m)}, \mathbf{x}) = G(\mathbf{x}^{(1)}, \mathbf{x}^{(2)}, \ldots, \mathbf{x}^{(m)}, \mathbf{y})$$
$$+ \begin{vmatrix} (\mathbf{x}^{(1)}, \mathbf{x}^{(1)}) & (\mathbf{x}^{(1)}, \mathbf{x}^{(2)}) & \cdots & (\mathbf{x}^{(1)}, \mathbf{x}^{(m)}) & (\mathbf{x}^{(1)}, \mathbf{y}) \\ (\mathbf{x}^{(2)}, \mathbf{x}^{(1)}) & (\mathbf{x}^{(2)}, \mathbf{x}^{(2)}) & \cdots & (\mathbf{x}^{(2)}, \mathbf{x}^{(m)}) & (\mathbf{x}^{(2)}, \mathbf{y}) \\ \cdots & \cdots & \cdots & \cdots & \cdots \\ (\mathbf{x}^{(m)}, \mathbf{x}^{(1)}) & (\mathbf{x}^{(m)}, \mathbf{x}^{(2)}) & \cdots & (\mathbf{x}^{(m)}, \mathbf{x}^{(m)}) & (\mathbf{x}^{(m)}, \mathbf{y}) \\ 0 & 0 & \cdots & 0 & \|\mathbf{z}\|^2 \end{vmatrix}. \tag{39}$$

The vector $\mathbf{y}$ belongs to the subspace spanned by the vectors $\mathbf{x}^{(1)}, \mathbf{x}^{(2)}, \ldots, \mathbf{x}^{(m)}$, i.e., $\mathbf{y}$ is a linear combination of the $\mathbf{x}^{(s)}$; therefore, by the theorem proved above, we have

$$G(\mathbf{x}^{(1)}, \mathbf{x}^{(2)}, \ldots, \mathbf{x}^{(m)}, \mathbf{y}) = 0.$$

Finally, expanding the determinant in (39) with respect to elements of the last row, we obtain (38).

An immediate consequence of this formula is the inequality

$$G(\mathbf{x}^{(1)}, \mathbf{x}^{(2)}, \ldots, \mathbf{x}^{(m)}, \mathbf{x}) \leq \|\mathbf{x}\|^2 G(\mathbf{x}^{(1)}, \mathbf{x}^{(2)}, \ldots, \mathbf{x}^{(m)}). \tag{40}$$

We note that

$$G(\mathbf{x}^{(1)}, \mathbf{x}^{(2)}, \ldots, \mathbf{x}^{(m)}, \mathbf{x}) = G(\mathbf{x}^{(1)}, \mathbf{x}^{(2)}, \ldots, \mathbf{x}^{(m)}) = 0$$

if the vectors $\mathbf{x}^{(s)}$ are linearly dependent. If the $\mathbf{x}^{(s)}$ are linearly independent, then the equality sign holds in the relation (40) if and only if $\mathbf{y} = 0$, i.e., if and only if $\mathbf{x}$ is orthogonal to all the $\mathbf{x}^{(s)}$. If we apply the inequality (40) several times in succession to the original Gram determinant, we obtain the following estimate:

$$G(\mathbf{x}^{(1)}, \mathbf{x}^{(2)}, \ldots, \mathbf{x}^{(m)}) \leq \|\mathbf{x}^{(1)}\|^2 \|\mathbf{x}^{(2)}\|^2 \cdots \|\mathbf{x}^{(m)}\|^2, \tag{41}$$

where we bear in mind that $G(\mathbf{x}^{(1)}) = \|\mathbf{x}^{(1)}\|^2$. The equality sign holds in (41) if and only if the vectors are orthogonal in pairs. (It is assumed that none of the vectors is the zero vector.)

From the inequality just proved, we can easily deduce an estimate for any determinant. Let Δ be a determinant of order n with elements a_{ik}. We shall regard the elements $a_{i1}, a_{i2}, \ldots, a_{in}$ of the ith row of Δ as the components of a vector $\mathbf{x}^{(i)}$ in R_n. Then we form the new determinant whose elements $\bar{a}_{ik}$ are the complex conjugates of the a_{ik}; this determinant is obviously equal to $\bar{\Delta}$. Multiplying Δ by $\bar{\Delta}$, using the "row by row" rule, we obtain the Gram determinant $G(\mathbf{x}^{(1)}, \mathbf{x}^{(2)}, \ldots, \mathbf{x}^{(n)})$. By the multiplication theorem for determinants,

$$G(\mathbf{x}^{(1)}, \mathbf{x}^{(2)}, \ldots, \mathbf{x}^{(n)}) = \Delta\bar{\Delta} = |\Delta|^2.$$

Applying the inequality (41), we obtain the following estimate of the absolute value of a determinant, known as *Hadamard's inequality:*

$$|\Delta|^2 \leq \sum_{k=1}^{n} |a_{1k}|^2 \sum_{k=1}^{n} |a_{2k}|^2 \cdots \sum_{k=1}^{n} |a_{nk}|^2. \tag{42}$$

If the elements of the determinant Δ are real, we can write

$$\Delta^2 \leq \sum_{k=1}^{n} a_{ik}^2 \sum_{k=1}^{n} a_{2k}^2 \cdots \sum_{k=1}^{n} a_{nk}^2. \tag{43}$$

If the estimate

$$|a_{ik}| \leq M \qquad (i, k = 1, 2, \ldots, n)$$

holds for the elements of Δ, then clearly

$$\sum_{k=1}^{n} |a_{ik}|^2 \leq nM^2,$$

and (42) gives the estimate

$$|\Delta| \leq n^{n/2}M^n. \tag{44}$$

By what was said above, the equality sign holds in (42) if and only if the vectors $\mathbf{x}^{(i)}$ are orthogonal in pairs.

One can also obtain further estimates for the Gram determinant, which are based on an inequality which is a generalization of the inequality (40). Let each of the letters X, Y, Z denote a sequence of vectors of R_n. The inequality we have in mind is the following:

$$G(X, Y, Z)G(X) \leq G(X, Z)G(X, Y). \tag{45}$$

(In this equation, we do not exclude the case of an "empty sequence" of vectors, i.e., a sequence containing no vectors at all; if W is such a sequence, we have to set $G(W) = 1$.) Using this inequality, one can obtain the following estimate for the Gram determinant:

$$G(\mathbf{x}^{(1)}, \mathbf{x}^{(2)}, \ldots, \mathbf{x}^{(m)}) \leq \left[\prod_{k=1}^{m} G(\mathbf{x}^{(1)}, \ldots, \mathbf{x}^{(k-1)}, \mathbf{x}^{(k+1)}, \ldots, \mathbf{x}^{(m)}) \right]^{\frac{1}{m-1}},$$

where $\prod$ is the multiplication sign. Repeated application of this estimate leads to new estimates, in which the Gram determinants contain fewer vectors. In all these estimates, the equality sign holds if and only if the vectors are orthogonal in pairs. (These estimates of the Gram determinant are due to Faguet.†)

17. Systems of Linear Differential Equations with Constant Coefficients. We now apply the results obtained above to the problem of integrating a system of linear differential equations with constant coefficients. Such a system can be written in the following form:‡

$$\begin{aligned}
\dot{x}_1 &= a_{11}x_1 + a_{12}x_2 + \cdots + a_{1n}x_n, \\
\dot{x}_2 &= a_{21}x_1 + a_{22}x_2 + \cdots + a_{2n}x_n, \\
&\cdots\cdots\cdots\cdots\cdots\cdots\cdots \\
\dot{x}_n &= a_{n1}x_1 + a_{n2}x_2 + \cdots + a_{nn}x_n.
\end{aligned} \tag{46}$$

† M. K. Faguet, "A Generalization of the Hadamard Determinant Inequality," *Doklady Akad. Nauk SSSR*, 54:761 (1946).

‡ In this section, the overdot means differentiation with respect to t.

Here the x_j are functions of t which we are trying to determine, the $\dot{x}_j$ are their derivatives, and the a_{ij} are given constants. We shall look for a solution of the form

$$x_1 = b_1 e^{\lambda t},\ x_2 = b_2 e^{\lambda t},\ \ldots,\ x_n = b_n e^{\lambda t}. \tag{47}$$

Substituting these expressions into the system (46) and dividing by the factor $e^{\lambda t}$, we obtain the system of equations

$$\begin{aligned} (a_{11} - \lambda)b_1 + \quad a_{12}b_2 + \cdots + \quad a_{1n}b_n &= 0, \\ a_{21}b_1 + (a_{22} - \lambda)b_2 + \cdots + \quad a_{2n}b_n &= 0, \\ \cdots\cdots\cdots\cdots\cdots\cdots\cdots\cdots \\ a_{n1}b_1 + \quad a_{n2}b_2 + \cdots + (a_{nn} - \lambda)b_n &= 0 \end{aligned} \tag{48}$$

for determining the constants $b_1, b_2, \ldots, b_n$. Since we must find a nontrivial solution for the unknowns b_j, the determinant of the system (48) has to vanish; thus, we obtain an equation of the form

$$\begin{vmatrix} a_{11} - \lambda & a_{12} & \cdots & a_{1n} \\ a_{21} & a_{22} - \lambda & \cdots & a_{2n} \\ \cdots & \cdots & \cdots & \cdots \\ a_{n1} & a_{n2} & \cdots & a_{nn} - \lambda \end{vmatrix} = 0, \tag{49}$$

which determines the constant λ. An equation of this kind is called a *secular equation*. Secular equations are encountered in the study of oscillations of mechanical systems without resistance. In this special case, the coefficient matrix is symmetric, i.e., $a_{ik} = a_{ki}$, and all the coefficients are real, a matter we shall discuss further when we study small oscillations (Sec. 38). At present, we study the system (48) in the general case. Equation (49) is an algebraic equation in λ of degree n, with leading term $(-\lambda)^n$. If this equation has n different roots $\lambda = \lambda_1$, $\lambda = \lambda_2, \ldots, \lambda = \lambda_n$, then, substituting each value $\lambda = \lambda_j$ in turn into the coefficients of the system (48), we obtain n homogeneous equations in the unknowns $b_1, b_2, \ldots, b_n$. Since these equations have nonvanishing determinants, we can find n nontrivial solutions for the unknowns $b_1, b_2, \ldots, b_n$. According to (47), we obtain in this way n linearly independent solutions of the system (46); the general integral of the system is a linear combination of these n solutions. However, if the secular equation (49) has multiple roots, the solution of the problem becomes more complicated. In this case, each root of (49) of multiplicity k is associated with k linearly independent solutions of (46); one of these solutions will certainly be of the form (47), but the rest will in general contain a polynomial in t as an extra factor. However, in the present case, unlike the case of a single differential equation with constant coefficients it may happen that not one but several (or even all) of the solutions corresponding to a multiple root will be of the form (47). We shall not go into this further now, since we shall solve the system (46) by another method in the Appendix, using the theory of functions of a complex variable.

We return to the secular equation (49), which plays a fundamental role in our problem. In the case of large n, even an approximate solution of this equation presents great practical difficulties, caused by the fact that the unknown λ appears in the diagonal terms of the determinant in (49) rather than in any single row or column. Thus, it requires a great deal of work to expand the determinant in powers of λ, as already remarked in Sec. 4. We now present a method for transforming (49) into a more practical form. In this method, due to Krylov,† the unknown λ appears in only one column.

First we form a linear combination

$$\xi = \alpha_{01}x_1 + \alpha_{02}x_2 + \cdots + \alpha_{0n}x_n \tag{50}$$

of the quantities x_j appearing in (46), where the α_{0j} are arbitrarily chosen numerical coefficients. We then differentiate (50) n times with respect to t, each time replacing the derivatives $\dot{x}_j$ appearing in the right-hand side by their expressions as given by the system (46). In this way, we obtain $n + 1$ equations of the form

$$\begin{aligned} \xi &= \alpha_{01}x_1 + \alpha_{02}x_2 + \cdots + \alpha_{0n}x_n, \\ \dot{\xi} &= \alpha_{11}x_1 + \alpha_{12}x_2 + \cdots + \alpha_{1n}x_n, \\ &\cdots\cdots\cdots\cdots\cdots\cdots\cdots\cdots \\ \xi^{(n-1)} &= \alpha_{n-1,1}x_1 + \alpha_{n-1,2}x_2 + \cdots + \alpha_{n-1,n}x_n, \\ \xi^{(n)} &= \alpha_{n1}x_1 + \alpha_{n2}x_2 + \cdots + \alpha_{nn}x_n. \end{aligned} \tag{51}$$

We now make the following assumption: *The determinant formed from the coefficients α_{ik} appearing in the first n equations is nonvanishing.* Then the first n equations give us expressions for the x_j in terms of $\xi, \dot{\xi}, \ldots, \xi^{(n-1)}$; substituting these expressions into the last equation, we obtain an nth order differential equation for ξ.

The elimination of the x_j from the $n + 1$ equations (51) can also be performed directly by using determinants. To do this, we rewrite the equations in the form

$$\begin{aligned} \xi x_0 + \alpha_{01}x_1 + \alpha_{02}x_2 + \cdots + \alpha_{0n}x_n &= 0, \\ \dot{\xi} x_0 + \alpha_{11}x_1 + \alpha_{12}x_2 + \cdots + \alpha_{1n}x_n &= 0, \\ \cdots\cdots\cdots\cdots\cdots\cdots\cdots\cdots \\ \xi^{(n)}x_0 + \alpha_{n1}x_1 + \alpha_{n2}x_2 + \cdots + \alpha_{nn}x_n &= 0, \end{aligned}$$

where $x_0 = -1$. Then we regard these new $n + 1$ equations as homo-

† A. N. Krylov, "On the Numerical Solution of the Equation Defining the Frequencies of Small Oscillations of Mechanical Systems," *Izvestiya Akad. Nauk SSSR, Otdel Mat. i Yest. Nauk,* 7:491 (1931).

geneous equations in the quantities $x_0, x_1, \ldots, x_n$. The determinant of the system must vanish, which gives us the equation

$$\begin{vmatrix} \xi & \alpha_{01} & \alpha_{02} & \cdots & \alpha_{0n} \\ \dot{\xi} & \alpha_{11} & \alpha_{12} & \cdots & \alpha_{1n} \\ \cdots & \cdots & \cdots & \cdots & \cdots \\ \xi^{(n)} & \alpha_{n1} & \alpha_{n2} & \cdots & \alpha_{nn} \end{vmatrix} = 0, \tag{52}$$

in which the elimination has been performed directly, as desired.

We now look for a solution of (52) in the form

$$\xi = e^{\lambda t}.$$

Substituting this into the first column of the determinant (52) and then factoring $e^{\lambda t}$ out of the first column, we obtain the following equation for λ:

$$\begin{vmatrix} 1 & \alpha_{01} & \alpha_{02} & \cdots & \alpha_{0n} \\ \lambda & \alpha_{11} & \alpha_{12} & \cdots & \alpha_{1n} \\ \cdots & \cdots & \cdots & \cdots & \cdots \\ \lambda^n & \alpha_{n1} & \alpha_{n2} & \cdots & \alpha_{nn} \end{vmatrix} = 0. \tag{53}$$

It is not hard to see that with our assumption, (53) has the same roots as (49). In fact, let λ_0 be a root of (53), so that we have a solution for (52) of the form

$$\xi = Ce^{\lambda_0 t}, \tag{54}$$

where C is an arbitrary constant. Then the first n equations of the system (51) give solutions for the x_j of the form (47) with $\lambda = \lambda_0$, i.e., λ_0 is a root of (49). Conversely, if λ_0 is a root of (49), then we have a solution of the form (47) for the x_j, where $\lambda = \lambda_0$ and the b_j are numerical coefficients which are not all zero. Substituting these expressions for the x_j into the first of the equations of the system (51), we find that ξ has a solution of the form (54). Here C is certainly not equal to zero, since otherwise we would have

$$\xi = \dot{\xi} = \cdots = \xi^{(n-1)} = 0,$$

and the first n equations of the system (51) would immediately imply

$$x_1 = x_2 = \cdots = x_n = 0,$$

contrary to hypothesis; therefore every root of (45) is actually a root of (53) as well. Thus, we finally see that (49) and (53) have the same roots (with the above assumption).

In Krylov's paper (cited above) the reader can find numerical examples of the application of this method as well as a treatment of the case where our assumption is not valid. The simplest calculations are obtained if we write (50) in the form $\xi = x_1$. Then (53) has the following appearance:

$$\begin{vmatrix} 1 & 1 & 0 & \cdots & 0 \\ \lambda & \alpha_{11} & \alpha_{12} & \cdots & \alpha_{1n} \\ \cdots & \cdots & \cdots & \cdots & \cdots \\ \lambda^n & \alpha_{n1} & \alpha_{n2} & \cdots & \alpha_{nn} \end{vmatrix} = 0.\dagger$$

Instead of the system (46), we now consider the following system of second order differential equations:

$$\begin{aligned} \ddot{x}_1 &= a_{11}x_1 + a_{12}x_2 + \cdots + a_{1n}x_n, \\ \ddot{x}_2 &= a_{21}x_1 + a_{22}x_2 + \cdots + a_{2n}x_n, \\ &\cdots\cdots\cdots\cdots\cdots\cdots \\ \ddot{x}_3 &= a_{n1}x_1 + a_{n2}x_2 + \cdots + a_{nn}x_n. \end{aligned} \tag{55}$$

Such systems are frequently encountered in mechanics. If we look for a solution of the form

$$x_j = b_j \cos (\lambda t + \varphi),$$

then we obtain the equation

$$\begin{vmatrix} a_{11} + \lambda^2 & a_{12} & \cdots & a_{1n} \\ a_{21} & a_{22} + \lambda^2 & \cdots & a_{2n} \\ \cdots & \cdots & \cdots & \cdots \\ a_{n1} & a_{n2} & \cdots & a_{nn} + \lambda^2 \end{vmatrix} = 0 \tag{56}$$

for λ. The constants b_j are determined from the system analogous to (48), and φ remains arbitrary.

Finally, we consider a system of the form

$$\begin{aligned} \ddot{x}_1 &= a_{11}x_1 + \cdots + a_{1n}x_n + c_{11}\dot{x}_1 + \cdots + c_{1n}\dot{x}_n, \\ \ddot{x}_2 &= a_{21}x_1 + \cdots + a_{2n}x_n + c_{21}\dot{x}_1 + \cdots + c_{2n}\dot{x}_n, \\ &\cdots\cdots\cdots\cdots\cdots\cdots \\ \ddot{x}_n &= a_{n1}x_1 + \cdots + a_{nn}x_n + c_{n1}\dot{x}_1 + \cdots + c_{nn}\dot{x}_n, \end{aligned} \tag{57}$$

where both the x_j and their first derivatives $\dot{x}_j$ appear in the right-hand side. Looking for a solution of the form (47), we arrive at the secular equation

$$\begin{vmatrix} a_{11} + c_{11}\lambda - \lambda^2 & a_{12} + c_{12}\lambda & \cdots & a_{1n} + c_{1n}\lambda \\ a_{21} + c_{21}\lambda & a_{22} + c_{22}\lambda - \lambda^2 & \cdots & a_{2n} + c_{2n}\lambda \\ \cdots & \cdots & \cdots & \cdots \\ a_{n1} + c_{n1}\lambda & a_{n2} + c_{n2}\lambda & \cdots & a_{nn} + c_{nn}\lambda - \lambda^2 \end{vmatrix} = 0. \tag{58}$$

† For another convenient method of transforming the secular equation, see A. Danilevski, "On the Numerical Solution of the Secular Equation," *Mat. Sbornik*, **2**:169 (1937).

Introducing the additional unknowns

$$x_{n+1} = \dot{x}_1,\ x_{n+2} = \dot{x}_2,\ \ldots,\ x_{2n} = \dot{x}_n, \tag{59}$$

we transform the system (57) into a system of $2n$ first order differential equations, where the first n equations are obtained from (57) by the substitutions

$$\ddot{x}_j = \dot{x}_{n+j},\ \dot{x}_j = x_{n+j} \qquad (j = 1, 2, \ldots, n),$$

and the remaining n are just the equations (59).

18. Jacobians. Suppose we have n functions

$$\varphi_1(x_1, x_2, \ldots, x_n),\ \varphi_2(x_1, x_2, \ldots, x_n),\ \ldots,\ \varphi_n(x_1, x_2, \ldots, x_n) \tag{60}$$

of n variables $x_1, x_2, \ldots, x_n$. By the *Jacobian* (or *functional determinant*) of these functions with respect to the variables x_s is meant the determinant of order n whose elements are given by the formula $a_{ik} = \partial\varphi_i/\partial x_k$. We introduce the following special notation for the Jacobian:

$$\frac{\partial(\varphi_1, \varphi_2, \ldots, \varphi_n)}{\partial(x_1, x_2, \ldots, x_n)} = \begin{vmatrix} \frac{\partial\varphi_1}{\partial x_1} & \frac{\partial\varphi_1}{\partial x_2} & \cdots & \frac{\partial\varphi_1}{\partial x_n} \\ \frac{\partial\varphi_2}{\partial x_1} & \frac{\partial\varphi_2}{\partial x_2} & \cdots & \frac{\partial\varphi_2}{\partial x_n} \\ \cdots & \cdots & \cdots & \cdots \\ \frac{\partial\varphi_n}{\partial x_1} & \frac{\partial\varphi_n}{\partial x_2} & \cdots & \frac{\partial\varphi_n}{\partial x_n} \end{vmatrix}. \tag{61}$$

We have already encountered such determinants previously in an earlier volume in connection with changes of variables in multiple integrals. For example, suppose we make the change of variables

$$x = \varphi(u, v),\ y = \psi(u, v) \tag{62}$$

in two dimensions, which carries the point (u, v) into the point (x, y). Then, under the point transformation (62), an element of area at the point (u, v) has to be multiplied by the absolute value of the Jacobian

$$\frac{\partial(\varphi, \psi)}{\partial(u, v)}. \tag{63}$$

Here we assume that in the region where the transformation (62) is applied, the partial derivatives of the functions (62) with respect to u and v are continuous and the determinant (63) does not vanish. Similarly, suppose we have a point transformation

$$x = \varphi(q_1, q_2, q_3),\ y = \psi(q_1, q_2, q_3),\ z = \omega(q_1, q_2, q_3)$$

in three dimensions, which carries the point (q_1, q_2, q_3) into the point (x, y, z) and the volume V_1 into the volume V. Then the formula for changing variables in a triple integral has the form

$$\iiint_V f(x, y, z)\, dx\, dy\, dz = \iiint_{V_1} f(\varphi, \psi, \omega)|D|\, dq_1, dq_2, dq_3.$$

Here

$$D = \frac{\partial(\varphi, \psi, \omega)}{\partial(q_1, q_2, q_3)},$$

and $|D|$ gives the relative change of the volume element $dq_1\, dq_2\, dq_3$ in going from (q_1, q_2, q_3) to (x, y, z). In just the same way, we may regard a function $u = f(x)$ of one variable x as a point transformation defined on the x-axis, carrying the point with abscissa x into the point with abscissa u. Clearly, in this case, $|f'(x)|$ characterizes the relative change of the line element dx in going from x to u. Thus, in the cases of two and three dimensions, there is an analogy between the Jacobian and the derivative; we now show that this analogy carries over to the formal properties of Jacobians and derivatives. (All that we have said can also be extended to the case of a point transformation in n-dimensional space and to the case of a change of variables in an n-fold integral.)

Consider the system of functions

$$\varphi_1(y_1, y_2, \ldots, y_n),\ \varphi_2(y_1, y_2, \ldots, y_n),\ \ldots,\ \varphi_n(y_1, y_2, \ldots, y_n),$$

where it is assumed that $y_1, y_2, \ldots, y_n$ are not independent variables, but are themselves functions of the variables $x_1, x_2, \ldots, x_n$; thus, the functions φ_i are also functions of the variables x_i. We form the Jacobians

$$\frac{\partial(\varphi_1, \varphi_2, \ldots, \varphi_n)}{\partial(y_1, y_2, \ldots, y_n)},\quad \frac{\partial(\varphi_1, \varphi_2, \ldots, \varphi_n)}{\partial(x_1, x_2, \ldots, x_n)},\quad \frac{\partial(y_1, y_2, \ldots, y_n)}{\partial(x_1, x_2, \ldots, x_n)},$$

whose elements are $\partial\varphi_i/\partial y_k$, $\partial\varphi_i/\partial x_k$, $\partial y_i/\partial x_k$, respectively. By the chain rule for differentiation of composite functions, we have

$$\frac{\partial\varphi_i}{\partial x_k} = \frac{\partial\varphi_i}{\partial y_1}\frac{\partial y_1}{\partial x_k} + \frac{\partial\varphi_i}{\partial y_2}\frac{\partial y_2}{\partial x_k} + \cdots + \frac{\partial\varphi_i}{\partial y_n}\frac{\partial y_n}{\partial x_k}.$$

Applying the theorem on multiplication of determinants (in the "row by column" form), we obtain the relation

$$\frac{\partial(\varphi_1, \varphi_2, \ldots, \varphi_n)}{\partial(x_1, x_2, \ldots, x_n)} = \frac{\partial(\varphi_1, \varphi_2, \ldots, \varphi_n)}{\partial(y_1, y_2, \ldots, y_n)}\frac{\partial(y_1, y_2, \ldots, y_n)}{\partial(x_1, x_2, \ldots, x_n)}. \tag{64}$$

This relation is analogous to the rule for differentiating a composite function of one independent variable.

Next we establish a related property of Jacobians. The system of functions

$$\varphi_i = \varphi_i(x_1, x_2, \ldots, x_n) \qquad (i = 1, 2, \ldots, n) \tag{65}$$

can be regarded as a transformation from the variables x_i to the new variables φ_i. In the special case of the so-called *identity transformation*

$$\varphi_1 = x_1, \varphi_2 = x_2, \ldots, \varphi_n = x_n,$$

the Jacobian is

$$\begin{vmatrix} 1 & 0 & 0 & \cdots & 0 \\ 0 & 1 & 0 & \cdots & 0 \\ 0 & 0 & 1 & \cdots & 0 \\ \cdot & \cdot & \cdot & \cdots & \cdot \\ 0 & 0 & 0 & \cdots & 1 \end{vmatrix} = 1.$$

Suppose now that we solve the equations (65) with respect to the x_i, so that the x_i are expressed in terms of the φ_i:

$$x_i = x_i(\varphi_1, \varphi_2, \ldots, \varphi_n) \qquad (i = 1, 2, \ldots, n). \tag{66}$$

It is natural to call the transformation (66) the *inverse* of the transformation (65). If we substitute (66) into the right-hand side of (65), we obtain the identities $\varphi_1 = \varphi_1, \varphi_2 = \varphi_2, \ldots, \varphi_n = \varphi_n$, i.e., the identity transformation. Thus, setting $y_i = x_i$ and $x_i = \varphi_i$ in (64), we obtain

$$\frac{\partial(\varphi_1, \varphi_2, \ldots, \varphi_n)}{\partial(\varphi_1, \varphi_2, \ldots, \varphi_n)} = \frac{\partial(\varphi_1, \varphi_2, \ldots, \varphi_n)}{\partial(x_1, x_2, \ldots, x_n)} \frac{\partial(x_1, x_2, \ldots, x_n)}{\partial(\varphi_1, \varphi_2, \ldots, \varphi_n)},$$

where the Jacobian of the identity transformation appears in the left-hand side. Thus, we have finally

$$\frac{\partial(\varphi_1, \varphi_2, \ldots, \varphi_n)}{\partial(x_1, x_2, \ldots, x_n)} \frac{\partial(x_1, x_2, \ldots, x_n)}{\partial(\varphi_1, \varphi_2, \ldots, \varphi_n)} = 1, \tag{67}$$

i.e., *the product of the Jacobians of the direct and the inverse transformations equals unity*, just as in the case of derivatives of direct and inverse functions of one variable.

Let

$$\frac{\partial(\varphi_1, \varphi_2, \ldots, \varphi_n)}{\partial(x_1, x_2, \ldots, x_n)} \tag{68}$$

be the Jacobian of the functions

$$\varphi_i(x_1, x_2, \ldots, x_n) \qquad (i = 1, 2, \ldots, n)$$

with respect to the variables $x_1, x_2, \ldots, x_n$. We now ask what it means for the Jacobian (68) to be identically zero. Suppose that there is a functional relationship

$$F(\varphi_1, \varphi_2, \ldots, \varphi_n) = 0 \tag{69}$$

between the function $\varphi_1, \varphi_2, \ldots, \varphi_n$, where (69) is an identity in the independent variables $x_1, x_2, \ldots, x_n$. Differentiating (69) with respect to all the independent variables, we obtain n identities:

$$
\begin{aligned}
\frac{\partial F}{\partial \varphi_1}\frac{\partial \varphi_1}{\partial x_1} + \frac{\partial F}{\partial \varphi_2}\frac{\partial \varphi_2}{\partial x_1} + \cdots + \frac{\partial F}{\partial \varphi_n}\frac{\partial \varphi_n}{\partial x_1} &= 0,\\
\frac{\partial F}{\partial \varphi_1}\frac{\partial \varphi_1}{\partial x_2} + \frac{\partial F}{\partial \varphi_2}\frac{\partial \varphi_2}{\partial x_2} + \cdots + \frac{\partial F}{\partial \varphi_n}\frac{\partial \varphi_n}{\partial x_2} &= 0,\\
\cdots\cdots\cdots\cdots\cdots\cdots&\cdots\\
\frac{\partial F}{\partial \varphi_1}\frac{\partial \varphi_1}{\partial x_n} + \frac{\partial F}{\partial \varphi_2}\frac{\partial \varphi_2}{\partial x_n} + \cdots + \frac{\partial F}{\partial \varphi_n}\frac{\partial \varphi_n}{\partial x_n} &= 0.
\end{aligned}
\tag{70}
$$

We can regard these identities as a system of linear equations in the n quantities $\partial F/\partial\varphi_1, \partial F/\partial\varphi_2, \ldots, \partial F/\partial\varphi_n$; clearly these quantities cannot all vanish identically, since then F would not depend on any of the functions φ_i. Consequently, the determinant of the homogeneous system (70) must vanish, which implies that the Jacobian (68) vanishes. (It should be noted that this argument is a formal one and does not purport to be a rigorous proof.) Thus the existence of the functional relationship (69) implies that the Jacobian (68) vanishes identically. The converse proposition can also be established, but we omit the proof. The final result is the following: *A necessary and sufficient condition for the Jacobian (68) to vanish identically is that there exist a functional relationship between the functions $\varphi_i(x_1, x_2, \ldots, x_n)$.*

As an example, consider the three functions

$$
\varphi_1 = x_1^2 + x_2^2 + x_3^2,\quad \varphi_2 = x_1 + x_2 + x_3,\quad \varphi_3 = x_1x_2 + x_1x_3 + x_2x_3 \tag{71}
$$

of three independent variables. It is easily verified that these functions satisfy the relation

$$
\varphi_2^2 - \varphi_1 - 2\varphi_3 = 0.
$$

The Jacobian of the system of functions (71) is

$$
\frac{\partial(\varphi_1, \varphi_2, \varphi_3)}{\partial(x_1, x_2, x_3)} = \begin{vmatrix} 2x_1 & 2x_2 & 2x_3 \\ 1 & 1 & 1 \\ x_2 + x_3 & x_1 + x_3 & x_1 + x_2 \end{vmatrix}.
$$

It is left to the reader to show that this Jacobian vanishes identically.

19. Implicit Functions. In an earlier volume, we proved the existence of a function defined implicitly by an equation. We now generalize this theorem to the case of a system of equations. Before stating the theorem, we first summarize our previous results.

Let $x = x_0$, $y = y_0$ be a solution of the equation

$$
F(x, y) = 0, \tag{72a}
$$

let $F(x, y)$ and its partial derivatives of the first order be continuous in a neighborhood of (x_0, y_0), and finally let the partial derivative $F_y(x, y)$ be different from zero at (x_0, y_0). Then, for x sufficiently close to x_0, (72a) defines a unique function $y(x)$ which is continuous and differentiable and satisfies the condition $y(x_0) = y_0$. Moreover, in just the same way we can prove the following result: If the equation

$$F(x, y, z) = 0 \tag{72b}$$

has the solution $x = x_0, y = y_0, z = z_0$, if $F(x, y, z)$ and its partial derivatives of the first order are continuous in a neighborhood of (x_0, y_0, z_0), and if $F_z(x_0, y_0, z_0) \neq 0$, then (72$b$) defines a unique function $z(x, y)$ which is continuous and differentiable with respect to x and y in a neighborhood of (x_0, y_0) and satisfies the condition $z(x_0, y_0) = z_0$.

We now consider the system of two equations

$$\varphi(x, y, z) = 0, \qquad \psi(x, y, z) = 0. \tag{73}$$

Let this system have the solution $x = x_0$, $y = y_0$, $z = z_0$, *let the functions* $\varphi(x, y, z)$, $\psi(x, y, z)$ *and their first partial derivatives be continuous in a neighborhood of* (x_0, y_0, z_0), *and let the Jacobian*

$$\frac{\partial(\varphi, \psi)}{\partial(y, z)} = \begin{vmatrix} \frac{\partial \varphi}{\partial y} & \frac{\partial \varphi}{\partial z} \\ \frac{\partial \psi}{\partial y} & \frac{\partial \psi}{\partial z} \end{vmatrix} = \frac{\partial \varphi}{\partial y}\frac{\partial \psi}{\partial z} - \frac{\partial \varphi}{\partial z}\frac{\partial \psi}{\partial y} \tag{74}$$

be different from zero at (x_0, y_0, z_0). *Then, for* x *sufficiently close to* x_0, *the system* (73) *defines a unique pair of functions* $y(x)$, $z(x)$ *which are continuous and differentiable and satisfy the conditions* $y(x_0) = y_0$, $z(x_0) = z_0$.

PROOF. Since the expression (74) is nonzero for $x = x_0, y = y_0, z = z_0$, then at least one of the partial derivatives $\partial\psi/\partial y$, $\partial\psi/\partial z$ is nonzero; suppose that this nonzero derivative is in fact $\partial\psi/\partial z$. By a theorem stated above, the second of the equations (73) uniquely defines a function $z(x, y)$. Substituting this function into the first equation of the system, we obtain an equation

$$\varphi[x, y, z(x, y)] \equiv g(x, y) = 0 \tag{75}$$

in the variables x and y. To prove the theorem, it suffices to show that the partial derivative of $g(x, y)$ with respect to y is nonzero at $x = x_0$, $y = y_0$. This partial derivative is given by the formula

$$\frac{\partial g}{\partial y} = \frac{\partial \varphi}{\partial y} + \frac{\partial \varphi}{\partial z}\frac{\partial z}{\partial y}, \tag{76}$$

where the function $z(x, y)$ is the solution of the second of the equations (73), i.e.,

$$\psi[x, y, z(x, y)] = 0$$

holds identically. Differentiating this identity with respect to y, we obtain

$$\frac{\partial\psi}{\partial y} + \frac{\partial\psi}{\partial z}\frac{\partial z}{\partial y} = 0. \tag{77}$$

We now multiply (76) by $\partial\psi/\partial z$ and (77) by $-\partial\varphi/\partial z$, and add the resulting equations, obtaining

$$\frac{\partial\psi}{\partial z}\frac{\partial g}{\partial y} = \frac{\partial(\varphi, \psi)}{\partial(y, z)}.$$

For $x = x_0$, $y = y_0$, the function $z(x, y)$ takes the value z_0. Moreover, for these values of the variables, $\partial\psi/\partial z$ and the Jacobian (74) are nonzero. It follows that $\partial g/\partial y$ is also nonzero for $x = x_0$, $y = y_0$. Consequently, (75) defines a unique function $y(x)$. Substituting $y(x)$ in $z(x, y)$, we obtain the other required function $z(x)$, QED. (In this proof, we can replace x by any number of independent variables.)

A more general version of the implicit function theorem is the following: *Let there be given a system of n equations*

$$F_1(x_1, \ldots, x_m, y_1, \ldots, y_n) = 0, \ldots, \qquad F_n(x_1, \ldots, x_m, y_1, \ldots, y_n) = 0, \tag{78}$$

with the solution

$$x_k = x_k^{(0)},\ y_l = y_l^{(0)} \qquad (k = 1, 2, \ldots, m;\ l = 1, 2, \ldots, n); \tag{79}$$

let the F_l be continuous functions with continuous first order partial derivatives in a neighborhood of the values (79), and finally let the Jacobian

$$\frac{\partial(F_1, F_2, \ldots, F_n)}{\partial(y_1, y_2, \ldots, y_n)} = \begin{vmatrix} \dfrac{\partial F_1}{\partial y_1} & \dfrac{\partial F_1}{\partial y_2} & \cdots & \dfrac{\partial F_1}{\partial y_n} \\ \dfrac{\partial F_2}{\partial y_1} & \dfrac{\partial F_2}{\partial y_2} & \cdots & \dfrac{\partial F_2}{\partial y_n} \\ \cdots & \cdots & \cdots & \cdots \\ \dfrac{\partial F_n}{\partial y_1} & \dfrac{\partial F_n}{\partial y_2} & \cdots & \dfrac{\partial F_n}{\partial y_n} \end{vmatrix} \tag{80}$$

be different from zero for the values (79). Then, for x_k close to $x_k^{(0)}$, (79) defines a unique system of functions $y_l(x_1, x_2, \ldots, x_m)$ which are continuous, have first order derivatives, and satisfy the conditions

$$y_l(x_1^{(0)}, x_2^{(0)}, \ldots, x_k^{(0)}) = y_l^{(0)}.$$

To prove the theorem, we proceed by induction. Assuming that the theorem holds for a system of $n - 1$ equations (it is certainly valid for $n = 1$ and $n = 2$), we shall show that it holds for a system of n equations. Expanding the determinant (80) with respect to elements of the first column, we see that at least one of the cofactors of these elements is nonzero for the values (79), since by hypothesis the determinant itself is nonzero for these values. By appropriately renumbering the functions F_l, we can assume that the cofactor of the element $\partial F_1/\partial y_1$ is nonzero. This coefficient is the Jacobian of the functions $F_2, \ldots, F_n$ with respect to the variables $y_2, \ldots, y_n$. By the theorem as applied to systems of $n - 1$ equations, the relations

$$F_2(x_1, \ldots, x_m, y_1, \ldots, y_n) = 0, \ldots, \quad F_n(x_1, \ldots, x_m, y_1, \ldots, y_n) = 0 \tag{81}$$

uniquely define the functions

$$y_2 = \varphi_2(x_1, \ldots, x_m, y_1), \ldots, y_n = \varphi_n(x_1, \ldots, x_m, y_1). \tag{82}$$

Substituting these functions into the first of the equations of the system (78), we obtain an equation

$$F_1(x_1, \ldots, x_m, y_1, \varphi_2, \ldots, \varphi_n) \equiv G(x_1, \ldots, x_m, y_1) = 0 \tag{83}$$

for determining y_1.

It remains to verify that the partial derivative of $G(x_1, \ldots, x_m, y_1)$ with respect to y_1 is nonzero for the values (79). This derivative is given by the formula

$$\frac{\partial G}{\partial y_1} = \frac{\partial F_1}{\partial y_1} + \sum_{s=2}^{n} \frac{\partial F_1}{\partial \varphi_s} \frac{\partial \varphi_s}{\partial y_1}. \tag{84}$$

Substituting the functions (82) into (81), we obtain identities which we then differentiate with respect to y_1; the result is

$$\frac{\partial F_l}{\partial y_1} + \sum_{s=2}^{n} \frac{\partial F_l}{\partial \varphi_s} \frac{\partial \varphi_s}{\partial y_1} = 0 \qquad (l = 2, \ldots, n). \tag{85}$$

We now denote the cofactors of the elements of the first column of the determinant (80) by $A_1, A_2, \ldots, A_n$. Multiplying (84) by A_1 and (85) by A_l $(l = 2, \ldots, n)$, and then subtracting the last $n - 1$ equations from the first, we obtain

$$A_1 \frac{\partial G}{\partial y_1} = \sum_{l=1}^{n} \frac{\partial F_l}{\partial y_1} A_l + \sum_{s=2}^{n} \left[\sum_{l=1}^{n} \frac{\partial F_l}{\partial \varphi_s} A_l \right] \frac{\partial \varphi_s}{\partial y_1}.$$

The first term in the right-hand side gives the determinant (80), which, for brevity, we denote simply by the letter D. The sum with respect to l appearing in the second term in the right-hand side is a sum of products of elements of a column of D, different from the first, with the cofactors of the elements of the first column, and therefore vanishes (Sec. 3). (In this regard, we note that differentiation with respect to φ_s is equivalent to differentiation with respect to y_s.) Thus, our equation can be written as

$$A_1 \frac{\partial G}{\partial y_1} = D.$$

Since A_1 and D are nonzero for the values (79), the same is true of $\partial G/\partial y_1$. Consequently, (83) defines a unique function $y_1(x_1, x_2, \ldots, x_m)$; substituting this function into the functions (82), we obtain the final result.

A special case of the implicit function theorem is the following theorem on the inversion of systems of functions: *Suppose that we are given the equations*

$$y_k = f_k(x_1, x_2, \ldots, x_n) \qquad (k = 1, 2, \ldots, n). \tag{86}$$

Assume that the functions f_k and their first order derivatives are continuous in a neighborhood of the values $x_k = x_k^{(0)}$ $(k = 1, 2, \ldots, n)$ and that the Jacobian

$$\frac{\partial(f_1, f_2, \ldots, f_n)}{\partial(x_1, x_2, \ldots, x_n)} \tag{87}$$

is nonzero for these values. Then, in a neighborhood of the values

$$y_k^{(0)} = f_k(x_1^{(0)}, x_2^{(0)}, \ldots, x_n^{(0)}),$$

the equations (86) *uniquely define $x_k(y_1, y_2, \ldots, y_n)$ as functions of $y_1, y_2, \ldots, y_n$; these functions are continuous, have first order derivatives, and satisfy the conditions $x_k(y_1^{(0)}, y_2^{(0)}, \ldots, y_n^{(0)}) = x_k^{(0)}$.*

To prove this theorem, it is sufficient to consider the equations

$$f_k(x_1, x_2, \ldots, x_n) - y_k = 0 \qquad (k = 1, 2, \ldots, n)$$

and use the implicit function theorem; here the x_k play the role of the y_l.

If the f_k are homogeneous linear functions of the variables x_k, then the system (86) has the form

$$y_k = a_{k1}x_1 + a_{k2}x_2 + \cdots + a_{kn}x_n \qquad (k = 1, 2, \ldots, n).$$

In this case, the determinant (87) reduces to the determinant $|a_{ik}|$ of the coefficients a_{ik}, and the possibility of uniquely solving the system is a consequence of Cramer's rule.

PROBLEMS†

1. Solve the following systems of linear equations by Cramer's rule:

(a) $$\begin{aligned} 2x \qquad - z &= 1, \\ 2x + 4y - z &= 1, \\ -x + 8y + 3z &= 2; \end{aligned}$$
(b) $$\begin{aligned} x + y + z &= a, \\ x + (1+a)y + z &= 2a, \\ x + y + (1+a)z &= 0; \end{aligned}$$

(c) $$\begin{aligned} 5x \qquad + 4z + 2t &= 3, \\ x - y + 2z + t &= 1, \\ 4x + y + 2z \qquad &= 1, \\ x + y + z + t &= 0; \end{aligned}$$
(d) $$\begin{aligned} 2x + y + 4z + 8t &= -1, \\ x + 3y - 6z + 2t &= 3, \\ 3x - 2y + 2z - 2t &= 8, \\ 2x - y + 2z \qquad &= 4. \end{aligned}$$

2. Two systems of linear equations in the same number of unknowns are said to be *equivalent* if they have the same solutions. Which of the following pairs are equivalent?

(a) $$\begin{aligned} w + x + y - z &= 1, \\ w + x - y + z &= -1, \\ w - x + y + z &= 1, \\ 3w + x + y + z &= 1; \end{aligned} \qquad \begin{aligned} w + z &= 0, \\ w + x &= 0, \\ w + y &= 1; \end{aligned}$$
(b) $$\begin{aligned} x + 2y &= 1, \\ x - y &= 2; \end{aligned} \qquad \begin{aligned} 2x + y &= 1, \\ x - 2y &= 2; \end{aligned}$$

(c) $$\begin{aligned} x + y + z &= 0, \\ x + y + t &= 0, \\ x + z + t &= 0, \\ y + z + t &= 0; \end{aligned} \qquad \begin{aligned} z - t &= 0, \\ y - t &= 0, \\ x - t &= 0, \\ x + y + z &= 0; \end{aligned}$$
(d) $$\begin{aligned} x + y + z &= 0, \\ x + y + t &= 0, \\ x + z + t &= 0, \\ y + z + t &= 0; \end{aligned} \qquad \begin{aligned} z - t &= 0, \\ y - t &= 0, \\ x - t &= 0, \\ x - z &= 0. \end{aligned}$$

3. Let $x_0 < x_1 < x_2 < \cdots < x_n$ and $y_0, y_1, y_2, \ldots, y_n$ be given numbers, and let $P(x) = a_0 + a_1x + a_2x^2 + \cdots + a_nx^n$ be a polynomial of degree n. Show that the coefficients $a_0, a_1, a_2, \ldots, a_n$ can be chosen in such a way that

$$P(x_i) = y_i \qquad (i = 0, 1, 2, \ldots, n).$$

4. Find the quadratic polynomial $f(x)$ for which $f(1) = -1, f(-1) = 9, f(2) = -3$, and the cubic polynomial $g(x)$ for which $g(-1) = 0, g(1) = 4, g(2) = 3, g(3) = 16$.

5. Solve the following systems of linear equations by means which are more appropriate than the use of Cramer's rule:

(a) $$\begin{aligned} -x + y + z + t &= a, \\ x - y + z + t &= b, \\ x + y - z + t &= c, \\ x + y + z - t &= d; \end{aligned}$$
(b) $$\begin{aligned} x + y + z &= a, \\ x + y + t &= b, \\ x + z + t &= c, \\ y + z + t &= d; \end{aligned}$$

(c) $$\begin{aligned} x_1 + 2x_2 + 3x_3 + \cdots + nx_n &= a_1, \\ x_2 + 2x_3 + 3x_4 + \cdots + nx_1 &= a_2, \\ \cdots\cdots\cdots\cdots\cdots\cdots\cdots \\ x_n + 2x_1 + 3x_2 + \cdots + nx_{n-1} &= a_n; \end{aligned}$$

† All equation numbers refer to the corresponding equations of Chap. 2. For hints and answers, see p. 433. These problems are by A. L. Shields and the translator, drawing largely from the sources mentioned in the Preface, especially Proskuryakov's book.

(d)
$$\begin{aligned} x_1 + x_2 + x_3 + \cdots + x_n &= 1, \\ x_1 + x_3 + x_4 + \cdots + x_n &= 2, \\ x_1 + x_2 + x_4 + \cdots + x_n &= 3, \\ \cdots\cdots\cdots\cdots\cdots\cdots & \\ x_1 + x_2 + x_3 + \cdots + x_{n-1} &= n; \end{aligned}$$

(e)
$$\begin{aligned} x_1 + x_2 &= a_1, \\ x_2 + x_3 &= a_2, \\ x_3 + x_4 &= a_3, \\ \cdots\cdots\cdots & \\ x_{n-1} + x_n &= a_{n-1}, \\ x_n + x_1 &= a_n. \end{aligned}$$

6. Using the method of characteristic determinants presented in Sec. 9, investigate the following systems and solve those which are *compatible*, i.e., which have at least one solution:†

(a)
$$\begin{aligned} 5x_1 - x_2 + 2x_3 + x_4 &= 7, \\ 2x_1 + x_2 + 4x_3 - 2x_4 &= 1, \\ x_1 - 3x_2 - 6x_3 + 5x_4 &= 0; \end{aligned}$$

(b)
$$\begin{aligned} 7x_1 + 3x_2 &= 2, \\ x_1 - 2x_2 &= -3, \\ 4x_1 + 9x_2 &= 11; \end{aligned}$$

(c)
$$\begin{aligned} x_1 + x_2 - 2x_3 - x_4 + x_5 &= 1, \\ 3x_1 - x_2 + x_3 + 4x_4 + 3x_5 &= 4, \\ x_1 + 5x_2 - 9x_3 - 8x_4 + x_5 &= 0; \end{aligned}$$

(d)
$$\begin{aligned} 4x_1 + x_2 - 2x_3 + x_4 &= 3, \\ x_1 - 2x_2 - x_3 + 2x_4 &= 2, \\ 2x_1 + 5x_2 \qquad - x_4 &= -1, \\ 3x_1 + 3x_2 - x_3 - 3x_4 &= 1; \end{aligned}$$

(e)
$$\begin{aligned} x_1 - 8x_2 &= 3, \\ 2x_1 + x_2 &= 1, \\ 4x_1 + 7x_2 &= -4. \end{aligned}$$

(These systems will be studied by another method in Prob. 25.)

7. Investigate the following homogeneous systems, and find the general solution and a complete system of solutions for the systems which have nontrivial solutions:

(a)
$$\begin{aligned} x_1 + 2x_2 + 4x_3 - 3x_4 &= 0, \\ 3x_1 + 5x_2 + 6x_3 - 4x_4 &= 0, \\ 4x_1 + 5x_2 - 2x_3 + 3x_4 &= 0, \\ 3x_1 + 8x_2 + 24x_3 - 19x_4 &= 0; \end{aligned}$$

(b)
$$\begin{aligned} 2x_1 - 4x_2 + 5x_3 + 3x_4 &= 0, \\ 3x_1 - 6x_2 + 4x_3 + 2x_4 &= 0, \\ 4x_1 - 8x_2 + 17x_3 - 11x_4 &= 0; \end{aligned}$$

(c)
$$\begin{aligned} 3x_1 + 2x_2 + x_3 + 3x_4 + 5x_5 &= 0, \\ 6x_1 + 4x_2 + 3x_3 + 5x_4 + 7x_5 &= 0, \\ 9x_1 + 6x_2 + 5x_3 + 7x_4 + 9x_5 &= 0, \\ 3x_1 + 2x_2 \qquad + 4x_4 + 8x_5 &= 0; \end{aligned}$$

(d)
$$\begin{aligned} 3x_1 + 5x_2 + 2x_3 &= 0, \\ 4x_1 + 7x_2 + 5x_3 &= 0, \\ x_1 + x_2 - 4x_3 &= 0, \\ 2x_1 + 9x_2 + 6x_3 &= 0; \end{aligned}$$

(e)
$$\begin{aligned} 6x_1 - 2x_2 + 2x_3 + 5x_4 + 7x_5 &= 0, \\ 9x_1 - 3x_2 + 4x_3 + 8x_4 + 9x_5 &= 0, \\ 6x_1 - 2x_2 + 6x_3 + 7x_4 + x_5 &= 0, \\ 3x_1 - x_2 + 4x_3 + 4x_4 - x_5 &= 0; \end{aligned}$$

(f)
$$\begin{aligned} x_1 - x_3 &= 0, \\ x_2 - x_4 &= 0, \\ -x_1 + x_3 - x_5 &= 0, \\ -x_2 + x_4 - x_6 &= 0, \\ -x_3 + x_5 &= 0, \\ -x_4 + x_6 &= 0; \end{aligned}$$

(g)
$$\begin{aligned} x_1 - x_3 + x_5 \qquad &= 0, \\ x_2 - x_4 + x_6 \qquad &= 0, \\ x_1 - x_2 + x_5 - x_6 &= 0, \\ x_2 - x_3 + x_6 \qquad &= 0, \\ x_1 - x_4 + x_5 \qquad &= 0; \end{aligned}$$

(h)
$$\begin{aligned} 5x_1 + 6x_2 - 2x_3 + 7x_4 + 4x_5 &= 0, \\ 2x_1 + 3x_2 - x_3 + 4x_4 + 2x_5 &= 0, \\ 7x_1 + 9x_2 - 3x_3 + 5x_4 + 6x_5 &= 0, \\ 5x_1 + 9x_2 - 3x_3 + x_4 + 6x_5 &= 0. \end{aligned}$$

8. Show that if a homogeneous system of linear equations has rational (in particular, integral) solutions, then one can always construct a complete system of solutions consisting entirely of integers.

† From a book by A. G. Kurosh.

9. Solve the following homogeneous system:

$$\begin{aligned} \lambda x + y + z + w &= 0, \\ x + \lambda y + z + w &= 0, \\ x + y + \lambda z + w &= 0, \\ x + y + z + \lambda w &= 0. \end{aligned}$$

For which values of λ are there nontrivial solutions?

10. Do the rows of the matrices

$$A = \begin{Vmatrix} 30 & -24 & 43 & -50 & -5 \\ 9 & -15 & 8 & 5 & 2 \\ 4 & 2 & 9 & -20 & -3 \end{Vmatrix} \quad \text{and} \quad B = \begin{Vmatrix} 4 & 2 & 9 & -20 & -3 \\ 1 & -11 & 2 & 13 & 4 \\ 9 & -15 & 8 & 5 & 2 \end{Vmatrix}$$

form a complete system of solutions for the system of equations

$$\begin{aligned} 3x_1 + 4x_2 + 2x_3 + x_4 + 6x_5 &= 0, \\ 5x_1 + 9x_2 + 7x_3 + 4x_4 + 7x_5 &= 0, \\ 4x_1 + 3x_2 - x_3 - x_4 + 11x_5 &= 0, \\ x_1 + 6x_2 + 8x_3 + 5x_4 - 4x_5 &= 0? \end{aligned}$$

11. Find the linear relations between the following sets of linear forms:

(a) $y_1 = 2x_1 + 2x_2 + 7x_3 - x_4,$
$y_2 = 3x_1 - x_2 + 2x_3 + 4x_4,$
$y_3 = x_1 + x_2 + 3x_3 + x_4;$

(b) $y_1 = 3x_1 + 2x_2 - 5x_3 + 4x_4,$
$y_2 = 3x_1 - x_2 + 3x_3 - 3x_4,$
$y_3 = 3x_1 + 5x_2 - 13x_3 + 11x_4;$

(c) $y_1 = 2x_1 + 3x_2 - 4x_3 - x_4,$
$y_2 = x_1 - 2x_2 + x_3 + 3x_4,$
$y_3 = 5x_1 - 3x_2 - x_3 + 8x_4,$
$y_4 = 3x_1 + 8x_2 - 9x_3 - 5x_4;$

(d) $y_1 = 2x_1 + x_2 - x_3 + x_4,$
$y_2 = x_1 + 2x_2 + x_3 - x_4,$
$y_3 = x_1 + x_2 + 2x_3 + x_4.$

12. Given the linear forms

$$\begin{aligned} y_1 &= x_1 + 2x_2 + x_3 - 3x_4 + 2x_5, \\ y_2 &= 2x_1 + x_2 + x_3 + x_4 - 3x_5, \\ y_3 &= x_1 + x_2 + 2x_3 + 2x_4 - 2x_5, \\ y_4 &= 2x_1 + 3x_2 - 5x_3 - 17x_4 + \lambda x_5, \end{aligned}$$

choose the parameter λ so that the fourth form is a linear combination of the other three.

13. Let R_n denote a real n-dimensional vector space referred to Cartesian axes. Do the following sets of vectors form a linear subspace of the corresponding vector spaces?

(a) All vectors of R_n whose components are integers;
(b) All vectors of R_2 lying along the x-axis and the y-axis;
(c) All vectors of R_2 whose end points lie on a line going through the origin;
(d) All vectors of R_2 whose end points lie on a line that does not go through the origin;
(e) All vectors of R_2 whose end points lie in the first quadrant;
(f) All vectors of R_n whose components satisfy the equation $x_1 + x_2 + \cdots + x_n = 0$;
(g) All vectors of R_n whose components satisfy the equation $x_1 + x_2 + \cdots + x_n = 1$.

(It is assumed that the initial points of all vectors are located at the origin.)

14. Which of the following sets of vectors are linearly dependent?

(a) $\mathbf{a}_1 = (1, 2, 3)$, $\mathbf{a}_2 = (3, 6, 7)$; (b) $\mathbf{a}_1 = (4, -2, 6)$, $\mathbf{a}_2 = (6, -3, 9)$;
(c) $\mathbf{a}_1 = (2, -3, 1)$, $\mathbf{a}_2 = (3, -1, 5)$, $\mathbf{a}_3 = (1, -4, 3)$;
(d) $\mathbf{a}_1 = (5, 4, 3)$, $\mathbf{a}_2 = (3, 3, 2)$, $\mathbf{a}_3 = (8, 1, 3)$;
(e) $\mathbf{a}_1 = (4, -5, 2, 6)$, $\mathbf{a}_2 = (2, -2, 1, 3)$, $\mathbf{a}_3 = (6, -3, 3, 9)$, $\mathbf{a}_4 = (4, -1, 5, 6)$;
(f) $\mathbf{a}_1 = (1, 0, 0, 2, 5)$, $\mathbf{a}_2 = (0, 1, 0, 3, 4)$, $\mathbf{a}_3 = (0, 0, 1, 4, 7)$, $\mathbf{a}_4 = (2, -3, 4, 11, 12)$.

Find the corresponding linear relations.

15. Show that a set of vectors is linearly dependent if

(a) It contains two equal vectors;
(b) It contains two vectors which differ only by a numerical factor;
(c) It contains the zero vector; (d) It contains a linearly dependent subset.

16. Suppose we have a set of vectors

$$\mathbf{a}_i = (a_{i1}, a_{i2}, \ldots, a_{in}) \qquad (i = 1, 2, \ldots, s;\ s \leq n),$$

such that

$$|a_{jj}| > \sum_{\substack{i=1\\ i\neq j}}^{s} |a_{ij}| \qquad (j = 1, 2, \ldots, n).$$

Show that this set of vectors is linearly independent.

17. Prove that the following sets of vectors form linear subspaces. Find a basis for each subspace and its dimension.

(a) All n-dimensional vectors whose first and last components are equal;
(b) All n-dimensional vectors whose even-numbered components vanish;
(c) All n-dimensional vectors whose even-numbered components are equal;
(d) All n-dimensional vectors of the form $(\alpha, \beta, \alpha, \beta, \ldots)$, where α and β are arbitrary numbers.

18. Let R_n be an n-dimensional vector space, in general complex. By the *sum* of two (linear) subspaces L_1 and L_2 of R_n is meant the set $S = L_1 + L_2$ of all vectors $\mathbf{x}$ of R_n which can be represented in the form $\mathbf{x} = \mathbf{x}_1 + \mathbf{x}_2$, where $\mathbf{x}_1$ is in L_1 and $\mathbf{x}_2$ is in L_2. By the *intersection* of two subspaces L_1 and L_2 of R_n is meant the set $D = L_1 \cap L_2$ of all vectors $\mathbf{x}$ which belong to both L_1 and L_2. Prove the following facts:

(a) The sum and intersection of two subspaces of R_n are themselves subspaces of R_n;
(b) The sum of two subspaces L_1 and L_2 of R_n equals the intersection of all subspaces of R_n containing both L_1 and L_2;
(c) The sum of the dimensions of two subspaces L_1 and L_2 of R_n equals the dimension of their sum plus the dimension of their intersection.

19. By a *hyperplane* in a vector space R is meant the set P of vectors in R obtained by adding the same fixed vector $\mathbf{x}_0$ to all the vectors of a subspace L of R; this relation is denoted by writing $P = L + \mathbf{x}_0$. Prove the following facts:

(a) Given two hyperplanes $P_1 = L_1 + \mathbf{x}_1$, $P_2 = L_2 + \mathbf{x}_2$, where L_1, L_2 are subspaces of R, and $\mathbf{x}_1$, $\mathbf{x}_2$ are fixed vectors in R, show that $P_1 = P_2$ if and only if $L_1 = L_2$ and $\mathbf{x}_1 - \mathbf{x}_2$ is in L_1;

(b) If $P = L + \mathbf{x}_0$, then the vector $\mathbf{x}_0$ belongs to P, and if $\mathbf{x}_0$ is replaced by any other vector $\mathbf{x}_1$ in P, we have $P = L + \mathbf{x}_1$;
(c) If a straight line has two points in common with a hyperplane, it is entirely contained in the hyperplane;
(d) If a hyperplane contains the vectors $\mathbf{x}$ and $\mathbf{y}$, it also contains the vector

$$\alpha\mathbf{x} + (1 - \alpha)\mathbf{y},$$

for any α.

20. Write the general equation of a straight line and of a plane in an n-dimensional space (using vectors). Then prove that

(a) In a space of dimension ≥ 3, any two straight lines lie in a three-dimensional hyperplane;
(b) In a space of dimension ≥ 5, any two planes lie in a five-dimensional hyperplane. (By the *dimension of the hyperplane* $P = L + \mathbf{x}_0$ is meant the dimension of the subspace L. Cf. Prob. 19*a*.)

21. Find the vectors which together with the following vectors form an orthonormal basis:

(a) $(\frac{2}{3}, \frac{1}{3}, \frac{2}{3})$, $(\frac{1}{3}, \frac{2}{3}, -\frac{2}{3})$; (b) $(\frac{1}{2}, \frac{1}{2}, \frac{1}{2}, \frac{1}{2})$, $(\frac{1}{2}, \frac{1}{2}, -\frac{1}{2}, -\frac{1}{2})$;
(c) $(\frac{1}{2}, \frac{1}{2}, \frac{1}{2}, \frac{1}{2})$, $(\frac{1}{6}, \frac{1}{6}, \frac{1}{2}, -\frac{5}{6})$.

22. The *angle* $\theta(\mathbf{x}, \mathbf{y})$ *between the vectors* $\mathbf{x}$ *and* $\mathbf{y}$ is defined by the formula

$$\cos\theta(\mathbf{x}, \mathbf{y}) = \frac{(\mathbf{x}, \mathbf{y})}{\|\mathbf{x}\| \, \|\mathbf{y}\|}.$$

Find the angles between the following pairs of vectors:

(a) $\mathbf{x} = (2, 1, 3, 2)$, $\mathbf{y} = (1, 2, -2, 1)$; (b) $\mathbf{x} = (1, 2, 2, 3)$, $\mathbf{y} = (3, 1, 5, 1)$;
(c) $\mathbf{x} = (1, 1, 1, 2)$, $\mathbf{y} = (3, 1, -1, 0)$.

23. Find the cosines of the angles between the line $x_1 = x_2 = \cdots = x_n$ and the coordinate axes.

24. What is the radius R of a sphere circumscribed about an n-dimensional cube of side a? Which of the quantities R and a is larger for different values of n?

25. Find the number of diagonals of an n-dimensional cube which are orthogonal to a given diagonal.

26. In an n-dimensional space, find n points with nonnegative coordinates such that the distance between each pair of points and the distance between each point and the origin is 1. Locate the first of these points on the first coordinate axis, the second in the plane spanned by the first two axes, etc.

Comment: These points together with the origin are said to form the vertices of a *regular simplex* of side 1.

27. In each of the following cases, represent $\mathbf{x}$ as the sum of a vector $\mathbf{y}$ in the subspace L and a vector $\mathbf{z}$ in the orthogonal complement of L:

(a) $\mathbf{x} = (4, -1, -3, 4)$, L spanned by the vectors $(1, 1, 1, 1,)$, $(1, 2, 2, -1)$, $(1, 0, 0, 3)$;
(b) $\mathbf{x} = (5, 2, -2, 2)$, L spanned by the vectors $(2, 1, 1, -1)$, $(1, 1, 3, 0)$, $(1, 2, 8, 1)$;

(c) $\mathbf{x} = (7, -4, -1, 2)$, L defined by the equations

$$\begin{aligned} 2x_1 + x_2 + x_3 + 3x_4 &= 0, \\ 3x_1 + 2x_2 + 2x_3 + x_4 &= 0, \\ x_1 + 2x_2 + 2x_3 - 9x_4 &= 0. \end{aligned}$$

28. By the *rank of a set of n vectors* $\mathbf{a}_1 = (a_{11}, a_{21}, \ldots, a_{m1})$, $\mathbf{a}_2 = (a_{12}, a_{22}, \ldots, a_{m2})$, $\ldots$, $\mathbf{a}_n = (a_{1n}, a_{2n}, \ldots, a_{mn})$ is meant either the maximum number of linearly independent vectors in the set or the rank of the matrix

$$\begin{Vmatrix} a_{11} & a_{12} & \cdots & a_{1n} \\ a_{21} & a_{22} & \cdots & a_{2n} \\ \cdots & \cdots & \cdots & \cdots \\ a_{m1} & a_{m2} & \cdots & a_{mn} \end{Vmatrix}$$

as previously defined (Sec. 7). These two definitions of rank are equivalent (Sec. 12). Show that the vector $\mathbf{b}$ can be expressed as a linear combination of the vectors $\mathbf{a}_1, \mathbf{a}_2, \ldots, \mathbf{a}_n$ if and only if the rank of the set $\mathbf{a}_1, \mathbf{a}_2, \ldots, \mathbf{a}_n$ is the same as the rank of the "augmented set" $\mathbf{a}_1, \mathbf{a}_2, \ldots, \mathbf{a}_n, \mathbf{b}$.

29. In Sec. 9, it is proved that a necessary and sufficient condition for the general linear system (6) to be *compatible*, i.e., to have at least one solution, is that all the characteristic determinants (8) vanish. Prove that an equivalent necessary and sufficient condition for the system (6) to be compatible is that the coefficient matrix and the *augmented matrix*, given by

$$\begin{Vmatrix} a_{11} & a_{12} & \cdots & a_{1n} \\ a_{21} & a_{22} & \cdots & a_{2n} \\ \cdots & \cdots & \cdots & \cdots \\ a_{m1} & a_{m2} & \cdots & a_{mn} \end{Vmatrix} \quad \text{and} \quad \begin{Vmatrix} a_{11} & a_{12} & \cdots & a_{1n} & b_1 \\ a_{21} & a_{22} & \cdots & a_{2n} & b_2 \\ \cdots & \cdots & \cdots & \cdots & \cdots \\ a_{m1} & a_{m2} & \cdots & a_{mn} & b_m \end{Vmatrix},$$

respectively, have the same rank.

30. Investigate the compatibility of the systems given in Prob. 6, this time using the augmented matrix approach.

31. By an *elementary transformation* of a matrix, we mean one of the following operations (cf. Sec. 56):

1. Interchange of two rows (or columns);
2. Multiplication of all the elements of a row (or column) by the same nonzero number;
3. Addition of the elements of a row (or column), multiplied by the same number, to the corresponding elements of another row (or column).

Show that elementary transformations do not change the rank of a matrix.

32. Show that by making suitable elementary transformations, any $m \times n$ matrix $\|a_{ik}\|$ of rank r $(0 \le r \le \min(m, n))$ can be brought into a form so that $a_{11} = a_{22} = \cdots = a_{rr} = 1$ and all the other elements vanish. Apply this method to find the rank of the matrix

$$\begin{Vmatrix} 0 & 2 & -4 \\ -1 & -4 & 5 \\ 3 & 1 & 7 \\ 0 & 5 & -10 \\ 2 & 3 & 0 \end{Vmatrix}.$$

33. Let the number of unknowns in a homogeneous system be one more than the number of equations. Show that the minors obtained by successively excluding the first column, the second column, etc., taken with alternating signs, give a solution of the system. (Thus, for example, the numbers

$$\begin{vmatrix} 3 & 4 \\ 5 & 6 \end{vmatrix}, \quad -\begin{vmatrix} 5 & 4 \\ 6 & 6 \end{vmatrix}, \quad \begin{vmatrix} 5 & 3 \\ 6 & 5 \end{vmatrix}$$

are a solution of the system

$$\begin{aligned} 5x + 3y + 4z &= 0, \\ 6x + 5y + 6z &= 0.) \end{aligned}$$

34. The coefficient matrices of the following inhomogeneous systems have vanishing determinants:

(a) $2x + 4y = 1$, $x + 2y = 3$; (b) $2x + 4y = 2$, $x + 2y = 1$; (c) $2x - y + 3z = 9$, $3x - 5y + z = -4$, $4x - 7y + z = 5$;

(d) $2x - y + 3z = 3$, $3x - 5y + z = 1$, $4x - 7y + z = 1$.

Use the adjoint matrix approach of Sec. 15 to find which of the systems are compatible.

35. A square matrix $\|a_{ij}\|$ is said to be *self-adjoint* if $a_{ij} = \bar{a}_{ji}$ for all i and j. (In particular, this implies that all the diagonal elements are real.) Using the criterion of Sec. 15, show that a self-adjoint system

$$\begin{aligned} a_{11}x_1 + a_{12}x_2 + \cdots + a_{1n}x_n &= b_1, \\ a_{21}x_1 + a_{22}x_2 + \cdots + a_{2n}x_n &= b_2, \\ \cdots\cdots\cdots\cdots\cdots\cdots\cdots \\ a_{n1}x_1 + a_{n2}x_2 + \cdots + a_{nn}x_n &= b_n \end{aligned}$$

has a solution if and only if the vector $\mathbf{b} = (b_1, b_2, \ldots, b_n)$ is a linear combination of the row vectors $\mathbf{a}_i = (a_{i1}, a_{i2}, \ldots, a_{in})$, $i = 1, 2, \ldots, n$.

36. For which choices of the parameter λ does the system

$$\begin{aligned} \lambda x + y + z + w &= 1, \\ x + \lambda y + z + w &= 1, \\ x + y + \lambda z + w &= 1, \\ x + y + z + \lambda w &= 1 \end{aligned}$$

have a solution? (Cf. Prob. 9.)

37. Find a necessary and sufficient condition for the three lines

$$\begin{aligned} a_1x + b_1y + c_1 &= 0, \\ a_2x + b_2y + c_2 &= 0, \\ a_3x + b_3y + c_3 &= 0 \end{aligned}$$

to pass through a common point.

38. What is a necessary and sufficient condition for n distinct planes $a_ix + b_iy + c_iz + d_i = 0$ $(i = 1, 2, \ldots, n)$ to have a line in common?

39. Find a necessary and sufficient condition for the kth unknown to vanish in any solution of a compatible system of linear equations.

40. Examine all possible cases encountered in solving systems of linear equations in two unknowns, and in each case interpret the system geometrically.

41. Calculate the Gram determinants of the following sets of vectors:

(a) $\mathbf{a}_1 = (1, 1, 0)$, $\mathbf{a}_2 = (1, 0, 1)$, $\mathbf{a}_3 = (0, 1, 1)$;
(b) $\mathbf{a}_1 = (1 + x, 1 - x, 0)$, $\mathbf{a}_2 = (1 - x, 0, 1 + x)$, $\mathbf{a}_3 = (0, 1 + x, 1 - x)$.

42. Show that all the principal minors of a Gram determinant are nonnegative.

43. Show that if

$$\Delta = \begin{vmatrix} 1 & b \\ \bar{b} & c \end{vmatrix} > 0,$$

then Δ is a Gram determinant.

44. Show that the inequality (44) can never be an equality for a real determinant $|a_{ik}|$ of odd order (>1). Show that the equality in (44) can be attained for a real determinant whose order is a power of 2.

45. Let L be a subspace and $\mathbf{x}$ a vector not in L. It is shown in Sec. 14 that $\mathbf{x} = \mathbf{y} + \mathbf{z}$, where $\mathbf{y}$ is in L and $\mathbf{z}$ is orthogonal to all the vectors in L. By the *distance from* $\mathbf{x}$ *to* L we mean

$$\inf_{\mathbf{w}} \|\mathbf{x} - \mathbf{w}\|,$$

where the infimum (greatest lower bound) is taken over all vectors $\mathbf{w}$ in L. Show that this distance is equal to $\|\mathbf{z}\|$.

46. Let the linearly independent set $\mathbf{x}^{(1)}, \mathbf{x}^{(2)}, \ldots, \mathbf{x}^{(m)}$ span a subspace L, and let $\mathbf{x}$ be a vector not in L. Show that the distance d from $\mathbf{x}$ to this subspace is given by

$$d^2 = \frac{G(\mathbf{x}^{(1)}, \ldots, \mathbf{x}^{(m)}, \mathbf{x})}{G(\mathbf{x}^{(1)}, \ldots, \mathbf{x}^{(m)})}.$$

47. Define the volume $V(\mathbf{a}_1, \mathbf{a}_2, \ldots, \mathbf{a}_n)$ of the n-dimensional parallelepiped spanned by the n linearly independent vectors $\mathbf{a}_1, \mathbf{a}_2, \ldots, \mathbf{a}_n$, using the following inductive approach:

1. $V(\mathbf{a}_1) = \|\mathbf{a}_1\|$,
2. $V(\mathbf{a}_1, \mathbf{a}_2, \ldots, \mathbf{a}_n) = V(\mathbf{a}_1, \mathbf{a}_2, \ldots, \mathbf{a}_{n-1})h_n$, where h_n is the length of the component of $\mathbf{a}_n$ orthogonal to the subspace spanned by $\mathbf{a}_1, \mathbf{a}_2, \ldots, \mathbf{a}_{n-1}$.

Then, prove that

$$V(\mathbf{a}_1, \mathbf{a}_2, \ldots, \mathbf{a}_n) = \sqrt{G(\mathbf{a}_1, \mathbf{a}_2, \ldots, \mathbf{a}_n)} = |D|,$$

where $G(\mathbf{a}_1, \mathbf{a}_2, \ldots, \mathbf{a}_n)$ is the Gram determinant of the vectors $\mathbf{a}_1, \mathbf{a}_2, \ldots, \mathbf{a}_n$, and D is the determinant of the components of these vectors relative to any orthonormal basis of the n-dimensional space spanned by them.

48. What is the geometrical meaning of Hadamard's inequality?

49. Let $\mathbf{a}_1, \mathbf{a}_2, \ldots, \mathbf{a}_k, \mathbf{b}, \mathbf{c}$ be linearly independent vectors in an n-dimensional space. Prove that

$$\frac{V(\mathbf{a}_1, \mathbf{a}_2, \ldots, \mathbf{a}_k, \mathbf{b}, \mathbf{c})}{V(\mathbf{a}_1, \mathbf{a}_2, \ldots, \mathbf{a}_k, \mathbf{b})} \le \frac{V(\mathbf{a}_1, \mathbf{a}_2, \ldots, \mathbf{a}_k, \mathbf{c})}{V(\mathbf{a}_1, \mathbf{a}_2, \ldots, \mathbf{a}_k)}.$$

Comment: This is a special case of formula (45) of Sec. 16.

50. Solve the following systems of first order differential equations:†

(a) $\dot{x} = z + y - x$, $\dot{y} = z + x - y$, $\dot{z} = x + y + z$, subject to the initial conditions $x(0) = 1$, $y(0) = z(0) = 0$;
(b) $\dot{x} = y - 7x$, $\dot{y} + 2x + 5y = 0$.

(The overdot denotes differentiation with respect to t.)

† From a problem collection by G. N. Berman.

51. Solve the secular equations of the preceding problem by Krylov's method.

Comment: No simplification is to be expected in such simple examples. As an example of a case where the power of Krylov's method begins to appear, consider the system $\dot{x} = y - z + w$, $\dot{y} = x + w + z$, $\dot{z} = x + w - y$, $\dot{w} = -2z - y + x$.

52. Solve the system $\ddot{x} = -2m^2 y$, $\ddot{y} = 2m^2 x$, where $m > 0$.

53. Find the Jacobian $\partial(\varphi_1, \varphi_2, \varphi_3)/\partial(x_1, x_2, x_3)$ of each of the following systems of functions:

(a) $\varphi_1(x_1, x_2, x_3) = x_1x_2x_3$, $\varphi_2(x_1, x_2, x_3) = x_1x_2 - x_1x_2x_3$, $\varphi_3(x_1, x_2, x_3) = x_2 - x_1x_2$;

(b) $\varphi_1(x_1, x_2, x_3) = x_1/d$, $\varphi_2(x_1, x_2, x_3) = x_2/d$, $\varphi_3(x_1, x_2, x_3) = x_3/d$, where $d = 1 + x_1 + x_2 + x_3$.

54. Prove that if

$$u_1 = \frac{x_1}{\sqrt{1 - r^2}}, \quad u_2 = \frac{x_2}{\sqrt{1 - r^2}}, \quad u_3 = \frac{x_3}{\sqrt{1 - r^2}},$$

where $r^2 = x_1^2 + x_2^2 + x_3^2$, then

$$\frac{\partial(u_1, u_2, u_3)}{\partial(x_1, x_2, x_3)} = (1 - r^2)^{-5/2}.$$

55. Consider the system of functions

$$h_i(x_1, x_2, \ldots, x_n) = \varphi_i(g_1(x_1), g_2(x_2), \ldots, g_n(x_n)) \qquad (i = 1, 2, \ldots, n).$$

Show that

$$\frac{\partial(h_1, h_2, \ldots, h_n)}{\partial(x_1, x_2, \ldots, x_n)} = \frac{\partial(\varphi_1, \varphi_2, \ldots, \varphi_n)}{\partial(g_1, g_2, \ldots, g_n)} g_1'(x_1)g_2'(x_2) \cdots g_n'(x_n),$$

where the derivatives appearing in this relation are assumed to exist.

56. Show that the Jacobian of the system of functions

$$\begin{aligned} \varphi_1(x_1, x_2, x_3) &= 2x_1^2 + 2x_2^2 + x_3^2, \\ \varphi_2(x_1, x_2, x_3) &= 2x_1 + x_3, \\ \varphi_3(x_1, x_2, x_3) &= x_1^2 - x_2^2 + 2x_1x_3 \end{aligned}$$

vanishes identically. Find a functional relationship between these three functions.

57. Consider the three-dimensional transformation

$$\begin{aligned} y_1 &= a_{11}x_1 + a_{12}x_2 + a_{13}x_3, \\ y_2 &= a_{21}x_1 + a_{22}x_2 + a_{23}x_3, \\ y_3 &= a_{31}x_1 + a_{32}x_2 + a_{33}x_3 \end{aligned}$$

taking the point (x_1, x_2, x_3) into the point (y_1, y_2, y_3). Show that the Jacobian $\partial(y_1, y_2, y_3)/\partial(x_1, x_2, x_3)$ vanishes if and only if there is a point different from the origin which is carried into the origin by the transformation.

58. Let $y_1 = \log(1 - x_2 - x_3)$, $y_2 = \log(1 - x_1 - x_3)$, $y_3 = \log(1 - x_2 - x_3)$. Can this system be inverted in a neighborhood of the origin? (Cf. the theorem associated with (86) and (87).)

PART II

Matrix Theory

Chapter 3

Linear Transformations

20. Coordinate Transformations in Three Dimensions. By a *linear transformation* in n variables is meant a transformation of the form

$$\begin{aligned} x_1' &= a_{11}x_1 + a_{12}x_2 + \cdots + a_{1n}x_n, \\ x_2' &= a_{21}x_1 + a_{22}x_2 + \cdots + a_{2n}x_n, \\ &\cdots\cdots\cdots\cdots\cdots\cdots \\ x_n' &= a_{n1}x_1 + a_{n2}x_2 + \cdots + a_{nn}x_n. \end{aligned} \tag{1}$$

This transformation can be interpreted in two ways. It can be regarded as taking one n-dimensional vector $(x_1, x_2, \ldots, x_n)$ into another vector $(x_1', x_2', \ldots, x_n')$, or alternatively, as taking one n-dimensional point with coordinates $x_1, x_2, \ldots, x_n$ into another point with coordinates $x_1', x_2', \ldots, x_n'$. On the other hand, if we regard $x_1, x_2, \ldots, x_n$ and $x_1', x_2', \ldots, x_n'$ as the components of the same vector (or the coordinates of the same point) with respect to two different choices of the axes, then the equations (1) give the transformation law of the components (or coordinates) in going from one coordinate system to the other. We have already frequently encountered formulas of the form (1) for the cases $n = 2$ and $n = 3$.

The first part of this chapter will be devoted to a detailed study of linear transformations of the form (1). For the sake of greater clarity, we begin with the case of real three-dimensional space, before considering the more general case of complex n-dimensional space. Moreover, we first consider the simplest three-dimensional case, i.e., the case where the transformation (1) corresponds to a transition from one set of rectangular axes to another. By locating the initial points of all vectors at the origin, we can either regard x_1, x_2, x_3 as the components of a vector or as the coordinates of its end point.

As we know from analytic geometry, the formulas for transforming from one set of rectangular Cartesian coordinates to another are

$$\begin{aligned} x_1' &= a_{11}x_1 + a_{12}x_2 + a_{13}x_3, \\ x_2' &= a_{21}x_1 + a_{22}x_2 + a_{23}x_3, \\ x_3' &= a_{31}x_1 + a_{32}x_2 + a_{33}x_3, \end{aligned} \tag{2}$$

where the a_{ik} are the cosines of the angles which the new axes make with the old axes. Suppose we write the a_{ik} in the form of a table:

	X_1	X_2	X_3
X'_1	a_{11}	a_{12}	a_{13}
X'_2	a_{21}	a_{22}	a_{23}
X'_3	a_{31}	a_{32}	a_{33}

, (3)

where X_1, X_2, X_3 denote the old axes and X'_1, X'_2, X'_3 denote the new axes. Then, as is well known, the sum of the squares of the elements of each row (and column) equals unity, while the sum of the products of the elements of one row (or column) with the corresponding elements of another row (or column) equals zero. The determinant $|a_{ik}|$ equals the volume of the rectangular parallelepiped whose sides have unit length and lie along the new coordinate axes (Sec. 5); this volume is $+1$ or -1, depending on whether or not the orientation of the new axes is the same as that of the old axes. Clearly, the inverse transformation from x'_1, x'_2, x'_3 to x_1, x_2, x_3 has the form

$$\begin{aligned} x_1 &= a_{11}x'_1 + a_{21}x'_2 + a_{31}x'_3, \\ x_2 &= a_{12}x'_1 + a_{22}x'_2 + a_{32}x'_3, \\ x_3 &= a_{13}x'_1 + a_{23}x'_2 + a_{33}x'_3. \end{aligned} \tag{2a}$$

In other words, the inverse of the transformation (2) is obtained by simply interchanging rows and columns in the coefficient array (3). The value of the determinant of the inverse transformation (2*a*) is obviously equal to that of the direct transformation (2).

We now show that the properties of the coefficients of the transformation (2) are actually a consequence of one simple requirement, which stems from the geometrical nature of the problem. Thus, suppose we set ourselves the problem of finding all real transformations of the form (2) such that

$$x_1'^2 + x_2'^2 + x_3'^2 = x_1^2 + x_2^2 + x_3^2. \tag{4}$$

(This way of posing the problem makes it possible to generalize our transformations to the case of spaces with an arbitrary number of dimensions.) In order to show that the transformations required by this new statement of the problem are the same as those already considered (i.e., that the requirement (4) leads to the properties of the coefficients a_{ik} already noted), we square the expressions (2) and substitute them into the lefts hand side of the relation (4). Then, setting the coefficients of the term-

x_k^2 $(k = 1, 2, 3)$ equal to unity and the coefficients of the terms x_kx_l $(k \neq l)$ equal to zero, we obtain the six relations

$$a_{1k}a_{1l} + a_{2k}a_{2l} + a_{3k}a_{3l} = \delta_{kl} \qquad (k, l = 1, 2, 3), \tag{5}$$

where

$$\delta_{kl} = 0 \qquad \text{for } k \neq l, \ \delta_{kk} = 1, \tag{6}$$

i.e., the sum of the squares of the elements of each column equals unity, and the sum of the products of elements of one column with the corresponding elements of another column equals zero. These conditions are called the *orthogonality conditions for columns.* Thus, the elements of each column are equal to the direction cosines of some direction, and the directions corresponding to different columns are orthogonal. It follows immediately that the transformations (2) satisfying the condition (4) are the same as those previously considered; in particular, the orthogonality conditions hold for rows as well as for columns.

In line with our previous remarks, (2) can be interpreted not only as a change of coordinates with the space remaining fixed (as we have just done), but also as a change of the space itself with the coordinate axes remaining fixed. We now consider this case. First suppose that the determinant of the transformation equals $+1$, i.e., that both sets of coordinate axes have the same orientation. Then we can make a rigid rotation of the space as a whole, together with the axes X_1', X_2', X_3', about the origin, until the axes X_1', X_2', X_3' coincide with the axes X_1, X_2, X_3. The axes X_1, X_2, X_3 are regarded as fixed during the rotation, and coordinates of all points are referred to them both before and after the rotation. If a point M had coordinates x_1, x_2, x_3 before the rotation, then as a result of the rotation it moves to a new position M' and has new coordinates x_1', x_2', x_3'. Since the point M and the coordinate axes X_1', X_2', X_3' move together, the coordinates x_1', x_2', x_3' of the point M' with respect to the axes X_1, X_2, X_3 after the rotation are the same as the coordinates of the point M with respect to the axes X_1', X_2', X_3' before the rotation. Thus, when the determinant equals $+1$, we see that the formulas (2) show how the coordinates of a point transform under a rotation.

Now suppose that the determinant $|a_{ik}|$ equals -1, and instead of the transformation (2), consider the transformation

$$x_i'' = -a_{i1}x_1 - a_{i2}x_2 - a_{i3}x_3 \qquad (i = 1, 2, 3).$$

These coefficients still have the property (5), but their determinant equals $+1$, i.e., this transformation corresponds to a spatial rotation about the origin. To obtain the coordinates x'_1, x'_2, x'_3, we have to perform another transformation:

$$x'_1 = -x''_1,\ x'_2 = -x''_2,\ x'_3 = -x''_3.$$

Since changing the signs of all the coordinates corresponds to performing a reflection in the origin, we see that when the determinant equals -1, the transformation (2) corresponds to a rotation about the origin followed by a reflection in the origin.

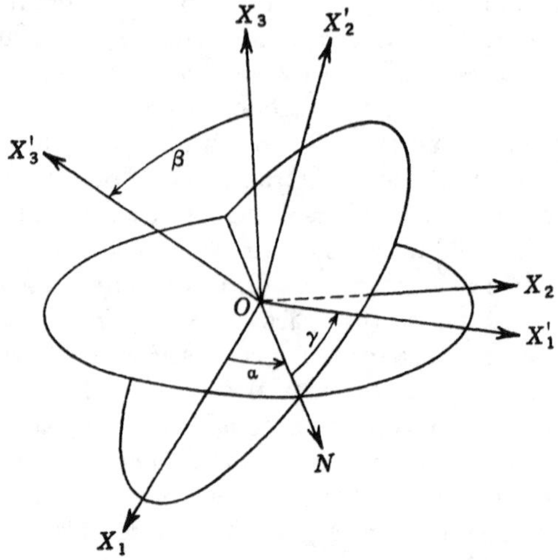

Fig. 1

We have just seen that the nine coefficients a_{ik} must satisfy the six relations (5). It follows that it must be possible to express these coefficients in terms of just three independent parameters. We now show one possible choice of these parameters in the case of a spatial rotation about the origin. Thus, consider two sets of coordinate axes, a fixed set X_1, X_2, X_3 to which all coordinates are referred, and another set X'_1, X'_2, X'_3 which is rigidly connected to the rotating space. To describe the rotation, we must specify three parameters which determine the position of the second coordinate system with respect to the first system. Referring to Fig. 1, let ON be the line of intersection of the plane $X'_1OX'_2$ with the plane X_1OX_2 (the *nodal line*). Choose a definite direction on this line, and let α be the angle made by ON with OX_1. Then, introduce the angle β made by OX'_3 with OX_3 and the angle γ made by OX'_1 with ON. These three angles, which we denote by the symbol $\{\alpha, \beta, \gamma\}$, completely characterize the position of the second system with respect to the first system, i.e., completely characterize the rotation. In fact, to obtain the rotation, we start from a position where the two sets of axes coincide and perform in succession the following three rotations: (*a*) a rotation through the angle α about the axis X_3, (*b*) a rotation through the angle β about the new position of the axis X_1', and (*c*) a rotation through the angle γ about the new position of the axis X_3'. These three angles are called the *Eulerian angles*. We note that they vary between the limits

$$0 \leq \alpha < 2\pi, \qquad 0 \leq \beta < \pi, \qquad 0 \leq \gamma < \pi.$$

If $\beta = 0$, then the rotation $\{\alpha, \beta, \gamma\}$ reduces to just a rotation through the angle $\alpha + \gamma$ about the axis X_3, so that we have the relation

$$\{\alpha, 0, \gamma\} = \{\alpha + \delta, 0, \gamma - \delta\}$$

for any δ. This shows that in some cases the correspondence between the parameters $\{\alpha, \beta, \gamma\}$ and rotations about the origin is not one-to-one, i.e., different values of the parameters can correspond to the same rotation. Finally, we note that it is not hard to derive formulas expressing the coefficients a_{ik} in terms of trigonometric functions of the angles α, β, γ (Sec. 75).

Subsequently, we shall introduce another choice of the parameters characterizing spatial rotations about the origin, and we shall return once more to the subject of Eulerian angles (Sec. 69).

21. General Linear Transformations in Three Dimensions. We now consider linear transformations of the form (2) with real coefficients which are such that the determinant of the transformation is nonvanishing, i.e.,

$$|a_{ik}| \neq 0, \tag{7}$$

but are otherwise arbitrary. A transformation satisfying the condition (7) is called a *nonsingular transformation*. If the transformation does not satisfy the condition (5) of the preceding section, then it leads to a deformation of space. The transformation (2) is characterized by the totality of its coefficients; they completely specify the law by which an arbitrary vector with components x_1, x_2, x_3 is transformed into a new vector with components x_1', x_2', x_3'. We denote the whole array or *matrix* of coefficients by one letter:

$$A = \begin{Vmatrix} a_{11} & a_{12} & a_{13} \\ a_{21} & a_{22} & a_{23} \\ a_{31} & a_{32} & a_{33} \end{Vmatrix}, \tag{8}$$

where, as before, we put the array between double lines to distinguish it from its determinant. The determinant of the matrix A will be denoted by det A and is a well-defined number. The transformation (2) can be written symbolically as

$$\mathbf{x}' = A\mathbf{x}, \tag{9}$$

where $\mathbf{x}'$ is the vector with components x_1', x_2', x_3', and $\mathbf{x}$ is the vector with components x_1, x_2, x_3. (Of course, instead of regarding the general linear transformation (2) as a spatial deformation carrying the vector with components x_1, x_2, x_3 into a new vector with components x_1', x_2', x_3', we can just as well treat (2) as a point transformation carrying the point with coordinates x_1, x_2, x_3 into the point with coordinates x_1', x_2', x_3', as already remarked in Sec. 20.)

The transformation which leaves every vector unchanged is called the *identity transformation.* It has the matrix

$$\begin{Vmatrix} 1 & 0 & 0 \\ 0 & 1 & 0 \\ 0 & 0 & 1 \end{Vmatrix}, \tag{10}$$

called the *unit matrix* and denoted by the symbol I. Assuming that det $A \neq 0$, we can solve the equations (2) for x_1, x_2, x_3, obtaining

$$\begin{aligned} x_1 &= \frac{A_{11}}{\det A} x_1' + \frac{A_{21}}{\det A} x_2' + \frac{A_{31}}{\det A} x_3', \\ x_2 &= \frac{A_{12}}{\det A} x_1' + \frac{A_{22}}{\det A} x_2' + \frac{A_{32}}{\det A} x_3', \\ x_3 &= \frac{A_{13}}{\det A} x_1' + \frac{A_{23}}{\det A} x_2' + \frac{A_{33}}{\det A} x_3', \end{aligned} \tag{11}$$

where A_{ik} is the cofactor of the element a_{ik} in the determinant det A. This linear transformation is called the *inverse* of the transformation (2); if the matrix of the direct transformation (2) is denoted by A, then the matrix of the inverse transformation (11) is denoted by A^{-1}.

We now introduce the important concept of the product of two transformations (or two matrices). Suppose we have two linear transformations $\mathbf{x}' = A\mathbf{x}$ and $\mathbf{x}'' = B\mathbf{x}'$, where the first takes $\mathbf{x} = (x_1, x_2, \ldots, x_n)$ into $\mathbf{x}' = (x_1', x_2', \ldots, x_n')$:

$$\begin{aligned} x_1' &= a_{11}x_1 + a_{12}x_2 + a_{13}x_3, \\ x_2' &= a_{21}x_1 + a_{22}x_2 + a_{23}x_3, \\ x_3' &= a_{31}x_1 + a_{32}x_2 + a_{33}x_3, \end{aligned} \tag{12}$$

and the second takes $\mathbf{x}' = (x_1', x_2', \ldots, x_n')$ into $\mathbf{x}'' = (x_1'', x_2'', \ldots, x_n'')$:

$$\begin{aligned} x_1'' &= b_{11}x_1' + b_{12}x_2' + b_{13}x_3', \\ x_2'' &= b_{21}x_1' + b_{22}x_2' + b_{23}x_3', \\ x_3'' &= b_{31}x_1' + b_{32}x_2' + b_{33}x_3'. \end{aligned} \tag{13}$$

Instead of first going from (x_1, x_2, x_3) to (x_1', x_2', x_3') and then from (x_1', x_2', x_3') to (x_1'', x_2'', x_3''), we can go directly from (x_1, x_2, x_3) to (x_1'', x_2'', x_3'') by another linear transformation:

$$x_k'' = c_{k1}x_1 + c_{k2}x_2 + c_{k3}x_3 \qquad (k = 1, 2, 3). \tag{14}$$

The linear transformation (14) is called the *product* of the linear transformations (12) and (13); here the order in which the transformations are applied is important. We obtain (14) by substituting (12) into the right-hand side of (13); the result is the following expression for the elements c_{ik} of the matrix of the product transformation in terms of the matrix elements of the transformations A and B:

$$c_{ik} = \sum_{s=1}^{3} b_{is}a_{sk} \qquad (i, k = 1, 2, 3). \tag{15}$$

The transformation (14) is usually written in the form

$$\mathbf{x}'' = BA\mathbf{x}. \tag{16}$$

The matrix C with the elements c_{ik} given by (15) is called the *product* of the matrices A and B and is written

$$C = BA. \tag{17}$$

To get the order in which the transformations are applied, we have to read this product from right to left. Comparing (15) with the multiplication theorem for determinants (Sec. 6), we can immediately write down the formula

$$\det C = \det B \det A, \tag{18}$$

relating the determinants of the transformations A, B, C. *Thus, the determinant of the product of two transformations equals the product of the determinants of the transformations.* It is not hard to prove the relation

$$AA^{-1} = A^{-1}A = I, \tag{19}$$

which has a simple geometrical meaning. We also note that the inverse of the transformation A^{-1} is the transformation A:

$$(A^{-1})^{-1} = A. \tag{20}$$

This follows from the way the inverse is constructed, since if we solve the system (11) for the x_k', we obviously get back the system (2).

The concept of product transformation can be extended to the case of an arbitrary number of factors; for example, the result of performing in succession the transformations with matrices A, B, and C is a new transformation, with matrix D given by

$$D = CBA. \tag{21}$$

If the matrices A, B, C have elements a_{ik}, b_{ik}, c_{ik}, respectively, then the product matrix D has elements given by the formula

$$d_{ik} = \sum_{p,q=1}^{3} c_{iq}b_{qp}a_{pk}. \tag{22}$$

This follows from the fact that by (15), the elements of the matrix $E = BA$ are

$$e_{ik} = \sum_{p=1}^{3} b_{ip}a_{pk},$$

while the elements of the matrix CE are

$$d_{ik} = \sum_{q=1}^{3} c_{iq}e_{qk},$$

which is equivalent to (22). (Subsequently, we shall often denote the elements of a matrix by $\{A\}_{ik}$.)

Matrix multiplication does not obey the *commutative law*, i.e., in general $BA \neq AB$. However, matrix multiplication does obey the *associative law*

$$C(BA) = (CB)A, \tag{23}$$

i.e., factors can be grouped in any way. To get $C(BA)$, we multiply A by B and then multiply the result by C; to get $(CB)A$, we multiply A by the result of multiplying B by C. It is easily verified that the elements of both $C(BA)$ and $(CB)A$ are given by (22); in fact, this has already been shown for $C(BA)$. As for $(CB)A$, performing in succession the required multiplications, we find

$$\{CB\}_{ik} = \sum_{q=1}^{3} \{C\}_{iq}\{B\}_{qk}$$

and

$$\{(CB)A\}_{ik} = \sum_{p=1}^{3} \{CB\}_{ip}\{A\}_{pk} = \sum_{p,q=1}^{3} \{C\}_{iq}\{B\}_{qp}\{A\}_{pk},$$

which is just (22), with our new notation.

An important special type of linear transformation is the following:

$$x_1' = k_1x_1, \qquad x_2' = k_2x_2, \qquad x_3' = k_3x_3. \tag{24}$$

This transformation represents expansion (or contraction) along each coordinate axis by a factor specified by the numerical coefficients k_1, k_2, k_3. The matrix of the transformation (24) is obviously

$$\left\| \begin{matrix} k_1 & 0 & 0 \\ 0 & k_2 & 0 \\ 0 & 0 & k_3 \end{matrix} \right\|,$$

where all elements except those lying on the principal diagonal vanish. Such a matrix is called a *diagonal matrix* and will be denoted by the symbol $[k_1, k_2, k_3]$. In particular, if the numbers are the same, i.e., $k_1 = k_2 = k_3 = k$, the transformation reduces to multiplying all the components of a vector by the same number k; this produces a so-called *homothetic transformation* with center at the origin of coordinates. If it is assumed that $k > 0$, then every vector has its length multiplied by k without changing its direction. A simpler way of writing this transformation is

$$\mathbf{x}' = k\mathbf{x}.$$

In this formula, k is regarded as being the special diagonal matrix

$$\left\| \begin{matrix} k & 0 & 0 \\ 0 & k & 0 \\ 0 & 0 & k \end{matrix} \right\|, \tag{25}$$

whose diagonal elements all equal the number k. Equation (15) shows that multiplication of matrices of this kind reduces to ordinary multiplication of numbers:

$$[k, k, k][l, l, l] = [kl, kl, kl].$$

More generally, it is easily verified that the multiplication law for arbitrary diagonal matrices is simply

$$[k_1, k_2, k_3][l_1, l_2, l_3] = [k_1l_1, k_2l_2, k_3l_3]. \tag{26}$$

Thus, two successive expansions along the coordinate axes are equivalent to a single expansion with coefficients equal to the product of the corresponding coefficients of the individual expansions. Equation (26) implies, among other things, that the product of two diagonal matrices does not change when the order of the factors is reversed. Using (15), together with the representation (25) of a number, as a diagonal matrix, we see that the product kA reduces to multiplying all the elements of the matrix A by the number k. This product does not depend on the order of the factors, and in fact

$$\{kA\}_{ij} = \{Ak\}_{ij} = k\{A\}_{ij}. \tag{27}$$

In defining the components of a vector, we can use any system of coordinate axes, i.e., any set of basis vectors. Thus, we can choose for the basis vectors any three vectors **i**, **j**, **k** which are not coplanar. Then, as is well known, every vector **x** can be uniquely represented in the form

$$\mathbf{x} = x_1\mathbf{i} + x_2\mathbf{j} + x_3\mathbf{k}. \tag{28}$$

The numbers x_1, x_2, x_3 are called the components of the vector **x** with respect to the coordinate system specified by the basis vectors **i**, **j**, **k**. We shall now investigate the influence of different choices of the basis vectors on the form of a linear transformation. More precisely, if a linear transformation has the form (12) in the coordinate system with basis vectors **i**, **j**, **k**, then what will the same linear transformation look like in another coordinate system with basis vectors $\mathbf{i}_1$, $\mathbf{j}_1$, $\mathbf{k}_1$? Suppose that the new basis vectors can be expressed in terms of the old basis vectors by the formulas

$$\begin{aligned} \mathbf{i}_1 &= t_{11}\mathbf{i} + t_{12}\mathbf{j} + t_{13}\mathbf{k}, \\ \mathbf{j}_1 &= t_{21}\mathbf{i} + t_{22}\mathbf{j} + t_{23}\mathbf{k}, \\ \mathbf{k}_1 &= t_{31}\mathbf{i} + t_{32}\mathbf{j} + t_{33}\mathbf{k}. \end{aligned} \tag{29}$$

(Note that the determinant formed from the coefficients t_{ik} must be nonvanishing, for otherwise the vectors would be linearly dependent and hence coplanar.) In the new coordinate system, the vector **x** defined by (28) has new components y_1, y_2, y_3:

$$\mathbf{x} = y_1\mathbf{i}_1 + y_2\mathbf{j}_1 + y_3\mathbf{k}_1. \tag{28a}$$

We begin by establishing a formula which expresses the new components of the vector **x** in terms of its old components. Thus, substituting the expressions (29) for the basis vectors into (28*a*), we obtain

$$\sum_{s=1}^{3} y_s(t_{s1}\mathbf{i} + t_{s2}\mathbf{j} + t_{s3}\mathbf{k}) = x_1\mathbf{i} + x_2\mathbf{j} + x_3\mathbf{k}.$$

Comparing coefficients of **i**, **j**, and **k**, we obtain the formulas

$$\begin{aligned} x_1 &= t_{11}y_1 + t_{21}y_2 + t_{31}y_3, \\ x_2 &= t_{12}y_1 + t_{22}y_2 + t_{32}y_3, \\ x_3 &= t_{13}y_1 + t_{23}y_2 + t_{33}y_3, \end{aligned} \tag{30}$$

which express the old components of **x** in terms of the new components of **x**. The difference between the matrix of this transformation and the matrix of the transformation (29) is that the rows and columns have been interchanged. In fact, in (29) the first indices are the same in each row, while in (30) the second indices are the same in each row. If the matrix of the transformation (29) is denoted by T then the matrix of the transformation (30) will be denoted by T^* and will be called the *transpose* of T.† Then (30) can be written compactly as

$$\mathbf{x} = T^*\mathbf{y}, \tag{31}$$

where $\mathbf{x} = (x_1, x_2, x_3)$ is the vector made up of the components of $\mathbf{x}$ relative to the old basis vectors, and $\mathbf{y} = (y_1, y_2, y_3)$ is the vector made up of the components of $\mathbf{x}$ relative to the new basis vectors. Conversely, the new components can be expressed in terms of the old by the formula

$$\mathbf{y} = (T^*)^{-1}\mathbf{x},$$

where $(T^*)^{-1}$ is the linear transformation which is the inverse of T^*; the transformation $(T^*)^{-1}$ is usually said to be *contragredient* to the transformation T. For brevity, we write

$$U = (T^*)^{-1}. \tag{32}$$

Then, we can assert that when the basis vectors are changed in accordance with (29), the components of every vector undergo the linear transformation with the matrix U defined by (32). Thus, after the transformation to new basis vectors, the two vectors $\mathbf{x} = (x_1, x_2, x_3)$ and $\mathbf{x}' = (x_1', x_2', x_3')$ appearing in the transformation (9) will have new components which are expressed in terms of the old components by the formulas

$$\mathbf{y} = U\mathbf{x}, \qquad \mathbf{y}' = U\mathbf{x}', \tag{33}$$

where $\mathbf{y} = (y_1, y_2, y_3)$ and $\mathbf{y}' = (y_1', y_2', y_3')$ are the vectors made up of the new components of $\mathbf{x}$ and $\mathbf{x}'$, respectively.

Our problem now is to find the linear transformation which takes the vector $\mathbf{y}$ into the vector $\mathbf{y}'$. We can go from the vector $\mathbf{y}$ to the new vector $\mathbf{y}'$ in the following steps: First we go from the vector $\mathbf{y}$ to the vector $\mathbf{x}$; according to (33), this is accomplished by using the transformation with matrix U^{-1}. Then we go from the vector $\mathbf{x}$ to the vector $\mathbf{x}'$ by using the transformation (9) with matrix A. Finally, we go from the vector $\mathbf{x}'$ to the vector $\mathbf{y}'$; according to (33), this is accomplished by using the transformation with matrix U. The result of applying these three transformations in succession is the transformation

$$\mathbf{y}' = UAU^{-1}\mathbf{y}. \tag{34}$$

This transformation is said to be *similar* to the transformation (9), and correspondingly, the matrices A and UAU^{-1} are said to be *similar matrices*. Thus, we can state the following final result: *If the components*

† Other commonly encountered notations for the transpose of T are T', T^t, T^T, $\tilde{T}$.

of a vector undergo the linear transformation (33) *when the basis vectors are changed, then every linear transformation which has the form*

$$\mathbf{x}' = A\mathbf{x}$$

in the old coordinate system has the form

$$\mathbf{y}' = UAU^{-1}\mathbf{y}$$

in the new coordinate system.

22. Covariant and Contravariant Affine Vectors. Consider the case where the linear transformation (9) represents a transformation from one Cartesian coordinate system to another; then the coefficients of the transformation are just the direction cosines given in the table (3). In this case, as we saw in Sec. 20, the transposed matrix A^* is the same as the inverse matrix A^{-1}, and consequently, the contragredient matrix $(A^*)^{-1}$ is the same as the original matrix A, i.e.,

$$A^* = A^{-1}, \qquad (A^*)^{-1} = A. \tag{35}$$

It follows that the components of a fixed vector transform by the same law as the law governing the transformation of the coordinates themselves. Thus, in any fixed Cartesian coordinate system, a vector is a triple of numbers (its components) which transforms by the same formulas

$$\begin{aligned} x_1' &= a_{11}x_1 + a_{12}x_2 + a_{13}x_3, \\ x_2' &= a_{21}x_1 + a_{22}x_2 + a_{23}x_3, \\ x_3' &= a_{31}x_1 + a_{32}x_2 + a_{33}x_3, \end{aligned} \tag{36}$$

as the coordinates themselves, when we go from one Cartesian coordinate system to another.

We now consider, in addition to transformations from one Cartesian coordinate system to another, the more general case of *all possible* linear coordinate transformations with nonvanishing determinants; as noted above, this case corresponds to choosing as the basis vectors an arbitrary set of three noncoplanar vectors. As before, we shall consider both the matrix A of the transformation (36) and the contragredient matrix $V = (A^*)^{-1}$. In general, these two matrices will be different, which leads to the following two possible definitions of a vector with regard to its behavior under linear coordinate transformations:

1. A vector is defined as a triple of numbers (its components) which transforms by the same law as the coordinates themselves, i.e., by the formula

$$\mathbf{x}' = A\mathbf{x}, \tag{37}$$

where $\mathbf{x} = (x_1, x_2, x_3)$, $\mathbf{x}' = (x_1', x_2', x_3')$, when we go from one coordinate system to another. Such a vector is called a *contravariant affine vector.*

(The general linear transformation (36) is sometimes called an *affine transformation*.)

2. A vector is defined as a triple of numbers (its components) which undergoes the contragredient transformation, i.e., transforms by the formula

$$\mathbf{x}' = V\mathbf{x}, \tag{38}$$

when we go from one coordinate system to another, where, as before, $\mathbf{x} = (x_1, x_2, x_3)$ and $\mathbf{x}' = (x_1', x_2', x_3')$. Such a vector is called a *covariant affine vector*.

In both cases, given the components of a vector in any coordinate system, we can determine its components in all other coordinate systems obtained from the first system by making an arbitrary affine transformation. We now give some examples of both kinds of vectors.

Consider first the radius vector joining two fixed points of space. This is obviously a contravariant vector, since its components in the above sense are just the differences between the coordinates of the two points, which transform by the same linear formulas as the coordinates themselves. As another example of a contravariant vector, consider the case where the coordinates of the point (x_1, x_2, x_3) are functions of a parameter t, and define the "velocity vector" with components

$$\frac{dx_1}{dt}, \frac{dx_2}{dt}, \frac{dx_3}{dt}.$$

Differentiating the basic equations (36) with respect to t, we find at once that the velocity vector is also a contravariant vector.

We now give an example of a covariant vector. Let $f(x_1, x_2, x_3)$ be a function of the point (x_1, x_2, x_3), and define in any coordinate system the vector with components

$$\frac{\partial f}{\partial x_1}, \frac{\partial f}{\partial x_2}, \frac{\partial f}{\partial x_3};$$

this vector is called the *gradient* of $f(x_1, x_2, x_3)$. According to (36) and the chain rule for differentiating composite functions, we have

$$\frac{\partial f}{\partial x_s} = a_{1s}\frac{\partial f}{\partial x_1'} + a_{2s}\frac{\partial f}{\partial x_2'} + a_{3s}\frac{\partial f}{\partial x_3'} \qquad (s = 1, 2, 3),$$

i.e., the components of the gradient in the old (unprimed) coordinate system are obtained from the components of the gradient in the new (primed) coordinate system by the linear transformation with matrix A^*. Consequently, the components in the new coordinate system are obtained from the components in the old coordinate system by the linear transformation with matrix $V = (A^*)^{-1}$; thus, the gradient of a function is in fact a covariant vector.

It is not hard to express (37) and (38) in terms of partial derivatives of the new coordinates with respect to the old coordinates, and conversely. We first introduce a notation which is different from that previously used, but which is customary in vector analysis, i.e., components of contravariant vectors will be labeled by superscripts, and components of covariant vectors will be labeled by subscripts. Then the coefficients of the transformation (36) can be represented as the partial derivatives

$$a_{ik} = \frac{\partial x'^i}{\partial x^k}. \tag{39}$$

The elements of the contragredient matrix V are

$$V_{ik} = \frac{A_{ik}}{\det A},$$

which are the same as the elements of the matrix $(A^{-1})^*$, i.e.,

$$(A^*)^{-1} = (A^{-1})^*.$$

It follows that we can first form the inverse matrix with coefficients $\partial x^i/\partial x'^k$ and then transpose rows and columns, obtaining

$$V_{ik} = \frac{\partial x^k}{\partial x'^i}. \tag{40}$$

Let u^s be the components of a contravariant vector in the coordinate system x^s, and let u'^s be its components in the coordinate system x'^s. By definition, we have

$$u'^i = \sum_{s=1}^{3} \frac{\partial x'^i}{\partial x^s} u^s \qquad (i = 1, 2, 3). \tag{41}$$

In just the same way, we have

$$u'_i = \sum_{s=1}^{3} \frac{\partial x^s}{\partial x'^i} u_s \tag{42}$$

for a covariant vector. Moreover, these formulas can be used for determining the coefficients of a vector not only in the case of linear coordinate transformations, but also in the case of the most general transformations, where the old coordinates are related to the new coordinates by arbitrary (nonlinear) functions. We leave it to the reader to show that the velocity vector is contravariant and that the gradient of a function is covariant under an arbitrary coordinate transformation (not just a linear transformation).

We now give another way of defining a covariant vector. Suppose that a contravariant vector has already been defined as a vector whose components have the same transformation law as the coordinates themselves, and suppose that we are given a contravariant vector u^s and a covariant vector v_s. Forming the expression

$$u^1v_1 + u^2v_2 + u^3v_3, \tag{43}$$

it is not hard to see that this expression is *invariant* (or, in other words, is a *scalar*) when u^s and v_s transform according to (41) and (42). In fact, it follows immediately from the chain rule for differentiation of composite functions that

$$\sum_{s=1}^{3} u'^s v'_s = \sum_{s=1}^{3} \left[\sum_{k=1}^{3} \frac{\partial x'^s}{\partial x^k} u^k \right] \left[\sum_{l=1}^{3} \frac{\partial x^l}{\partial x'^s} v_l \right] = u^1v_1 + u^2v_2 + u^3v_3.$$

Thus, given our previous definition of a contravariant vector, we can define a covariant vector by the requirement that the expression (43) be invariant. By just the same kind of argument as that given in the previous section, we find that if (43) is to be invariant, the components v_s must undergo a linear transformation which is contragredient to that experienced by the components u^s.

Finally, we make a remark concerning the difference between contravariant and covariant vectors, which so far have been defined in a purely formal way, i.e., by the way they transform in going from one coordinate system to another. Let $\mathbf{x}$ be a vector specified by a length and a direction. Given a set of basis vectors $\mathbf{i}$, $\mathbf{j}$, and $\mathbf{k}$, we form the components of $\mathbf{x}$ as given by (28). We then call these components the *contravariant components* of $\mathbf{x}$ and write (28) in the form

$$\mathbf{x} = x^1\mathbf{i} + x^2\mathbf{j} + x^3\mathbf{k}. \tag{44}$$

By the *covariant component* of the vector $\mathbf{x}$ with respect to the basis vector $\mathbf{i}$, we mean the orthogonal projection of $\mathbf{x}$ on $\mathbf{i}$, multiplied by the length of $\mathbf{i}$; the other covariant components of $\mathbf{x}$ are defined similarly. In this way, we obtain three covariant components x_1, x_2, x_3 for every system of basis vectors. It can be shown that in going from one set of basis vectors to another, these components transform like the components of a covariant vector (as previously defined). In fact, it can be shown that with our present definitions, the sum

$$x^1x_1 + x^2x_2 + x^3x_3$$

gives the square of the length of the vector $\mathbf{x}$ and is therefore invariant under changes of the basis vectors. We omit the proof.

23. The Tensor Concept. We now turn to a generalization of the vector concept, restricting ourselves for the time being to the case of linear coordinate transformations. Suppose we are given a set of nine numbers

$$b_{ik} \qquad (i, k = 1, 2, 3).$$

We form the expression

$$\sum_{i,k=1}^{3} b_{ik}u^i v^k, \tag{45}$$

where u^i and v^k are the components of two contravariant vectors. If we transform to a new coordinate system and express u^i and v^k in terms of the new components u'^i and v'^k, then (45) becomes

$$\sum_{i,k=1}^{3} b_{ik}u^i v^k = \sum_{i,k=1}^{3} b'_{ik}u'^i v'^k. \tag{46}$$

Thus, in the new coordinate system, we also have a set of numbers, this time consisting of the nine elements b'_{ik}. Such a set, defined by the requirement that (45) be invariant in any coordinate system, is called a *covariant tensor of the second rank.* In just the same way, we can take two covariant vectors with components u_i and v_k and form the expression

$$\sum_{i,k=1}^{3} b^{ik}u_i v_k. \tag{47}$$

Then, if the set consisting of the nine elements b^{ik} is specified in one coordinate system, it can be defined in any other coordinate system by the requirement that (47) remain invariant; such a set is called a *contravariant tensor of the second rank.* Finally, if we take a contravariant vector with components u^i and a covariant vector with components v_k and form the expression

$$\sum_{i,k=1}^{3} b_i^k u^i v_k, \tag{48}$$

we are led in just the same way to the concept of a *mixed tensor of the second rank.*

We now show how to express the components of a tensor in the new coordinate system in terms of its components in the old coordinate system, given the coefficients of the linear coordinate transformation (36). Beginning with the case of a covariant tensor of the second rank, we examine the transformation law of the contravariant vectors u^i and v^k appearing in (45). The components u^i and v^k in the old coordinate system are expressed in terms of the components u'^i and v'^k in the new coordinate system by using the linear transformation with matrix A^{-1}. Thus, denoting the elements of this matrix by $\{A^{-1}\}_{ik}$, we have

$$u^i = \sum_{k=1}^{3} \{A^{-1}\}_{ik} u'^k, \qquad v^i = \sum_{k=1}^{3} \{A^{-1}\}_{ik} v'^k.$$

Substituting these expressions into (45) and finding the coefficient of the product $u'^i v'^k$, we obtain the expression

$$b'_{ik} = \sum_{p,q=1}^{3} b_{pq} \{A^{-1}\}_{pi} \{A^{-1}\}_{qk} \tag{49}$$

for the component b'_{ik} of the tensor in the new coordinate system. In just the same way, in the case of a contravariant tensor of the second rank, we can express the old components u_i and v_k of the covariant vectors appearing in (47) in terms of their new components u'_i and v'_k. According to the definition of a covariant vector, u'_i is obtained from the u_i by using the matrix $(A^*)^{-1}$, so that u_i is expressed in terms of the u'_i by using A^*, the transpose of A. Since the same is true for v_i, we have

$$u_i = \sum_{k=1}^{3} \{A\}_{ki} u'_k, \qquad v_i = \sum_{k=1}^{3} \{A\}_{ki} v'_k.$$

Substituting these expressions into (47), we obtain the transformation formula

$$b'^{ik} = \sum_{p,q=1}^{3} b^{pq} \{A\}_{ip} \{A\}_{kq} \tag{50}$$

for the components of a contravariant tensor of the second rank. Similarly, we have the transformation formula

$$b'^{k}_{i} = \sum_{p,q=1}^{3} b^{q}_{p} \{A^{-1}\}_{pi} \{A\}_{kq} \tag{51}$$

for the components of a mixed tensor of the second rank.

If we express the coefficients of the linear transformation A in terms of the partial derivatives $\partial x'^i/\partial x^k$, $\partial x^i/\partial x'^k$ and substitute them into the preceding formulas, we obtain the transformation law for tensors of the second rank for the case of *arbitrary* coordinate transformations. Moreover, in complete analogy with the case of second rank tensors, we can introduce the concept of tensors of rank higher than 2; we omit the details.

Suppose now that we have a matrix B which expresses a linear transformation in three dimensions relative to a given coordinate system, and suppose that we perform the affine coordinate transformation described by the formula

$$\mathbf{y} = A\mathbf{x},$$

where $\mathbf{x} = (x_1, x_2, x_3)$, $\mathbf{y} = (y_1, y_2, y_3)$, and A is a matrix whose determinant does not vanish. As shown in Sec. 21, our transformation has the matrix ABA^{-1} in the new coordinate system. It is not hard to see that this transformation law for the matrix B is the same as the transformation law given above for a mixed tensor of the second rank. In fact, applying the multiplication formula for matrices, we obtain

$$\{BA^{-1}\}_{qi} = \sum_{p=1}^{3} \{B\}_{qp}\{A^{-1}\}_{pi},$$

whence $$\{A(BA^{-1})\}_{ki} = \sum_{q=1}^{3} \{A\}_{kq}\{BA^{-1}\}_{qi} = \sum_{p,q=1}^{3} \{B\}_{qp}\{A^{-1}\}_{pi}\{A\}_{kq}.$$

Denoting $\{B\}_{ik}$ by b_k^i, we obtain at once a formula of the form (51), as asserted.

We now consider some special tensors. Suppose that a covariant tensor has the property that

$$b_{ik} = b_{ki} \qquad (i, k = 1, 2, 3) \tag{52}$$

in some coordinate system. Then it is not hard to see that it will have the same property in any other coordinate system as well. In fact, according to (49), we have

$$b'_{ki} = \sum_{p,q=1}^{3} b_{pq}\{A^{-1}\}_{pk}\{A^{-1}\}_{qi},$$

or by (52),

$$b'_{ki} = \sum_{p,q=1}^{3} b_{qp}\{A^{-1}\}_{pk}\{A^{-1}\}_{qi}.$$

Interchanging the summation variables, we have

$$b'_{ki} = \sum_{p,q=1}^{3} b_{pq}\{A^{-1}\}_{qk}\{A^{-1}\}_{pi}.$$

Comparing this with the first expression for b'_{ki}, we see that $b'_{ki} = b'_{ik}$, as asserted. A tensor of this kind is called a *symmetric covariant tensor*. The definition of a *symmetric contravariant tensor* is completely analogous. In just the same way, if $b_{ik} = -b_{ki}$ or $b^{ik} = -b^{ki}$ in one coordinate system, then the same relations hold in any other coordinate system; the corresponding tensors are called *skew-symmetric.* However, this property does *not* hold for mixed tensors; for example, the relation $b_i^k = b_k^i$ is not invariant under coordinate transformations.

In the next section, we shall study another special kind of tensor.

24. Cartesian Tensors. In the examples that follow, we confine ourselves to the case of linear coordinate transformations of the type studied in Sec. 20, i.e., transformations from one Cartesian coordinate system to another. Such transformations are usually called *orthogonal* transformations of three-dimensional space. As we have already seen, in this case the contragredient transformation $(A^*)^{-1}$ coincides with A, and the difference between contravariant and covariant vectors disappears; for just the same reason, there will only be one kind of second rank tensor. We denote the coefficient matrix of the orthogonal transformation by $\{A\}_{ik}$, as before. Then, an immediate consequence of the formulas of the preceding section is the formula

$$b'_{ik} = \sum_{p,q=1}^{3} b_{pq}\{A\}_{ip}\{A\}_{kq} \tag{53}$$

describing the transformation law of a second rank tensor. The elements of each column of the matrix $\|b_{ik}\|$ will be regarded as the components of a vector. Thus, we have the three vectors

$$\mathbf{b}^{(1)} = (b_{11}, b_{21}, b_{31}),\ \mathbf{b}^{(2)} = (b_{12}, b_{22}, b_{32}),\ \mathbf{b}^{(3)} = (b_{13}, b_{23}, b_{33}),$$

which will be said to correspond to the x_1-axis, the x_2-axis, and the x_3-axis, respectively. Using these vectors, we associate with every direction n the vector $\mathbf{b}^{(n)}$ defined by the formula

$$\mathbf{b}^{(n)} = \cos(n, x_1)\mathbf{b}^{(1)} + \cos(n, x_2)\mathbf{b}^{(2)} + \cos(n, x_3)\mathbf{b}^{(3)}. \tag{54}$$

Suppose now that instead of the Cartesian coordinate system x_1, x_2, x_3, we take any other Cartesian system x'_1, x'_2, x'_3 and use (54) to form the vectors corresponding to the directions of the new coordinate axes:

$$\mathbf{b}'^{(k)} = \cos(x'_k, x_1)\mathbf{b}^{(1)} + \cos(x'_k, x_2)\mathbf{b}^{(2)} + \cos(x'_k, x_3)\mathbf{b}^{(3)}. \tag{55}$$

Then the projections of these vectors onto the new coordinate axes x'_1, x'_2, x'_3 give us a new matrix $\|b'_{ik}\|$, with nine elements just like the matrix $\|b_{ik}\|$. We now show that the expression for the elements of this new matrix in terms of the elements of the old matrix $\|b_{ik}\|$ is just the transformation law for the components of a second rank tensor. To see this, consider the element b'_{12}, say. By definition, b'_{12} is the component of the vector $\mathbf{b}'^{(2)}$ along the x'_1-axis. According to (55),

$$\mathbf{b}'^{(2)} = \cos(x'_2, x_1)\mathbf{b}^{(1)} + \cos(x'_2, x_2)\mathbf{b}^{(2)} + \cos(x'_2, x_3)\mathbf{b}^{(3)}, \tag{56}$$

from which it is clear that $\mathbf{b}'^{(2)}$ is a linear function of the vectors $\mathbf{b}^{(i)}$. Thus, to obtain b'_{12} it is sufficient to replace the vectors $\mathbf{b}^{(i)}$ in the right-hand side of (56) by their projections onto the x'_1-axis, i.e., to replace $\mathbf{b}^{(i)}$ by

$$b_{1i} \cos (x'_1, x_1) + b_{2i} \cos (x'_1, x_2) + b_{3i} \cos (x'_1, x_3) \qquad (i = 1, 2, 3).$$

Furthermore, we note that

$$\cos (x'_i, x_k) = a_{ik} = \{A\}_{ik},$$

according to the table (3) of Sec. 20. Replacing the vectors $\mathbf{b}^{(i)}$ in the right-hand side of (56) by these expressions, we find that

$$b'_{12} = \sum_{p,q=1}^{3} b_{pq}\{A\}_{1p}\{A\}_{2q},$$

which is the same as (53) for the case $i = 1$, $k = 2$. Thus, we can state the following result: Given three mutually orthogonal directions, associate three vectors $\mathbf{b}^{(1)}, \mathbf{b}^{(2)}, \mathbf{b}^{(3)}$ with these directions and use (54) to define a vector for any direction n. Let x'_1, x'_2, x'_3 be an arbitrary Cartesian coordinate system. Then the array of nine numbers $\|b_{ik}\|$ giving the projections of the vectors $\mathbf{b}'^{(i)}$ onto the x'_k-axis $(i, k = 1, 2, 3)$ defines a *Cartesian tensor* of the second rank, i.e., a tensor of the second rank defined for all possible orthogonal transformations. (Note that when we say that $\mathbf{b}^{(1)}$ corresponds to the x_1-axis, we do not mean that $\mathbf{b}^{(1)}$ must have the same direction as the x_1-axis. The important thing is the formula (54), which associates with every direction n a vector $\mathbf{b}^{(n)}$, whose direction is in general not the same as n.)

We now give two examples of Cartesian tensors of the second rank. The first example is the familiar *stress tensor* of elasticity theory. Through a fixed point M of an elastic body under deformation we pass an infinitely small surface element $d\sigma$ with the normal n. It is assumed in elastic theory that the force exerted on $d\sigma$ by the part of the elastic medium lying on the same side of $d\sigma$ as the normal n is equal to the product of the surface element $d\sigma$ and a vector $\mathbf{b}^{(n)}$ which depends on the direction of the normal n. An analysis of the equilibrium conditions for an infinitely small tetrahedron formed from the elastic medium then gives (54), which shows at once that the stress is a second rank tensor. In any Cartesian coordinate system, this tensor is characterized by an array of nine numbers $\|b_{ik}\|$; moreover, as shown in elastic theory, $b_{ik} = b_{ki}$, i.e., the stress tensor is symmetric. In other words, the projection on the x_i-axis of the stress acting on a surface element perpendicular to the x_k-axis equals the projection on the x_k-axis of the stress acting on a surface element perpendicular to the x_i-axis.

We now consider another example of a tensor. Let $\mathbf{c}(P)$ be a vector field with components c_1, c_2, c_3. If we choose a Cartesian coordinate system x_1, x_2, x_3 and then differentiate the field components with respect to the x_i, we obtain the following array of nine quantities:

$$\left\| \begin{matrix} \frac{\partial c_1}{\partial x_1} & \frac{\partial c_1}{\partial x_2} & \frac{\partial c_1}{\partial x_3} \\ \frac{\partial c_2}{\partial x_1} & \frac{\partial c_2}{\partial x_2} & \frac{\partial c_2}{\partial x_3} \\ \frac{\partial c_3}{\partial x_1} & \frac{\partial c_3}{\partial x_2} & \frac{\partial c_3}{\partial x_3} \end{matrix} \right\|. \tag{57}$$

Given any direction n, we define the vector corresponding to n to be the derivative $\partial \mathbf{c}/\partial n$; thus, the elements of the kth column of the matrix (57) give the components of the vector associated with the x_k-axis. For any n, we have the formula

$$\frac{\partial c_i}{\partial n} = \cos(n, x_1) \frac{\partial c_i}{\partial x_1} + \cos(n, x_2) \frac{\partial c_i}{\partial x_2} + \cos(n, x_3) \frac{\partial c_i}{\partial x_3} \qquad (i = 1, 2, 3),$$

i.e., the matrix (57) is a second rank tensor. In general, this tensor is neither symmetric nor skew-symmetric. However, as we shall now show, it can easily be represented as the sum of a symmetric and a skew-symmetric tensor. (By the sum of two matrices, we mean the matrix whose elements are the sums of the corresponding elements of the individual matrices.)

First of all, we make some general remarks. It follows from the linear character of (53) that if $\|b_{ik}\|$ and $\|c_{ik}\|$ are two tensors, then the sum $\|b_{ik} + c_{ik}\|$ is also a tensor. Moreover, the same equation is valid if we interchange the indices i and k:

$$b'_{ki} = \sum_{p,q=1}^{3} b_{qp}\{A\}_{ip}\{A\}_{kq}.$$

Thus, if a certain matrix (defined for all coordinate systems) represents a tensor, then the transposed matrix also represents a tensor. Suppose now that we have a tensor $\|b_{ik}\|$. We can represent $\|b_{ik}\|$ as the sum

$$\|b_{ik}\| = \left\| \frac{b_{ik} + b_{ki}}{2} \right\| + \left\| \frac{b_{ik} - b_{ki}}{2} \right\|.$$

Obviously, the first summand is a symmetric tensor, and the second summand is a skew-symmetric tensor. Applying this decomposition to the tensor defined by the matrix (57), we find that its symmetric part is

$$\left\|\begin{matrix} \dfrac{\partial c_1}{\partial x_1} & \dfrac{1}{2}\left(\dfrac{\partial c_1}{\partial x_2}+\dfrac{\partial c_2}{\partial x_1}\right) & \dfrac{1}{2}\left(\dfrac{\partial c_1}{\partial x_3}+\dfrac{\partial c_3}{\partial x_1}\right) \\ \dfrac{1}{2}\left(\dfrac{\partial c_1}{\partial x_2}+\dfrac{\partial c_2}{\partial x_1}\right) & \dfrac{\partial c_2}{\partial x_2} & \dfrac{1}{2}\left(\dfrac{\partial c_2}{\partial x_3}+\dfrac{\partial c_3}{\partial x_2}\right) \\ \dfrac{1}{2}\left(\dfrac{\partial c_1}{\partial x_3}+\dfrac{\partial c_3}{\partial x_1}\right) & \dfrac{1}{2}\left(\dfrac{\partial c_2}{\partial x_3}+\dfrac{\partial c_3}{\partial x_2}\right) & \dfrac{\partial c_3}{\partial x_3} \end{matrix}\right\|. \tag{58}$$

If $\mathbf{c}(P)$ is the vector by which the point P of a continuous medium is displaced when the medium is deformed, then the matrix (58) defines the so-called *strain tensor*. The skew-symmetric part of the tensor (57) is

$$\left\|\begin{matrix} 0 & \dfrac{1}{2}\left(\dfrac{\partial c_1}{\partial x_2}-\dfrac{\partial c_2}{\partial x_1}\right) & \dfrac{1}{2}\left(\dfrac{\partial c_1}{\partial x_3}-\dfrac{\partial c_3}{\partial x_1}\right) \\ \dfrac{1}{2}\left(\dfrac{\partial c_2}{\partial x_1}-\dfrac{\partial c_1}{\partial x_2}\right) & 0 & \dfrac{1}{2}\left(\dfrac{\partial c_2}{\partial x_3}-\dfrac{\partial c_3}{\partial x_2}\right) \\ \dfrac{1}{2}\left(\dfrac{\partial c_3}{\partial x_1}-\dfrac{\partial c_1}{\partial x_3}\right) & \dfrac{1}{2}\left(\dfrac{\partial c_3}{\partial x_2}-\dfrac{\partial c_2}{\partial x_3}\right) & 0 \end{matrix}\right\|. \tag{59}$$

In an earlier volume, we have already encountered this decomposition of a tensor into two parts for the special case of a linear homogeneous deformation. In that case, the skew-symmetric part corresponds to a rotation of the space as a whole about some axis (without deformation).

25. The n-Dimensional Case. We now turn to the general case of an n-dimensional space. In Sec. 12, we defined a vector in such a space to be an ordered set

$$\mathbf{x} = (x_1, x_2, \ldots, x_n)$$

of n real or complex numbers called the *components* of $\mathbf{x}$; moreover, we defined the simplest operations on vectors and gave the condition for two vectors to be equal. We shall assume that our space is referred to the basis vectors

$$\mathbf{a}^{(1)} = (1, 0, \ldots, 0), \mathbf{a}^{(2)} = (0, 1, \ldots, 0), \ldots, \mathbf{a}^{(n)} = (0, 0, \ldots, 1),$$

so that

$$\mathbf{x} = x_1\mathbf{a}^{(1)} + x_2\mathbf{a}^{(2)} + \cdots + x_n\mathbf{a}^{(n)}. \tag{60}$$

The transformation

$$y_i = a_{i1}x_1 + a_{i2}x_2 + \cdots + a_{in}x_n \qquad (i = 1, 2, \ldots, n), \tag{61}$$

or simply

$$\mathbf{y} = A\mathbf{x}, \tag{62}$$

carrying one vector $\mathbf{x} = (x_1, x_2, \ldots, x_n)$ into another vector $\mathbf{y} = (y_1, y_2, \ldots, y_n)$, is called a *linear transformation* of n-dimensional space. Here A denotes the transformation matrix $\|a_{ik}\|$; if $\det A$ is nonvanishing, then the transformation (62) is called a *nonsingular* transformation and the matrix A is called a nonsingular matrix. In this case, solving (61) for the x_i, we obtain the transformation

$$\mathbf{x} = A^{-1}\mathbf{y} \tag{63}$$

inverse to (61) and (62). Here the matrix A^{-1} has the elements

$$\{A^{-1}\}_{ik} = \frac{A_{ki}}{\det A}, \tag{64}$$

$\det A$ denotes the determinant of the matrix A, and A_{ik} denotes the cofactor of a_{ik} in $\det A$.

Just as before (Sec. 21), we define the product of two transformations as follows: Performing the two transformations

$$\mathbf{y} = A\mathbf{x}, \qquad \mathbf{z} = B\mathbf{y}$$

in succession is equivalent to one linear transformation

$$\mathbf{z} = BA\mathbf{x};$$

BA is called the *product* of the transformations A and B, and has matrix elements given by

$$\{BA\}_{ik} = \sum_{s=1}^{n} \{B\}_{is}\{A\}_{sk}. \tag{65}$$

This product depends in general on the order of the factors, i.e., apart from certain exceptional cases, we have

$$BA \neq AB.$$

It is easy to define the product of any number of linear transformations. Then the associative law holds, i.e., factors can be grouped in any way; for example, we have

$$(CB)A = C(BA). \tag{66}$$

The inverse transformation satisfies the relations

$$AA^{-1} = A^{-1}A = I, \qquad (A^{-1})^{-1} = A, \tag{67}$$

where the symbol I denotes the so-called *unit matrix*, which has 1's along the principal diagonal and 0's elsewhere. This matrix corresponds to the identity transformation

$$y_i = x_i \qquad (i = 1, 2, \ldots, n).$$

Just as before, we define a *diagonal matrix* of order n as

$$[k_1, k_2, \ldots, k_n] = \begin{Vmatrix} k_1 & 0 & \cdots & 0 \\ 0 & k_2 & \cdots & 0 \\ \cdots & \cdots & \cdots & \cdots \\ 0 & 0 & \cdots & k_n \end{Vmatrix}. \tag{68}$$

This matrix corresponds to a transformation of the form

$$y_i = k_i x_i \qquad (i = 1, 2, \ldots, n).$$

The product of diagonal matrices is defined by the formula

$$\begin{aligned}[k_1, k_2, \ldots, k_n][l_1, l_2, \ldots, l_n] &= [l_1, l_2, \ldots, l_n][k_1, k_2, \ldots, k_n] \\ &= [k_1 l_1, k_2 l_2, \ldots, k_n l_n]\end{aligned}$$

and does not depend on the order of the factors. In the special case $k_1 = k_2 = \cdots = k_n = k$, we obtain the matrix

$$[k, k, \ldots, k] = \begin{Vmatrix} k & 0 & \cdots & 0 \\ 0 & k & \cdots & 0 \\ \cdots & \cdots & \cdots & \cdots \\ 0 & 0 & \cdots & k \end{Vmatrix}, \tag{69}$$

which corresponds to multiplying all the components of a vector by the number k. We shall regard the matrix (69) as being equivalent to the number k, i.e., the number k is regarded as a special case of a matrix. By using (65), it is easy to see that the product of the matrix A and the number k (treated as a matrix) does not depend on the order of the factors and reduces to multiplying all the elements of A by the number k:

$$\{[k, k, \ldots, k]A\}_{ik} = \{kA\}_{ik} = k\{A\}_{ik}. \tag{70}$$

Suppose now that instead of the basis vectors $\mathbf{a}^{(k)}$, we choose new basis vectors $\mathbf{b}^{(k)}$ related to the $\mathbf{a}^{(k)}$ by the formulas

$$\begin{aligned} \mathbf{b}^{(1)} &= t_{11}\mathbf{a}^{(1)} + t_{12}\mathbf{a}^{(2)} + \cdots + t_{1n}\mathbf{a}^{(n)}, \\ \mathbf{b}^{(2)} &= t_{21}\mathbf{a}^{(1)} + t_{22}\mathbf{a}^{(2)} + \cdots + t_{2n}\mathbf{a}^{(n)}, \\ &\cdots\cdots\cdots\cdots\cdots\cdots\cdots \\ \mathbf{b}^{(n)} &= t_{n1}\mathbf{a}^{(1)} + t_{n2}\mathbf{a}^{(2)} + \cdots + t_{nn}\mathbf{a}^{(n)}, \end{aligned} \tag{71}$$

where the determinant formed by the elements t_{ik} is nonvanishing. Then the vectors $\mathbf{a}^{(k)}$ can themselves be expressed in terms of the vectors $\mathbf{b}^{(k)}$; moreover, every linear combination of the vectors $\mathbf{a}^{(k)}$ is a linear combination of the vectors $\mathbf{b}^{(k)}$, and conversely. In other words, the vectors $\mathbf{b}^{(k)}$ *span the same space* as the vectors $\mathbf{a}^{(k)}$. If a vector $\mathbf{x}$ has components $x_1, x_2, \ldots, x_n$ in the coordinate system with basis vectors $\mathbf{a}^{(k)}$, then it will have other components $x'_1, x'_2, \ldots, x'_n$ in the coordinate system with basis vectors $\mathbf{b}^{(k)}$. These new components can be expressed in terms of the old components by using the contragredient transformation

$$\mathbf{x}' = (T^*)^{-1}\mathbf{x}, \tag{72}$$

where $\mathbf{x} = (x_1, x_2, \ldots, x_n)$, $\mathbf{x}' = (x_1', x_2', \ldots, x_n')$, and T^* is the transpose of the matrix T corresponding to the transformation (71). The transformation expressed by (62) in the original coordinate system is expressed by

$$\mathbf{y}' = UAU^{-1}\mathbf{x}' \tag{73}$$

in the new coordinate system, where $U = (T^*)^{-1}$. The matrix UAU^{-1} is said to be *similar* to the matrix A.

So far, we have been concerned with the two basic notions of vectors and matrices. In this regard, it should be noted that it is sometimes convenient to consider the vector $\mathbf{x} = (x_1, x_2, \ldots, x_n)$ to be a matrix itself, i.e., a matrix which has the numbers $x_1, x_2, \ldots, x_n$ as one of its columns (it does not matter which one), while the other elements are all zero. For example, suppose we represent the vector $\mathbf{x}$ by the matrix

$$\begin{Vmatrix} x_1 & 0 & \cdots & 0 \\ x_2 & 0 & \cdots & 0 \\ \cdot & \cdot & \cdot & \cdot \\ x_n & 0 & \cdots & 0 \end{Vmatrix}. \tag{74}$$

Such a matrix, with only one column containing nonzero elements, is sometimes denoted by the symbol

$$\begin{pmatrix} x_1 \\ x_2 \\ \cdot \\ \cdot \\ \cdot \\ x_n \end{pmatrix}.$$

We now show that the linear transformation (62) can be written as the product of the matrix A with the matrix (74). In fact, if we multiply the matrix (74) by the matrix A, using the rule (65), and bear in mind that the matrix (74) has nonzero elements only in the first column, we find that the product is also a matrix which has nonzero elements only in the first column. Moreover, it is easily seen that these elements are just

$$y_i = a_{i1}x_1 + a_{i2}x_2 + \cdots + a_{in}x_n \qquad (i = 1, 2, \ldots, n),$$

i.e., they correspond to the linear transformation (62), which can therefore be written in the form

$$\begin{pmatrix} y_1 \\ y_2 \\ \cdot \\ \cdot \\ \cdot \\ y_n \end{pmatrix} = A \begin{pmatrix} x_1 \\ x_2 \\ \cdot \\ \cdot \\ \cdot \\ x_n \end{pmatrix}, \tag{75}$$

where the right-hand side is the product of two matrices.

We conclude this section by again noting some general laws governing operations on n-dimensional vectors:

1. $\mathbf{x} + \mathbf{y} = \mathbf{y} + \mathbf{x}$, $(\mathbf{x} + \mathbf{y}) + \mathbf{z} = \mathbf{x} + (\mathbf{y} + \mathbf{z})$.
2. If $\mathbf{x}$ and $\mathbf{y}$ are any two vectors, then the vector $\mathbf{z} = \mathbf{y} - \mathbf{x}$, with components $y_k - x_k$, is the only vector satisfying the relation $\mathbf{x} + \mathbf{z} = \mathbf{y}$.
3. If a and b are any numbers, then $(a + b)\mathbf{x} = a\mathbf{x} + b\mathbf{x}$, $a(b\mathbf{x}) = (ab)\mathbf{x}$, and $a(\mathbf{x} + \mathbf{y}) = a\mathbf{x} + a\mathbf{y}$.
4. $1\mathbf{x} = \mathbf{x}$ and $0\mathbf{x} = 0$, where the symbol 0 appearing in the right-hand side of the last relation denotes the vector with all its components equal to zero.

26. Elements of Matrix Algebra. In the formulas of the preceding section, matrices appear as symbols on which we can perform operations resembling operations on ordinary numbers. This leads naturally to the idea of constructing a new algebra, suitable for calculations involving matrices. In other words, we want to interpret matrices as a new kind of number, a so-called *hypercomplex number*. Just as starting with two real numbers a and b, we are led to construct a new kind of number, namely, the complex number of the form $a + ib$, in the same way, starting with n^2 complex numbers a_{ik} arranged in a square array, we are led to a new kind of number, namely, a matrix. However, the following important difference between the two cases must be noted: All the familiar operations of algebra are also valid for the symbols representing complex numbers. On the other hand, for matrices we obtain an algebra which differs significantly from the familiar algebra of complex numbers in that *multiplication is noncommutative*, i.e., the result of multiplying matrices depends on the order of the factors.

We now establish the basic rules of matrix algebra. In many cases, we shall be guided by the results which were obtained above by interpreting a matrix as an array of coefficients associated with a linear transformation. Henceforth, unless a statement is made to the contrary, we shall always be concerned with square matrices, which are all of the same order n. As before, we denote the elements of the matrix A by $\{A\}_{ik}$.

Two matrices A and B are regarded as equal if and only if the corresponding elements of the two matrices are equal, i.e.,

$$\{A\}_{ik} = \{B\}_{ik} \qquad (i, k = 1, 2, \ldots, n). \tag{76}$$

Addition of two matrices is defined by the formula

$$\{A + B\}_{ik} = \{A\}_{ik} + \{B\}_{ik}, \tag{77}$$

which reduces to addition of corresponding elements of the individual matrices. Multiplication is defined by the formula

$$\{BA\}_{ik} = \sum_{s=1}^{n} \{B\}_{is}\{A\}_{sk}. \tag{78}$$

As we have already seen, in general,

$$BA \neq AB,$$

but the associative law (Sec. 21)

$$(CB)A = C(BA) \tag{79}$$

holds. It is clear that the distributive law

$$(A + B)C = AC + BC \qquad \text{and} \qquad C(A + B) = CA + CB \tag{80}$$

also holds. The determinant of the product of two matrices is equal to the product of the determinants of the factors:

$$\det AB = \det A \det B. \tag{81}$$

We now note another peculiarity of matrix multiplication: A product of two matrices can vanish, i.e., can equal the matrix all of whose elements are zero, even though both factors are nonvanishing. As an example, consider the following product of two identical second order matrices:

$$\begin{Vmatrix} 0 & 0 \\ 1 & 0 \end{Vmatrix} \begin{Vmatrix} 0 & 0 \\ 1 & 0 \end{Vmatrix} = \begin{Vmatrix} 0 & 0 \\ 0 & 0 \end{Vmatrix}.$$

If A is nonsingular, i.e., if $\det A \neq 0$, then, just as in the preceding section, we introduce the notion of the inverse matrix A^{-1}. If $C = BA$ and R_A, R_B, R_C are the ranks of the matrices A, B, C, respectively, then we saw in Sec. 7 that $R_C \leq R_A$. If B is a nonsingular matrix, then $A = B^{-1}C$, and in the same way we can assert that $R_A \leq R_C$, so that $R_C = R_A$. Thus, *if a matrix A is multiplied by a nonsingular matrix B, its rank does not change.* For the unit matrix I, the relation

$$BI = IB = B \tag{82}$$

holds, where B is an arbitrary matrix. It is not hard to see that the matrix A^{-1} is the unique solution of the equations

$$AX = I, \qquad XA = I, \tag{83}$$

where I is the unit matrix. For example, suppose we multiply the first of these equations by A^{-1} on the left. Then, using (79) and (67), we obtain $X = A^{-1}$. We can treat the second equation similarly. We note that if $\det A = 0$, then the equations (83) have no solutions at all, i.e., the matrix A has no inverse. For if X were the inverse of A, (83) would imply

$$\det A \det X = 1,$$

which contradicts the hypothesis that $\det A = 0$. From the preceding section, we recall the concept of a diagonal matrix and the fact that every number k can be regarded as a special case of a matrix.

A positive integral power of a matrix A is readily defined as

$$A^p = A \cdot A \cdot \cdots \cdot A \qquad (p \text{ times}).$$

A negative integral power of a matrix is defined as the corresponding positive integral power of the inverse matrix:

$$A^{-p} = (A^{-1})^p. \tag{84}$$

We obviously have

$$A^{-p} = (A^p)^{-1}, \text{ i.e., } A^{-p}A^p = A^pA^{-p} = I. \tag{85}$$

The ratio A/B of two matrices has no definite meaning. We can interpret the ratio in two different ways, either as the product AB^{-1} or as the product $B^{-1}A$. In general, the two products are different, and it is only in the special cases where they are the same that the symbol A/B has any unique meaning. Another basic notion, which was also introduced in the preceding section, is the notion of similar matrices.

Next we note the following formulas, which are very easily proved:

$$(CBA)^{-1} = A^{-1}B^{-1}C^{-1}, \tag{86}$$
$$CBAC^{-1} = (CBC^{-1})(CAC^{-1}). \tag{87}$$

If A^* denotes the transpose of the matrix A, then we also have the formula

$$(CBA)^* = A^*B^*C^*, \tag{88}$$

which is easily verified by using the definition of multiplication. Now we introduce two further items of notation: We denote by $\bar{A}$ the matrix whose elements are the complex conjugates of the elements of the matrix A:

$$\{\bar{A}\}_{ik} = \overline{\{A\}_{ik}}; \tag{89}$$

here, as always, we denote the complex conjugate of the number α by $\bar{\alpha}$. Moreover, we denote by $\tilde{A}$ the matrix which is obtained from the matrix A by interchanging rows and columns and then taking the complex conjugates of all elements:

$$\{\tilde{A}\}_{ik} = \overline{\{A\}_{ki}}. \tag{90}$$

The matrix $\tilde{A}$ is called the *Hermitian conjugate*† or *adjoint* of the matrix A. It is easy to verify the formula.

$$\widetilde{CBA} = \tilde{A}\tilde{B}\tilde{C}. \tag{91}$$

The elementary formula

$$(A^*)^{-1} = (A^{-1})^*$$

is also easily verified, i.e., the operations of taking the inverse and transposing can be interchanged, as already noted (Sec. 22). We also note another formula which will be useful later. It is an immediate consequence of (67) that

$$\det A \det A^{-1} = 1$$

or

$$\det A^{-1} = [\det A]^{-1}. \tag{92}$$

In other words, the determinant of an inverse matrix is the reciprocal of the determinant of the original matrix.

Finally, we introduce the concept of a *quasi-diagonal matrix*, which is a generalization of the concept of a diagonal matrix. As an illustration, consider the following matrix of order 7:

$$\begin{Vmatrix} b_{11} & b_{12} & b_{13} & 0 & 0 & 0 & 0 \\ b_{21} & b_{22} & b_{23} & 0 & 0 & 0 & 0 \\ b_{31} & b_{32} & b_{33} & 0 & 0 & 0 & 0 \\ 0 & 0 & 0 & c_{11} & c_{12} & 0 & 0 \\ 0 & 0 & 0 & c_{21} & c_{22} & 0 & 0 \\ 0 & 0 & 0 & 0 & 0 & d_{11} & d_{12} \\ 0 & 0 & 0 & 0 & 0 & d_{21} & d_{22} \end{Vmatrix}.$$

Denote by B the third order matrix with elements b_{ik}, and denote by C and D the second order matrices with elements c_{ik} and d_{ik}. This matrix of order 7 is called *quasi-diagonal with the structure* $\{3, 2, 2\}$ and is denoted by the symbol

$$[B, C, D].$$

In general, suppose that the principal diagonal of a matrix A of order n, consisting of the elements a_{ii}, is divided into m parts, where the first part consists of the first k_1 elements, the second part of the next k_2 elements, etc., so that $k_1 + k_2 + \cdots + k_m = n$. We can regard the first k_1 ele-

† Named after C. Hermite, a French mathematician of the second half of the nineteenth century. Other commonly encountered notations for the Hermitian conjugate of A are A^* and $A†$.

ments as the principal diagonal of a matrix X_1 of order k_1, the next k_2 elements as the principal diagonal of a matrix X_2 of order k_2, etc. Suppose now that all the elements of the matrix A which do not belong to the matrices X_k vanish. Then the matrix A is called a *quasi-diagonal matrix with the structure* $\{k_1, k_2, \ldots, k_m\}$ and is written as

$$A = [X_1, X_2, \ldots, X_m].$$

The rules for operating on quasi-diagonal matrices of identical structure are particularly simple. We cite the relevant formulas without proof; the proofs are elementary consequences of the definitions of the operations. First we have the formula

$$[X_1, X_2, \ldots, X_m] + [Y_1, Y_2, \ldots, Y_m] = [X_1 + Y_1, X_2 + Y_2, \ldots, X_m + Y_m] \quad (93)$$

for adding quasi-diagonal matrices of identical structure (i.e., the order of every matrix X_k is the same as the order of the corresponding matrix Y_k). Then, in just the same way, we have the formulas

$$[Y_1, Y_2, \ldots, Y_m][X_1, X_2, \ldots, X_m] = [Y_1X_1, Y_2X_2, \ldots, Y_mX_m] \quad (94)$$

$$[X_1, X_2, \ldots, X_m]^p = [X_1^p, X_2^p, \ldots, X_m^p] \quad (95)$$

for multiplication and raising to a power. Here p is any positive or negative integer; of course, if p is a negative integer, we must require that the determinants det X_k be different from zero. The rule for performing a similarity transformation on the matrix $[X_1, X_2, \ldots, X_m]$ by using a matrix of the same structure is given by the formula

$$[Y_1, Y_2, \ldots, Y_m][X_1, X_2, \ldots, X_m][Y_1, Y_2, \ldots, Y_m]^{-1} = [Y_1X_1Y_1^{-1}, \ldots, Y_mX_mY_m^{-1}]. \quad (96)$$

The geometrical meaning of the linear transformations corresponding to quasi-diagonal matrices should be noted. Consider for simplicity the case given above of the quasi-diagonal matrix of order 7 with the structure specified by the numbers $\{3, 2, 2\}$, and examine the linear transformation corresponding to this matrix. If we have

$$x_4 = x_5 = x_6 = x_7 = 0$$

in the original vector $\mathbf{x} = (x_1, x_2, \ldots, x_7)$, then obviously we also have

$$y_4 = y_5 = y_6 = y_7 = 0$$

in the vector $\mathbf{y} = (y_1, y_2, \ldots, y_7)$ obtained as a result of the transformation. In other words, all vectors belonging to the subspace spanned by the first three basis vectors continue to belong to the same subspace after the transformation, and for these vectors the transformation is specified by the third order matrix B. Similar remarks pertain to the subspace spanned by the fourth and fifth basis vectors and the subspace spanned by the sixth and seventh basis vectors. (In this connection, we recall that the *subspace spanned by the vectors* $\mathbf{x}^{(1)}, \mathbf{x}^{(2)}, \ldots, \mathbf{x}^{(l)}$ means the set of vectors given by the formula

$$c_1\mathbf{x}^{(1)} + c_2\mathbf{x}^{(2)} + \cdots + c_l\mathbf{x}^{(l)},$$

where $c_1, c_2, \ldots, c_l$ are arbitrary constants.)

27. Eigenvalues of a Matrix. Reduction of a Matrix to Canonical Form. It goes without saying that similar matrices are not equal in the sense of (76). However, they are equivalent in the geometrical sense in that they bring about the same linear transformation of space, expressed in different coordinate systems. We now look for invariants of these matrices, i.e., expressions formed from their elements which have the same value for all similar matrices. It is not hard to see that the determinant is one such invariant. In fact, if A is a matrix and UAU^{-1} is a matrix equivalent to A, where U is any matrix with a nonvanishing determinant, then by (81) and (92) we have

$$\det UAU^{-1} = \det U \det A \det U^{-1} = \det U \det A\ [\det U]^{-1} = \det A.$$

To construct further invariants, we proceed as follows: We subtract the parameter λ from all the diagonal elements of the matrix A and then take the determinant of the resulting matrix. The result is the polynomial

$$\varphi(\lambda) = \begin{vmatrix} a_{11} - \lambda & a_{12} & \cdots & a_{1n} \\ a_{21} & a_{22} - \lambda & \cdots & a_{2n} \\ \cdot & \cdot & \cdot & \cdot \\ a_{n1} & a_{n2} & \cdots & a_{nn} - \lambda \end{vmatrix} \tag{97}$$

of degree n in λ; here the a_{ik} are the elements of the matrix A. We can also write $\varphi(\lambda)$ as

$$\varphi(\lambda) = \det (A - \lambda) = \det (A - \lambda I), \tag{98}$$

since λ, or λI, is the diagonal matrix with all its diagonal elements equal to λ. Replacing A by UAU^{-1} and bearing in mind that $U\lambda U^{-1} = \lambda$, since any matrix commutes with the matrix λ, we have

$$\det (UAU^{-1} - \lambda) = \det [U(A - \lambda)U^{-1}] = \det (A - \lambda). \tag{99}$$

Thus, we see that the polynomials of the form (97) associated with the matrices UAU^{-1} and A are identical. In other words, all the coefficients of the polynomial (97) are invariant with respect to similarity transformations. It is easy to see that the leading coefficient of the polynomial $\varphi(\lambda)$ is $(-1)^n$. Two other coefficients of $\varphi(\lambda)$ are of particular interest, namely, the constant term and the coefficient of $(-1)^{n-1}\lambda^{n-1}$. The first of these coefficients is clearly the determinant $|a_{ik}|$, an invariant which we have already mentioned. As for the coefficient of $(-1)^{n-1}\lambda^{n-1}$, using the result of Sec. 5, we see that it equals the sum of the diagonal elements. This sum is usually called the *trace of the matrix A* and is written

$$\operatorname{tr} A = \{A\}_{11} + \{A\}_{22} + \cdots + \{A\}_{nn} = a_{11} + a_{22} + \cdots + a_{nn};$$

the German word "Spur" is often used instead of "trace." Thus, similar matrices have the same determinant and the same trace. Consider now the equation

$$\det (A - \lambda) = 0, \tag{100}$$

which is called the *characteristic equation of the matrix A:* The roots of this equation are called the *eigenvalues* (or *characteristic numbers*) *of the matrix A*. Then, by the argument just given, we can assert that *similar matrices have the same eigenvalues.* (We have already encountered equations of the form (100) in Sec. 17.)

We now ask the following question: Given a matrix A, can we find a matrix V such that the similar matrix $V^{-1}AV$ is a diagonal matrix? (Here, for convenience, we denote the matrix similar to A by $V^{-1}AV$, i.e., we write the inverse matrix first instead of last; obviously, these two ways of writing similar matrices are entirely equivalent.) In other words, from the point of view of linear transformations, can we choose a new coordinate system such that the linear transformation characterized by the matrix A in the original coordinate system becomes simply a transformation of the form $y_k = \lambda_k x_k$ ($k = 1, 2, \ldots, n$) in the new coordinate system? We can write this condition in the form

$$V^{-1}AV = [\lambda_1, \lambda_2, \ldots, \lambda_n], \tag{101}$$

where we wish to determine the numbers λ_k and the elements of the matrix V. Multiplying both sides of (101) by V from the left, we can rewrite the condition in the form†

$$AV = V[\lambda_1, \lambda_2, \ldots, \lambda_n]. \tag{102}$$

Using (65) to calculate the element with indices i and k in both sides of (102), we obtain n^2 equations:

$$\sum_{s=1}^{n} a_{is}v_{sk} = v_{ik}\lambda_k,$$

where a_{ik} and v_{ik} are the elements of the matrices A and V. Holding the

† To avoid confusion in expressions like (102), we observe that functions of one or several arguments will always be denoted by $f(\ \ )$ and not by $f[\ \ ]$.

second index k fixed and setting $i = 1, 2, \ldots, n$ in turn, we obtain n equations:

$$\sum_{s=1}^{n} a_{is}v_{sk} = \lambda_k v_{ik} \qquad (i = 1, 2, \ldots, n), \tag{103}$$

each containing only λ_k and the elements of the kth column of the matrix V. If we now regard the elements $v_{1k}, v_{2k}, \ldots, v_{nk}$ as components of a vector $\mathbf{v}^{(k)}$, then we can write (103) in the form of a vector equation

$$A\mathbf{v}^{(k)} = \lambda_k \mathbf{v}^{(k)}. \tag{104}$$

Thus, we see that finding the matrix V which transforms the matrix A to diagonal form reduces to finding vectors $\mathbf{v}^{(k)}$, which, when acted upon by the linear transformation A, are just multiplied by a numerical factor. This parallelism is the algebraic analog of a proposition of modern quantum mechanics which states that Heisenberg's matrix mechanics is fundamentally the same as Schrödinger's wave mechanics. In matrix mechanics, the main problem is to transform a certain (infinite) matrix into diagonal form, while in wave mechanics the problem is to find vectors (in an infinite-dimensional space) which when acted upon by a certain linear transformation are just multiplied by a numerical factor. In our case, since we consider only finite-dimensional spaces, the problem is a purely algebraic analog of the quantum mechanical situation. However, in the more complicated case of infinite-dimensional space, we are compelled for basic reasons to abandon the context of ordinary algebra, and instead we have to adopt techniques belonging to analysis.

All these matters will be studied in more detail later. For the present, we note that for the purposes of physical applications, we can restrict ourselves (in the finite-dimensional case under consideration) to matrices A of a special type, i.e., *Hermitian matrices*, for which $a_{ki} = \bar{a}_{ik}$; moreover, the matrices V also have to be of a certain type, i.e., *unitary matrices*, to be defined below. Here we shall consider the general problem for any finite matrix, but only final results will be given, with proofs being deferred to Chap. 6. However, the problem will be solved in complete detail for the cases of interest in the applications.

We now solve the system (103) or (104), which can be written in expanded form as

$$\begin{aligned}
(a_{11} - \lambda_k)v_{1k} + a_{12}v_{2k} + \cdots + a_{1n}v_{nk} &= 0,\\
a_{21}v_{1k} + (a_{22} - \lambda_k)v_{2k} + \cdots + a_{2n}v_{nk} &= 0,\\
\cdots\cdots\cdots\cdots\cdots\cdots\cdots\cdots\cdots\cdots\cdots\cdots&\\
a_{n1}v_{1k} + a_{n2}v_{2k} + \cdots + (a_{nn} - \lambda_k)v_{nk} &= 0.
\end{aligned} \tag{105}$$

To obtain a solution for $(v_{1k}, v_{2k}, \ldots, v_{nk})$ which is different from zero, it is necessary and sufficient that the determinant of the system (105) vanish, i.e., that the number λ_k be a root of the characteristic equation. We shall consider in detail only the case where all the roots of the characteristic equation are different; we denote these roots by $\lambda_1, \lambda_2, \ldots, \lambda_n$. Substituting the first root λ_1 for λ_k in the coefficients of the system (105), we can determine the elements of the first column of the matrix V. Here we are not concerned with the question of how many different choices of the quantities v_{i1} are possible; the solution can be chosen in any way, provided only that it is different from zero. Similarly, substituting $\lambda_k = \lambda_2$ in the coefficients of the system (105), we can determine the elements of the second column of the matrix V, and so on, until finally we have found the nth column of V. Since (105) and (102) are equivalent, to go over to the basic equation (101) it is necessary only that the matrix V have an inverse V^{-1}, i.e., that the determinant of V be nonzero. We shall prove that this is the case by contradiction. Suppose that the determinant of V vanishes. Then, as we have seen (Sec. 12), this implies the existence of a linear relation between the vectors $\mathbf{v}^{(k)}$ which determine the columns of the matrix V, i.e., a relation of the form

$$C_1\mathbf{v}^{(1)} + C_2\mathbf{v}^{(2)} + \cdots + C_n\mathbf{v}^{(n)} = 0,$$

where not all the coefficients C_k are zero. Applying the transformation A to both sides of this equation $n - 1$ times and using (104), we obtain the n equations

$$\begin{array}{rcccccl} C_1\mathbf{v}^{(1)} &+& C_2\mathbf{v}^{(2)} &+ \cdots +& C_n\mathbf{v}^{(n)} &=& 0, \\ \lambda_1 C_1\mathbf{v}^{(1)} &+& \lambda_2 C_2\mathbf{v}^{(2)} &+ \cdots +& \lambda_n C_n\mathbf{v}^{(n)} &=& 0, \\ \multicolumn{7}{c}{\cdots\cdots\cdots\cdots\cdots\cdots\cdots\cdots\cdots} \\ \lambda_1^{n-1} C_1\mathbf{v}^{(1)} &+& \lambda_2^{n-1} C_2\mathbf{v}^{(2)} &+ \cdots +& \lambda_n^{n-1} C_n\mathbf{v}^{(n)} &=& 0. \end{array}$$

Bearing in mind that not all the vectors $C_k\mathbf{v}^{(k)}$ are zero, we can assert that the determinant of this system must vanish, i.e.,

$$\begin{vmatrix} 1 & 1 & \cdots & 1 \\ \lambda_1 & \lambda_2 & \cdots & \lambda_n \\ \cdots & \cdots & \cdots & \cdots \\ \lambda_1^{n-1} & \lambda_2^{n-1} & \cdots & \lambda_n^{n-1} \end{vmatrix} = 0,$$

where by hypothesis the numbers λ_k are all different. But this last equation contradicts the fact that a Vandermonde determinant (Sec. 5) formed from unequal numbers cannot vanish. Thus, we have proved that in the case where all the eigenvalues of the matrix A are different, it is possible to reduce A to diagonal form by making a similarity transformation.

In the case where some of the eigenvalues are equal, it may happen that the matrix A cannot be reduced to diagonal form by a similarity transformation. Nevertheless, in this case A still has very simple representation, the so-called *canonical representation* or *Jordan form*. In the special case where the matrix can be reduced to diagonal form, the canonical representation can be written as

$$[\lambda_1, \lambda_2, \ldots, \lambda_n],$$

where the λ_k are the eigenvalues of the matrix. We shall now discuss the more general case of unequal eigenvalues; we shall state only results, deferring proofs to Chap. 6.

Let $\lambda = a$ be a root of (100) of multiplicity k. Suppose that all the minors of order $n - 1$ of the matrix $A - \lambda$ have $\lambda = a$ as a root of at most multiplicity k_1, i.e., all these minors are divisible by $(\lambda - a)^{k_1}$, but at least one of them is not divisible by $(\lambda - a)^{k_1+1}$. Suppose further that all the minors of order $n - 2$ of the matrix $A - \lambda$ have $\lambda = a$ as a root of at most multiplicity k_2, etc., and finally, suppose that all the minors of order $n - m$ have $\lambda = a$ as a root of at most multiplicity k_m, while at least one minor of order $n - m - 1$ does not vanish for $\lambda = a$. Then, of course, at least one minor of any order less than $n - m - 1$ will not vanish. It can be shown that the numbers k_s form a decreasing sequence

$$k > k_1 > k_2 > \cdots > k_m.$$

We now introduce the positive integers

$$l_1 = k - k_1, \qquad l_2 = k_1 - k_2, \ \ldots, \ l_m = k_{m-1} - k_m, \qquad l_{m+1} = k_m;$$

clearly $l_1 + l_2 + \cdots + l_{m+1} = k$. The binomials

$$(\lambda - a)^{l_1}, (\lambda - a)^{l_2}, \ldots, (\lambda - a)^{l_{m+1}}$$

are called the *elementary divisors* of the matrix A for the root $\lambda = a$. In the same way, we can determine the elementary divisors for all the eigenvalues of the matrix A, finally obtaining the *set of all elementary divisors*

$$(\lambda - \lambda_1)^{p_1}, (\lambda - \lambda_2)^{p_2}, \ldots, (\lambda - \lambda_p)^{p_p}, \tag{106}$$

where now some of the numbers λ_k will in general be the same, and

$$p_1 + p_2 + \cdots + p_p = n. \tag{107}$$

We have seen above that the eigenvalues are invariant under similarity transformations. It turns out that the set of all elementary divisors of a matrix has the same property.

We now introduce some new matrices $I_\rho(a)$ of particularly simple form: $I_\rho(a)$ will denote a matrix of order ρ which has the number a everywhere along the principal diagonal and the number 1 everywhere along the diagonal just below the principal diagonal:

$$I_\rho(a) = \begin{Vmatrix} a & 0 & 0 & \cdots & 0 & 0 \\ 1 & a & 0 & \cdots & 0 & 0 \\ 0 & 1 & a & \cdots & 0 & 0 \\ \cdots & \cdots & \cdots & \cdots & \cdots & \cdots \\ 0 & 0 & 0 & \cdots & a & 0 \\ 0 & 0 & 0 & \cdots & 1 & a \end{Vmatrix}. \tag{108}$$

We can now state the basic result on the representation of matrices in canonical form: *If the matrix A has the elementary divisors* (106), *then there exists a matrix U with a nonvanishing determinant such that*

$$UAU^{-1} = [I_{\rho_1}(\lambda_1), I_{\rho_2}(\lambda_2), \ldots, I_{\rho_p}(\lambda_p)]. \tag{109}$$

Thus, once the eigenvalues of the matrix A are known, finding the matrix U reduces to a sequence of elementary algebraic operations. If $\rho = 1$, then $I_\rho(a)$ means just the number a itself. Even when some of the eigenvalues are the same, it may turn out that all the elementary divisors are simple, i.e., have the form

$$(\lambda - \lambda_1), (\lambda - \lambda_2), \ldots, (\lambda - \lambda_n).$$

In this case, the quasi-diagonal matrix

$$[I_{\rho_1}(\lambda_1), I_{\rho_2}(\lambda_2), \ldots, I_{\rho_p}(\lambda_p)]$$

reduces to just the diagonal matrix $[\lambda_1, \lambda_2, \ldots, \lambda_n]$, and the matrix A can actually be reduced to diagonal form. It is not hard to prove that a matrix can be reduced to diagonal form if and only if for every root λ_k of the characteristic equation, the rank of the coefficient matrix of the system (105) is equal to $n - \mu_k$, where μ_k is the multiplicity of the root λ_k. When this condition is satisfied, the system (105) defines μ_k linearly independent vectors $(v_{1k}, v_{2k}, \ldots, v_{nk})$ for each μ_k (see Sec. 14). Finally, we note that the matrix U appearing in (109) is not uniquely determined. In particular, if d is the value of the determinant of the matrix U, then in (105) we can replace U by $(1/\sqrt[n]{d})\, U$ and U^{-1} by $\sqrt[n]{d}\, U$. Thus, we might as well assume that the determinant of the matrix U equals unity.

This concludes what we shall say now about the general problem of reducing a matrix to canonical form, a problem which will be considered further in Chap. 6. As already noted, a detailed study will also be made of the problem of reducing matrices of a special type (Hermitian matrices and unitary matrices) to canonical form (see Chap. 4).

28. Unitary and Orthogonal Transformations. In this and subsequent sections, we shall make use of the concepts of the scalar product and the norm (length) of a vector, which were introduced in Sec. 13. We recall that the square of the norm (length) of a vector is defined by the formula

$$\|\mathbf{x}\|^2 = (\mathbf{x}, \mathbf{x}) = \sum_{s=1}^{n} |x_s|^2, \tag{110}$$

or in the case of real components, by

$$\|\mathbf{x}\|^2 = \sum_{s=1}^{n} x_s^2.$$

(This definition of the norm presupposes an appropriate choice of the basis vectors or coordinate axes; a coordinate system with a norm defined in this way will be called a *Cartesian* coordinate system.) In addition to the length of a vector, we also define the *scalar product of two vectors* by the formula

$$(\mathbf{x}, \mathbf{y}) = x_1\bar{y}_1 + x_2\bar{y}_2 + \cdots + x_n\bar{y}_n. \tag{111}$$

In the case of real vectors, this formula takes the more symmetric form

$$(\mathbf{x}, \mathbf{y}) = x_1y_1 + x_2y_2 + \cdots + x_ny_n.$$

It follows from (111) that when the order of the vectors is reversed, the scalar product is replaced by its complex conjugate, i.e.,

$$(\mathbf{y}, \mathbf{x}) = \overline{(\mathbf{x}, \mathbf{y})}. \tag{112}$$

Two vectors are said to be *orthogonal* (or *perpendicular*) if their scalar product vanishes.

Henceforth, unless the contrary is explicitly stated, we shall always assume that we are dealing with a Cartesian coordinate system. Thus, the linear transformations corresponding to a change from one Cartesian coordinate system to another will be of particular importance. As we know, every transformation from one set of basis vectors to another corresponds to a linear transformation of the components of a vector. Suppose that we have a transformation

$$\mathbf{y} = U\mathbf{x} \tag{113}$$

from the vector $\mathbf{x} = (x_1, x_2, \ldots, x_n)$ to the vector $\mathbf{y} = (y_1, y_2, \ldots, y_n)$, where the original coordinate system is Cartesian. A necessary and sufficient condition for the new coordinate system to be Cartesian also is that the length of a vector be equal to the sum of the squares of its components in the new coordinate system as well as in the old, i.e.,

$$|y_1|^2 + |y_2|^2 + \cdots + |y_n|^2 = |x_1|^2 + |x_2|^2 + \cdots + |x_n|^2. \quad (114)$$

We now show that if the condition (114) is met, then the scalar product in the new coordinate system is given by a formula analogous to (111). Suppose that we have two vectors

$$\mathbf{x} = (x_1, x_2, \ldots, x_n), \mathbf{x}' = (x_1', x_2', \ldots, x_n')$$

in the original coordinate system, which become

$$\mathbf{y} = (y_1, y_2, \ldots, y_n), \mathbf{y}' = (y_1', y_2', \ldots, y_n')$$

in the new coordinate system. We form two new vectors $\mathbf{z} = \mathbf{x} + \mathbf{x}'$ and $\mathbf{u} = \mathbf{x} + i\mathbf{x}'$, with components $x_k + x_k'$ and $x_k' + ix_k'$. If the condition (114) holds, we have†

$$\sum_{k=1}^{n} (y_k + y_k')(\bar{y}_k + \bar{y}_k') = \sum_{k=1}^{n} (x_k + x_k')(\bar{x}_k + \bar{x}_k').$$

Then, since

$$\sum_{k=1}^{n} |y_k|^2 = \sum_{k=1}^{n} |x_k|^2$$

and

$$\sum_{k=1}^{n} |y_k'|^2 = \sum_{k=1}^{n} |x_k'|^2,$$

we obtain

$$\sum_{k=1}^{n} (y_k\bar{y}_k' + y_k'\bar{y}_k) = \sum_{k=1}^{n} (x_k\bar{x}_k' + x_k'\bar{x}_k). \quad (115a)$$

In just the same way, we have

$$\sum_{k=1}^{n} (y_k + iy_k')(\bar{y}_k - i\bar{y}_k') = \sum_{k=1}^{n} (x_k + ix_k')(\bar{x}_k - i\bar{x}_k'),$$

and hence

$$\sum_{k=1}^{n} (y_k'\bar{y}_k - y_k\bar{y}_k') = \sum_{k=1}^{n} (x_k'\bar{x}_k - x_k\bar{x}_k'). \quad (115b)$$

Subtracting (115b) from (115a), we obtain

$$\sum_{k=1}^{n} y_k\bar{y}_k' = \sum_{k=1}^{n} x_k\bar{x}_k', \quad (116)$$

† There is no distinction between $\bar{x}_k'$ and $\overline{x_k'}$, between $\bar{y}_k'$ and $\overline{y_k'}$, etc.

so that the scalar product is in fact given by the same formula as before. Thus, if the transformation (113) satisfies the condition (114), it also satisfies the condition (116), i.e., the scalar product is invariant under the transformation. Conversely, substituting $x'_k = x_k$ in (116), we see that (116) implies (114), since obviously the scalar product of two identical vectors is just the square of the length of the vector. Linear transformations satisfying the condition (114) or the condition (116) are called *unitary transformations.* (We note that the conditions (114) and (116) can be written as $\|U\mathbf{x}\|^2 = \|\mathbf{x}\|^2$ and $(U\mathbf{x}, U\mathbf{x}') = (\mathbf{x}, \mathbf{x}')$, respectively, where $\mathbf{x}$ and $\mathbf{x}'$ are arbitrary vectors.) In the case of real space and linear transformations with real matrices, the condition (114) reduces simply to

$$y_1^2 + y_2^2 + \cdots + y_n^2 = x_1^2 + x_2^2 + \cdots + x_n^2, \tag{117}$$

and the corresponding real transformations are called *orthogonal transformations.* They are clearly a special case of unitary transformations-

Next we establish the fundamental properties of unitary transformations. Denoting the elements of the matrix U by u_{ik}, we write out the condition (114) on the transformation (113) explicitly; the result is

$$\sum_{k=1}^{n} |u_{k1}x_1 + u_{k2}x_2 + \cdots + u_{kn}x_n|^2 = \sum_{k=1}^{n} |x_k|^2$$

or

$$\sum_{k=1}^{n} (u_{k1}x_1 + u_{k2}x_2 + \cdots + u_{kn}x_n)(\bar{u}_{k1}\bar{x}_1 + \bar{u}_{k2}\bar{x}_2 + \cdots + \bar{u}_{kn}\bar{x}_n) = \sum_{k=1}^{n} x_k\bar{x}_k. \tag{118}$$

Expanding the left-hand side of this equation and setting the coefficients of $x_p\bar{x}_p$ equal to unity and those of $x_p\bar{x}_q$ $(p \neq q)$ equal to zero, we obtain the following sufficient conditions for a transformation U to be unitary:

$$\begin{aligned} &\sum_{k=1}^{n} |u_{kp}|^2 = 1 \qquad (p = 1, 2, \ldots, n) \\ &\sum_{k=1}^{n} u_{kp}\bar{u}_{kq} = 0 \qquad (p \neq q), \end{aligned} \tag{119}$$

i.e., the sum of the squares of the absolute values of each column must equal unity, and the sum of the products of the elements of one column with the complex conjugates of the corresponding elements of any other column must vanish. Sometimes these conditions are written in the form

$$\sum_{k=1}^{n} u_{kp}\bar{u}_{kq} = \delta_{pq}, \tag{120}$$

where $\|\delta_{pq}\|$ is the unit matrix, i.e.,

$$\delta_{pq} = \begin{cases} 0 & (p \neq q) \\ 1 & (p = q). \end{cases} \tag{121}$$

By choosing special values for the x_k in (118), it is not hard to show that the conditions (119) are necessary as well as sufficient conditions for U to be unitary.

Consider the determinant det U and the determinant det $\bar{U}$ which consists of the complex conjugates of the elements of det U. Multiplying these determinants together by the "column by column" rule, we obtain the determinant of the unit matrix, which of course equals unity. On the other hand, since the two determinants in question are complex conjugates of each other, it follows at once that

$$|\det U|^2 = 1.$$

Thus *the square of the absolute value of the determinant of a unitary matrix equals unity.* In other words, the determinant of a unitary matrix has absolute value 1, i.e., is a complex number of the form $e^{i\varphi}$, where φ is real.

Next, consider the matrix U^* which is the transpose of U. The conditions (119), usually referred to as the *column orthogonality conditions,* can be written in the form of the matrix equality

$$\bar{U}^*U = I, \tag{122}$$

which is equivalent to

$$U^{-1} = \bar{U}^* = \tilde{U}. \tag{123}$$

Thus, the inverse of a unitary matrix U is the same as the Hermitian conjugate of U. The transformation U^{-1}, which is the inverse of U, transforms the vector **y** back into the vector **x** and obviously also satisfies the unitary conditions (114), i.e., *the inverse of a unitary matrix is also unitary.* It follows from (123) that the matrix $\tilde{U}$ is unitary as well, so that the columns of $\tilde{U}$ satisfy the orthogonality conditions. However, since the columns of $\tilde{U}$ are the same as the rows of U, we can assert that not only the columns but also the rows of a unitary matrix satisfy the orthogonality conditions, i.e., together with (120) we have the formula

$$\sum_{k=1}^{n} u_{pk}\bar{u}_{qk} = \delta_{pq}. \tag{124}$$

Moreover, if two matrices U_1 and U_2 satisfy the condition (114), then their product U_2U_1 clearly satisfies the same condition. Thus, *the product of two unitary matrices is also unitary.*

We now study the situation that occurs when a unitary matrix has real elements. In this case, as already noted, the unitary matrix is called *orthogonal*, and the corresponding transformation is called an *orthogonal transformation.* Then, instead of the formulas (120) and (124), we have the formulas

$$\sum_{k=1}^{n} u_{kp}u_{kq} = \delta_{pq}, \qquad \sum_{k=1}^{n} u_{pk}u_{qk} = \delta_{pq}. \tag{125}$$

Moreover, since the determinant of the transformation must obviously be a real number, it can only take the values -1 and $+1$. These real orthogonal transformations of n-dimensional space are the natural generalization of the transformations of three-dimensional space studied in Sec. 20. In the real case, $\tilde{U}$ equals U^*, i.e., the inverse U^{-1} of an orthogonal transformation U is obtained from U by interchanging rows and columns. Finally, we note that every complex number $e^{i\varphi}$, where φ is real, can be regarded as a matrix $[e^{i\varphi}, e^{i\varphi}, \ldots, e^{i\varphi}]$ and as such is a unitary matrix. Thus, if U is a unitary matrix, so is the product $e^{i\varphi}U$. (The meaning of the product of a matrix and a number was given in Sec. 25.)

29. Schwarz's Inequality. In this section we establish an inequality which will be of great use later. This inequality, which has already been derived in a previous volume, goes as follows: If $\alpha_1, \alpha_2, \ldots, \alpha_m$ and $\beta_1, \beta_2, \ldots, \beta_m$ are arbitrary real numbers, then

$$\left(\sum_{k=1}^{m} \alpha_k\beta_k\right)^2 \leq \sum_{k=1}^{m} \alpha_k^2 \sum_{k=1}^{m} \beta_k^2, \tag{126}$$

where the equality sign holds if and only if α_k and β_k are proportional:

$$\frac{\beta_1}{\alpha_1} = \frac{\beta_2}{\alpha_2} = \cdots = \frac{\beta_m}{\alpha_m}. \tag{127}$$

To review the derivation, let ξ be an arbitrary real number and form the sum

$$S = \sum_{k=1}^{m} (\xi\alpha_k - \beta_k)^2 \geq 0,$$

which is obviously nonnegative. The equality sign holds if and only if

$$\frac{\beta_1}{\alpha_1} = \frac{\beta_2}{\alpha_2} = \cdots = \frac{\beta_m}{\alpha_m} = \xi,$$

and in this case, we obviously have

$$\left(\sum_{k=1}^{m} \alpha_k\beta_k\right)^2 = \sum_{k=1}^{m} \alpha_k^2 \sum_{k=1}^{m} \beta_k^2.$$

However, in general, expanding the expression for S, we obtain a polynomial

$$S = A\xi^2 - 2B\xi + C$$

of the second degree, where

$$A = \sum_{k=1}^{m} \alpha_k^2,\ B = \sum_{k=1}^{m} \alpha_k\beta_k,\ C = \sum_{k=1}^{m} \beta_k^2.$$

This polynomial is ≥ 0 for all real ξ, whence it follows that $AC - B^2 \geq 0$, i.e., $B^2 \leq AC$, which proves the inequality (126). If $AC - B^2 = 0$, the polynomial must vanish for some real ξ; then, as we have seen, the condition (127) must hold. Conversely, if the condition (127) is met, the equality sign holds in (126).

We now assume that the α_k and β_k are complex numbers. Bearing in mind that

$$\left|\sum_{k=1}^{m} \alpha_k\beta_k\right| \leq \sum_{k=1}^{m} |\alpha_k|\,|\beta_k|$$

and applying the inequality (126) to the right-hand side, which is a sum of positive terms, we obtain

$$\left|\sum_{k=1}^{m} \alpha_k\beta_k\right|^2 \leq \sum_{k=1}^{m} |\alpha_k|^2 \sum_{k=1}^{m} |\beta_k|^2. \tag{126a}$$

In this case, where α_k and β_k are complex, it is not hard to show that the equality sign holds if and only if $|\alpha_k|$ and $|\beta_k|$ are proportional and all the products $\alpha_k\beta_k$ have the same argument.

The inequality (126) is applicable not only to sums but also to integrals, as observed in an earlier volume. Thus, if $f_1(x)$ and $f_2(x)$ are two real functions defined on the interval $a \leq x \leq b$, the inequality for integrals takes the form

$$\left[\int_a^b f_1(x)f_2(x)\,dx\right]^2 \leq \int_a^b f_1^2(x)\,dx \int_a^b f_2^2(x)\,dx. \tag{126b}$$

To prove this, form the expression

$$\int_a^b [\xi f_1(x) - f_2(x)]^2\,dx$$
$$= \xi^2 \int_a^b f_1^2(x)\,dx - 2\xi \int_a^b f_1(x)f_2(x)\,dx + \int_a^b f_2^2(x)\,dx,$$

where ξ is an arbitrary real number. Because of the form of the left-hand side, this expression is nonnegative. But if the polynomial $A\xi^2 - 2B\xi + C$ is nonnegative for all ξ, then, as is familiar from elementary algebra, $AC - B^2 \geq 0$; with the present choice of A, B, and C, this gives the inequality (126*b*).

The inequality (126*b*) for integrals was first proved by Bunyakovski, and the inequality (126*a*) for sums was first proved by Cauchy.† However, in the English-language literature, both cases are usually called *Schwarz's inequality.*

30. Properties of the Scalar Product and Norm. We now note a few more properties of the scalar product and norm. First, applying the inequality (126*a*) and bearing in mind that $|\bar{y}_k| = |y_k|$, we are able to write

$$|(\mathbf{x}, \mathbf{y})|^2 = \left|\sum_{k=1}^{n} x_k\bar{y}_k\right|^2 \leq \sum_{k=1}^{n} |x_k|^2 \sum_{k=1}^{n} |y_k|^2,$$

i.e.,

$$|(\mathbf{x}, \mathbf{y})| \leq \|\mathbf{x}\|\,\|\mathbf{y}\|. \tag{128}$$

Then, we prove the so-called *triangle inequality*,

$$\|\mathbf{x} + \mathbf{y}\| \leq \|\mathbf{x}\| + \|\mathbf{y}\|. \tag{129}$$

Since

$$\|\mathbf{x} + \mathbf{y}\|^2 = (\mathbf{x} + \mathbf{y}, \mathbf{x} + \mathbf{y}) = (\mathbf{x}, \mathbf{x}) + (\mathbf{y}, \mathbf{y}) + (\mathbf{x}, \mathbf{y}) + (\mathbf{y}, \mathbf{x}),$$

taking into account (128), we have

$$\|\mathbf{x} + \mathbf{y}\|^2 \leq \|\mathbf{x}\|^2 + \|\mathbf{y}\|^2 + 2\|\mathbf{x}\|\,\|\mathbf{y}\| = (\|\mathbf{x}\| + \|\mathbf{y}\|)^2,$$

which implies (129).

Next, we investigate the influence which the choice of a coordinate system has on the *metric* of the space, i.e., on the expression for the square of the length of a vector. Suppose that we replace the original Cartesian coordinate system by a new coordinate system which is in general not Cartesian and has the linearly independent vectors $\mathbf{z}^{(1)}, \mathbf{z}^{(2)}, \ldots, \mathbf{z}^{(n)}$ as basis vectors. Then, for any vector x, we have

$$\mathbf{x} = z_1\mathbf{z}^{(1)} + z_2\mathbf{z}^{(2)} + \cdots + z_n\mathbf{z}^{(n)},$$

where the z_k are the components of $\mathbf{x}$ in the new coordinate system. The square of the length of $\mathbf{x}$ is given by the scalar product of $\mathbf{x}$ with itself:

$$\|\mathbf{x}\|^2 = (z_1\mathbf{z}^{(1)} + \cdots + z_n\mathbf{z}^{(n)}, z_1\mathbf{z}^{(1)} + \cdots + z_n\mathbf{z}^{(n)}).$$

Expanding this expression, we find that

$$\|\mathbf{x}\|^2 = \sum_{i,k=1}^{n} a_{ik}z_i\bar{z}_k, \tag{130}$$

† See Probs. 59, 60 of Chap. 1.

where the coefficients α_{ik} are given by

$$\alpha_{ik} = (\mathbf{z}^{(i)}, \mathbf{z}^{(k)}).$$

Reversing indices obviously changes $(\mathbf{z}^{(i)}, \mathbf{z}^{(k)})$ into its complex conjugate, so that

$$\alpha_{ki} = \bar{\alpha}_{ik}. \tag{131}$$

A sum of the form (130), with coefficients satisfying the condition (131), is called a *Hermitian form*. It is immediately clear that every expression of the form (130) with the condition (131) can take only real values for any choice of the complex numbers z_k, since the terms $\alpha_{ik}z_i\bar{z}_k$ and $\alpha_{ki}z_k\bar{z}_i$ $(i \neq k)$ are complex conjugates of each other, while the coefficients α_{kk} of the terms $\alpha_{kk}|z_k|^2$ are real because of the condition (131). Moreover, by the very construction of the Hermitian form (130), we can assert that the sum (130) is nonnegative and vanishes only if all the z_k vanish. *The metric in the new coordinate system is defined by* (130). The metric (130) will coincide with the metric (110) of a Cartesian coordinate system if

$$\alpha_{ik} = 0 \qquad \text{for } i \neq k,$$
$$\alpha_{kk} = 1$$

or

$$(\mathbf{z}^{(i)}, \mathbf{z}^{(k)}) = 0 \qquad \text{for } i \neq k,$$
$$(\mathbf{z}^{(k)}, \mathbf{z}^{(k)}) = 1,$$

i.e., if the vectors $\mathbf{z}^{(k)}$ which we have chosen for the new basis vectors are (pairwise) orthogonal unit vectors (vectors of unit length).

Finally, we note that if (113) defines a unitary transformation of the components of a vector, then the corresponding transformation from the old basis vectors to the new basis vectors is given by the matrix $(U^*)^{-1}$ which is contragredient to U. In the present case, because of (123), this matrix is just $\bar{U}$, while for real orthogonal transformations, it is simply U.

31. The Orthogonalization Process for Vectors. Suppose that we are given any m linearly independent vectors $\mathbf{x}^{(1)}, \mathbf{x}^{(2)}, \ldots, \mathbf{x}^{(m)}$. The set of all vectors of the form

$$C_1\mathbf{x}^{(1)} + C_2\mathbf{x}^{(2)} + \cdots + C_m\mathbf{x}^{(m)},$$

where the C_k are arbitrary coefficients, gives the whole space if $m = n$, and some subspace R_m of dimension m if $m < n$. We now show that we can always construct a set of m orthogonal unit vectors $\mathbf{z}^{(k)}$ (a so-called *orthonormal system*) spanning the same subspace R_m as the vectors $\mathbf{x}^{(k)}$. In other words, the new orthonormal vectors $\mathbf{z}^{(k)}$ must be linear combinations of the $\mathbf{x}^{(k)}$, and conversely, the $\mathbf{x}^{(k)}$ must be linear combinations of the $\mathbf{z}^{(k)}$. We construct the vectors $\mathbf{z}^{(k)}$ by writing

$$\begin{aligned}
\mathbf{y}^{(1)} &= \mathbf{x}^{(1)}, \\
\mathbf{y}^{(2)} &= \mathbf{x}^{(2)} - (\mathbf{x}^{(2)}, \mathbf{z}^{(1)})\mathbf{z}^{(1)}, \\
\mathbf{y}^{(3)} &= \mathbf{x}^{(3)} - (\mathbf{x}^{(3)}, \mathbf{z}^{(1)})\mathbf{z}^{(1)} - (\mathbf{x}^{(3)}, \mathbf{z}^{(2)})\mathbf{z}^{(2)}, \\
&\cdots\cdots\cdots\cdots\cdots\cdots\cdots\cdots
\end{aligned} \tag{132}$$

where the $\mathbf{z}^{(k)}$ are given by

$$\mathbf{z}^{(1)} = \frac{\mathbf{y}^{(1)}}{\|\mathbf{y}^{(1)}\|}, \quad \mathbf{z}^{(2)} = \frac{\mathbf{y}^{(2)}}{\|\mathbf{y}^{(2)}\|}, \quad \ldots, \quad \mathbf{z}^{(m)} = \frac{\mathbf{y}^{(m)}}{\|\mathbf{y}^{(m)}\|}. \tag{133}$$

Thus, first the vector $\mathbf{z}^{(1)}$ is obtained by simply dividing $\mathbf{y}^{(1)}$ by its length, so that $\mathbf{z}^{(1)}$ has unit length. Then, the vector $\mathbf{y}^{(2)}$ is constructed by the second of the formulas (132); it follows immediately from this construction that $\mathbf{y}^{(2)}$ is orthogonal to $\mathbf{z}^{(1)}$, since

$$(\mathbf{y}^{(2)}, \mathbf{z}^{(1)}) = (\mathbf{x}^{(2)}, \mathbf{z}^{(1)}) - (\mathbf{x}^{(2)}, \mathbf{z}^{(1)})(\mathbf{z}^{(1)}, \mathbf{z}^{(1)}) = 0.$$

Dividing $\mathbf{y}^{(2)}$ by its length, we obtain $\mathbf{z}^{(2)}$. Then, we construct $\mathbf{y}^{(3)}$ by the third of the formulas (132); it follows immediately that $\mathbf{y}^{(3)}$ is orthogonal to $\mathbf{z}^{(1)}$ and $\mathbf{z}^{(2)}$. For example, since $\mathbf{z}^{(1)}$ and $\mathbf{z}^{(2)}$ are orthogonal, we have

$$(\mathbf{y}^{(3)}, \mathbf{z}^{(2)}) = (\mathbf{x}^{(3)}, \mathbf{z}^{(2)}) - (\mathbf{x}^{(3)}, \mathbf{z}^{(2)})(\mathbf{z}^{(2)}, \mathbf{z}^{(2)}) = 0.$$

Dividing the vector $\mathbf{y}^{(3)}$ by its length, we obtain $\mathbf{z}^{(3)}$, and so on. Clearly, all the new vectors $\mathbf{z}^{(k)}$ are linear combinations of the $\mathbf{x}^{(k)}$. Conversely, it is not hard to see that the $\mathbf{x}^{(k)}$ are linear combinations of the $\mathbf{z}^{(k)}$. To see this, it is enough just to solve the equations (132) in succession for $\mathbf{x}^{(1)}, \mathbf{x}^{(2)}$, etc.

We note that none of the vectors $\mathbf{y}^{(k)}$ can vanish. Indeed, if at some step of the calculation we obtained a vector $\mathbf{y}^{(k)}$ equal to zero, then, since $\mathbf{y}^{(k)}$ is a linear combination of the $\mathbf{x}^{(s)}$, with the coefficient of $\mathbf{x}^{(k)}$ equal to unity, we would have a linear relation between the vectors $\mathbf{x}^{(s)}$, which would contradict the assumption that these vectors are linearly independent. We also note that a set of orthogonal vectors like the $\mathbf{z}^{(k)}$, none of which is the zero vector, must be linearly independent.

If $m = n$, the $\mathbf{z}^{(k)}$ form a complete orthonormal system (complete in the sense that every vector is a linear combination of the $\mathbf{z}^{(k)}$), and the $\mathbf{z}^{(k)}$ can be chosen as the basis vectors of a Cartesian coordinate system. However, if $m < n$, to obtain a complete orthonormal system, we must supplement the vectors $\mathbf{z}^{(k)}$ just constructed by constructing $n - m$ more unit vectors which are orthogonal both to each other and to the $\mathbf{z}^{(k)}$. These new unit vectors must therefore span a subspace R'_{n-m} of dimension

$n - m$ which is orthogonal to the subspace R_m (see Sec. 12). Each new vector $\mathbf{u}$ must satisfy the system of equations

$$(\mathbf{u}, \mathbf{x}^{(1)}) = 0, (\mathbf{u}, \mathbf{x}^{(2)}) = 0, \ldots, (\mathbf{u}, \mathbf{x}^{(m)}) = 0.$$

This is a system of m homogeneous equations in n unknowns, where the rank of the system is m, since the vectors $\mathbf{x}^{(k)}$ are linearly independent (Sec. 12). Therefore, the system has $n - m$ linearly independent solutions, i.e., we obtain $n - m$ linearly independent vectors. Orthogonalizing these vectors by the procedure just given and then normalizing them to unit length, we obtain a complete system of linearly independent vectors, consisting of these vectors and the vectors $\mathbf{z}^{(k)}$.

We make one final remark. The subspace R_m spanned by the orthonormal vectors $\mathbf{z}^{(k)}$ can also be spanned by any other system of orthonormal vectors. In fact, to get a new system of orthonormal vectors, one has only to apply a unitary transformation to the system of vectors $\mathbf{z}^{(k)}$. Thus, we see that the process of orthogonalizing a system of vectors can be done in various ways and that the method given above is just one possible way.

PROBLEMS†

1. Show that the transformation

$$x' = \frac{1}{\sqrt{3}}(x - y + z),$$
$$y' = \frac{1}{\sqrt{6}}(x + 2y + z),$$
$$z' = \frac{1}{\sqrt{2}}(x - z)$$

satisfies (4). What are the new coordinates of the point $x = y = z = a$?

2. Let $\|a_{ij}\|$ and $\|b_{ij}\|$ $(i, j = 1, 2, 3)$ be two arrays of coefficients which when written in (2) give values of x_1', x_2', x_3' satisfying (4). Show that $\|c_{ij}\|$, where

$$c_{ij} = \sum_{k=1}^{3} a_{ik}b_{kj} \qquad (i, j = 1, 2, 3),$$

is another such array.

3. Write the array (3) corresponding to each of the following three-dimensional transformations:

(**a**) Reflection in the yz-plane; (**b**) Reflection in the plane $x + y + z = 0$;
(**c**) Rotation through $\pi/2$ about the z-axis;
(**d**) Rotation through $\pi/2$ about the axis $x = y = z$.

† All equation numbers (unless otherwise specified) refer to the corresponding equations of Chap. 3. For hints and answers, see p. 436. These problems are by A. L. Shields and the translator, making considerable use of the books by Proskuryakov and by Faddeyev and Sominski (cited in the Preface).

4. Let x, y, x′, y′ be vectors in three dimensions, and let

$$\|\mathbf{x}\| = \|\mathbf{y}\|, \qquad \|\mathbf{x}'\| = \|\mathbf{y}'\|, \qquad (\mathbf{x}, \mathbf{x}') = (\mathbf{y}, \mathbf{y}')$$

(Sec. 13). Show that there exists a linear transformation (2) satisfying the condition (4) and carrying x into y and x′ into y′.

5. Calculate the following matrix products:

(a) $\begin{Vmatrix} 2 & 1 \\ 3 & 2 \end{Vmatrix}\begin{Vmatrix} 1 & -1 \\ 1 & 1 \end{Vmatrix}$; (b) $\begin{Vmatrix} 3 & 5 \\ 6 & -1 \end{Vmatrix}\begin{Vmatrix} 2 & 1 \\ -3 & 2 \end{Vmatrix}$;

(c) $\begin{Vmatrix} 3 & 1 & 1 \\ 2 & 1 & 2 \\ 1 & 2 & 3 \end{Vmatrix}\begin{Vmatrix} 1 & 1 & -1 \\ 2 & -1 & 1 \\ 1 & 0 & 1 \end{Vmatrix}$; (d) $\begin{Vmatrix} 1 & 2 & 3 \\ 2 & 4 & 6 \\ 3 & 6 & 9 \end{Vmatrix}\begin{Vmatrix} -1 & -2 & -4 \\ -1 & -2 & -4 \\ 1 & 2 & 4 \end{Vmatrix}$;

(e) $\begin{Vmatrix} 1 & 2 & 1 \\ 0 & 1 & 2 \\ 3 & 1 & 1 \end{Vmatrix}\begin{Vmatrix} 2 & 3 & 1 \\ -1 & 1 & 0 \\ 1 & 2 & -1 \end{Vmatrix}$; (f) $\begin{Vmatrix} a & b & c \\ c & b & a \\ 1 & 1 & 1 \end{Vmatrix}\begin{Vmatrix} 1 & a & c \\ 1 & b & b \\ 1 & c & a \end{Vmatrix}$.

6. Carry out the following operations:

(a) $\begin{Vmatrix} 2 & 1 & 1 \\ 3 & 1 & 0 \\ 0 & 1 & 2 \end{Vmatrix}^2$; (b) $\begin{Vmatrix} 2 & 1 \\ 1 & 3 \end{Vmatrix}^3$; (c) $\begin{Vmatrix} 3 & 2 \\ -4 & -2 \end{Vmatrix}^5$; (d) $\begin{Vmatrix} 1 & 1 \\ 0 & 1 \end{Vmatrix}^n$;

(e) $\begin{Vmatrix} \cos\alpha & -\sin\alpha \\ \sin\alpha & \cos\alpha \end{Vmatrix}^n$.

7. Find the limit

$$\lim_{n\to\infty} \begin{Vmatrix} 1 & \frac{\alpha}{n} \\ -\frac{\alpha}{n} & 1 \end{Vmatrix}^n,$$

where α is a real number.

8. Calculate $AB - BA$ where

(a) $A = \begin{Vmatrix} 1 & 2 & 1 \\ 2 & 1 & 2 \\ 1 & 2 & 3 \end{Vmatrix}, \quad B = \begin{Vmatrix} 4 & 1 & 1 \\ -4 & 2 & 0 \\ 1 & 2 & 1 \end{Vmatrix}$;

(b) $A = \begin{Vmatrix} 2 & 1 & 0 \\ 1 & 1 & 2 \\ -1 & 2 & 1 \end{Vmatrix}, \quad B = \begin{Vmatrix} 3 & 1 & -2 \\ 3 & -2 & 4 \\ -3 & 5 & -1 \end{Vmatrix}$.

9. Find all the matrices commuting with the matrix A:

(a) $A = \begin{Vmatrix} 1 & 2 \\ -1 & -1 \end{Vmatrix}$; (b) $A = \begin{Vmatrix} 1 & 1 \\ 0 & 1 \end{Vmatrix}$; (c) $A = \begin{Vmatrix} 1 & 0 & 0 \\ 0 & 1 & 0 \\ 3 & 1 & 2 \end{Vmatrix}$.

10. Find $f(A)$, where

(a) $f(x) = x^2 - x - 1, \quad A = \begin{Vmatrix} 2 & 1 & 1 \\ 3 & 1 & 2 \\ 1 & -1 & 0 \end{Vmatrix}$;

(b) $f(x) = x^2 - 5x + 3, \quad A = \begin{Vmatrix} 2 & -1 \\ -3 & 3 \end{Vmatrix}$.

(Cf. Sec. 44.)

11. Show that every second order matrix

$$A = \begin{Vmatrix} a & b \\ c & d \end{Vmatrix}$$

satisfies the matrix equation

$$x^2 - (a + d)x + (ad - bc) = 0.$$

Comment: This is a special case of the *Hamilton-Cayley theorem* proved in Sec. 52.

12. Find all second order matrices

(a) Whose squares equal the zero matrix; (b) Whose cubes equal the zero matrix; (c) Whose squares equal the unit matrix.

13. Find all second order matrices X such that

(a) $XA = 0$, where A is a given second order matrix;
(b) $X^2 = \begin{Vmatrix} 1 & 0 \\ 0 & 1 \end{Vmatrix}$; (c) $X^2 = \begin{Vmatrix} 1 & 1 \\ 0 & 1 \end{Vmatrix}$; (d) $X^2 = \begin{Vmatrix} 1 & 1 \\ 1 & 1 \end{Vmatrix}$; (e) $X^2 = \begin{Vmatrix} 2 & 1 \\ 1 & 1 \end{Vmatrix}$.

14. Find the inverse matrix A^{-1} of the following matrices:

(a) $A = \begin{Vmatrix} a & b \\ c & d \end{Vmatrix}$; (b) $A = \begin{Vmatrix} 1 & 2 & -3 \\ 0 & 1 & 2 \\ 0 & 0 & 1 \end{Vmatrix}$; (c) $A = \begin{Vmatrix} 2 & 2 & 3 \\ 1 & -1 & 0 \\ -1 & 2 & 1 \end{Vmatrix}$.

15. Find the unknown matrix X satisfying each of the following matrix equations:

(a) $\begin{Vmatrix} 2 & 5 \\ 1 & 3 \end{Vmatrix} X = \begin{Vmatrix} 4 & -6 \\ 2 & 1 \end{Vmatrix}$; (b) $X \begin{Vmatrix} 1 & 1 & -1 \\ 2 & 1 & 0 \\ 1 & -1 & 1 \end{Vmatrix} = \begin{Vmatrix} 1 & -1 & 3 \\ 4 & 3 & 2 \\ 1 & -2 & 5 \end{Vmatrix}$;

(c) $\begin{Vmatrix} 2 & 1 \\ 3 & 2 \end{Vmatrix} X \begin{Vmatrix} -3 & 2 \\ 5 & -3 \end{Vmatrix} = \begin{Vmatrix} -2 & 4 \\ 3 & -1 \end{Vmatrix}$; (d) $\begin{Vmatrix} 2 & 1 \\ 2 & 1 \end{Vmatrix} X = \begin{Vmatrix} 2 & 1 \\ 2 & 1 \end{Vmatrix}$;

(e) $X \begin{Vmatrix} 2 & 1 \\ 2 & 1 \end{Vmatrix} = \begin{Vmatrix} 1 & 0 \\ 0 & 1 \end{Vmatrix}$.

16. Show that the one-to-one correspondence

$$\begin{Vmatrix} a & b \\ -b & a \end{Vmatrix} \leftrightarrow a + ib$$

preserves products ($i = \sqrt{-1}$).

In the following four problems, A, B, and X denote third order matrices and I denotes the unit matrix (10). All the results generalize at once to the case of matrices of order n.

17. By the *trace* of a matrix is meant the sum of its diagonal elements (cf. Sec. 27). Show that AB and BA have the same trace.

18. Show that the relation $AB - BA = I$ is impossible.

19. If $AX = XA$ for all X, then $A = cI$, where c is a constant.

20. If $AX = XA$ for all diagonal matrices X, then A is a diagonal matrix.

21. Without calculating A^{-1}, find $\det A^{-1}$, where

(a) $A = \begin{Vmatrix} 0 & 1 & 1 \\ 1 & 0 & 1 \\ 0 & 0 & 1 \end{Vmatrix}$; (b) $A = \begin{Vmatrix} -1 & 1 & 1 \\ 1 & -1 & 1 \\ 1 & 1 & -1 \end{Vmatrix}$.

In problems involving tensors, the following *summation convention* is very convenient (and often indispensable): If a certain index occurs twice in an expression, then it is understood that the expression is summed with respect to that index for all admissible values of the index. Thus, for example, in terms of the coefficients a_{ik}, b_i and the variables x_i $(i, j = 1, 2, \ldots, n)$, we can write a system of n linear equations in n unknowns as

$$a_{ik}x_k = b_i \qquad (i = 1, 2, \ldots, n).$$

In what follows, the summation convention will be assumed to hold in any relevant context.

22. Let $\Delta = |a_{ik}|$ be a determinant of order n, with cofactors A_{ik}. Use the summation convention to summarize the formulas (18), (19), (21*a*), and (22*b*) of Sec. 3. Then give a compact proof of Cramer's rule (Sec. 8).

23. Let **i**, **j**, and **k** be three linearly independent three-dimensional vectors which are in general not orthogonal. Define the *contravariant components* x^1, x^2, x^3 of a vector **x** with respect to this basis by the formula

$$\mathbf{x} = x^1\mathbf{i} + x^2\mathbf{j} + x^3\mathbf{k}$$

and the *covariant components* of **x** by the formulas

$$x_1 = (\mathbf{x}, i),\ x_2 = (\mathbf{x}, j),\ x_3 = (\mathbf{x}, k).$$

Then, prove the two assertions made at the end of Sec. 22, i.e., (x_1, x_2, x_3) transforms like a covariant vector (as defined earlier) and

$$x^1x_1 + x^2x_2 + x^3x_3 = \|\mathbf{x}\|^2.$$

24. Write the transformation laws for the three types of second rank tensors (T^{ij}, T^i_j, T_{ij}) for an arbitrary coordinate transformation.

25. The *Kronecker delta* δ^i_j is defined as equal to 1 if $i = j$ and 0 if $i \neq j$ in *every* coordinate system. Show that δ^i_j is a mixed second rank tensor, as anticipated in the notation.

26. Let T_{ij} and U^m_n be second rank tensors of the types indicated by the indices. Show that the *outer product* $T_{ij}U^m_n$ is a tensor of rank 4, covariant in the indices i, j, n, and contravariant in the index m. Generalize this result to tensors of any rank and any type.

27. Let T^m_{nij} be a tensor of rank 4, covariant in the indices n, i, j, and contravariant in the index m. Consider the *contraction* of T^m_{nij}, obtained by setting $m = n$ and invoking the summation convention, i.e., consider the quantity T^m_{mij}. Show that T^m_{mij} is a covariant tensor of rank 2. Generalize this result to tensors of any rank and any type.

28. How many different tensors can be obtained by contraction of the tensor T^{ij}_{kmn}?

29. Show that the *inner products* $T_{im}U^m_n$ and $T_{mj}U^m_n$, obtained by contraction of the outer product $T_{ij}U^m_n$, are covariant tensors of rank 2. Generalize this result to tensors of any rank and any type.

30. Let δ^i_j be the Kronecker delta of Prob. 25. What are the values in an n-dimensional space of the contractions

(a) δ^m_m; (b) $\delta^m_n\delta^n_m$; (c) $\delta^m_n\delta^n_r\delta^r_m$?

31. Prove that the set of n^3 quantities T^{ijk} form the components of a contravariant tensor of rank 3 (in an n-dimensional space), if we have

$$T^{ijk}X^p_{ij} = U^{kp},$$

where U^{kp} is a contravariant tensor of rank 2 and X^p_{ij} is an *arbitrary* tensor of rank 3 of the type indicated. Generalize this *quotient law* to tensors of any rank and type.

Comment: This law is used in Sec. 23 to define the three types of second rank tensors.

32. Show that if X^i is an arbitrary contravariant tensor and if $T_{ij}X^iX^j$ is an invariant, we can only infer that the *sum* $T_{ij} + T_{ji}$ is a covariant tensor of rank 2, but not T_{ij} itself.

33. Show that the partial derivatives of a vector (contravariant or covariant) do not form a tensor, but that they do form a *Cartesian tensor.*

Comment: Thus, every tensor is a Cartesian tensor, but the converse is not true. There is a slight abuse of language involved here. See remarks in J. L. Synge and A. Schild, "Tensor Calculus," p. 128, University of Toronto Press, Toronto, 1949.

34. Let $A = \|a_{ik}\|$ be the matrix of a transformation from one n-dimensional Cartesian coordinate system to another. Then either $\det A = +1$ or $\det A = -1$ (Secs. 20, 28). By an *oriented Cartesian tensor* of rank r, we mean a set of n^r quantities which transform like a Cartesian tensor (Sec. 24) under transformations for which $\det A = +1$, but not under transformations for which $\det A = -1$.

Consider the *permutation symbol* e_{ijk} defined in every coordinate system as follows: $e_{ijk} = 0$ if any two subscripts are equal; $e_{ijk} = +1$ if i, j, k is an *even* permutation† of 1, 2, 3; $e_{ijk} = -1$ if i, j, k is an *odd* permutation of 1, 2, 3. Express an arbitrary determinant of order 3 in terms of e_{ijk}, and show that e_{ijk} is an oriented Cartesian tensor (of dimension and rank 3).

35. Use the permutation symbol e_{ijk} of the preceding problem to write expressions for

(a) The vector product $\mathbf{A} \times \mathbf{B}$ of two vectors $\mathbf{A}$ and $\mathbf{B}$;
(b) The triple product $\mathbf{A} \cdot (\mathbf{B} \times \mathbf{C})$ of three vectors $\mathbf{A}$, $\mathbf{B}$, and $\mathbf{C}$;
(c) The curl of a vector field $\mathbf{A}(\mathbf{r})$.

Show that all these quantities change sign when we change the orientation of the coordinate system.

36. Prove the formulas

(a) $e_{ijk}e_{imn} = \delta_{jm}\delta_{kn} - \delta_{jn}\delta_{km}$;
(b) $e_{ijk}e_{imn} + e_{ijn}e_{ikm} = e_{ijm}e_{ikn}$.

37. A Cartesian tensor which transforms into itself under any three-dimensional rotation is called an *isotropic tensor.* Prove that the most general isotropic tensors of ranks 2 and 3 are $\lambda\delta_{ij}$ and λe_{ijk}, respectively, where λ is an invariant (scalar).

† The identity permutation is regarded as even.

38. Evaluate the following matrix products:

(a) $$\begin{Vmatrix} 2 & -1 & 3 & -4 \\ 3 & -2 & 4 & -3 \\ 5 & -3 & -2 & 1 \\ 3 & -3 & -1 & 2 \end{Vmatrix} \begin{Vmatrix} 7 & 8 & 6 & 9 \\ 5 & 7 & 4 & 5 \\ 3 & 4 & 5 & 6 \\ 2 & 1 & 1 & 2 \end{Vmatrix};$$

(b) $$\begin{Vmatrix} 5 & 7 & -3 & -4 \\ 7 & 6 & -4 & -5 \\ 6 & 4 & -3 & -2 \\ 8 & 5 & -6 & -1 \end{Vmatrix} \begin{Vmatrix} 1 & 2 & 3 & 4 \\ 2 & 3 & 4 & 5 \\ 1 & 3 & 5 & 7 \\ 2 & 4 & 6 & 8 \end{Vmatrix};$$ (c) $$\begin{Vmatrix} 1 & 1 & 1 & \cdots & 1 \\ 0 & 1 & 1 & \cdots & 1 \\ 0 & 0 & 1 & \cdots & 1 \\ \cdots & \cdots & \cdots & \cdots & \cdots \\ 0 & 0 & 0 & \cdots & 1 \end{Vmatrix}^2.$$

In the last example, the matrix is of order n.

39. Find all matrices which commute with the matrix

$$\begin{Vmatrix} 0 & 1 & 0 & 0 \\ 0 & 0 & 1 & 0 \\ 0 & 0 & 0 & 1 \\ 0 & 0 & 0 & 0 \end{Vmatrix}.$$

40. How does the product AB of two matrices change if

(a) Rows i and j of A are interchanged;
(b) Row j of A is multiplied by a number c and added to row i;
(c) Columns i and j of B are interchanged;
(d) Column j of B is multiplied by a number c and added to column i?

41. An $n \times n$ matrix is said to be *triangular* if all the elements above the principal diagonal are zero. Show that the product of two triangular matrices is triangular.

42. A square matrix A is said to be *idempotent* if $A^2 = A$. Show that an idempotent matrix A cannot have an inverse unless A is the identity matrix.

43. Prove that if A and B commute and are idempotent, then AB is idempotent. Give an example to show that if A and B are idempotent but do not commute, then AB need not be idempotent.

44. Find the inverses of the following matrices of order n:

(a) $$\begin{Vmatrix} 1 & 1 & 1 & \cdots & 1 \\ 0 & 1 & 1 & \cdots & 1 \\ 0 & 0 & 1 & \cdots & 1 \\ \cdots & \cdots & \cdots & \cdots & \cdots \\ 0 & 0 & 0 & \cdots & 1 \end{Vmatrix};$$ (b) $$\begin{Vmatrix} 1 & 1 & 0 & \cdots & 0 \\ 0 & 1 & 1 & \cdots & 0 \\ 0 & 0 & 1 & \cdots & 0 \\ \cdots & \cdots & \cdots & \cdots & \cdots \\ 0 & 0 & 0 & \cdots & 1 \end{Vmatrix};$$ (c) $$\begin{Vmatrix} 1 & 1 & 1 & \cdots & 1 \\ 1 & 0 & 1 & \cdots & 1 \\ 1 & 1 & 0 & \cdots & 1 \\ \cdots & \cdots & \cdots & \cdots & \cdots \\ 1 & 1 & 1 & \cdots & 0 \end{Vmatrix};$$

(d) $$\begin{Vmatrix} 0 & 1 & 1 & \cdots & 1 \\ 1 & 0 & 1 & \cdots & 1 \\ 1 & 1 & 0 & \cdots & 1 \\ \cdots & \cdots & \cdots & \cdots & \cdots \\ 1 & 1 & 1 & \cdots & 0 \end{Vmatrix}.$$

45. Consider the matrix

$$A = \begin{Vmatrix} 1 & 1 & 1 & 1 \\ 1 & 1 & 1 & 1 \\ 1 & 1 & 1 & 1 \\ 1 & 1 & 1 & 1 \end{Vmatrix}.$$

Show that the matrix $I - A$ has an inverse of the form $I + cA$ for some constant c (I is the unit matrix).

46. A square matrix is called *nilpotent* if one of its powers equals zero. Show that a triangular matrix T (cf. Prob. 41) is nilpotent if and only if all its diagonal elements vanish, and show that the smallest integer k for which $T^k = 0$ does not exceed the order of A.

Comment: Nilpotent matrices play an important role in Chap. 6.

47. Let A be any square matrix, and let $\tilde{A}$ be its Hermitian conjugate (Sec. 26). Show that the principal minors (Sec. 5) of the matrix $A\tilde{A}$ are nonnegative.

48. Show that the rank of $A\tilde{A}$ is the same as that of A.

49. Let A and B be $n \times n$ matrices. Show that the matrix equation $AX = B$ has a solution if and only if the rank of the matrix A is equal to the rank of the $2n \times n$ matrix (A, B) obtained by writing the matrix B to the right of the matrix A.

50. Suppose A and B are rectangular matrices with the same number of rows. As in Prob. 49, let (A, B) denote the matrix obtained by writing the matrix B to the right of the matrix A. Show that

$$\text{rank }(A, B) \le \text{rank } A + \text{rank } B.$$

51. Show that the *elementary transformations* on the rows of a matrix A (defined in Prob. 31 of Chap. 2) can be achieved by multiplying A from the *left* by suitable nonsingular matrices. Show that the elementary transformations on the columns of A can be achieved by multiplying A from the *right* by suitable nonsingular matrices. (The matrix A need not be square.)

52. Show that any rectangular matrix A of rank r can be represented in the form of a product $A = PRQ$, where P and Q are nonsingular matrices, and R is a matrix of the same form as A, all of whose elements vanish except the first r elements along the principal diagonal.

53. Show that if a matrix A contains m rows (columns) and has rank r, then any s rows (columns) of A form a matrix of rank not less than $r + s - m$.

54. Let A be an $m \times n$ matrix of rank r, let $P = \|p_{ij}\|$ be an $s \times m$ matrix for which $p_{11} = p_{22} = \cdots = p_{kk} = 1$ and all other elements vanish, and let $Q = \|q_{ij}\|$ be an $n \times t$ matrix for which $q_{11} = q_{22} = \cdots = q_{ll} = 1$ and all other elements vanish. Prove the following inequalities:

(**a**) rank $PA \ge k + r - m$; (**b**) rank $AQ \ge l + r - n$;
(**c**) rank $PAQ \ge k + l + r - m - n$.

55. Let A be a rectangular matrix with n columns and rank r_A, and let B be a rectangular matrix with n rows and rank r_B. Let r_{AB} be the rank of the product AB. Prove *Sylvester's inequality:*

$$r_A + r_B - n \le r_{AB} \le r_A, r_B.$$

56. A matrix is said to be in *block form* if it is divided into submatrices by one or more horizontal or vertical lines. Show that multiplication of two matrices in block form reduces to matrix multiplication *with the submatrices playing the role of matrix elements* if and only if the vertical subdivisions in the first factor correspond to the horizontal subdivisions in the second factor. Use this rule to evaluate the product AB, where

$$A = \left\| \begin{array}{c:cc} 1 & -2 & 3 \\ 3 & -1 & 2 \\ \hdashline 4 & -2 & 1 \end{array} \right\|, \quad B = \left\| \begin{array}{c:cc} 2 & 3 & 1 \\ \hdashline 1 & 2 & 3 \\ 2 & 1 & 3 \end{array} \right\|.$$

57. Find the eigenvalues of A, and find a matrix V such that $V^{-1}AV$ is diagonal. where A is the matrix

$$\text{(a)} \begin{Vmatrix} 2 & 1 \\ 1 & 2 \end{Vmatrix}; \quad \text{(b)} \begin{Vmatrix} 3 & 4 \\ 5 & 2 \end{Vmatrix}; \quad \text{(c)} \begin{Vmatrix} 0 & a \\ -a & 0 \end{Vmatrix}; \quad \text{(d)} \begin{Vmatrix} 0 & 2 & 1 \\ -2 & 0 & 3 \\ -1 & -3 & 0 \end{Vmatrix};$$

$$\text{(e)} \begin{Vmatrix} 2 & 5 & -6 \\ 4 & 6 & -9 \\ 3 & 6 & -8 \end{Vmatrix}.$$

Comment: In all these cases, the eigenvalues are distinct. The case of multiple eigenvalues will be considered later.

58. Given the eigenvalues of the matrix A, find the eigenvalues of the matrix A^{-1} and the matrix A^2 (cf. end of Sec. 52).

59. Find the eigenvalues of the matrix

$$\begin{Vmatrix} a_0 & a_1 & a_2 & \cdots & a_{n-1} \\ a_{n-1} & a_0 & a_1 & \cdots & a_{n-2} \\ \cdots & \cdots & \cdots & \cdots & \cdots \\ a_1 & a_2 & a_3 & \cdots & a_0 \end{Vmatrix}$$

whose determinant is the *circulant* of Prob. 52 of Chap. 1.

60. Show that the conditions (120) are necessary for a matrix U to be unitary.

61. Prove that the sum of the squares of the absolute values of all the minors of order k appearing in any k rows (or columns) of a unitary matrix is 1.

62. Let U be a unitary matrix, let M be any minor of U, and let C be the cofactor of M. Prove that $C = \bar{M} \det U$.

63. Show that

$$\text{(a)} \left| \sum_{k=1}^{n} a_k \right| \leq \sqrt{n} \left(\sum_{k=1}^{n} |a_k|^2 \right)^{1/2}; \quad \text{(b)} \left| \sum_{k=1}^{n} a_k \right| \leq \sqrt{\frac{n(n+1)}{2}} \left(\sum_{k=1}^{n} \frac{|a_k|^2}{k} \right)^{1/2}.$$

64. Use (126) to estimate the sum

$$s_n = \sum_{k=0}^{n} \frac{1}{\sqrt{k+1}\sqrt{n-k+1}}.$$

Comment: Actually, $s_n < \pi$, but this result does not follow from (126).

65. Show that the Schwarz inequality for integrals (126*b*) can be derived from the inequality

$$0 \leq \int_a^b \int_a^b (f(x)g(y) - f(y)g(x))^2 \, dx \, dy$$

by squaring and separating the integrals.

66. Show that the equality is attained in (129) if and only if $y = cx$ for some constant $c > 0$.

67. Using (129), show that

$$\|\mathbf{x} - \mathbf{y}\| \geq | \, \|\mathbf{x}\| - \|\mathbf{y}\| \, |.$$

68. If $\mathbf{x}$ and $\mathbf{y}$ are any two vectors, show that

$$\|\mathbf{x} + \mathbf{y}\|^2 + \|\mathbf{x} - \mathbf{y}\|^2 = 2(\|\mathbf{x}\|^2 + \|\mathbf{y}\|)^2.$$

What is the geometrical interpretation of this formula?

69. Define the *distance* between two vectors $\mathbf{x}$ and $\mathbf{y}$ to be

$$d(\mathbf{x}, \mathbf{y}) = \|\mathbf{x} - \mathbf{y}\|.$$

Show that

$$d(\mathbf{x}, \mathbf{y}) \leq d(\mathbf{x}, \mathbf{z}) + d(\mathbf{y}, \mathbf{z})$$

for any three vectors $\mathbf{x}$, $\mathbf{y}$, and $\mathbf{z}$.

70. Orthonormalize the following sets of vectors:

(**a**) (1, 1, 0), (1, 0, 1), (0, 1, 1); (**b**) (1, 1, 1), (1, 0, −1), (1, −1, 0).

71. Show that the Gram determinant (Sec. 16) is invariant under the orthogonalization process, i.e., if the systems of vectors $\mathbf{x}^{(1)}, \ldots, \mathbf{x}^{(m)}$ and $\mathbf{y}^{(1)}, \ldots, \mathbf{y}^{(m)}$ are related by (132) and (133), then

$$G(\mathbf{x}^{(1)}, \ldots, \mathbf{x}^{(m)}) = G(\mathbf{y}^{(1)}, \ldots, \mathbf{y}^{(m)}) = \|\mathbf{y}^{(1)}\|^2 \cdots \|\mathbf{y}^{(m)}\|^2.$$

72. Prove that the Gram determinant satisfies the inequality

$$G(\mathbf{a}_1, \ldots, \mathbf{a}_k, \mathbf{b}_1, \ldots, \mathbf{b}_l) \leq G(\mathbf{a}_1, \ldots, \mathbf{a}_k)G(\mathbf{b}_1, \ldots, \mathbf{b}_l),$$

where the equality holds if and only if either

$$(\mathbf{a}_i, b_j) = 0, \qquad (i = 1, \ldots, k; j = 1, \ldots, l),$$

or at least one of the systems $\mathbf{a}_1, \ldots, \mathbf{a}_k$ and $\mathbf{b}_1, \ldots, \mathbf{b}_l$ is linearly dependent.

Comment: This is a special case of formula (45) of Sec. 16.

Chapter 4

Quadratic Forms

32. Reduction of a Quadratic Form to a Sum of Squares. Let

$$Ax^2 + By^2 + Cz^2 + 2Dxy + 2Exz + 2Fyz + G = 0 \qquad (1)$$

be the equation of a quadric surface in three dimensions, which has its center at the origin of coordinates. The following problem then arises: Choose new coordinates x', y', z' in such a way that the mixed terms in (1) disappear, i.e., find an orthogonal transformation which takes the variables x, y, z into x', y', z' and which is such that all the second degree terms in the left-hand side of (1) reduce to just a sum of squares. In three dimensions, this problem can always be solved, and the transformed equation takes the form

$$\lambda_1 x'^2 + \lambda_2 y'^2 + \lambda_3 z'^2 + G = 0.$$

We now study the analogous problem in a real n-dimensional space.

Let

$$\varphi(x_1, x_2, \ldots, x_n) = \sum_{i,k=1}^{n} a_{ik} x_i x_k \qquad (2)$$

be a real quadratic form in n variables, where the a_{ik} are real coefficients satisfying the condition

$$a_{ki} = a_{ik}. \qquad (3)$$

(For example, in the case of (1), we have $x = x_1$, $y = x_2$, $z = x_3$ and $a_{11} = A$, $a_{22} = B$, $a_{33} = C$, $a_{12} = a_{21} = D$, $a_{13} = a_{31} = E$, $a_{23} = a_{32} = F$.) We call the matrix consisting of the elements a_{ik} the *matrix of the quadratic form* (2). This matrix is *symmetric*, i.e., equals its own transpose. We now transform the expression (2) by replacing the variables x_k by new variables x'_k, writing the transformation in the form

$$\mathbf{x} = B\mathbf{y}, \qquad (4)$$

where $\quad \mathbf{x} = (x_1, x_2, \ldots, x_n), \qquad \mathbf{y} = (y_1, y_2, \ldots, y_n).$

Substituting the expression (4) into (2), we obtain

$$\varphi = \sum_{i,k=1}^{n} a_{ik}(b_{i1}x_1' + b_{i2}x_2' + \cdots + b_{in}x_n')(b_{k1}x_1' + b_{k2}x_2' + \cdots + b_{kn}x_n').$$

Expanding this expression, we find that the coefficient of $x_p'x_q'$ $(p \neq q)$ is

$$\sum_{i,k=1}^{n} a_{ik}(b_{ip}b_{kq} + b_{iq}b_{kp}).$$

Using (3), we easily see that this expression is just twice the sum

$$\sum_{i=1}^{n} b_{ip} \sum_{k=1}^{n} a_{ik}b_{kq}.$$

Treating the case $p = q$ similarly, we see that in the new variables the quadratic form becomes

$$\varphi = \sum_{i,k=1}^{n} c_{ik}x_i'x_k', \tag{5}$$

where $$c_{ik} = c_{ki} = \sum_{t=1}^{n} b_{ti} \sum_{s=1}^{n} a_{ts}b_{sk}.$$

The summation over s gives $\{AB\}_{tk}$. If we regard t as a column index and i as a row index, then b_{ti} is the element $\{B^*\}_{it}$ of the transposed matrix B^*. Then we have

$$c_{ik} = c_{ki} = \sum_{t=1}^{n} \{B^*\}_{it}\{AB\}_{tk},$$

i.e., the new form (5) has a matrix defined by

$$C = B^*AB \tag{6}$$

in terms of the matrix A of the original form and the matrix B of the transformation (4). If the transformation (4) is orthogonal, then the transpose B^* of the orthogonal matrix B coincides with the inverse matrix B^{-1}. In this case, we have the formula

$$C = B^{-1}AB \tag{7}$$

instead of the formula (6).

Thus, the problem of constructing an orthogonal transformation (4) which reduces the quadratic form (2) to a sum of squares is equivalent to the problem of constructing an orthogonal matrix B such that the matrix (7) is just a diagonal matrix $[\lambda_1, \lambda_2, \ldots, \lambda_n]$, since the matrix of a quadratic form which has been reduced to diagonal form is in fact a diagonal matrix, whose elements λ_k are the coefficients of the terms $x_k'^2$. Thus, we must have, as before (Sec. 27),

$$B^{-1}AB = [\lambda_1, \lambda_2, \ldots, \lambda_n]$$

or

$$AB = B[\lambda_1, \lambda_2, \ldots, \lambda_n]. \tag{8}$$

We note that in the present case, A is not an arbitrary matrix, but a real symmetric matrix, and B has to be an orthogonal matrix. We now proceed just as we did previously (Sec. 27) in studying the general case. We rewrite (8) in the form

$$\sum_{s=1}^{n} a_{is}b_{sk} = \lambda_k b_{ik}, \tag{9}$$

which gives n equations for the elements of the kth column of the matrix B. Introducing the vector $\mathbf{x}^{(k)}$ with components $b_{1k}, b_{2k}, \ldots, b_{nk}$, we can write this last equation as

$$A\mathbf{x}^{(k)} = \lambda_k \mathbf{x}^{(k)}. \tag{10}$$

Transposing all the terms of (9) to one side, we obtain the following system of n homogeneous equations for determining $b_{1k}, b_{2k}, \ldots, b_{nk}$:

$$\begin{aligned}
(a_{11} - \lambda_k)b_{1k} + a_{12}b_{2k} + \cdots + a_{1n}b_{nk} &= 0,\\
a_{21}b_{1k} + (a_{22} - \lambda_k)b_{2k} + \cdots + a_{2n}b_{nk} &= 0,\\
\cdots\cdots\cdots\cdots\cdots\cdots\cdots\cdots\cdots\cdots&\\
a_{n1}b_{1k} + a_{n2}b_{2k} + \cdots + (a_{nn} - \lambda_k)b_{nk} &= 0.
\end{aligned} \tag{11}$$

Since the determinant of this system must vanish, we obtain an algebraic equation of degree n for the numbers λ_k:

$$\begin{vmatrix} a_{11} - \lambda & a_{12} & \cdots & a_{1n} \\ a_{21} & a_{22} - \lambda & \cdots & a_{2n} \\ \cdots & \cdots & \cdots & \cdots \\ a_{n1} & a_{n2} & \cdots & a_{nn} - \lambda \end{vmatrix} = 0. \tag{12}$$

As we know, this is just the characteristic equation of the matrix A.†

We now show that all the roots of (12) are real in the case of a real symmetric matrix A. We begin by writing the quadratic form slightly differently and in a somewhat more general form. Let $\mathbf{x}$ be the vector with components $x_1, x_2, \ldots, x_n$ (real or complex), and let A be the matrix with elements a_{ik}. We form the scalar product

$$(A\mathbf{x}, \mathbf{x}) = \sum_{i=1}^{n} \bar{x}_i(a_{i1}x_1 + a_{i2}x_2 + \cdots + a_{in}x_n),$$

† The left-hand side of (12) is called the *characteristic polynomial* of A.

which can also be written in the form

$$(A\mathbf{x}, \mathbf{x}) = \sum_{i,k=1}^{n} a_{ik}\bar{x}_i x_k. \tag{13}$$

If the condition

$$a_{ki} = \bar{a}_{ik} \qquad (a_{kk} \text{ real}) \tag{14}$$

is met, then the expression (13) is a Hermitian form (Sec. 30), whose values must be real; the case where A is a real symmetric matrix is a special case of (14). If in addition to A being real and symmetric, the components of $\mathbf{x}$ are real, then (13) reduces to just the quadratic form (2). To proceed with the proof that the roots of (12) are real, let λ_k be a root of this equation. Then the system (11) gives us the components of the vector $\mathbf{x}^{(k)}$ (real or complex) which satisfies (10). We take the scalar product of both sides of this equation with $\mathbf{x}^{(k)}$ from the right, obtaining

$$\|\mathbf{x}^{(k)}\|^2\lambda_k = (A\mathbf{x}^{(k)}, \mathbf{x}^{(k)}).$$

As we have seen, the right-hand side is a real number, and consequently λ_k is also a real number. Thus, we have proved that the roots of (12) are real not only for the case of real symmetric matrices, but also for matrices whose elements satisfy the condition (14). Such matrices are called *Hermitian matrices*. In the present case, the coefficients of the system (11) are real, and we can assume that the components of the vector $\mathbf{x}^{(k)}$ are also real.

We now show that if λ_p and λ_q are two different roots of (12), then the corresponding vectors $\mathbf{x}^{(p)}$ and $\mathbf{x}^{(q)}$ satisfying (10) are orthogonal. By hypothesis we have

$$A\mathbf{x}^{(p)} = \lambda_p\mathbf{x}^{(p)}, \quad A\mathbf{x}^{(q)} = \lambda_q\mathbf{x}^{(q)}.$$

Taking the scalar product of the first equation with $\mathbf{x}^{(q)}$ and the scalar product of the second equation with $\mathbf{x}^{(p)}$, and then subtracting, we obtain

$$(A\mathbf{x}^{(p)}, \mathbf{x}^{(q)}) - (\mathbf{x}^{(p)}, A\mathbf{x}^{(q)}) = (\lambda_p - \lambda_q)(\mathbf{x}^{(p)}, \mathbf{x}^{(q)}). \tag{15}$$

Next, we show that the formula

$$(A\mathbf{x}, \mathbf{y}) = (\mathbf{x}, A\mathbf{y}) \tag{16}$$

holds for any two vectors $\mathbf{x}$ and $\mathbf{y}$ (real or complex), provided only that the elements of the matrix satisfy the condition (14). In fact, the left-hand side of (16) gives

$$(A\mathbf{x}, \mathbf{y}) = \sum_{k=1}^{n} (a_{k1}x_1 + a_{k2}x_2 + \cdots + a_{kn}x_n)\bar{y}_k = \sum_{i,k=1}^{n} a_{ki}x_i\bar{y}_k,$$

or in view of the condition (14),

$$(A\mathbf{x}, \mathbf{y}) = \sum_{i,k=1}^{n} \bar{a}_{ik} x_i \bar{y}_k,$$

and the right-hand side of (16) gives just the same result, as required. Equation (16) is also valid for the case of real orthogonal matrices, which are a special case of Hermitian matrices. Because of (16), the left-hand side of (15) is zero, and since $\lambda_p \neq \lambda_q$, we have $(\mathbf{x}^{(p)}, \mathbf{x}^{(q)}) = 0$, i.e., the vectors $\mathbf{x}^{(p)}$ and $\mathbf{x}^{(q)}$ are in fact orthogonal, as asserted. In the present case, these vectors are real, and their orthogonality means that the sum of the products of their components vanishes.

Thus, if the roots of (12) are all different, we have n pairwise orthogonal vectors $\mathbf{x}^{(k)}$. Since (10) is linear and homogeneous in $\mathbf{x}^{(k)}$, we can multiply a solution of this equation by any constant; hence, we can assume that the vectors $\mathbf{x}^{(k)}$ have unit length. The components of these vectors form the columns of the matrix B. It follows that B satisfies the column orthogonality conditions and is therefore actually an orthogonal matrix (as assumed). Thus, under the assumption that all the roots of (12) are different, we have solved the problem of reducing a quadratic form to a sum of squares by an orthogonal transformation, or what amounts to the same thing, the problem of reducing the matrix A to diagonal form. The numbers λ_k are usually called the *eigenvalues of the matrix* A and the vectors $\mathbf{x}^{(k)}$ *the eigenvectors of* A.

33. Multiple Roots of the Characteristic Equation. In the general case where the characteristic equation (12) can have multiple roots, we use the following procedure for reducing a quadratic form to a sum of squares: We take a root $\lambda = \lambda_1$ of (12) and construct the corresponding solution of (10); this solution is a certain real vector $\mathbf{x}^{(1)}$, which we normalize to unit length. We then take $n - 1$ other real unit vectors which, together with $\mathbf{x}^{(1)}$, form a complete system of orthonormal vectors (Sec. 31). As we know, the transformation from the old basis vectors to this new system of basis vectors is accomplished by using an orthogonal matrix; the matrix A thereby goes into a similar matrix $A_1 = B_1^{-1} A B_1$, where B_1 is an orthogonal matrix. For this new matrix, the equation corresponding to (10) is

$$A_1 \mathbf{x} = \lambda \mathbf{x}. \tag{17}$$

The solution of (17) for the eigenvalue $\lambda = \lambda_1$ (the eigenvalues are invariant under similarity transformations) is just the vector $\mathbf{x}^{(1)}$, which we took as the first of the new basis vectors, and which therefore has the components 1, 0, . . . , 0 in the new basis. Thus, when A_1 operates on

$\mathbf{x}^{(1)} = (1, 0, \ldots, 0)$, the result is the vector $(\lambda_1, 0, \ldots, 0)$. From this it follows at once that the elements of the first column of A_1 are

$$\{A_1\}_{11} = \lambda_1, \{A_1\}_{21} = \{A_1\}_{31} = \cdots = \{A_1\}_{n1} = 0. \tag{18}$$

Since A is a real symmetric matrix, so is A_1, i.e., A_1 coincides with its own transpose. To see this, we write

$$A_1^* = (B_1^{-1}AB_1)^* = B_1^*A^*(B_1^{-1})^*.$$

Since B_1 is orthogonal and A is symmetric,

$$B_1^* = B_1^{-1}, (B_1^{-1})^* = B_1, A^* = A,$$

from which it follows that

$$A_1^* = A_1.$$

Using (18) and the fact that A_1 is symmetric, we find

$$\{A_1\}_{11} = \lambda_1, \{A_1\}_{21} = \{A_1\}_{12} = \cdots = \{A_1\}_{n1} = \{A_1\}_{1n} = 0,$$

i.e., all the elements of the first row and first column of the matrix A_1 vanish, with the possible exception of the element $\{A_1\}_{11} = \lambda_1$. Thus, the matrix A_1 has the form

$$A_1 = \left\| \begin{matrix} \lambda_1 & 0 & \cdots & 0 \\ 0 & a_{22}^{(1)} & \cdots & a_{2n}^{(1)} \\ \cdot & \cdot & \cdot & \cdot \\ 0 & a_{n2}^{(1)} & \cdots & a_{nn}^{(1)} \end{matrix} \right\|,$$

where we denote the elements of A_1 by $a_{ik}^{(1)}$. In the new variables, the quadratic form φ becomes

$$\varphi = \lambda_1 y_1'^2 + \sum_{i,k=2}^{n} a_{ik}^{(1)} y_i' y_k'.$$

In this way, we have eliminated one of the square terms and have thereby reduced our problem to a study of the quadratic form

$$\sum_{i,k=2}^{n} a_{ik}^{(1)} y_i' y_k'$$

in $n - 1$ variables, or equivalently, to a study of the corresponding matrix C_1 of order $n - 1$ (a submatrix of A_1). We can now repeat the argument just given, i.e., in the subspace of dimension $n - 1$ spanned by the last $n - 1$ basis vectors, we choose a unit vector $\mathbf{x}^{(2)}$, which is a solution of the equation

$$C_1\mathbf{x}^{(2)} = \lambda_2\mathbf{x}^{(2)};$$

by construction, this vector is orthogonal to $\mathbf{x}^{(1)}$. We now make a transformation which preserves the basis vector $\mathbf{x}^{(1)}$ and carries the other basis vectors into a new orthonormal basis which has $\mathbf{x}^{(2)}$ as its first basis vector. In the new variables, the quadratic form φ becomes

$$\varphi = \lambda_1 y_1''^2 + \lambda_2 y_2''^2 + \sum_{i,k=3}^{n} a_{ik}^{(2)} y_i'' y_k''.$$

Continuing in this way, we finally reduce the quadratic form to a sum of squares and the corresponding matrix to diagonal form. This is accomplished by successive application of several orthogonal transformations, which is clearly equivalent to application of a single orthogonal product transformation B. The final diagonal matrix

$$B^{-1}AB = [\lambda_1, \lambda_2, \ldots, \lambda_n] \tag{19}$$

is similar to the original matrix A, and therefore its characteristic equation

$$\begin{vmatrix} \lambda_1 - \lambda & 0 & \cdots & 0 \\ 0 & \lambda_2 - \lambda & \cdots & 0 \\ \cdots & \cdots & \cdots & \cdots \\ 0 & 0 & \cdots & \lambda_n - \lambda \end{vmatrix} = 0$$

coincides with the equation (12). In other words, the coefficients λ_k of the square terms in the final reduced quadratic form

$$\varphi = \lambda_1 x_1'^2 + \lambda_2 x_2'^2 + \cdots + \lambda_n x_n'^2 \tag{20}$$

are the roots of the equation (12), and every multiple root of multiplicity k appears k times in (20).

As we know, every column of the final orthogonal matrix B gives a vector which is a solution of (10); it follows from the construction that the value of λ_k corresponding to a given solution of (10) is the same as the coefficient of the corresponding variable in the quadratic form (20). More precisely, we mean the following: On the one hand, the vector $\mathbf{x}^{(k)}$ with components $b_{1k}, b_{2k}, \ldots, b_{nk}$ is a solution of (10) for $\lambda = \lambda_k$. On the other hand, since by (4) the orthogonal transformation B carries the variables $x_1', x_2', \ldots, x_n'$ into the variables $x_1, x_2, \ldots, x_n$, and since the inverse matrix B^{-1} is the transpose of B, we have

$$x_k' = b_{1k}x_1 + b_{2k}x_2 + \cdots + b_{nk}x_n \qquad (k = 1, 2, \ldots, n), \tag{21}$$

i.e., x_k' is the new variable corresponding to the vector $\mathbf{x}^{(k)}$.

Finally, we prove that the above procedure effectively gives *all solutions* of (10). We start from the characteristic equation (12), since the values λ_k appearing in (10) must be roots of (12). To be specific, suppose, for example, that the first three roots $\lambda_1, \lambda_2, \lambda_3$ of (12) are the same, i.e., $\lambda_1 = \lambda_2 = \lambda_3$. Then our procedure gives us three solutions:

$$\mathbf{x}^{(1)} = (b_{11}, b_{21}, \ldots, b_{n1}), \mathbf{x}^{(2)} = (b_{12}, b_{22}, \ldots, b_{n2}),$$
$$\mathbf{x}^{(3)} = (b_{13}, b_{23}, \ldots, b_{n3})$$

of the equation

$$A\mathbf{x} = \lambda_1\mathbf{x}. \tag{22}$$

We wish to show that any solution of (22) is a linear combination of $\mathbf{x}^{(1)}, \mathbf{x}^{(2)}, \mathbf{x}^{(3)}$. In fact, suppose that this were not the case. Then we would have a solution $\mathbf{y}$ of (22) which is linearly independent of $\mathbf{x}^{(1)}$, $\mathbf{x}^{(2)}, \mathbf{x}^{(3)}$. Moreover, we can assume that $\mathbf{y}$ is a real vector. (If $\mathbf{y}$ were complex, its real and imaginary parts would separately satisfy (22), since (22) has real coefficients; therefore, either the real or the imaginary part of $\mathbf{y}$ would have to be linearly independent of $\mathbf{x}^{(1)}, \mathbf{x}^{(2)}, \mathbf{x}^{(3)}$.) As we have already shown, $\mathbf{y}$ has to be orthogonal to all the vectors $\mathbf{x}^{(k)}$ for $k > 3$, since these vectors correspond to values of λ_k which are different from λ_1. Thus, $\mathbf{y}$ has to be linearly independent of the whole set of vectors $\mathbf{x}^{(1)}, \mathbf{x}^{(2)}, \mathbf{x}^{(3)}, \ldots, \mathbf{x}^{(n)}$, i.e., we have $n + 1$ linearly independent vectors in an n-dimensional space. This contradiction shows that for every root of (12) with multiplicity m, there are exactly m linearly independent real solutions.

We now give more details on the construction of solutions corresponding to multiple roots of (12). Suppose that in the coefficients of the system (11) we replace λ_k by a root $\lambda = \lambda_0$ of (12) of multiplicity m. Then we obtain a homogeneous system with m linearly independent solutions. Thus, the rank of the system must equal $n - m$, i.e., the system reduces to just $n - m$ equations. Take any solution of this system, and normalize it to unity by multiplying it by an appropriate factor. This gives a unit vector $\mathbf{x}^{(1)}$ corresponding to the root λ_0. To find the next vector $\mathbf{x}^{(2)}$ corresponding to λ_0, we supplement our system of $n - m$ equations with another equation, expressing the requirement that the new vector $\mathbf{x}^{(2)}$ be orthogonal to the vector $\mathbf{x}^{(1)}$ already found. Thus, to find the components of $\mathbf{x}^{(2)}$, we have a homogeneous system of $n - m + 1$ equations. The vector $\mathbf{x}^{(2)}$ is obtained by taking any solution of this new system and normalizing it to unity. To find the next vector $\mathbf{x}^{(3)}$ corresponding to λ_0, we supplement the original $n - m$ equations with two more equations, expressing the requirement that the new vector $\mathbf{x}^{(3)}$ be orthogonal to the vectors $\mathbf{x}^{(1)}$ and $\mathbf{x}^{(2)}$ already found. We continue in this way until we have constructed the whole set of m orthonormal vectors corresponding to the root λ_0 of multiplicity m.

A certain arbitrariness in constructing the solutions of (10) is implicit in this procedure. If all the roots of (10) are simple, then this arbitrariness consists only in the fact that all the components of the vectors

$\mathbf{x}^{(k)}$ can be multiplied by -1. However, suppose that the characteristic equation (12) has a root of multiplicity m. Then the corresponding m orthonormal vectors which are solutions of (10) span a certain m-dimensional subspace R_m. Clearly, any orthonormal basis that we choose in this subspace will also be a solution of (10) corresponding to $\lambda = \lambda_0$, i.e., we can go from one set of orthonormal solutions to another such set by performing an orthogonal transformation in R_m. (All this applies equally well to any other multiple root of (12).)

To clarify the situation, we return to the problem discussed at the beginning of Sec. 32, i.e., the problem of reducing a quadric surface to principal axes. To be explicit, we assume that the surface is an ellipsoid. Then, the case where all the roots of (12) are different corresponds to all the semiaxes of the ellipsoid being different, and the only arbitrariness in the choice of the principal axes is the choice of directions along these axes. If the equation (12), which in this case is a third degree equation, has two identical roots, then the ellipsoid is an ellipsoid of revolution, and we can choose for the principal axes any two axes lying in the plane P through the origin perpendicular to the axis of rotation, provided only that they are orthogonal axes. Thus, in this case, the arbitrariness in the choice of principal axes corresponds to making an arbitrary orthogonal transformation in the plane P. Finally, if all the roots of (12) are the same, our ellipsoid becomes a sphere, and the equation defining the surface no longer has any terms containing products of coordinates. In this case, we can choose the axes of any Cartesian coordinate system whatsoever for our principal axes.

34. Examples. We now consider two numerical examples.

1. *Reduce the surface*

$$x_1^2 + 5x_2^2 + x_3^2 + 2x_1x_2 + 6x_1x_3 + 2x_2x_3 = 5$$

to principal axes. The corresponding quadratic form is

$$\begin{aligned} \varphi = {} & x_1^2 + x_1x_2 + 3x_1x_3 \\ & + x_2x_1 + 5x_2^2 + x_2x_3 \\ & + 3x_3x_1 + x_3x_2 + x_3^2. \end{aligned}$$

The characteristic equation is

$$\begin{vmatrix} 1-\lambda & 1 & 3 \\ 1 & 5-\lambda & 1 \\ 3 & 1 & 1-\lambda \end{vmatrix} = 0.$$

Expanding with respect to elements of the first row, we obtain

$$(1-\lambda)[(5-\lambda)(1-\lambda) - 1] - (1 - \lambda - 3) + 3[1 - 3(5-\lambda)] = 0$$

or

$$\lambda^3 - 7\lambda^2 + 36 = 0.$$

It is easily verified that this equation has the roots

$$\lambda_1 = -2,\ \lambda_2 = 3,\ \lambda_3 = 6,$$

and referred to principal axes, the equation of our surface is

$$-2x_1'^2 + 3x_2'^2 + 6x_3'^2 = 5.$$

We now determine the elements of the orthogonal matrix

$$B = \begin{Vmatrix} b_{11} & b_{12} & b_{13} \\ b_{21} & b_{22} & b_{23} \\ b_{31} & b_{32} & b_{33} \end{Vmatrix},$$

which satisfy the system

$$\begin{aligned} (1-\lambda)b_{1k} + \qquad b_{2k} + \qquad 3b_{3k} &= 0, \\ b_{1k} + (5-\lambda)b_{2k} + \qquad b_{3k} &= 0, \\ 3b_{1k} + \qquad b_{2k} + (1-\lambda)b_{3k} &= 0. \end{aligned} \tag{23}$$

We first make the substitution $\lambda = \lambda_1 = -2$, which leads us to two equations

$$\begin{aligned} 3b_{11} + b_{21} + 3b_{31} &= 0, \\ b_{11} + 7b_{21} + b_{31} &= 0. \end{aligned}$$

The solution of this system has the form

$$b_{11} = -k_1,\ b_{21} = 0,\ b_{31} = k_1,$$

where k_1 is an arbitrary number, which we choose so as to make the sum $b_{11}^2 + b_{21}^2 + b_{31}^2$ equal to 1. Thus, we obtain

$$b_{11} = \frac{1}{\sqrt{2}},\ b_{21} = 0,\ b_{31} = -\frac{1}{\sqrt{2}},$$

where we can also take the solution with the opposite sign.

Next we substitute $\lambda = \lambda_2 = 3$ in the coefficients of the system (23). We obtain a system in which the third equation is the difference between the first two; thus we are left with the two equations

$$\begin{aligned} -2b_{12} + b_{22} + 3b_{32} &= 0, \\ b_{12} + 2b_{22} + b_{32} &= 0. \end{aligned}$$

The normalized solution of this system is easily found to be

$$b_{12} = \frac{1}{\sqrt{3}},\ b_{22} = -\frac{1}{\sqrt{3}},\ b_{32} = \frac{1}{\sqrt{3}}.$$

Finally, we substitute the third root $\lambda = \lambda_3 = 6$ in the coefficients of the system (23). The result is again a system in which one of the equations is a consequence of the other two. Solving the two equations which are left, and normalizing the solution obtained, we have

$$b_{13} = \frac{1}{\sqrt{6}},\ b_{23} = \frac{2}{\sqrt{6}},\ b_{33} = \frac{1}{\sqrt{6}}.$$

Thus, in this example, the transformation from the old to the new variables is given by the formulas

$$\begin{aligned} x_1' &= \frac{1}{\sqrt{2}} x_1 && - \frac{1}{\sqrt{2}} x_3, \\ x_2' &= \frac{1}{\sqrt{3}} x_1 - \frac{1}{\sqrt{3}} x_2 && + \frac{1}{\sqrt{3}} x_3, \\ x_3' &= \frac{1}{\sqrt{6}} x_1 + \frac{2}{\sqrt{6}} x_2 && + \frac{1}{\sqrt{6}} x_3. \end{aligned}$$

2. *Reduce the surface*

$$2x_1^2 + 6x_2^2 + 2x_3^2 + 8x_1x_3 = 1$$

to principal axes. In this example, the quadratic form is

$$\begin{aligned} \varphi = \quad & 2x_1^2 + 0x_1x_2 + 4x_1x_3 \\ & + 0x_2x_1 + 6x_2^2 + 0x_2x_3 \\ & + 4x_3x_1 + 0x_3x_2 + 2x_3^2, \end{aligned}$$

whose characteristic equation is

$$\begin{vmatrix} 2-\lambda & 0 & 4 \\ 0 & 6-\lambda & 0 \\ 4 & 0 & 2-\lambda \end{vmatrix} = 0.$$

Expanding this determinant, we obtain the equation

$$\lambda^3 - 10\lambda^2 + 12\lambda + 72 = 0.$$

Its roots are

$$\lambda_1 = -2,\ \lambda_2 = \lambda_3 = 6,$$

i.e., the equation has a double root.

To determine the coefficients of the orthogonal transformation, we have the system

$$\begin{aligned} (2-\lambda)b_{1k} \qquad\qquad & + \qquad 4b_{3k} = 0, \\ (6-\lambda)b_{2k} \qquad\qquad & = 0, \\ 4b_{1k} \qquad\qquad & + (2-\lambda)b_{3k} = 0. \end{aligned} \tag{24}$$

Substituting $\lambda = -2$, we easily obtain the normalized solution

$$b_{11} = \frac{1}{\sqrt{2}},\ b_{21} = 0,\ b_{31} = -\frac{1}{\sqrt{2}}.$$

Next we substitute the double root $\lambda = 6$ into the coefficients of the system (24); corresponding to this root, we should obtain two linearly independent orthogonal solutions. As a result of the substitution, the system reduces to the single equation

$$-b_{12} + b_{32} = 0.$$

For one normalized solution of this equation, we take

$$b_{12} = \frac{1}{\sqrt{2}},\ b_{22} = 0,\ b_{32} = \frac{1}{\sqrt{2}}.$$

To find the other solution, we note that this solution must satisfy both (24) and the condition that it be orthogonal to the solution already found, i.e., it must satisfy the two equations

$$\begin{aligned} -b_{13} + \qquad b_{33} &= 0, \\ \frac{1}{\sqrt{2}} b_{13} + \frac{1}{\sqrt{2}} b_{33} &= 0, \end{aligned}$$

or

$$b_{13} = b_{33} = 0,$$

which has the normalized solution

$$b_{13} = 0,\ b_{23} = 1,\ b_{33} = 0.$$

Finally, the orthogonal transformation is

$$\begin{aligned} x_1' &= \frac{1}{\sqrt{2}} x_1 - \frac{1}{\sqrt{2}} x_3, \\ x_2' &= \frac{1}{\sqrt{2}} x_1 + \frac{1}{\sqrt{2}} x_3, \\ x_3' &= \qquad x_2, \end{aligned}$$

and referred to principal axes, the equation of the surface becomes

$$-2x_1'^2 + 6(x_2'^2 + x_3'^2) = 1.$$

35. Classification of Quadratic Forms. We have stated the problem of reducing a quadratic form to a sum of squares with the restriction that the linear transformation relating the old to the new variables be orthogonal. More generally, the problem can be stated as follows: Reduce the real quadratic form (2) to the form

$$\varphi = \mu_1 X_1^2 + \mu_2 X_2^2 + \cdots + \mu_n X_n^2, \tag{25}$$

where the X_k are any n linearly independent real linear forms in the variables x_k. When the problem is stated in this way, the coefficients μ_k are no longer uniquely defined numbers such as we had before. Nevertheless, we can still make the following statement about the coefficients

μ_k: *The number of nonzero coefficients is always equal to the rank of the coefficient matrix* $\|a_{ik}\|$ *of the quadratic form, regardless of the way in which the quadratic form is represented as a sum of squares of linearly independent linear forms.* Moreover, the following property, usually called the *law of inertia for quadratic forms,* is also valid: *Regardless of the way in which a real quadratic form is reduced to the form* (25), *where the linear forms* X_k *are themselves real, the number of positive coefficients and the number of negative coefficients always remain the same.* The proof of these facts will be given at the end of this section.

The general problem of reducing a quadratic form to the form (25) can be solved very simply by completing squares. We illustrate this procedure with the example

$$\varphi = x_1^2 + 4x_2^2 + x_3^2 + 2x_1x_2 - 6x_1x_3 + 8x_2x_3.$$

Adding the terms $x_2^2 + 9x_3^2 - 6x_2x_3$ to the terms $x_1^2 + 2x_1x_2 - 6x_1x_3$, we obtain a complete square, and we can write φ in the form

$$\varphi = (x_1 + x_2 - 3x_3)^2 + 3x_2^2 - 8x_3^2 + 14x_2x_3.$$

Then, completing another square, we reduce φ to the final form (25):

$$\varphi = (x_1 + x_2 - 3x_3)^2 - 2(2x_3 - \tfrac{7}{4}x_2)^2 + \tfrac{73}{8}(x_2)^2.$$

The linear forms appearing in parentheses are obviously linearly independent.

In the case where there are missing squares in the expression for φ, the calculation has to be carried out somewhat differently. Suppose we have

$$\varphi = ax_1x_2 + Px_1 + Qx_2 + R,$$

where a is a nonzero numerical coefficient, P and Q are linear forms in the variables other than x_1 and x_2, and R is a quadratic form which likewise does not contain x_1 and x_2. We can write

$$\varphi = a\left(x_1 + \frac{Q}{a}\right)\left(x_2 + \frac{P}{a}\right) + R - \frac{PQ}{a}.$$

If we set

$$X_1 = \tfrac{1}{2}\left(x_1 + x_2 + \frac{P+Q}{a}\right), \quad X_2 = \tfrac{1}{2}\left(x_1 - x_2 - \frac{P-Q}{a}\right)$$

and

$$\varphi_1 = R - \frac{PQ}{a},$$

then we obtain

$$\varphi = aX_1^2 - aX_2^2 + \varphi_1,$$

where φ_1 is a quadratic form which does not contain x_1 and x_2. Thus, by completing two squares, we have eliminated two variables.

Reduction of a quadratic form to the form (25) leads to a natural classification of quadratic forms. We now consider the various cases that arise:

1. If all the coefficients μ_k in (25) are positive, the quadratic form is called *positive definite*. In this case, it is easy to see that the quadratic form takes positive values for all real nonzero values of x_k and vanishes if and only if all the x_k vanish. In fact, since all the μ_k are positive, a necessary and sufficient condition for the right-hand side of (25) to vanish is that all the linear forms X_k vanish. Since the X_k are linearly independent, this gives a system of n homogeneous equations for x_k with a nonvanishing determinant, and therefore the system has only the trivial solution.

2. If all the coefficients μ_k are negative, the quadratic form is called *negative definite*. Just as before, it can be shown that the form takes only negative values for all real nonzero values of x_k and vanishes if and only if all the x_k vanish.

3. Next we consider the case where some of the coefficients μ_k vanish, and all the nonzero coefficients have a definite sign, e.g., are positive. In this case, φ has the representation

$$\varphi = \mu_1 X_1^2 + \mu_2 X_2^2 + \cdots + \mu_m X_m^2 \qquad (m < n) \tag{26}$$

where all the μ_k are positive, and again the form cannot be negative for any values of x_k. However, the form can vanish for nonzero values of x_k. In fact, φ vanishes when the following system of m homogeneous equations is satisfied:

$$X_1 = X_2 = \cdots = X_m = 0;$$

since $m < n$, this system must have a nontrivial solution (Sec. 10). In just the same way, if all the coefficients μ_k in (26) are negative, the quadratic form cannot take positive values, but it can vanish for nonzero values of x_k. In this case, the form is called *semidefinite* (*positive semidefinite* or *negative semidefinite*).

4. Finally, if some of the coefficients μ_k in (25) are positive and others are negative, it is easily seen that the quadratic form can take both positive and negative values for real values of x_k. In this case, the form is called *indefinite*.

This classification of quadratic forms has an immediate application to the problem of maxima and minima of functions of several variables. Suppose we have a function $\psi(x_1, x_2, \ldots, x_n)$ of several independent variables $x_1, x_2, \ldots, x_n$, and suppose that the necessary conditions for the existence of a maximum or a minimum are met at the point $x_1 = x_2 = \cdots = x_n = 0$, i.e., the partial derivatives of the function with respect to all the independent variables vanish at this point. Expanding ψ in a Maclaurin series, we obtain

$$\psi(x_1, x_2, \ldots, x_n) - \psi(0, 0, \ldots, 0) = \varphi(x_1, x_2, \ldots, x_n) + \omega,$$

where $\varphi(x_1, x_2, \ldots, x_n)$ denotes a quadratic form in the variables x_k and ω denotes all the terms in the x_k of degree higher than 2. If the quadratic form φ is positive definite, then we have a minimum of the function ψ at the point $x_1 = x_2 = \cdots = x_n = 0$, and if it is negative definite we have a maximum. If it is indefinite, we have neither a maximum nor a minimum, and finally, if it is semidefinite, the question of whether we have a maximum or a minimum cannot be answered in this way. (These results are a natural extension of those given in an earlier volume, for the case of a function of two variables.)

We now prove the statements made at the beginning of this section. Suppose we have a quadratic form

$$\varphi = \sum_{i,k=1}^{n} a_{ik}x_ix_k \qquad (a_{ik} = a_{ki}),$$

where r is the rank of the coefficient matrix $\|a_{ik}\|$. We write down the following system of n linear forms:

$$\frac{1}{2}\frac{\partial \varphi}{\partial x_s} = \sum_{l=1}^{n} a_{sl}x_l \qquad (s = 1, 2, \ldots, n), \tag{27}$$

where we have used the condition $a_{ik} = a_{ki}$. The number r is clearly the rank of the system of forms (27) in the sense of Sec. 11. Suppose that φ can be reduced to a sum of squares of m linearly independent forms

$$y_s = \beta_{s1}x_1 + \beta_{s2}x_2 + \cdots + \beta_{sn}x_n \qquad (s = 1, 2, \ldots, m), \tag{28}$$

so that

$$\varphi = \mu_1y_1^2 + \mu_2y_2^2 + \cdots + \mu_my_m^2, \tag{29}$$

where the μ_s are nonzero. We have to show that $m = r$. Using (29), we find that the linear forms (27) can be written as

$$\frac{1}{2}\frac{\partial \varphi}{\partial x_s} = \mu_1\beta_{1s}y_1 + \mu_2\beta_{2s}y_2 + \cdots + \mu_m\beta_{ms}y_m \qquad (s = 1, 2, \ldots, n). \tag{30}$$

The variables y_s can take any values, since the forms (28) are linearly independent (Sec. 11). Therefore, in studying the linear dependence of the forms (30), the y_s can be regarded as independent variables, and the largest number of linearly independent forms in the system (30) equals the rank of the matrix $\|\mu_k\beta_{ki}\|$, where the row index i ranges from 1 to n

and the column index k ranges from 1 to m. The elements of each column of this matrix are multiplied by a common nonzero factor μ_k; therefore, the rank of the matrix $\|\mu_k\beta_{ki}\|$ is the same as that of the matrix $\|\beta_{ki}\|$. Since (28) is a system of m linearly independent forms, this rank equals m, i.e., the largest number of linearly independent forms in the system (30), or equivalently in the system (27), equals m. On the other hand, by hypothesis, this number equals r, from which it follows that $m = r$, QED.

We now prove the number of positive coefficients μ_k (and the number of negative coefficients) is always the same, for any representation of φ in the form (29), where the real linear forms y_s are linearly independent. The proof will be by contradiction. Suppose that φ has two representations of the form (29), i.e.,

$$\begin{aligned}\varphi &= \lambda_1 y_1^2 + \cdots + \lambda_p y_p^2 - \lambda_{p+1} y_{p+1}^2 - \cdots - \lambda_m y_m^2,\\ \varphi &= \lambda_1' y_1'^2 + \cdots + \lambda_q' y_q'^2 - \lambda_{q+1}' y_{q+1}'^2 - \cdots - \lambda_m' y_m'^2,\end{aligned} \tag{31}$$

where the number of positive coefficients λ_s and λ_s' is different in the two representations. We assume that the forms $y_1, y_2, \ldots, y_m$ are linearly independent and that the same is true of the forms $y_1', y_2', \ldots, y_m'$. Suppose that $p \neq q$; we show that this leads to an absurdity. Without loss of generality, we can assume that $p < q$. We supplement the forms $y_1, \ldots, y_m$ with the forms $y_{m+1}, \ldots, y_n$ in such a way as to make a complete system of linearly independent forms (Sec. 11), and then write the following homogeneous system of linear equations in the variables $x_1, x_2, \ldots, x_n$:

$$y_1 = 0, \ldots, y_p = 0, y_{q+1}' = 0, \ldots, y_m' = 0, y_{m+1} = 0, \ldots, y_n = 0. \tag{32}$$

There are

$$p + (m - q) + (n - m) = n - (q - p),$$

of these homogeneous equations; since $q - p > 0$, this number is less than n. Therefore, the system (32) has nontrivial real solutions. Let $x_s = x_s^{(0)}$ $(s = 1, 2, \ldots, n)$ be any of these solutions. By (32), we have

$$\varphi = -\lambda_{p+1} y_{p+1}^2 - \cdots - \lambda_m y_m^2 = \lambda_1' y_1'^2 + \cdots + \lambda_q' y_q'^2$$

for these values of x_s. From this it is clear that the quadratic form must vanish for $x_s = x_s^{(0)}$. Therefore, $x_s = x_s^{(0)}$ must satisfy the equations

$$y_{p+1} = 0, \ldots, y_m = 0,$$

as well as the equations (32). Thus, we finally find that all the forms of a complete system of linear forms must vanish for $x_s = x_s^{(0)}$. But this is impossible, because the homogeneous system

$$y_1 = 0, y_2 = 0, \ldots, y_n = 0$$

in the variables $x_1, x_2, \ldots, x_n$ has a nonvanishing determinant (since the forms y_s are linearly independent). This contradiction proves the law of inertia.

36. Jacobi's Formula. We now present without proof a formula due to Jacobi, which gives a convenient method for reducing a quadratic form to a sum of squares. First we need some notation. We write

$$A_i(x) = \sum_{k=1}^{n} a_{ik}x_k \qquad (i = 1, 2, \ldots, n),$$

$$\Delta_0 = 1,\ \Delta_1 = a_{11},\ \Delta_k = \begin{vmatrix} a_{11} & a_{12} & \cdots & a_{1k} \\ a_{21} & a_{22} & \cdots & a_{2k} \\ \cdots & \cdots & \cdots & \cdots \\ a_{k1} & a_{k2} & \cdots & a_{kk} \end{vmatrix} \qquad (k = 2, 3, \ldots, n),$$

$$X_1 = A_1(x),\ X_k = \begin{vmatrix} a_{11} & \cdots & a_{1,k-1} & A_1(x) \\ a_{21} & \cdots & a_{2,k-1} & A_2(x) \\ \cdots & \cdots & \cdots & \cdots \\ a_{k1} & \cdots & a_{k,k-1} & A_k(x) \end{vmatrix} \qquad (k = 2, 3, \ldots, n).$$

If the rank of the coefficient matrix $\|a_{ik}\|$ equals r and the determinants $\Delta_1, \Delta_2, \ldots, \Delta_r$ are nonvanishing, then Jacobi's formula has the form

$$\varphi = \sum_{i,k=1}^{n} a_{ik}x_ix_k = \sum_{k=1}^{r} \frac{X_k^2}{\Delta_k\,\Delta_{k-1}}, \tag{33}$$

where the linear forms X_k $(k = 1, 2, \ldots, r)$ are linearly independent. This formula allows us to determine the category to which the form φ belongs, insofar as the law of inertia is concerned. In particular, if all the determinants $\Delta_1, \Delta_2, \ldots, \Delta_n$ are positive (in which case $r = n$), then it follows from (33) that φ is positive definite. The converse statement can also be proved, i.e., if φ is a positive definite form, then all the determinants $\Delta_1, \Delta_2, \ldots, \Delta_n$ must be positive.

Of course, in applying (33), we can number the variables x_s in any order. For different enumerations, the determinants Δ_k will be different, and every principal minor of the matrix $\|a_{ik}\|$ appears in the sequence $\Delta_1, \Delta_2, \ldots, \Delta_n$ for some enumeration of the variables x_s. It follows from the foregoing that all the principal minors of a positive definite form are positive; however, it is sufficient to verify that just the determinants $\Delta_1, \Delta_2, \ldots, \Delta_n$ are positive. It can be shown that a necessary and sufficient condition for the form φ to be *positive* (i.e., positive definite or positive semidefinite) is that all the principal minors be nonnegative. In this case, it is not enough to determine the signs of just the determinants $\Delta_1, \Delta_2, \ldots, \Delta_n$, but it is also necessary to determine the signs of *all* the principal minors.

A necessary and sufficient condition for a form to be negative definite is that the inequalities

$$(-1)^k \Delta_k > 0 \qquad (k = 1, 2, \ldots, n)$$

hold. A necessary and sufficient condition for φ to be *negative* (i.e., negative definite or negative semidefinite) is that the principal minors either have the sign $(-1)^k$, where k is the order of the minor, or else vanish.

Proofs of the statements made in this section can be found in Gantmacher's book.†

37. Simultaneous Reduction of Two Quadratic Forms to Sums of Squares. Suppose we have two quadratic forms

$$\varphi_1 = \sum_{i,k}^{n} a_{ik} x_i x_k, \qquad \varphi_2 = \sum_{i,k}^{n} b_{ik} x_i x_k,$$

where φ_1 is positive definite, i.e., reduces to a sum of n squares with positive coefficients. It is required to find a linear transformation (not necessarily orthogonal) which reduces both forms to sums of squares. We begin by introducing new variables y_k which reduce the form φ_1 to a sum of squares. This can be done, for example, by the elementary method indicated in the preceding section. In the new variables, the two quadratic forms become

$$\varphi_1 = \sum_{k=1}^{n} \mu_k y_k^2, \qquad \varphi_2 = \sum_{i,k=1}^{n} b'_{ik} y_i y_k.$$

By hypothesis, all the numbers μ_k are positive, and we can introduce new (real) variables $z_k = \sqrt{\mu_k}\, y_k$, obtaining expressions of the form

$$\varphi_1 = \sum_{k=1}^{n} z_k^2, \qquad \varphi_2 = \sum_{i,k=1}^{n} b''_{ik} z_i z_k.$$

We now carry out the orthogonal transformation from the variables z_k to the new variables z'_k, which reduces the form φ_2 to a sum of squares. Since this transformation is orthogonal, the form φ_1 remains a sum of squares, so that both forms have finally been reduced to sums of squares:

$$\varphi_1 = \sum_{k=1}^{n} z_k'^2, \qquad \varphi_2 = \sum_{k=1}^{n} \lambda_k z_k'^2.$$

The numbers λ_k are sometimes called the *characteristic numbers* (or *eigenvalues*) *of the form* φ_2 *with respect to the form* φ_1.

† F. R. Gantmacher, "The Theory of Matrices," translated by K. A. Hirsch, vol. 1, pp. 304–308, Chelsea Publishing Company, New York, 1959 (in two volumes).

We now find the equation which must be satisfied by the numbers λ_k; this equation is completely analogous to the characteristic equation (12) of Sec. 32. First we prove a simple property of the *discriminant* of a quadratic form, which is defined as the determinant of the coefficient matrix of the form. Let φ be a quadratic form with coefficient matrix A, and suppose we use the transformation

$$\mathbf{x} = B\mathbf{x}',$$

where $\mathbf{x} = (x_1, x_2, \ldots, x_n)$, $\mathbf{x}' = (x_1', x_2', \ldots, x_n')$ to transform φ to new variables. As we know (Sec. 32), the matrix of the new quadratic form is

$$C = B^*AB,$$

and its determinant is given by

$$\det C = \det B^* \det A \det B;$$

$\det B^*$ and $\det B$ are obviously equal, since the corresponding matrices are transposes of each other. Thus we have

$$\det C = \det A(\det B)^2,$$

i.e., under a linear transformation of variables, the discriminant of a quadratic form is multiplied by the square of the determinant of the matrix which transforms from the new to the old variables.

Returning to the quadratic forms φ_1 and φ_2, we write the new quadratic form

$$\omega = \varphi_2 - \lambda\varphi_1 = \sum_{i,k=1}^{n} (b_{ik} - \lambda a_{ik})x_i x_k,$$

whose coefficients contain the parameter λ. When transformed to the new variables which simultaneously reduce φ_1 and φ_2 to sums of squares, this form becomes

$$\omega = \sum_{k=1}^{n} (\lambda_k - \lambda)z_k'^2;$$

clearly, the discriminant of this form is the product

$$(\lambda_1 - \lambda)(\lambda_2 - \lambda) \cdots (\lambda_n - \lambda) \tag{34}$$

in the new variables and the determinant of the matrix $\|b_{ik} - \lambda a_{ik}\|$ in the old variables. As we have just shown, these two discriminants differ only by a factor equal to the square of the determinant of the transformation matrix, a nonzero factor which is independent of λ. It follows immediately that, regarded as functions of λ, both discriminants have the same roots. Taking account of (34), we see that the numbers λ_k are the roots of the equation

$$\begin{vmatrix} b_{11} - \lambda a_{11} & b_{12} - \lambda a_{12} & \cdots & b_{1n} - \lambda a_{1n} \\ b_{21} - \lambda a_{21} & b_{22} - \lambda a_{22} & \cdots & b_{2n} - \lambda a_{2n} \\ \cdots & \cdots & \cdots & \cdots \\ b_{n1} - \lambda a_{n1} & b_{n2} - \lambda a_{n2} & \cdots & b_{nn} - \lambda a_{nn} \end{vmatrix} = 0.$$

38. Small Oscillations. In an earlier volume, we saw that the motion of a mechanical system with n degrees of freedom and time-independent constraints, which is acted upon by forces derivable from a potential, is described by a system of differential equations (the *Lagrange equations*) of the form†

$$\frac{d}{dt}\left(\frac{\partial T}{\partial \dot{q}_k}\right) - \frac{\partial T}{\partial q_k} = \frac{\partial U}{\partial q_k} \qquad (k = 1, 2, \ldots, n); \tag{35}$$

here T is the kinetic energy of the system, and U is a given function of the q_k (the potential) which we assume to be independent of t. As already noted, T is a quadratic function

$$T = \sum_{i,k=1}^{m} a_{ik}\dot{q}_i\dot{q}_k \qquad (a_{ki} = a_{ik}), \tag{36}$$

where $\dot{q}_k$ is the time derivative of q_k, and the coefficients a_{ik} are given functions of the q_k. Suppose that

$$\frac{\partial U}{\partial q_k} = 0 \qquad \text{for } q_1 = q_2 = \cdots = q_n = 0. \tag{37}$$

Then, it is obvious physically that the system of differential equations (35) has the solution $q_1 = q_2 = \cdots = q_n = 0$, corresponding to an equilibrium position of the system. The function U is defined only to within an additive constant, and we can always assume that it vanishes for $q_1 = q_2 = \cdots = q_n = 0$. Then, because of the condition (37), we can assert that the expansion of the function U in powers of the q_k begins with terms of the second degree in the q_k. Now suppose that the quadratic form consisting of these second degree terms is negative definite, which implies that U has a maximum at

$$q_1 = q_2 = \cdots = q_n = 0$$

(Sec. 35), or equivalently, that the potential energy $-U$ has a minimum. In this case the equilibrium position $q_1 = q_2 = \cdots = q_n = 0$ is stable; thus, when subjected to small initial disturbances, the system undergoes

† In this section, as in Sec. 17, the overdot means differentiation with respect to t.

small oscillations about the equilibrium position, i.e., the q_k remain small during the entire course of the motion. Therefore, in studying these small oscillations, we can assume that U reduces to just the second degree terms and has the form

$$-U = \sum_{i,k=1}^{n} b_{ik}q_iq_k \qquad (b_{ik} = b_{ki}). \tag{38}$$

In just the same way, it is a good approximation to set $q_k = 0$ in the coefficients a_{ik} of the expression (36), so that the a_{ik} can be regarded as constants. Making these substitutions in the system (35), we obtain a system of n linear differential equations with constant coefficients:

$$\begin{aligned}
a_{11}\ddot{q}_1 + a_{12}\ddot{q}_2 + \cdots + a_{1n}\ddot{q}_n + b_{11}q_1 + b_{12}q_2 + \cdots + b_{1n}q_n &= 0,\\
a_{21}\ddot{q}_1 + a_{22}\ddot{q}_2 + \cdots + a_{2n}\ddot{q}_n + b_{21}q_1 + b_{22}q_2 + \cdots + b_{2n}q_n &= 0,\\
\cdots\cdots\cdots\cdots\cdots\cdots\cdots\cdots\cdots\cdots\cdots\cdots\cdots\cdots\cdots\cdots\cdots\cdots\cdots&\\
a_{n1}\ddot{q}_1 + a_{n2}\ddot{q}_2 + \cdots + a_{nn}\ddot{q}_n + b_{n1}q_1 + b_{n2}q_2 + \cdots + b_{nn}q_n &= 0.
\end{aligned} \tag{39}$$

We now look for a solution of this system in the form of harmonic oscillations of the same frequency and initial phase, but with different amplitudes, i.e., we look for a solution of the form

$$q_k = A_k \cos(\lambda t + \varphi) \qquad (k = 1, 2, \ldots, n). \tag{40}$$

Substituting (40) into (39), we obtain the following system of equations for A_k and λ:

$$\begin{aligned}
(b_{11} - \lambda^2 a_{11})A_1 + (b_{12} - \lambda^2 a_{12})A_2 + \cdots + (b_{1n} - \lambda^2 a_{1n})A_n &= 0,\\
(b_{21} - \lambda^2 a_{21})A_1 + (b_{22} - \lambda^2 a_{22})A_2 + \cdots + (b_{2n} - \lambda^2 a_{2n})A_n &= 0,\\
\cdots\cdots\cdots\cdots\cdots\cdots\cdots\cdots\cdots\cdots\cdots\cdots\cdots\cdots\cdots\cdots\cdots\cdots\cdots&\\
(b_{n1} - \lambda^2 a_{n1})A_1 + (b_{n2} - \lambda^2 a_{n2})A_2 + \cdots + (b_{nn} - \lambda^2 a_{nn})A_n &= 0.
\end{aligned} \tag{41}$$

If this system is to have a nontrivial solution for the A_k, its determinant must vanish, i.e.,

$$\begin{vmatrix}
b_{11} - \lambda^2 a_{11} & b_{12} - \lambda^2 a_{12} & \cdots & b_{1n} - \lambda^2 a_{1n}\\
b_{21} - \lambda^2 a_{21} & b_{22} - \lambda^2 a_{22} & \cdots & b_{2n} - \lambda^2 a_{2n}\\
\cdots & \cdots & \cdots & \cdots\\
b_{n1} - \lambda^2 a_{n1} & b_{n2} - \lambda^2 a_{n2} & \cdots & b_{nn} - \lambda^2 a_{nn}
\end{vmatrix} = 0. \tag{42}$$

Taking a root of this equation and substituting it into the coefficients of the system (41), we obtain one or several solutions for the A_k, which we can then multiply by arbitrary constants. Moreover, the equations (40) still contain the arbitrary constant φ.

A more elegant solution of the problem can be obtained by applying the theory of quadratic forms. First we observe that by its very meaning, the quadratic form (36) in the variables $\dot{q}_k$ must be positive definite, since it expresses the kinetic energy of the system. Moreover, in the present case the form (38) is also positive definite, by hypothesis. As we have seen (Sec. 37), we can replace the variables q_k by new variables p_k, related to the old variables by a linear transformation with constant coefficients, which are such that in the new variables the quadratic forms T and $-U$ are simultaneously reduced to sums of squares, with the form T simply becoming a sum of squares with coefficients equal to 1. In this regard, we note that the linear dependence between the $\dot{p}_k$ and the $\dot{q}_k$ is precisely the same as that between the p_k and the q_k. Thus, we have

$$T = \sum_{s=1}^{n} \dot{p}_s^2, \qquad -U = \sum_{s=1}^{n} \lambda_s^2 p_s^2, \tag{43}$$

where all the coefficients of the p_s^2 are positive and hence can be written as squares λ_s^2. In the new variables, the Lagrange equations (35) become

$$\frac{d}{dt}\left(\frac{\partial T}{\partial \dot{p}_k}\right) = \frac{\partial U}{\partial p_k} \qquad (k = 1, 2, \ldots, n).$$

Substituting (43) in these equations, we obtain the remarkably simple system

$$\ddot{p}_k + \lambda_k^2 p_k = 0 \qquad (k = 1, 2, \ldots, n).$$

The solutions of this system are

$$p_k = C_k \cos(\lambda_k t + \psi_k) \qquad (k = 1, 2, \ldots, n), \tag{44}$$

where the C_k and ψ_k are arbitrary constants. The generalized coordinates p_k are called the *normal coordinates* of the mechanical system under consideration. The original coordinates q_k are expressed in terms of the normal coordinates by linear relations with constant coefficients. It follows immediately from the results of the preceding section that the numbers λ_k, called the *natural frequencies* of the system, must be the roots of the equation (42). We note that even if some of these roots are the same, (44) still gives the general solution of the problem of small oscillations as formulated here.

39. Extremal Properties of the Eigenvalues of a Quadratic Form. We now examine from another point of view the problem of reducing a real quadratic form to a sum of squares. For simplicity, we begin by considering the case of three variables:

$$\varphi = \sum_{i,k=1}^{3} a_{ik} x_i x_k = \sum_{k=1}^{3} \lambda_k x_k'^2, \tag{45}$$

where the x'_k are related to the x_k by an orthogonal transformation:

$$\begin{aligned} x_1 &= b_{11}x'_1 + b_{12}x'_2 + b_{13}x'_3, \\ x_2 &= b_{21}x'_1 + b_{22}x'_2 + b_{23}x'_3, \\ x_3 &= b_{31}x'_1 + b_{32}x'_2 + b_{33}x'_3. \end{aligned} \tag{46}$$

To be explicit, we assume that the numbers λ_k are arranged in descending order, so that

$$\lambda_1 > \lambda_2 > \lambda_3. \tag{47}$$

Our problem is to determine the numbers λ_k and the coefficients b_{ik} from a knowledge of the values of the form φ on the unit sphere K, i.e., the sphere

$$x_1^2 + x_2^2 + x_3^2 = 1 \qquad \text{or} \qquad x_1'^2 + x_2'^2 + x_3'^2 = 1, \tag{48}$$

with center at the origin and unit radius. Every point P of this sphere characterizes a direction in space, determined by the unit vector drawn from the origin to P.

We now write (45) in the form

$$\varphi = \lambda_1(x_1'^2 + x_2'^2 + x_3'^2) + (\lambda_2 - \lambda_1)x_2'^2 + (\lambda_3 - \lambda_1)x_3'^2,$$

from which it is apparent that

$$\varphi = \lambda_1 + (\lambda_2 - \lambda_1)x_2'^2 + (\lambda_3 - \lambda_1)x_3'^2$$

on the sphere K. It follows at once that λ_1 is the maximum of φ on K; this maximum is obviously achieved at the point

$$x'_1 = 1, \; x'_2 = x'_3 = 0.$$

According to (46), this point has the coordinates

$$x_1 = b_{11}, \; x_2 = b_{21}, \; x_3 = b_{31}$$

in the original coordinates and is therefore the point determined by the vector corresponding to the first column of the matrix of the orthogonal transformation (46), i.e., by the vector which is the solution of the equation

$$A\mathbf{x} = \lambda\mathbf{x} \tag{49}$$

for $\lambda = \lambda_1$ (Sec. 32). Thus, we see that the largest eigenvalue of the quadratic form (45) is the maximum value of this form on the unit sphere, and moreover, the corresponding eigenvector $\mathbf{x}^{(1)}$ (which is a solution of (49)) is just the vector drawn from the origin to the point of the unit sphere where this maximum is achieved.

To determine the second eigenvalue, and the corresponding eigenvector, we set $x'_1 = 0$, which defines a plane through the origin perpendicular to the vector $\mathbf{x}^{(1)}$. The intersection of this plane and the unit sphere is the circle

$$x_2'^2 + x_3'^2 = 1.$$

On this circle, we have

$$\varphi = \lambda_2 x_2'^2 + \lambda_3 x_3'^2,$$

from which it is immediately clear that λ_2 is the maximum value of φ on the unit sphere K, under the condition that the variable point on K always corresponds to a vector which is orthogonal to the vector $\mathbf{x}^{(1)}$ already found. In just the same way as before, we show that the eigenvector $\mathbf{x}^{(2)}$, i.e., the solution of (49) for $\lambda = \lambda_2$, is the vector drawn from the origin to the point where this maximum is achieved. Once we have the two eigenvectors $\mathbf{x}^{(1)}$ and $\mathbf{x}^{(2)}$, we obtain the third eigenvector $\mathbf{x}^{(3)}$ from the condition that $\mathbf{x}^{(3)}$ be orthogonal to both $\mathbf{x}^{(1)}$ and $\mathbf{x}^{(2)}$, and the eigenvalue λ_3 is the value of the form φ at the point where $\mathbf{x}^{(3)}$ intersects the unit sphere. It should be noted that if $\lambda_1 = \lambda_2$, for example, then in looking for the first maximum of the form on the unit sphere, we would have found that this maximum is obtained not at a point but on a whole circle.

The preceding argument can easily be extended to n dimensions. In this general case, we shall state only the results, which are completely analogous to those obtained in three dimensions. Let

$$\varphi = \sum_{i,k=1}^{n} a_{ik} x_i x_k \tag{50}$$

be a real quadratic form in n variables. The unit vectors in a real n-dimensional space are just the vectors $\mathbf{x} = (x_1, x_2, \ldots, x_n)$, for which

$$x_1^2 + x_2^2 + \cdots + x_n^2 = 1, \tag{51}$$

i.e., the end points of all the unit vectors lie on the unit n-dimensional sphere defined by (51). The largest eigenvalue of the form φ equals the maximum of φ on the sphere (51), and the corresponding eigenvector is the vector $\mathbf{x}^{(1)}$ drawn from the origin to the point on the sphere where φ achieves its maximum. To find the second largest eigenvalue, we consider the unit vectors which are orthogonal to the vector $\mathbf{x}^{(1)}$ already found. Among these vectors, there is a vector $\mathbf{x}^{(2)}$ for which φ takes the largest value. This second maximum λ_2 is just the second eigenvalue of the form φ, and the vector $\mathbf{x}^{(2)}$ is the corresponding eigenvector. Next, we consider the unit vectors which are orthogonal to $\mathbf{x}^{(1)}$ and $\mathbf{x}^{(2)}$; this is equivalent to imposing the two constraints

$$(\mathbf{x}^{(1)}, \mathbf{x}) = 0, \qquad (\mathbf{x}^{(2)}, \mathbf{x}) = 0$$

on (51). Among these vectors, there is again a vector $\mathbf{x}^{(3)}$ which gives φ its largest value. This value is the third largest eigenvalue λ_3 of the form φ, and $\mathbf{x}^{(3)}$ is the corresponding eigenvalue, and so on.

We could have arranged the eigenvalues of the quadratic form φ in ascending order instead of descending order, in which case the first eigenvalue would be the smallest, the second eigenvalue the next smallest, etc. Then we would have a situation completely analogous to that just discussed, except that the phrases "minimum" and "smallest value" would have to be substituted for "maximum" and "largest value" in all the appropriate places.

We can also generalize all the above considerations to the problem of simultaneous reduction of two quadratic forms to sums of squares. Let the two quadratic forms

$$\varphi = \sum_{i,k=1}^{n} a_{ik}x_ix_k, \qquad \psi = \sum_{i,k=1}^{n} b_{ik}x_ix_k$$

be reduced to the sums of squares

$$\varphi = \sum_{k=1}^{n} x_k'^2, \qquad \psi = \sum_{k=1}^{n} \lambda_k x_k'^2$$

by using the linear transformation

$$\mathbf{x} = B\mathbf{x}',$$

where $\mathbf{x} = (x_1, x_2, \ldots, x_n)$, $\mathbf{x}' = (x_1', x_2', \ldots, x_n')$, and it is assumed that the numbers λ_k are in descending order. Then λ_1 is the largest value of ψ under the condition that $\varphi = 1$, and this largest value is achieved for

$$x_1 = b_{11},\ x_2 = b_{21},\ \ldots,\ x_n = b_{n1}.$$

The subsequent eigenvalues are defined similarly.

40. Hermitian Matrices and Hermitian Forms. In the preceding sections, we studied real symmetric matrices and noted that they are special cases of Hermitian matrices, whose elements are complex numbers satisfying the relation

$$a_{ki} = \bar{a}_{ik}. \tag{52}$$

For $i = k$, this relation shows that the diagonal elements a_{kk} must be real. A Hermitian matrix can also be defined as a matrix which does not change if we interchange its rows and its columns and then take the complex conjugates of all its elements; thus, with the notation of Sec. 26, we have

$$\bar{A}^* = A \qquad \text{or} \qquad \tilde{A} = A. \tag{53}$$

As already noted, the matrix $\tilde{A}$ is called the *Hermitian conjugate*, or *adjoint*, of A. Therefore, Hermitian matrices are also said to be *self-adjoint*. Moreover, as shown above (Sec. 32), a Hermitian matrix A satisfies the relation

$$(A\mathbf{x}, \mathbf{y}) = (\mathbf{x}, A\mathbf{y}) \tag{54}$$

for any vectors $\mathbf{x}$ and $\mathbf{y}$. This relation, like the two preceding ones, can serve as a definition of a Hermitian matrix.

We now note another property of Hermitian matrices. Let A be a Hermitian matrix, and let U be any unitary matrix. Then it is easy to see that $U^{-1}AU$ is also a Hermitian matrix, i.e., is self-adjoint. In fact, we have (Sec. 26)

$$(\overline{U^{-1}AU})^* = \bar{U}^* \bar{A}^* (\bar{U}^*)^{-1}.$$

Since $\bar{A}^* = A$, by hypothesis, and $\bar{U}^* = U^{-1}$, because U is unitary (Sec. 28), this relation becomes

$$(\overline{U^{-1}AU})^* = U^{-1}AU,$$

as was to be proved. If we make the unitary transformation

$$\mathbf{x} = U\mathbf{x}', \tag{55}$$

where $\mathbf{x} = (x_1, x_2, \ldots, x_n)$ and $\mathbf{x}' = (x_1', x_2', \ldots, x_n')$, then the Hermitian matrix A will take the form $U^{-1}AU$ in the new coordinates. Thus, the result just proved can be stated as follows: *A unitary transformation does not change the Hermitian character of a matrix.*

We now study the problem of reducing a Hermitian matrix to diagonal form by using a unitary transformation:

$$U^{-1}AU = [\lambda_1, \lambda_2, \ldots, \lambda_n]. \tag{56}$$

As was the case for real symmetric matrices, this problem is equivalent to solving an equation of the form

$$A\mathbf{x} = \lambda\mathbf{x}, \tag{57}$$

where λ is one of the numbers λ_k, and the components of the vector $\mathbf{x}$ give the elements of the corresponding column of the matrix U. The numbers λ_k and the corresponding vectors $\mathbf{x}^{(k)}$ are called the *eigenvalues* and *eigenvectors* of the matrix A. As we know, the eigenvalues must be roots of the equation

$$\begin{vmatrix} a_{11} - \lambda & a_{12} & \cdots & a_{1n} \\ a_{21} & a_{22} - \lambda & \cdots & a_{2n} \\ \cdots & \cdots & \cdots & \cdots \\ a_{n1} & a_{n2} & \cdots & a_{nn} - \lambda \end{vmatrix} = 0. \tag{58}$$

Let $\lambda = \lambda_1$ be a root of (58) and let $\mathbf{x}^{(1)}$ be a solution of (57) for $\lambda = \lambda_1$. Since (57) is linear, we can multiply its solutions by any constant; therefore, we can assume that the length of the vector $\mathbf{x}^{(1)}$ is 1. We choose this vector as the first basis vector of a new coordinate system, and we then construct any $n - 1$ additional unit vectors such that these vectors together with $\mathbf{x}^{(1)}$ constitute a system of n orthonormal vectors. We take these n vectors as new basis vectors, and we let U_1 be the corresponding unitary transformation (55). In the new coordinate system, our Hermitian matrix A becomes the new Hermitian matrix $A_1 = U_1^{-1}AU_1$, and for $\lambda = \lambda_1$, the corresponding equation

$$A_1\mathbf{x} = \lambda\mathbf{x}$$

must have the vector $(1, 0, \ldots, 0)$ as a solution. Just as in Sec. 33, this fact shows us that all the elements of the first row and first column of the matrix A_1 vanish, except the element λ_1 appearing in the first row and the first column. Thus, the matrix A_1 has the form

$$\begin{Vmatrix} \lambda_1 & 0 & \cdots & 0 \\ 0 & a_{22}^{(1)} & \cdots & a_{2n}^{(1)} \\ \cdots & \cdots & \cdots & \cdots \\ 0 & a_{n2}^{(1)} & \cdots & a_{nn}^{(1)} \end{Vmatrix},$$

i.e., A_1 is a quasi-diagonal matrix of the form $[\lambda_1, C_1]$, where C_1 denotes a Hermitian matrix of order $n - 1$ with elements $a_{ik}^{(1)}$. (As we have seen above (Sec. 32), it is an immediate consequence of the Hermitian character of the matrix A_1 that the element λ_1, and in fact all the other roots of the equation (58), are real.) Repeating the argument just given, we can find a unitary transformation U_2 which acts on all the new basis vectors except the first one and reduces C_1 to a form in which all the elements of the first row and first column of C_1 are zero except the element in the upper left-hand corner of C_1. If this unitary transformation is regarded as a transformation acting in the *whole* space, it is a quasi-diagonal unitary matrix of the form

$$[1, U_2].$$

As a result of this transformation, our Hermitian matrix is transformed into the new Hermitian matrix

$$[1, U_2]^{-1}[\lambda_1, C_1][1, U_2] = [\lambda_1, U_2^{-1}C_1U_2],$$

which when written out in full has the form

$$\begin{Vmatrix} \lambda_1 & 0 & 0 & \cdots & 0 \\ 0 & \lambda_2 & 0 & \cdots & 0 \\ 0 & 0 & a_{33}^{(2)} & \cdots & a_{3n}^{(2)} \\ \cdots & \cdots & \cdots & \cdots & \cdots \\ 0 & 0 & a_{n3}^{(2)} & \cdots & a_{nn}^{(2)} \end{Vmatrix}.$$

Continuing this construction, we ultimately reduce the original Hermitian matrix A to diagonal form, and the overall unitary transformation U which appears in (56) is the product of the separate unitary transformations which were required at each step of the diagonalization procedure just described.

In Sec. 32, we showed that the solutions of (57) corresponding to different values of λ must be orthogonal to one another. Moreover, just as in Sec. 33, we can show that the vectors making up the columns of the matrix U (together with the corresponding values of λ) give *all* the solutions of (57). However, in this regard, we must keep in mind the following important fact concerning multiple roots of (58): If (58) has a root $\lambda = \lambda_1$ of multiplicity m, say, then (58) has m linearly independent solutions $\mathbf{x}^{(1)}, \mathbf{x}^{(2)}, \ldots, \mathbf{x}^{(m)}$ for $\lambda = \lambda_1$. Every linear combination

$$\mathbf{x} = C_1\mathbf{x}^{(1)} + C_2\mathbf{x}^{(2)} + \cdots + C_m\mathbf{x}^{(m)}$$

of these solutions with arbitrary coefficients $C_1, C_2, \ldots, C_m$ is also a solution of (57); thus, the equation

$$A\mathbf{x} = \lambda_1\mathbf{x}$$

has a family of solutions corresponding to the subspace spanned by the vectors $\mathbf{x}^{(1)}, \mathbf{x}^{(2)}, \ldots, \mathbf{x}^{(m)}$. We can choose as a basis in this subspace any system of m orthonormal vectors, whose components will then give us the columns of a unitary matrix U corresponding to the eigenvalue λ_1, i.e., we have the same arbitrariness in choosing the matrix U as we had in Sec. 33 in choosing the matrix B. Moreover, we can obviously multiply the components of every vector $\mathbf{x}^{(s)}$ obtained as a solution of (57) by any numerical factor of absolute value 1, i.e., by any (phase) factor of the form $e^{i\varphi}$; if we do this, the length of $\mathbf{x}^{(s)}$ will still be equal to 1 and $\mathbf{x}^{(s)}$ will still be orthogonal to all the other vectors in the complete system of solutions of (57). Finally, we can rearrange the columns of U in any way we please, since obviously this transformation leads only to a renumbering of the basis vectors in the new coordinate system, with a consequent rearrangement of the numbers λ_k in the diagonal form of the matrix A. Thus, we can henceforth assume that these numbers appear in increasing order.

We now consider Hermitian forms. The *Hermitian form* corresponding to the Hermitian matrix A is defined as

$$A(\mathbf{x}) = (A\mathbf{x}, \mathbf{x}) = \sum_{i,k=1}^{n} a_{ik}x_k\bar{x}_i, \tag{59}$$

where $x_1, x_2, \ldots, x_n$ are the components of the vector $\mathbf{x}$. As before, A is regarded as a linear transformation which when applied to a vector $\mathbf{x}$ gives a new vector $\mathbf{x}' = A\mathbf{x}$, and (59) associates a real number $A(\mathbf{x})$ with every vector $\mathbf{x}$. Suppose we carry out a unitary transformation U such that the old components of a vector are expressed in terms of the new components by the formula $\mathbf{x} = U\mathbf{x}'$. Then, in the new coordinates, the Hermitian form (59) becomes

$$(AU\mathbf{x}', U\mathbf{x}').$$

Using the property

$$(U\mathbf{x}, U\mathbf{x}') = (\mathbf{x}, \mathbf{x}')$$

of unitary transformation (Sec. 29), we can multiply both vectors in the scalar product by the unitary matrix U^{-1} from the left; the result is the expression

$$(U^{-1}AU\mathbf{x}', \mathbf{x}') \tag{60}$$

for the Hermitian form (59) in the new coordinates. In particular, if U transforms the matrix A into diagonal form, i.e., if (56) holds, then the only terms left in our Hermitian form when it is expressed in the new variables are the terms containing the products $\bar{x}_i'x_i' = |x_i'|^2$, so that we have reduced the form to a sum of squares:

$$(\mathbf{x}', U^{-1}AU\mathbf{x}') = \lambda_1|x_1|^2 + \lambda_2|x_2|^2 + \cdots + \lambda_n|x_n|^2.$$

Thus here, as in Sec. 32, the problem of reducing the matrix A to diagonal form is equivalent to the problem of reducing the corresponding Hermitian form to a sum of squares.

Sometimes, instead of Hermitian forms, one studies so-called *bilinear forms,* defined by

$$(A\mathbf{x}, \mathbf{y}) = \sum_{i,k=1}^{n} a_{ik}x_k\bar{y}_i.$$

If we again apply the unitary transformation $\mathbf{x} = U\mathbf{x}'$, we find that

$$(A\mathbf{x}, \mathbf{y}) = (AU\mathbf{x}', U\mathbf{y}')$$

in the new coordinates, and by the same argument as before, we have

$$(A\mathbf{x}, \mathbf{y}) = (U^{-1}AU\mathbf{x}', \mathbf{y}').$$

Thus, if U reduces A to diagonal form, the bilinear form becomes simply

$$\sum_{k=1}^{n} \lambda_k x_k'\bar{y}_k'$$

in the new coordinates.

We note that every diagonal matrix with real elements is Hermitian, and therefore the matrix $U^{-1}[\lambda_1, \ldots, \lambda_n]U$ is also Hermitian, where U

is an arbitrary unitary matrix. Conversely, as we have just seen, *every* Hermitian matrix can be written in the form $U^{-1}[\lambda_1, \ldots, \lambda_n]U$.

Just like real quadratic forms, Hermitian forms can be classified according to the sign of the eigenvalues λ_k. For example, if all the λ_k are positive, a Hermitian form is called *positive definite* and is characterized by the fact that it is positive for all nonzero x_k and vanishes only for $x_1 = x_2 = \cdots = x_n = 0$. *Semidefinite* and *indefinite* Hermitian forms are defined similarly and are studied by analogy with the case of real quadratic forms, i.e., by using the diagonal representation

$$(A\mathbf{x}, \mathbf{x}) = \lambda_1|x_1'|^2 + \lambda_2|x_2'|^2 + \cdots + \lambda_n|x_n'|^2.$$

Finally, we note the generalization of the relation (54) for matrices which are not Hermitian. If A is any matrix and $\tilde{A} = \bar{A}^*$ is its adjoint, then instead of (54) we have

$$(A\mathbf{x}, \mathbf{y}) = (\mathbf{x}, \tilde{A}\mathbf{y}). \tag{61}$$

If A has the elements a_{ik}, then $\tilde{A}$ has the elements $\{\tilde{A}\}_{ik} = \bar{a}_{ki}$, and (61) is verified by direct substitution, just as in the case of (54).

41. Commuting Hermitian Matrices. Let A and B be two Hermitian matrices. To find a condition for the product matrix BA to be Hermitian also, we write the adjoint of BA, obtaining

$$(\overline{BA})^* = \bar{A}^* \bar{B}^*.$$

Since A and B are Hermitian, this becomes

$$(\overline{BA})^* = AB,$$

so that a necessary and sufficient condition for BA to be Hermitian is that $AB = BA$, i.e., that A and B *commute.*

Suppose that the Hermitian matrices A and B are reduced to diagonal form by the same unitary transformation U:

$$A = U^{-1}[\lambda_1, \ldots, \lambda_n]U, \qquad B = U^{-1}[\mu_1, \ldots, \mu_n]U.$$

Then, it is easy to see that they commute, since

$$AB = BA = U^{-1}[\lambda_1\mu_1, \ldots, \lambda_n\mu_n]U,$$

i.e., a *necessary* condition for the matrices A and B to be simultaneously diagonalizable by the same unitary transformation is that they commute. We now prove that this condition is also *sufficient*, i.e., *if two Hermitian matrices commute, then they can be simultaneously reduced to diagonal form by the same unitary transformation.* The proof is as follows: If $AB = BA$, the similar matrices $C^{-1}AC$ and $C^{-1}BC$ also commute, since

$$(C^{-1}AC)(C^{-1}BC) = C^{-1}ABC = C^{-1}BAC = (C^{-1}BC)(C^{-1}AC).$$

Thus, if we take C to be the unitary transformation which diagonalizes A and subject B to the same transformation, the new matrices will still commute, so that we can assume that A is already in diagonal form, i.e., that the elements a_{ik} of A satisfy the condition

$$a_{ik} = 0 \qquad \text{for } i \neq k. \tag{62}$$

We denote the elements of B by b_{ik} and write the condition that A and B commute:

$$\sum_{s=1}^{n} a_{is}b_{sk} = \sum_{s=1}^{n} b_{is}a_{sk} \qquad (i, k = 1, 2, \ldots, n).$$

By (62), this condition becomes

$$(a_{ii} - a_{kk})b_{ik} = 0 \qquad (i, k = 1, 2, \ldots, n). \tag{63}$$

If the numbers a_{ii} are all different, then it follows at once from (63) that $b_{ik} = 0$ if $i \neq k$, i.e., B is also diagonal, and our assertion is proved.

Consider now the more general case where some of the a_{ii} are the same. To be explicit, we assume that the a_{ii} separate into two groups, each consisting of identical numbers:

$$a_{11} = \cdots = a_{mm}, \; a_{m+1,m+1} = \cdots = a_{nn}.$$

It follows at once from (63) that in this case the elements b_{ik} can be nonzero only when both indices i and k are greater than m or neither is greater than m. Thus, in this case, the matrix B has the quasi-diagonal form

$$B = [B_1, B_2],$$

where B_1 is a Hermitian matrix of order m, and B_2 is a Hermitian matrix of order $n - m$. In expanded form, the matrix B is

$$\begin{Vmatrix} b_{11} & \cdots & b_{1m} & 0 & \cdots & 0 \\ \cdot & \cdot & \cdot & \cdot & \cdot & \cdot \\ b_{m1} & \cdots & b_{mm} & 0 & \cdots & 0 \\ 0 & \cdots & 0 & b_{m+1,m+1} & \cdots & b_{m+1,n} \\ \cdot & \cdot & \cdot & \cdot & \cdot & \cdot \\ 0 & \cdots & 0 & b_{n,m+1} & \cdots & b_{nn} \end{Vmatrix}.$$

Without changing the diagonal form of A, we can subject the subspace spanned by the first m basis vectors to any unitary transformation, and the same applies to the subspace spanned by the last $n - m$ basis vectors. Choose these unitary transformations V_1 and V_2 in such a way as to diagonalize the matrices B_1 and B_2, and consider the overall transformation of the whole n-dimensional space which has the quasi-diagonal form

$[V_1, V_2]$. As already remarked, the matrix A preserves its diagonal form in the new coordinates, while the matrix B becomes

$$[V_1, V_2]^{-1}[B_1, B_2][V_1, V_2] = [V_1^{-1}B_1V_1, V_2^{-1}B_2V_2],$$

i.e., B is also reduced to diagonal form, which completes the proof of our assertion.

Now let A and B commute, and write the equations

$$A\mathbf{x} = \lambda\mathbf{x}, \qquad B\mathbf{x} = \mu\mathbf{x}.$$

Then it follows at once from what has just been proved that we can construct the same set of n linearly independent solutions for both of these equations. Moreover, these solutions will give the columns of the unitary matrix U, which reduces both matrices to diagonal form. In other words, we can construct the same system of n linearly independent eigenvectors for two commuting Hermitian matrices. As for the eigenvalues (i.e., the values of the parameters λ and μ), they will in general be different. We note that this still does not imply that *every* eigenvector of the matrix A is also an eigenvector of the matrix B. If all the eigenvalues of A and B are different, so that every value of λ_k and μ_k corresponds to just one vector (to within a numerical factor), then in fact every eigenvector of A is an eigenvector of B. However, this is in general no longer the case if some of the eigenvalues are the same. To see this, let $\mathbf{x}^{(k)}$ $(k = 1, \ldots, n)$ be a complete system of eigenvectors of the matrices A and B, let λ_k and μ_k be the corresponding eigenvalues, and assume, for example, that $\lambda_1 = \lambda_2$ while $\mu_1 \neq \mu_2$. Then, the vectors $C_1\mathbf{x}^{(1)} + C_2\mathbf{x}^{(2)}$ will be eigenvectors of A for any choice of the constants C_1 and C_2, but they will not be eigenvectors of B.

All the above considerations can easily be generalized to the case of several matrices. Thus, if we have several Hermitian matrices $A_1, A_2, \ldots, A_l$, then they are all simultaneously diagonalizable by the same unitary transformation if and only if they commute in pairs, i.e., $A_iA_k = A_kA_i$ for any i and k from 1 to l.

42. Reduction of Unitary Matrices to Diagonal Form. As regards reduction to diagonal form, unitary matrices have a property completely analogous to Hermitian matrices, i.e., if V is a unitary matrix, then another unitary matrix U can always be found such that the matrix $U^{-1}VU$ is diagonal. The problem of finding this matrix can be posed by writing

$$VU = U[\lambda_1, \lambda_2, \ldots, \lambda_k], \tag{64}$$

where the matrix U and the numbers λ_k are to be determined. As in the case of a Hermitian matrix (Sec. 40), the columns of the matrix U will correspond to certain vectors $\mathbf{x}^{(k)}$ which have to be solutions of the equation

$$V\mathbf{x} = \lambda\mathbf{x}, \tag{65}$$

for the values $\lambda = \lambda_k$. Just as before, this implies at once that the numbers λ_k must be roots of the characteristic equations

$$\begin{vmatrix} v_{11} - \lambda & v_{12} & \cdots & v_{1n} \\ v_{21} & v_{22} - \lambda & \cdots & v_{2n} \\ \cdots & \cdots & \cdots & \cdots \\ v_{n1} & v_{n2} & \cdots & v_{nn} - \lambda \end{vmatrix} = 0, \tag{66}$$

where the v_{ik} are the elements of the matrix V. Before proceeding, we note that if the matrices V_1 and U_1 are unitary, then so is the matrix $U_1^{-1}V_1U_1$. In fact, since U_1 is unitary, so is U_1^{-1}, and a product of unitary matrices is also unitary.

Now let $\lambda = \lambda_1$ be a root of (66). Substituting λ_1 for λ in (65), we determine a unit vector $\mathbf{x}^{(1)}$ satisfying (65). We then choose a new set of basis vectors consisting of n orthonormal vectors, the first of which is $\mathbf{x}^{(1)}$. The transformation from the old basis vectors to the new basis vectors is equivalent to a unitary transformation U_1, which transforms our unitary matrix V into the similar matrix

$$V_1 = U_1^{-1}VU_1.$$

For $\lambda = \lambda_1$, the corresponding equation

$$V_1\mathbf{x} = \lambda\mathbf{x}$$

has for a solution the vector with components $1, 0, \ldots, 0$. From this, it immediately follows, just as before, that the elements of the first column of the matrix V_1 are all zero except for the first element which equals λ_1. Since, in a unitary matrix, the sum of the squares of the absolute values of the elements in each *column* equals 1 (Sec. 28), we can assert that λ_1 has absolute value 1. Next, we recall that the sum of the squares of the absolute values of the elements in each *row* of the unitary matrix V_1 also equals 1. But, as just shown, the first element of the first row has absolute value 1. Therefore, all the other elements of this row must be zero. Thus, as a result of the first unitary transformation U_1, we have reduced our unitary matrix V to the form

$$\begin{Vmatrix} \lambda_1 & 0 & \cdots & 0 \\ 0 & v_{22}^{(1)} & \cdots & v_{2n}^{(1)} \\ \cdots & \cdots & \cdots & \cdots \\ 0 & v_{n2}^{(1)} & \cdots & v_{nn}^{(1)} \end{Vmatrix},$$

in which every element of the first row and first column vanishes except the first element. This is completely analogous to the situation which we had previously for Hermitian matrices. Moreover, the elements $v_{ik}^{(1)}$ themselves form a unitary matrix of order $n - 1$. Applying another unitary transformation, we can also transform this matrix into a matrix in which every element of the first row and first column vanishes except the first element, which has absolute value 1. Then, as a result of applying both of the above unitary transformations, our unitary matrix V is reduced to the form

$$\left\| \begin{matrix} \lambda_1 & 0 & 0 & \cdots & 0 \\ 0 & \lambda_2 & 0 & \cdots & 0 \\ 0 & 0 & v_{33}^{(2)} & \cdots & v_{3n}^{(2)} \\ \cdots & \cdots & \cdots & \cdots & \cdots \\ 0 & 0 & v_{n3}^{(2)} & \cdots & v_{nn}^{(2)} \end{matrix} \right\|.$$

Continuing this procedure, we finally reduce our unitary matrix to diagonal form by using several unitary transformations in succession. We note that it follows from these considerations that *all the eigenvalues of a unitary matrix have absolute value unity.* Moreover, just as in Sec. 41, we can show that if several unitary matrices commute in pairs, they can be simultaneously diagonalized by the same unitary transformation.

The following fact should be noted: Suppose the unitary matrix U diagonalizes a matrix A, i.e., suppose $U^{-1}AU$ is a diagonal matrix. As we know, the absolute value of the determinant of U equals 1, so that we can select a real number ω such that the determinant of the unitary matrix $e^{i\omega}U$ is unity. Then the unitary matrix $e^{i\omega}U$ will also reduce A to diagonal form, since

$$(e^{i\omega}U)^{-1}A(e^{i\omega}U) = e^{i\omega}e^{-i\omega}U^{-1}AU = U^{-1}AU.$$

Thus, we can always assume that the determinant of a unitary matrix diagonalizing another matrix equals 1.

EXAMPLE. As an example, we consider the reduction to diagonal form of a real orthogonal matrix:

$$V = \left\| \begin{matrix} v_{11} & v_{12} & v_{13} \\ v_{21} & v_{22} & v_{23} \\ v_{31} & v_{32} & v_{33} \end{matrix} \right\| \tag{67}$$

of order 3. We assume that the determinant of V equals $+1$, so that V corresponds to a rotation of the three-dimensional space as a whole about the origin. By hypothesis, the characteristic equation of the matrix (67) has a constant term equal to 1, since clearly this term is just the determinant of the matrix. On the other hand, we have seen that all the roots of our characteristic equation must have absolute value 1. The leading term of the characteristic equation is $(-\lambda)^3 = -\lambda^3$, and therefore the constant term 1 must equal the product of all the roots of the characteristic equation. Since this equation has real coefficients, only two cases are possible: Either one root equals 1 and the other two roots are purely imaginary and complex conjugates of each other, i.e., the other two roots are $e^{\pm i\varphi}$, or one root equals 1 and the other two roots equal -1. (The second case reduces to the first case for $\varphi = \pi$.)

Corresponding to the eigenvalue $\lambda = 1$ is the real vector $\mathbf{x}^{(1)}$, which must be a solution of the equation

$$V\mathbf{x}^{(1)} = \mathbf{x}^{(1)}, \tag{68}$$

i.e., $\mathbf{x}^{(1)}$ is invariant under the spatial rotation determined by the matrix V. Clearly, this real vector $\mathbf{x}^{(1)}$, corresponding to the value $\lambda = 1$, defines the axis about which the rotation is performed. (Every rotation about the origin is equivalent to a rotation about an axis passing through the origin.) To find the components of $\mathbf{x}^{(1)}$ in terms of the elements of the matrix V, we rewrite (68) in the form

$$V^{-1}\mathbf{x}^{(1)} = \mathbf{x}^{(1)};$$

since V is real and unitary, this can be written as

$$V^*\mathbf{x}^{(1)} = \mathbf{x}^{(1)}. \tag{69}$$

Subtracting (69) from (68), we obtain

$$(V - V^*)\mathbf{x}^{(1)} = 0. \tag{70}$$

We now denote the components of $\mathbf{x}^{(1)}$ by (u_{11}, u_{21}, u_{31}) and write (70) in expanded form. The result is the system of equations

$$\begin{aligned} (v_{12} - v_{21})u_{21} + (v_{13} - v_{31})u_{31} &= 0, \\ (v_{21} - v_{12})u_{11} \qquad\qquad\qquad + (v_{23} - v_{32})u_{31} &= 0, \\ (v_{31} - v_{13})u_{11} + (v_{32} - v_{23})u_{21} \qquad\qquad\qquad &= 0, \end{aligned}$$

which immediately lead to the following formula determining the direction of the axis of rotation:

$$u_{11}:u_{21}:u_{31} = (v_{23} - v_{32}):(v_{31} - v_{13}):(v_{12} - v_{21}).$$

The other two eigenvectors $\mathbf{x}^{(2)}$ and $\mathbf{x}^{(3)}$ must satisfy the equations

$$V\mathbf{x}^{(2)} = e^{i\varphi}\mathbf{x}^{(2)}, \quad V\mathbf{x}^{(3)} = e^{-i\varphi}\mathbf{x}^{(3)}, \tag{71}$$

and these will now be vectors with complex components. We can determine φ from the condition that the sum of the roots of the characteristic equation must equal the sum of the diagonal elements of (67), i.e., the trace of the matrix V (Sec. 27). Thus, we have

$$1 + e^{i\varphi} + e^{-i\varphi} = 1 + 2\cos\varphi = v_{11} + v_{22} + v_{33},$$

where it can be assumed that φ lies between 0 and π. It follows from (71) that the components of the vector $\mathbf{x}^{(2)}$ are the complex conjugates of those of $\mathbf{x}^{(3)}$, since the two values of λ in (71) are complex conjugates. We now form a new unitary matrix

$$U_0 = \left\| \begin{matrix} 1 & 0 & 0 \\ 0 & \frac{1}{\sqrt{2}} & \frac{i}{\sqrt{2}} \\ 0 & \frac{1}{\sqrt{2}} & -\frac{i}{\sqrt{2}} \end{matrix} \right\|. \tag{72}$$

It is easy to verify by direct calculation that the elements of the columns of the matrix $W = UU_0$ are just the components of the vectors

$$\mathbf{x}^{(1)}, \qquad \frac{\mathbf{x}^{(2)} + \mathbf{x}^{(3)}}{\sqrt{2}}, \qquad i\,\frac{\mathbf{x}^{(2)} - \mathbf{x}^{(3)}}{\sqrt{2}},$$

and are therefore real. Moreover, the matrix W is unitary, since it is the product of two unitary matrices. It follows that W is an orthogonal matrix, being real and unitary. We now apply a similarity transformation to the matrix V, using the orthogonal matrix W. The result is

$$W^{-1}VW = U_0^{-1}U^{-1}VUU_0 = U_0^{-1}[1, e^{i\varphi}, e^{-i\varphi}]U_0.$$

Calculating the matrix product explicitly, we obtain

$$W^{-1}VW = \left\| \begin{matrix} 1 & 0 & 0 \\ 0 & \cos\varphi & -\sin\varphi \\ 0 & \sin\varphi & \cos\varphi \end{matrix} \right\|. \tag{73}$$

We can always assume that the determinant of the orthogonal matrix W equals $+1$, for otherwise we can always multiply W by -1, which does not alter the relation (73). Thus, W also corresponds to a three-dimensional rotation. The matrix (73), obtained as a result of the coordinate transformation $\mathbf{x} = W\mathbf{x}'$, is similar to the matrix V and gives the same transformation relative to the new coordinates that V gives relative to the old coordinates. From the form of the matrix (73), it is immediately clear that this matrix corresponds to a rotation through the angle φ about the new axis $\mathbf{x}'^{(1)}$, and the significance of the transformation W is that it corresponds to taking the axis of rotation determined by the vector $\mathbf{x}^{(1)}$ as the first basis vector of a new coordinate system.

The following important fact is an immediate consequence of the considerations of this section: All real matrices corresponding to a spatial rotation through a certain definite angle φ can be reduced by a similarity transformation (which is different for different matrices) to the same form (73); therefore, all these matrices are similar to one another. However, matrices corresponding to different angles of rotation cannot be similar, since their eigenvalues 1, $e^{i\varphi}$, $e^{-i\varphi}$ are certainly different for different values of the angle φ. All these properties have a very simple geometrical meaning.

43. Projection Matrices. We now consider Hermitian matrices of a special type, the so-called *projection matrices.* Let R_m be a subspace of dimension m spanned by m linearly independent vectors $\mathbf{y}^{(1)}, \ldots, \mathbf{y}^{(m)}$, i.e., the subspace consisting of all vectors of the form

$$C_1\mathbf{y}^{(1)} + \cdots + C_m\mathbf{y}^{(m)},$$

where $C_1, \ldots, C_m$ are arbitrary numerical coefficients. By orthogonalizing the vectors $\mathbf{y}^{(k)}$, we can construct m orthonormal vectors

$$\mathbf{x}^{(1)}, \ldots, \mathbf{x}^{(m)},$$

which span the same subspace R_m. This system of vectors can be extended to a complete system of n orthonormal vectors by constructing $n - m$ additional unit vectors

$$\mathbf{x}^{(m+1)}, \ldots, \mathbf{x}^{(n)}$$

which span a subspace R'_{n-m} of dimension $n - m$. The two subspaces R_m and R'_{n-m} are orthogonal in the sense that any vector in R_m is orthogonal to any vector in R'_{n-m} (Sec. 14). The expansion

$$\mathbf{x} = x_1\mathbf{x}^{(1)} + \cdots + x_n\mathbf{x}^{(n)}$$

of an arbitrary vector $\mathbf{x}$ with respect to the basis vectors $\mathbf{x}^{(k)}$ can be written in the form

$$\mathbf{x} = [x_1\mathbf{x}^{(1)} + \cdots + x_m\mathbf{x}^{(m)}] + [x_{m+1}\mathbf{x}^{(m+1)} + \cdots + x_n\mathbf{x}^{(n)}] = \mathbf{u} + \mathbf{v}, \tag{74}$$

i.e., as the sum of a vector $\mathbf{u}$ in R_m and a vector $\mathbf{v}$ in R'_{n-m}.

It is easy to see that the representation (74) of the vector $\mathbf{x}$ is unique; for suppose that in addition to (74), we have another representation $\mathbf{x} = \mathbf{u}' + \mathbf{v}'$, where $\mathbf{u}' \in R_m$ and $\mathbf{v}' \in R_{n-m}$.† This implies that

$$\mathbf{u} + \mathbf{v} = \mathbf{u}' + \mathbf{v}'$$

or

$$\mathbf{u} - \mathbf{u}' = \mathbf{v} - \mathbf{v}',$$

where $\mathbf{u} - \mathbf{u}'$ and $\mathbf{v} - \mathbf{v}'$ are orthogonal, since $\mathbf{u} - \mathbf{u}' \in R_m$, $\mathbf{v} - \mathbf{v}' \in R'_{n-m}$. Therefore,

$$0 = (\mathbf{u} - \mathbf{u}', \mathbf{v} - \mathbf{v}') = \|\mathbf{u} - \mathbf{u}'\|^2 = \|\mathbf{v} - \mathbf{v}'\|^2,$$

so that $\mathbf{u} = \mathbf{u}'$, $\mathbf{v} = \mathbf{v}'$, i.e., the vector $\mathbf{x}$ uniquely determines the vectors $\mathbf{u}$ and $\mathbf{v}$. The vector $\mathbf{u}$ is called the *projection of the vector* $\mathbf{x}$ *onto the subspace* R_m; the matrix which transforms $\mathbf{x}$ into $\mathbf{u}$ is called a *projection*

† The symbol $\in$ means *is a member of.*

matrix (corresponding to projection onto the subspace R_m) and is denoted by P_{R_m}.

The form of a projection matrix obviously depends on the choice of coordinate axes. If we choose the $\mathbf{x}^{(k)}$ as basis vectors, then $\mathbf{x}$ has the representation (74) and $\mathbf{u}$ has the representation

$$\mathbf{u} = x_1\mathbf{x}^{(1)} + \cdots + x_m\mathbf{x}^{(m)};$$

thus, in this case, the operation of projection reduces to just leaving the first m components of $\mathbf{x}$ unchanged and setting the other components equal to zero. The corresponding projection matrix is obviously a diagonal matrix of the form

$$P_{R_m} = [1, \ldots, 1, 0, \ldots, 0],$$

where the first m entries are all 1's and the rest are all 0's. If we renumber the basis vectors, we get a different order of the elements of P_{R_m}, but we still have a diagonal matrix consisting only of 1's and 0's. More generally, when the choice of Cartesian axes is arbitrary, the projection matrix has the form

$$P_{R_m} = U^{-1}[1, \ldots, 1, 0, \ldots, 0]U, \tag{75}$$

where U is a unitary matrix; thus, the eigenvalues of P_{R_m} are either 0 or 1. Conversely, every Hermitian matrix of this type is a projection matrix corresponding to projection onto a subspace whose dimension is just the number of eigenvalues of P_{R_m} which equal 1.

An alternative definition of a projection matrix is the following: *A projection matrix is a Hermitian matrix which satisfies the condition*

$$P^2 = P. \tag{76}$$

In fact, since $1^2 = 1$ and $0^2 = 0$, it is easy to see that a matrix of the form (75) satisfies the relation (76). Conversely, let

$$P = U^{-1}[\lambda_1, \ldots, \lambda_n]U$$

be a Hermitian matrix satisfying (76). Then

$$U^{-1}[\lambda_1^2, \ldots, \lambda_n^2]U = U^{-1}[\lambda_1, \ldots, \lambda_n]U,$$

i.e., $\lambda_k^2 = \lambda_k$, from which it follows at once that $\lambda_k = 1$ or 0. If all the eigenvalues of a matrix equal 1, then the matrix is the unit matrix corresponding to the identity transformation, which leaves vectors unchanged, i.e., projects them onto the whole space. With the exception of this trivial case, a projection matrix P has at least one eigenvalue equal to zero; therefore, its determinant, which equals the product of its eigenvalues, is also zero, and the inverse matrix P^{-1} does not exist. We also note that it is an immediate consequence of the definition of the projection matrix P_{R_m} that P_{R_m} leaves vectors which belong to R_m unchanged and decreases the lengths of vectors which do not belong to R_m.

Having dispensed with these preliminaries, we now consider some operations involving projection matrices. Suppose P_R and P_S are two projection matrices whose product is zero, so that

$$P_S P_R = 0, \tag{77}$$

where the right-hand side denotes the matrix all of whose elements are zero. If $\mathbf{x}$ is a vector in the subspace R, then $P_R\mathbf{x} = \mathbf{x}$ and (77) gives

$$P_S\mathbf{x} = 0. \tag{78}$$

This implies that $\mathbf{x}$ is orthogonal to any vector in the subspace S. Otherwise, we could choose as our first basis vector a unit vector $\mathbf{y}$ in S which is not orthogonal to $\mathbf{x}$; then the first component of $\mathbf{x}$, which remains unchanged when $\mathbf{x}$ is projected onto S, would be nonzero, contradicting (78). Thus, we see that if (77) holds, every vector in R is orthogonal to every vector in S; then we have

$$P_R P_S = 0, \tag{79}$$

as well as (77). In fact, $P_S\mathbf{y}$ belongs to S for any $\mathbf{y}$, and therefore $P_S\mathbf{y}$ is orthogonal to all the vectors in R, i.e.,

$$P_R P_S \mathbf{y} = 0$$

for any vector $\mathbf{y}$, which is equivalent to (79). Conversely, if two subspaces R and S are orthogonal (in the sense indicated above), then (77) and (79) hold.

Next, we consider the sum

$$P = P_R + P_S \tag{80}$$

of two projection matrices, assuming that the conditions (77) and (79) hold. The matrix (80), which is obviously Hermitian, is also a projection matrix, i.e., equals its own square. To see this, we write

$$P^2 = (P_R + P_S)(P_R + P_S) = P_R^2 + P_R P_S + P_S P_R + P_S^2.$$

In view of (77) and (79) and the fact that P_R and P_S are projection matrices, we have

$$P^2 = P_R + P_S = P.$$

It is not hard to show that in this case the matrix P corresponds to the operation of projection onto the subspace $R + S$, which is the sum of the subspaces R and S in the sense that $R + S$ is spanned by all the vectors spanning R and S separately. In other words, if the vectors $\mathbf{x}^{(1)}, \ldots,$

$\mathbf{x}^{(p)}$ span R and the vectors $\mathbf{y}^{(1)}, \ldots, \mathbf{y}^{(q)}$ span S, then $R + S$ consists of all vectors of the form

$$C_1\mathbf{x}^{(1)} + \cdots + C_p\mathbf{x}^{(p)} + D_1\mathbf{y}^{(1)} + \cdots + D_q\mathbf{y}^{(q)},$$

where the C_k and D_k are arbitrary constants. This property can be generalized to the case of any number of summands. Thus, if the subspaces $S_1, \ldots, S_m$ are pairwise orthogonal, i.e., if any vector in S_i is orthogonal to any vector in S_j for all $i \neq j$, then the sum

$$P = P_{S_1} + \cdots + P_{S_m}$$

represents the projection matrix onto the subspace $S_1 + \cdots + S_m$, spanned by all the vectors spanning the separate subspaces $S_1, \ldots, S_m$. In particular, this sum can equal the unit matrix, i.e.,

$$P_{S_1} + \cdots + P_{S_m} = I,$$

and then one usually speaks of a resolution of the unit matrix into projection matrices, or simply of a *resolution of the identity*.

Next, we consider a product

$$P = P_S P_R \tag{81}$$

of two projection matrices. In order for this product to be also a projection matrix, it is first necessary that it be a Hermitian matrix, and for this to be the case, it is necessary that the matrices commute (Sec. 41):

$$P_R P_S = P_S P_R. \tag{82}$$

Thus, (82) is a necessary condition for (81) to be a projection matrix. Moreover, (82) is also a sufficient condition for (81) to be a projection matrix, i.e., if (82) holds, then $P^2 = P$. To see this, we write

$$P^2 = P_S P_R P_S P_R$$

and use (82) to interchange matrices, obtaining

$$P^2 = P_S^2 P_R^2 = P_S P_R = P,$$

as required. It is not hard to see that if P_R and P_S commute, then $P = P_S P_R$ corresponds to projection onto the subspace which is the intersection of the subspaces R and S, i.e., which consists of the vectors belonging to both R and S.

Finally, we mention another result, without giving its proof (which is quite simple): If the subspace S is contained in the subspace R, then the difference

$$P = P_R - P_S \tag{83}$$

is also a projection matrix. If $\mathbf{x}^{(k)}$ $(k = 1, \ldots, q)$ are vectors spanning S, then to obtain vectors spanning R, we have to supplement the $\mathbf{x}^{(k)}$ with one or more additional linearly independent vectors. These additional vectors themselves span a subspace T, and the matrix (83) is the projection matrix corresponding to projection onto T.

Using projection matrices, we can give a unique formulation of the problem of reducing a Hermitian matrix to diagonal form, even when multiple roots occur. Suppose, for example, that we have a Hermitian matrix

$$A = U[\lambda_1, \ldots, \lambda_n]U^{-1},$$

where U is some unitary matrix. To be explicit, we assume that the λ_k are divided into two groups of identical numbers, i.e.,

$$A = U[\mu, \ldots, \mu, \nu, \ldots, \nu]U^{-1},$$

where the first m numbers equal μ and the remaining $n - m$ numbers equal ν. Clearly, we can write this matrix in the form

$$A = \mu U[1, \ldots, 1, 0, \ldots, 0]U^{-1} + \nu U[0, \ldots, 0, 1, \ldots, 1]U^{-1}.$$

We now introduce the projection matrices

$$P_R = U[1, \ldots, 1, 0, \ldots, 0]U^{-1}, P_S = U[0, \ldots, 0, 1, \ldots, 1]U^{-1},$$

whose sum is the unit matrix; the corresponding subspaces R and S are obviously orthogonal. Then we have

$$A = \mu P_R + \nu P_S,$$

where $\lambda_1 = \cdots = \lambda_m = \mu, \qquad \lambda_{m+1} = \cdots = \lambda_n = \nu.$

More generally, the problem of reducing a Hermitian form to diagonal form reduces to finding a resolution of the identity

$$I = P_{S_1} + \cdots + P_{S_m}, \tag{84}$$

such that our matrix A can be represented in the form

$$A = \mu_1 P_{S_1} + \cdots + \mu_m P_{S_m},$$

where the μ_k are the different eigenvalues of A.

It is not hard to restate all these results in the language of Hermitian forms. Every projection matrix P_R, with elements p_{ik}, corresponds to a Hermitian form

$$P_R(\mathbf{x}) = (P_R\mathbf{x}, \mathbf{x}) = \sum_{i,k=1}^{n} p_{ik}x_k\bar{x}_i, \tag{85}$$

which is sometimes called an *Einzelform*. Let the subspace R be m-dimensional, and choose m orthonormal vectors in R as the first m basis vectors. Then, in this coordinate system, (85) takes the form

$$(P_R\mathbf{x}', \mathbf{x}') = |x_1'|^2 + |x_2'|^2 + \cdots + |x_m'|^2.$$

Similarly, given a resolution of the identity (84) and choosing orthonormal vectors in each subspace S_k for the basis vectors, we clearly have

$$\sum_{k=1}^{m} P_{S_k}(\mathbf{x}') = \sum_{i=1}^{n} |x_i'|^2,$$

and therefore, the sum

$$\sum_{k=1}^{m} P_{S_k}(\mathbf{x})$$

equals the square of the length of the vector $\mathbf{x}$ in *any* coordinate system. Thus, we see that the problem of reducing a Hermitian form to a sum of squares is described by the two equations

$$A(\mathbf{x}) = \sum_{k=1}^{m} \mu_k P_{S_k}(\mathbf{x}),$$

$$\|\mathbf{x}\|^2 = \sum_{k=1}^{m} P_{S_k}(\mathbf{x}),$$

so that the introduction of projection matrices allows us to formulate the problem of reducing a Hermitian matrix to diagonal form without recourse to any special coordinate systems. In particular, this approach allows us to extend our results with appropriate modifications to the case of infinite-dimensional spaces. Indeed, this leads us to the basic mathematical problem of modern quantum mechanics, with which we shall be concerned in a later volume. This extension to the infinite-dimensional case cannot be made in an algebraic context, but rather is intimately connected with the introduction of analytic techniques.

44. Functions of Matrices. We now study functions of matrix arguments, restricting ourselves to the simplest cases, i.e., *polynomial and rational functions of matrices.* We defer a more detailed treatment of the theory of matrix functions until after we have introduced the theory of functions of a complex variable (in a later volume).† A polynomial $f(A)$ of degree m in the variable matrix A has the form

$$f(A) = c_0 + c_1A + \cdots + c_mA^m, \tag{86}$$

where the c_k are numerical coefficients. In this case, the value of the

† See, however, Probs. 6, 7 of Chap. 6, and Probs. 4 to 6 of the Appendix.

function is also a matrix, whose elements are obviously given by the formula

$$\{f(A)\}_{ik} = c_0\delta_{ik} + c_1\{A\}_{ik} + \cdots + c_m\{A^m\}_{ik},$$

where

$$\delta_{ik} = 0 \quad \text{for } i \neq k,\ \delta_{ii} = 1.$$

We can also consider polynomials in several matrices, but here we must keep in mind the fact that matrix multiplication is noncommutative. The general form of a second degree polynomial in two variable matrices A and B is

$$f(A, B) = c_0 + c_1A + c_2B + c_3A^2 + c_4B^2 + c_5AB + c_6BA.$$

Suppose that we replace the matrix A in (86) by a similar matrix $U^{-1}AU$. Bearing in mind that $(U^{-1}AU)^k = U^{-1}A^kU$, we have

$$f(U^{-1}AU) = c_0 + c_1U^{-1}AU + \cdots + c_mU^{-1}A^mU$$
$$= U^{-1}(c_0 + c_1A + \cdots + c_mA^m)U,$$

i.e.,

$$f(U^{-1}AU) = U^{-1}f(A)U. \tag{87}$$

A similar formula, which also holds for a polynomial in several matrices, is

$$f(U^{-1}AU, U^{-1}BU) = U^{-1}f(A, B)U.$$

We now treat the case of Hermitian matrices in somewhat more detail. It follows at once from the definition of a Hermitian matrix A that any positive power A^k and any product cA, where c is real, are also Hermitian matrices; moreover, a sum of Hermitian matrices is also Hermitian. This implies that if the matrix A in (86) is Hermitian and if the coefficients c_k are real, then the function $f(A)$ is also a Hermitian matrix. This Hermitian matrix $f(A)$ obviously commutes with A, so that both A and $f(A)$ can be diagonalized by the same unitary transformation. In fact, let V be the unitary transformation which diagonalizes A, i.e.,

$$A = V[\lambda_1, \ldots, \lambda_n]V^{-1}.$$

By (87), we have

$$f(A) = Vf([\lambda_1, \ldots, \lambda_n])V^{-1}. \tag{88}$$

If we replace A in (86) by the diagonal matrix $[\lambda_1, \ldots, \lambda_n]$, then the result is obviously the diagonal matrix

$$\sum_{k=0}^{m} c_k[\lambda_1^k, \ldots, \lambda_n^k] = [f(\lambda_1), \ldots, f(\lambda_n)],$$

where $f(\lambda_k)$ is the value of the polynomial $f(A)$ when the matrix A is replaced by the number λ_k. Therefore, (88) becomes

$$f(A) = V[f(\lambda_1), \ldots, f(\lambda_n)]V^{-1},$$

i.e., V diagonalizes $f(A)$ and the numbers $f(\lambda_k)$ are the eigenvalues of $f(A)$.

Next we consider rational functions of a matrix argument. Let $f_1(A)$ and $f_2(A)$ be two polynomials in the matrix A, and consider their ratio

$$\frac{f_1(A)}{f_2(A)}. \tag{89}$$

As we have seen previously (Sec. 26), in general, the ratio of two matrices has no definite meaning, but in the present case the ratio does have a unique meaning, provided that the determinant of the matrix $f_2(A)$ does not vanish. To see this, we equate the two different versions of (89), obtaining

$$f_1(A)[f_2(A)]^{-1} = [f_2(A)]^{-1}f_1(A),$$

or equivalently,

$$f_2(A)f_1(A) = f_1(A)f_2(A). \tag{90}$$

The polynomials $f_1(A)$ and $f_2(A)$ commute, since they both contain the same matrix A, i.e., (90) is in fact valid and the ratio (89) has a definite meaning. Moreover, it is easily verified that rational functions of one matrix argument are multiplied by the same rule as ordinary ratios. In fact, we have

$$\begin{aligned}\frac{f_1(A)}{f_2(A)}\cdot\frac{f_3(A)}{f_4(A)} = f_1(A)[f_2(A)]^{-1}f_3(A)[f_4(A)]^{-1} &= f_1(A)f_3(A)[f_2(A)f_4(A)]^{-1} \\ &= \frac{f_1(A)f_3(A)}{f_2(A)f_4(A)},\end{aligned}$$

where we have made free use of the commutativity of the various matrices.

As an example, consider the rational function of the form

$$U = \frac{1 + iA}{1 - iA}, \tag{91}$$

where A is a Hermitian matrix, i.e., $\bar{A}^* = A$. It is easily verified that the matrix U is unitary, i.e., that

$$\bar{U}^* = U^{-1}. \tag{92}$$

To see this, we first write

$$\bar{U} = \frac{1 - i\bar{A}}{1 + i\bar{A}} = (1 - i\bar{A})(1 + i\bar{A})^{-1},$$

and then take transposes (Sec. 26), obtaining

$$\bar{U}^* = [(1 + i\bar{A})^*]^{-1}(1 - i\bar{A})^* = (1 + i\bar{A}^*)^{-1}(1 - i\bar{A}^*).$$

Since $\bar{A}^* = A$, this becomes

$$\bar{U}^* = (1 + iA)^{-1}(1 - iA) = \frac{1 - iA}{1 + iA} = U^{-1},$$

i.e., (92) is satisfied and U is unitary, as asserted. Since U commutes with A, we can solve (91) for A, with the result

$$A = i\frac{1 - U}{1 + U}; \tag{93}$$

in just the same way as before, it can be shown that the matrix A defined by (93) is Hermitian, if U is unitary and if the determinant of $1 + U$ is nonvanishing. Thus, any unitary matrix U for which $\det(1 + U) \neq 0$ has the representation (91) in terms of the Hermitian matrix (93).

PROBLEMS†

1. By a *quadric surface* in a real n-dimensional space is meant the locus of all points $\mathbf{x} = (x_1, x_2, \ldots, x_n)$ satisfying an equation of the form

$$\sum_{i,k=1}^{n} a_{ik}x_ix_k + 2\sum_{i=1}^{n} b_ix_i + c = 0$$

or

$$(A\mathbf{x}, \mathbf{x}) + 2(\mathbf{b}, \mathbf{x}) + c = 0,$$

where A is the matrix $\|a_{ik}\|$ (which we assume to be symmetric), $\mathbf{b}$ is the vector $(b_1, b_2, \ldots, b_n)$, and c is a constant. By a *center* of the surface is meant a point $\mathbf{x}_0$ such that if $\mathbf{x}_0 + \mathbf{x}$ lies on the surface, so does $\mathbf{x}_0 - \mathbf{x}$. Show that the surface $(A\mathbf{x}, \mathbf{x}) + 2(\mathbf{b}, \mathbf{x}) + c = 0$ has a center if and only if the rank of the matrix A equals the rank of the augmented matrix obtained by writing the components of $\mathbf{b}$ to the right of A.

2. Reduce each of the following quadratic forms to a sum of squares by making an orthogonal transformation (which is not uniquely defined):

(a) $6x_1^2 + 5x_2^2 + 7x_3^2 - 4x_1x_2 + 4x_1x_3$;
(b) $11x_1^2 + 5x_2^2 + 2x_3^2 + 16x_1x_2 + 4x_1x_2 - 20x_2x_3$;
(c) $x_1^2 + x_2^2 + 5x_3^2 - 6x_1x_2 - 2x_1x_3 + 2x_2x_3$;
(d) $x_1^2 + x_2^2 + x_3^2 + 4x_1x_2 + 4x_1x_3 + 4x_2x_3$;
(e) $17x_1^2 + 14x_2^2 + 14x_3^2 - 4x_1x_2 - 4x_1x_3 - 8x_2x_3$;
(f) $x_1^2 - 5x_2^2 + x_3^2 + 4x_1x_2 + 2x_1x_3 + 4x_2x_3$;
(g) $8x_1^2 - 7x_2^2 + 8x_3^2 + 8x_1x_2 - 2x_1x_3 + 8x_2x_3$;
(h) $2x_1x_2 - 6x_1x_3 - 6x_2x_4 + 2x_3x_4$;
(i) $5x_1^2 + 5x_2^2 + 5x_3^2 + 5x_4^2 - 10x_1x_2 + 2x_1x_3 + 6x_1x_4 + 6x_2x_3 + 2x_2x_4 - 10x_3x_4$;
(j) $3x_1^2 + 8x_1x_2 - 3x_2^2 + 4x_3^2 - 4x_3x_4 + x_4^2$;
(k) $x_1^2 + 2x_1x_2 + x_2^2 - 2x_3^2 - 4x_3x_4 - 2x_4^2$;
(l) $9x_1^2 + 5x_2^2 + 5x_3^2 + 8x_4^2 + 8x_2x_3 - 4x_2x_4 + 4x_3x_4$;
(m) $4x_1^2 - 4x_1x_2 + x_2^2 + 5x_3^2 - 4x_4^2 + 12x_4x_5 + x_5^2$;

† For hints and answers see page 439. These problems are by A. L. Shields and the translator, making some use of the books by Proskuryakov and by Faddeyev and Sominski (cited in the Preface).

(n) $4x_1^2 - 4x_2^2 - 8x_2x_3 + 2x_3^2 - 5x_4^2 + 6x_4x_5 + 3x_5^2$;
(o) $3x_1^2 + 8x_1x_2 - 3x_2^2 + 4x_3^2 - 6x_3x_4 - 4x_4^2 + 4x_5^2 + 4x_5x_6 + x_6^2$;

(p) $\sum_{i=1}^{n} x_i^2 + \sum_{i<j}^{n} x_ix_j$; (q) $\sum_{i<j}^{n} x_ix_j$.

3. Two quadratic forms are said to be *orthogonally equivalent* if we can go from each to the other by means of an orthogonal transformation. Prove that two quadratic forms are orthogonally equivalent if and only if the characteristic polynomials of their matrices are the same.
4. Which of the following quadratic forms are orthogonally equivalent?

(a) $f = 9x_2^2 + 9x_3^2 + 12x_1x_2 + 12x_1x_3 - 6x_2x_3$,
$g = -3y_1^2 + 6y_2^2 + 6y_3^2 - 12y_1y_2 + 12y_1y_3 + 6y_2y_3$,
$h = 11z_1^2 - 47z_2^2 + 11z_3^2 + 8z_1z_2 - 2z_1z_3 + 8z_2z_3$;
(b) $f = 7x_1^2 + x_2^2 + x_3^2 - 8x_1x_2 - 8x_1x_3 - 16x_2x_3$,
$g = \frac{2}{3}y_1^2 - \frac{1}{3}y_2^2 - \frac{1}{3}y_3^2 - \frac{4}{3}y_1y_2 + \frac{4}{3}y_1y_3 + \frac{8}{3}y_2y_3$,
$h = z_2^2 - z_3^2 + 2\sqrt{2}\,z_1z_3$.

5. Find the orthogonal matrix B which diagonalizes each of the following real symmetric matrices A, i.e., such that $B^{-1}AB$ is diagonal. Find the corresponding diagonal matrices:

(a) $A = \begin{Vmatrix} a & b \\ b & a \end{Vmatrix}$; (b) $A = \begin{Vmatrix} 3 & 2 & 0 \\ 2 & 4 & -2 \\ 0 & -2 & 5 \end{Vmatrix}$; (c) $A = \begin{Vmatrix} 2 & 2 & -2 \\ 2 & 5 & -4 \\ -2 & -4 & 5 \end{Vmatrix}$.

6. Let $A = \|a_{ik}\|$ be a real symmetric matrix. Show that any matrix C obtained by applying the *same* elementary transformations (see Prob. 31 of Chap. 2) to the rows and columns of A can be written in the form $C = B^*AB$, where B is a nonsingular matrix which is not necessarily orthogonal. (Cf. (6) of Sec. 32.) Show that A can be brought into a form where $a_{11}, a_{22}, \ldots, a_{rr} = \pm 1$ (r = rank A) and all other elements vanish.
7. Use the method of Prob. 6 or the method of Sec. 33 to reduce each of the following quadratic forms to a sum of squares, with coefficients ± 1 (the corresponding transformation is not unique):

(a) $x_1^2 + 5x_2^2 - 4x_3^2 + 2x_1x_2 - 4x_1x_3$;
(b) $4x_1^2 + x_2^2 + x_3^2 - 4x_1x_2 + 4x_1x_3 - 3x_2x_3$;
(c) $x_1x_2 + x_1x_3 + x_2x_3$;
(d) $2x_1^2 + 18x_2^2 + 8x_3^2 - 12x_1x_2 + 8x_1x_3 - 27x_2x_3$;
(e) $-12x_1^2 - 3x_2^2 - 12x_3^2 + 12x_1x_2 - 24x_1x_3 + 8x_2x_3$;
(f) $x_1x_2 + x_2x_3 + x_3x_4 + x_4x_1$;
(g) $3x_1^2 + 2x_2^2 - x_3^2 - 2x_4^2 + 2x_1x_2 - 4x_2x_3 + 2x_2x_4$.

8. Two quadratic forms with matrices A and C, respectively, are said to be *equivalent* if we can go from each to the other by introducing new variables as in Sec. 35, i.e., if A and C are related by formula (6) of Sec. 32, where B is nonsingular. In each of the following examples, find a nonsingular transformation (which is not uniquely defined), taking f into the equivalent form g:

(a) $f = 2x_1^2 + 9x_2^2 + 3x_3^2 + 8x_1x_2 - 4x_1x_3 - 10x_2x_3$,
$g = 2y_1^2 + 3y_2^2 + 6y_3^2 - 4y_1y_2 - 4y_1y_3 + 8y_2y_3$;

(b) $f = 3x_1^2 + 10x_2^2 + 25x_3^2 - 12x_1x_2 - 18x_1x_3 + 40x_2x_3$,
$g = 5y_1^2 + 6y_2^2 + 12y_1y_2$;
(c) $f = 5x_1^2 + 5x_2^2 + 2x_3^2 + 8x_1x_2 + 6x_1x_3 + 6x_2x_3$,
$g = 4y_1^2 + y_2^2 + 9y_3^2 - 12y_1y_3$.

9. Prove that a real matrix A is symmetric and positive† (i.e., $(A\mathbf{x}, \mathbf{x}) \geq 0$ for all real $\mathbf{x}$) if and only if

(a) $A = C^*C$ for some real matrix C (so that det A is a Gram determinant);
(b) $A = D^2$ for some positive symmetric matrix D.

10. Find the values of the parameter λ for which the following quadratic forms are positive definite:

(a) $5x_1^2 + x_2^2 + \lambda x_3^2 + 4x_1x_2 - 2x_1x_3 - 2x_2x_3$;
(b) $2x_1^2 + x_2^2 + 3x_3^2 + 2\lambda x_1x_2 + 2x_1x_3$;
(c) $x_1^2 + x_2^2 + 5x_3^2 + 2\lambda x_1x_2 - 2x_1x_3 + 4x_2x_3$;
(d) $x_1^2 + 4x_2^2 + x_3^2 + 2\lambda x_1x_2 + 10x_1x_3 + 6x_2x_3$;
(e) $2x_1^2 + 2x_2^2 + x_3^2 + 2\lambda x_1x_2 + 6x_1x_3 + 2x_2x_3$.

11. By the *composition* of two quadratic forms

$$f = \sum_{i,j=1}^{n} a_{ij}x_ix_j, \qquad g = \sum_{i,j=1}^{n} b_{ij}x_ix_j$$

is meant the quadratic form

$$[f, g] = \sum_{i,j=1}^{n} a_{ij}b_{ij}x_ix_j.$$

Show that

(a) If the forms f and g are positive, then the form $[f, g]$ is positive;
(b) If the forms f and g are positive definite, then the form $[f, g]$ is positive definite.

Comment: The terminology adopted in this book is the following: A quadratic form $(A\mathbf{x}, \mathbf{x})$, or the corresponding matrix A, is *positive* if $(A\mathbf{x}, \mathbf{x}) \geq 0$ for all (real) $\mathbf{x}$, *positive definite* if $(A\mathbf{x}, \mathbf{x}) > 0$ for all $\mathbf{x}$, and *positive semidefinite* if $(A\mathbf{x}, \mathbf{x}) \geq 0$ for all $\mathbf{x}$, but $(A\mathbf{x}, \mathbf{x}) = 0$ for some nonzero $\mathbf{x}$. (Similar remarks apply *mutatis mutandis* to negative, negative definite, and negative semidefinite forms.)

12. Prove that if the quadratic form with coefficient matrix $\|a_{ik}\|$ is positive definite, then the determinants $\Delta_1, \Delta_2, \ldots, \Delta_n$ of Sec. 36 must be positive.

13. Show by an example that the nonnegativity of the determinants $\Delta_1, \Delta_2, \ldots, \Delta_n$ of Sec. 36 does not necessarily imply that the quadratic form with coefficient matrix $\|a_{ik}\|$ is positive.

14. Let $f(x_1, x_2, \ldots, x_n)$ be a positive definite quadratic form. Show that the result of adding the square of a nonzero linear form in the same variables to f is to increase the *discriminant* of f, i.e., the determinant of its coefficient matrix (Sec. 37).

15. Let $f(x_1, x_2, \ldots, x_n)$ be a positive definite quadratic form with coefficient matrix $\|a_{ik}\|$, and let $g(x_2, x_3, \ldots, x_n) = f(0, x_2, \ldots, x_n)$. Show that the discriminants D_f and D_g of these forms satisfy the inequality

$$D_f \leq a_{11}D_g.$$

† Cf. comment to Prob. 11.

16. Show that if $A = \|a_{ik}\|$ is a positive definite matrix of order n, i.e., if $(A\mathbf{x}, \mathbf{x}) > 0$ for all real $\mathbf{x}$, then

$$\det A \le a_{11}a_{22} \cdots a_{nn}.$$

17. In each of the following examples, one of the two quadratic forms f and g is positive definite. Find a nonsingular linear transformation (not uniquely defined) which simultaneously reduces both f and g to sums of squares, with the coefficients of the positive definite form being all equal to 1.

(a) $f = -4x_1x_2,$
$g = x_1^2 - 2x_1x_2 + 4x_2^2;$
(b) $f = x_1^2 + 26x_2^2 + 10x_1x_2,$
$g = x_1^2 + 56x_2^2 + 16x_1x_2;$
(c) $f = -4x_1^2 + 3x_2^2 + 3x_3^2 - 4x_1x_2 - 6x_2x_3,$
$g = 4x_1^2 + 3x_2^2 + x_3^2 + 4x_1x_2 - 2x_2x_3;$
(d) $f = 7x_1^2 + 2x_2^2 + 4x_3^2 + 10x_1x_2 + 4x_2x_3,$
$g = 5x_1^2 + 3x_2^2 + 4x_3^2 + 2x_1x_2 + 4x_2x_3;$
(e) $f = 23x_1^2 + 23x_2^2 + 48x_3^2 - 46x_1x_2 - 66x_1x_3 + 66x_2x_3,$
$g = 9x_1^2 + 6x_2^2 + 11x_3^2 - 6x_1x_2 - 10x_1x_3 + 16x_2x_3;$
(f) $f = x_1^2 - 15x_2^2 + 4x_1x_2 - 2x_1x_3 + 6x_2x_3,$
$g = x_1^2 + 17x_2^2 + 3x_3^2 + 4x_1x_2 - 2x_1x_3 - 14x_2x_3.$

18. Can either of the following pairs of quadratic forms be simultaneously reduced to sums of squares by a nonsingular real transformation:

(a) $f = x_1^2 + 4x_1x_2 - x_2^2,\ g = x_1^2 + 6x_1x_2 + 5x_2^2;$
(b) $f = x_1^2 + x_1x_2 - x_2^2,\ g = x_1^2 - 2x_1x_2?$

19. Suppose a mechanical system has kinetic energy

$$T = \frac{m_A}{2}(\dot{q}_1^2 + \dot{q}_3^2) + \frac{m_B}{2}\dot{q}_2^2$$

and potential energy

$$V = \frac{k}{2}(q_2 - q_1)^2 + \frac{k}{2}(q_3 - q_2)^2,$$

where the overdot denotes differentiation with respect to time. (This corresponds to the longitudinal oscillations of a symmetric linear molecule ABA consisting of a central atom B of mass m_B between two identical atoms of mass m_A, with a restoring force acting on each atom A equal to k times the displacement of A from its equilibrium position.) Find the *natural frequencies* λ_k and the normal coordinates p_k of this system ($k = 1, 2, 3$). To what kind of motion does the natural frequency zero correspond?

20. Suppose a mechanical system has kinetic energy

$$T = \frac{m_A}{2}\dot{q}_1^2 + \frac{m_B}{2}\dot{q}_2^2 + \frac{m_C}{2}\dot{q}_3^2$$

and potential energy

$$V = \frac{k_A}{2}(q_2 - q_1)^2 + \frac{k_C}{2}(q_3 - q_2)^2.$$

(This corresponds to the longitudinal oscillations of an asymmetric linear molecule ABC consisting of a central atom B of mass m_B between two different atoms A and C of masses m_A and m_C, acted upon by different restoring forces with "spring constants" k_A and k_C.) Find the natural frequencies λ_k and the normal coordinates p_k of this system ($k = 1, 2, 3$), and show that they reduce to the corresponding quantities of Prob. 19 when $m_A = m_C$ and $k_A = k_C$.

21.† Let A and C be real symmetric matrices, and let C be positive definite. Let $\lambda_1 \geq \lambda_2 \geq \cdots \geq \lambda_n$ be the roots of the equation $\det(A - \lambda C) = 0$. Show that

$$\lambda_1 = \max_{\mathbf{x}\neq 0} \frac{(A\mathbf{x}, \mathbf{x})}{(C\mathbf{x}, \mathbf{x})}, \qquad \lambda_n = \min_{\mathbf{x}\neq 0} \frac{(Ax, \mathbf{x})}{(Cx, \mathbf{x})},$$

where the maximum and minimum are taken over all nonzero vectors. Show that the maximum is attained for a vector $\mathbf{x}$ if and only if $A\mathbf{x} = \lambda_1\mathbf{x}$ and that the minimum is attained if and only if $A\mathbf{x} = \lambda_n\mathbf{x}$.

22. Let A be a real symmetric matrix. Let λ_1 and λ_n be the eigenvalues of A with the largest and smallest absolute value, respectively (λ_1 and λ_n may not be unique). Show that

$$|\lambda_1| = \max_{\|\mathbf{x}\|=1} \|A\mathbf{x}\|, \qquad |\lambda_n| = \min_{\|\mathbf{x}\|=1} \|A\mathbf{x}\|,$$

where the maximum and minimum are taken over all vectors $\mathbf{x}$ *of norm* 1. Show that the maximum is attained if and only if $A\mathbf{x} = \lambda_1\mathbf{x}$ and that the minimum is attained if and only if $A\mathbf{x} = \lambda_n\mathbf{x}$.

23. Let A be a real symmetric matrix. Show that

$$\max_{\|\mathbf{x}\|=1, \|\mathbf{y}\|=1} |(A\mathbf{x}, \mathbf{y})| = \max_{\|\mathbf{x}\|=1} |(A\mathbf{x}, \mathbf{x})|.$$

24. For a real matrix A (not necessarily symmetric), define the *norm* as

$$\|A\| = \max_{\|\mathbf{x}\|=1} \|A\mathbf{x}\|.$$

Show that

(a) $\|A\| = \max_{\mathbf{x}\neq 0} \dfrac{\|A\mathbf{x}\|}{\|\mathbf{x}\|}$, so that $\|A\mathbf{x}\| \leq \|A\|\,\|\mathbf{x}\|$ for all $\mathbf{x}$;
(b) $\|A + B\| \leq \|A\| + \|B\|$; (c) $\|AB\| \leq \|A\|\,\|B\|$;
(d) $\|A\|^2 = \|A^*A\|$ = largest eigenvalue of A^*A; (e) $\|A^*\| = \|A\|$.

25. Let A be a real symmetric matrix, and let λ_n be the eigenvalue of A of smallest absolute value (λ_n may not be unique). Show that A has an inverse if and only if $\lambda_n \neq 0$, and show that in this case

$$\|A^{-1}\| = \frac{1}{|\lambda_n|}.$$

26. Let A be a symmetric matrix all of whose elements are positive, with eigenvalues $\lambda_1 \geq \lambda_2 \geq \cdots \geq \lambda_n$. Let $\mathbf{z} = (z_1, z_2, \ldots, z_n)$, $\|\mathbf{z}\| = 1$, be an eigenvector of unit length corresponding to the eigenvalue λ_1, and let $\mathbf{u}$ be the vector with components $u_i = |z_i|$, $i = 1, 2, \ldots, n$. Show that

(a) $A\mathbf{u} = \lambda_1\mathbf{u}$, where $\lambda_1 > 0$; (b) $\lambda_1 > |\lambda_k|$, if $\lambda_1 \neq \lambda_k$;
(c) The components z_i of $\mathbf{z}$ are either all positive or all negative;
(d) The set of eigenvectors associated with the eigenvalue λ_1 is one-dimensional.

Comment: These results are due to O. Perron, *Math. Annalen*, **64**, 1–76 (1907) and are valid even if A is not symmetric.

† Unless otherwise stated, the matrices in the following problems are $n \times n$.

27. Find $\|A\|$ and $\|A^{-1}\|$ (if A^{-1} exists):

(a) $A = \begin{Vmatrix} 3 & 4 \\ 5 & 2 \end{Vmatrix}$; (b) $A = \begin{Vmatrix} \alpha & 1 & \cdots & 1 \\ 1 & \alpha & \cdots & 1 \\ \cdots & \cdots & \cdots & \cdots \\ 1 & 1 & \cdots & \alpha \end{Vmatrix}$.

28. Find the maximum and minimum values of the following expressions, when the variables are restricted by the condition $x^2 + y^2 + z^2 = 1$:

(a) $x^2 + y^2 + z^2 + xy + xz + yz$; (b) $x^2 + y^2 + z^2 - xy + xz - yz$.

29. A complex matrix A is said to be *normal* if $\tilde{A}A = A\tilde{A}$, i.e., if A commutes with its adjoint $\tilde{A}$. Show that A is normal if and only if

$$\|A\mathbf{x}\| = \|\tilde{A}\mathbf{x}\|$$

for all vectors $\mathbf{x}$.

30. Show that if A is a normal matrix and if $A\mathbf{x} = \lambda\mathbf{x}$, then $\tilde{A}\mathbf{x} = \bar{\lambda}\mathbf{x}$.

31. Show that if A is a normal matrix and if $A\mathbf{x}_1 = \lambda_1\mathbf{x}_1$, $A\mathbf{x}_2 = \lambda_2\mathbf{x}_2$ ($\lambda_1 \neq \lambda_2$), then the vectors $\mathbf{x}_1$ and $\mathbf{x}_2$ are orthogonal.

32. Show that every matrix A can be written in the form $A = B + iC$, where B and C are Hermitian matrices uniquely determined by A. Show that A is normal if and only if $BC = CB$.

33. Show that Hermitian matrices and unitary matrices are normal. Show that if A is normal and U is unitary, then $U^{-1}AU$ is normal.

34. Let A be an $n \times n$ normal matrix, let λ be an eigenvalue of A, and let $A\mathbf{x} = \lambda\mathbf{x}$, where $\|\mathbf{x}\| = 1$. Let U be a unitary matrix whose first column is the vector $\mathbf{x}$. Show that

$$U^{-1}AU = \begin{Vmatrix} \lambda & 0 \\ 0 & A_1 \end{Vmatrix},$$

where A_1 is an $(n - 1) \times (n - 1)$ normal matrix.

35. Show that a matrix A can be reduced to diagonal form by a unitary transformation (i.e., $U^{-1}AU$ is diagonal, where U is unitary) if and only if A is normal.

36. Which of the following matrices are normal? Reduce the normal matrices to diagonal form. (Give the unitary matrix as well as the diagonal form.)

(a) $\begin{Vmatrix} 1 & 2 \\ 3 & 4 \end{Vmatrix}$; (b) $\begin{Vmatrix} a & b \\ -b & a \end{Vmatrix}$; (c) $\begin{Vmatrix} a & b & c \\ -b & a & d \\ -c & -d & a \end{Vmatrix}$.

37. Show that an $n \times n$ matrix A is normal if and only if it has n pairwise orthogonal eigenvectors.

38. Suppose that an $n \times n$ matrix A has k pairwise orthogonal eigenvectors. Show that there exists a unitary matrix U such that

$$U^{-1}AU = \begin{Vmatrix} \lambda_1 & 0 & \cdots & 0 & 0 \\ 0 & \lambda_2 & \cdots & 0 & 0 \\ \cdots & \cdots & \cdots & \cdots & \cdots \\ 0 & 0 & \cdots & \lambda_k & 0 \\ 0 & 0 & \cdots & 0 & B \end{Vmatrix},$$

where B is an $(n - k) \times (n - k)$ matrix.

39. Show that if A is normal with eigenvalues $\lambda_1, \lambda_2, \ldots, \lambda_n$, then the eigenvalues of $\tilde{A}A$ are $|\lambda_1|^2, |\lambda_2|^2, \ldots, |\lambda_n|^2$.

40. Show that if $A = \|a_{ij}\|$ is any $n \times n$ matrix and if $\mu_1, \mu_2, \ldots, \mu_n$ are the eigenvalues of $\tilde{A}A$, then

$$\sum_{i,j=1}^{n} |a_{ij}|^2 = \sum_{i=1}^{n} \mu_i.$$

41. Let $A = \|a_{ij}\|$ be an $n \times n$ matrix, and let

$$\rho_k = \sum_{j \neq k} |a_{kj}|.$$

Show that every eigenvalue of A lies in one of the circles $|z - a_{ii}| \leq \rho_i$ $(i = 1, 2, \ldots, n)$. (Cf. Prob. 16 of Chap. 2.)

42. Let A and B be $n \times n$ matrices, and let A have n *distinct* eigenvalues. Show that $AB = BA$ if and only if every eigenvector of A is also an eigenvector of B.

43. Show that if A and B are commuting normal matrices, then they can be simultaneously reduced to diagonal form by a unitary transformation.

44. Let A be a normal matrix. Show that

(a) A is Hermitian if and only if all its eigenvalues are real;
(b) A is unitary if and only if all its eigenvalues have absolute value 1.

45. For each of the following Hermitian matrices H, find a unitary matrix reducing H to diagonal form, and find the corresponding diagonal form:

(a) $H = \begin{Vmatrix} 3 & 2+2i \\ 2-2i & 1 \end{Vmatrix}$; (b) $H = \begin{Vmatrix} 3 & 2-i \\ 2+i & 7 \end{Vmatrix}$;

(c) $H = \begin{Vmatrix} 2 & i & 1 \\ -i & 1 & -i \\ 1 & i & 0 \end{Vmatrix}$.

46. Which of the following pairs of Hermitian matrices commute? For the commuting pairs, find a unitary matrix which simultaneously reduces them to diagonal form, and find the corresponding forms:

(a) $\begin{Vmatrix} i & 2 \\ 2 & 1+i \end{Vmatrix}$, $\begin{Vmatrix} i & 2 \\ 2 & 1-i \end{Vmatrix}$; (b) $\begin{Vmatrix} 1 & \sqrt{3} \\ \sqrt{3} & -1 \end{Vmatrix}$, $\begin{Vmatrix} 2 & \sqrt{3} \\ \sqrt{3} & 0 \end{Vmatrix}$;

(c) $\begin{Vmatrix} 1 & 2i & 0 & 0 \\ -2i & 1 & 0 & 0 \\ 0 & 0 & 1 & 0 \\ 0 & 0 & 0 & 1 \end{Vmatrix}$, $\begin{Vmatrix} 1 & 0 & 0 & 0 \\ 0 & 1 & 0 & 0 \\ 0 & 0 & 2 & -i \\ 0 & 0 & i & 2 \end{Vmatrix}$.

47. For each of the following orthogonal matrices V, find an orthogonal matrix reducing V to the canonical form (73) of Sec. 42, and find the corresponding canonical form:

(a) $V = \begin{Vmatrix} \frac{1}{2} & \frac{1}{2} & -\frac{1}{2}\sqrt{2} \\ \frac{1}{2} & \frac{1}{2} & \frac{1}{2}\sqrt{2} \\ \frac{1}{2}\sqrt{2} & -\frac{1}{2}\sqrt{2} & 0 \end{Vmatrix}$; (b) $V = \begin{Vmatrix} \frac{2}{3} & -\frac{1}{3} & \frac{2}{3} \\ \frac{2}{3} & \frac{2}{3} & -\frac{1}{3} \\ -\frac{1}{3} & \frac{2}{3} & \frac{2}{3} \end{Vmatrix}$;

(c) $V = \begin{Vmatrix} \frac{3}{4} & \frac{1}{4} & -\frac{1}{4}\sqrt{6} \\ \frac{1}{4} & \frac{3}{4} & \frac{1}{4}\sqrt{6} \\ \frac{1}{4}\sqrt{6} & -\frac{1}{4}\sqrt{6} & \frac{1}{2} \end{Vmatrix}$.

48. Let U be any unitary matrix diagonalizing a rotation matrix R, i.e., such that $U^{-1}RU = [1, e^{i\varphi}, e^{-i\varphi}]$. Show that $U^{-1}[0, i\alpha, -i\alpha]U$, where U is any real number, is a real skew-symmetric matrix. (A is *skew-symmetric* means that $A^* = -A$.)

49. For each of the following unitary matrices U, find a unitary matrix reducing U to diagonal form, and find the corresponding diagonal form:

(a) $$U = \begin{Vmatrix} \frac{1+i}{2} & \frac{1-i}{2} \\ \frac{1-i}{2} & \frac{1+i}{2} \end{Vmatrix};$$ (b) $$U = \frac{1}{9}\begin{Vmatrix} 4+3i & 4i & -6-2i \\ -4i & 4-3i & -2-6i \\ 6+2i & -2-6i & 1 \end{Vmatrix};$$

(c) $$U = \begin{Vmatrix} \frac{1}{3}\sqrt{3} & \frac{i}{3}\sqrt{3} & -\frac{i}{3}\sqrt{3} \\ 0 & \frac{1}{2}\sqrt{2} & \frac{1}{2}\sqrt{2} \\ \frac{1}{3}\sqrt{6} & -\frac{i}{6}\sqrt{6} & \frac{i}{6}\sqrt{6} \end{Vmatrix}.$$

50. Show that every nonsingular matrix A has a unique representation $A = UB$, where U is a unitary matrix and B is a positive Hermitian matrix.

51. Let $P = P_R + P_S$ be the sum of two (Hermitian) projection matrices. It is shown in Sec. 43 that if $P_RP_S = P_SP_R = 0$, then P is also a projection matrix. Prove the converse.

52.† Prove that every symmetric unitary matrix is the square of a symmetric unitary matrix.

† Suggested by J. S. Lomont.

Chapter 5

Infinite-Dimensional Spaces

45. Infinite-Dimensional Spaces. Before defining an infinite-dimensional space, we begin by introducing the notion of a limit of a sequence of complex numbers. Let the complex variable $z = x + iy$ take the sequence of values

$$z_1 = x_1 + iy_1,\ z_2 = x_2 + iy_2,\ \ldots,\ z_n = x_n + iy_n,\ \ldots \tag{1}$$

We say that the complex number $\alpha = a + ib$ is the limit of the sequence (1), and we write $\alpha = \lim z_n$ or $z_n \to \alpha$, if the absolute value of the difference $\alpha - z_n$ goes to zero as n increases without limit, i.e., if $|\alpha - z_n| \to 0$ as $n \to \infty$. The condition $|\alpha - z_n| \to 0$ is equivalent to the two conditions $x_n \to a$, $y_n \to b$, since

$$|\alpha - z_n| = |(a - x_n) + i(b - y_n)| = \sqrt{(a - x_n)^2 + (b - y_n)^2}$$

and both terms under the radical are nonnegative; thus $x_n + iy_n \to a + ib$ is equivalent to $x_n \to a$, $y_n \to b$.

Consider now an infinite series of complex terms

$$\sum_{k=1}^{\infty} (a_k + ib_k). \tag{2}$$

This series is said to be *convergent* if the sum of its first n terms

$$S_n = \sum_{k=1}^{n} (a_k + ib_k) = (a_1 + a_2 + \cdots + a_n) + i(b_1 + b_2 + \cdots + b_n)$$

has a limit, i.e., if $S_n \to a + ib$ as $n \to \infty$; this limit $a + ib$ is called the *sum* of the series. It follows from the definition of a limit that the convergence of the series (2) is equivalent to the convergence of the series

$$a = \sum_{k=1}^{\infty} a_k, \qquad b = \sum_{k=1}^{\infty} b_k, \tag{3}$$

consisting of the real and imaginary parts of (2).

Suppose that the series

$$\sum_{k=1}^{\infty} |a_k + ib_k| = \sum_{k=1}^{\infty} \sqrt{a_k^2 + b_k^2}, \tag{4}$$

consisting of the absolute values of the terms of the series (2), converges. Then, it follows from the obvious inequalities

$$|a_k| \le \sqrt{a_k^2 + b_k^2}, \qquad |b_k| \le \sqrt{a_k^2 + b_k^2}$$

that the series (3) will be absolutely convergent and hence convergent. Therefore, the series (2) will also be convergent, i.e., if (4) converges, then (2) converges *a fortiori;* in this case, the series (2) is said to be *absolutely convergent.* Applying the usual Cauchy criterion, we can state the following necessary and sufficient condition for absolute convergence: Given $\epsilon > 0$, however small, there exists an N such that

$$\sum_{k=n}^{n+p} |a_k + ib_k| < \epsilon \tag{5}$$

for any positive integer p, provided that $n > N$.

We now apply the above considerations to some special cases which are of great importance in what follows. Consider the series

$$\sum_{k=1}^{\infty} \alpha_k \beta_k, \tag{6}$$

where α_k and β_k are complex numbers for which the series

$$\sum_{k=1}^{\infty} |\alpha_k|^2, \qquad \sum_{k=1}^{\infty} |\beta_k|^2 \tag{7}$$

are known to converge. By applying the inequality

$$\left\{ \sum_{k=n}^{n+p} |\alpha_k \beta_k| \right\}^2 \le \sum_{k=n}^{n+p} |\alpha_k|^2 \sum_{k=n}^{n+p} |\beta_k|^2,$$

proved in Sec. 29, it follows from the convergence of the series (7) that the sum

$$\sum_{k=n}^{n+p} |\alpha_k \beta_k|$$

can be made arbitrarily small for any p, provided that n is sufficiently large. Thus, the convergence of the series (7) guarantees the absolute convergence of the series (6). Now consider the series

$$\sum_{k=1}^{\infty} |\alpha_k + \beta_k|^2 = \sum_{k=1}^{\infty} (\alpha_k + \beta_k)(\bar{\alpha}_k + \bar{\beta}_k), \tag{8}$$

where, as before, we assume that the series (7) converge. The series (8) can be written as the sum of the four series

$$\sum_{k=1}^{\infty} |\alpha_k|^2, \qquad \sum_{k=1}^{\infty} |\beta_k|^2, \qquad \sum_{k=1}^{\infty} \alpha_k\bar{\beta}_k, \qquad \sum_{k=1}^{\infty} \bar{\alpha}_k\beta_k.$$

The first two series converge by hypothesis, and the last two converge by what was just proved, i.e., the convergence of the series (7) guarantees the convergence of the series (8).

We now introduce the concept of an infinite-dimensional space. By a *vector* $\mathbf{x}$ *in an infinite-dimensional space,* we mean an infinite sequence of complex numbers

$$\mathbf{x} = (x_1, x_2, \ldots),$$

where we always assume that these numbers obey the condition that the series

$$\sum_{k=1}^{\infty} |x_k|^2 \tag{9}$$

converges. The set of all such vectors is usually called a *Hilbert space,* after D. Hilbert, who was first to introduce such spaces; for brevity, this space will henceforth be referred to as the space H. The basic operations on vectors, i.e., multiplication of vectors by numbers and addition of vectors, are defined for the vectors of H in just the same way as for ordinary vectors. If the components of the vector $\mathbf{x}$ are x_k, then the components of the vector $c\mathbf{x}$ (where c is any complex number) are defined as cx_k; if the components of the vectors $\mathbf{x}$, $\mathbf{y}$ are x_k, y_k, then the components of the vector $\mathbf{x} + \mathbf{y}$ are defined as $x_k + y_k$. The difference $\mathbf{x} - \mathbf{y}$ is defined as the sum of $\mathbf{x}$ and $(-1)\mathbf{y}$ (cf. Sec. 12). If the series (9) converges, then obviously so does the series

$$\sum_{k=1}^{\infty} |cx_k|^2;$$

in just the same way, if the series

$$\sum_{k=1}^{\infty} |x_k|^2, \qquad \sum_{k=1}^{\infty} |y_k|^2$$

converge, then it follows by the argument given above that the series

$$\sum_{k=1}^{\infty} |x_k + y_k|^2$$

also converges, i.e., if $\mathbf{x}$ and $\mathbf{y}$ belong to H, the sequences of numbers $(cx_1, cx_2, \ldots)$ and $(x_1 + y_1, x_2 + y_2, \ldots)$ define vectors $c\mathbf{x}$ and $\mathbf{x} + \mathbf{y}$ in H. The zero vector is defined as the vector all of whose components are zero and is denoted by 0 in vector equations. The operations on vectors obey the usual rules (cf. Sec. 12):

$$\mathbf{x} + \mathbf{y} = \mathbf{y} + \mathbf{x}, \qquad (\mathbf{x} + \mathbf{y}) + \mathbf{z} = \mathbf{x} + (\mathbf{y} + \mathbf{z}),$$
$$(a + b)\mathbf{x} = a\mathbf{x} + b\mathbf{x}, \qquad a(\mathbf{x} + \mathbf{y}) = a\mathbf{x} + a\mathbf{y}, \qquad a(b\mathbf{x}) = (ab)\mathbf{x}.$$

In view of the foregoing, we can define the *scalar product*

$$(\mathbf{x}, \mathbf{y}) = \sum_{k=1}^{\infty} x_k \bar{y}_k$$

of two vectors $\mathbf{x}$ and $\mathbf{y}$ in H. The sum

$$(\mathbf{x}, \mathbf{x}) = \sum_{k=1}^{\infty} |x_k|^2 \tag{10}$$

defines the square of the *norm* (*length*) of the vector $\mathbf{x}$; we introduce the special notation

$$\|\mathbf{x}\|^2 = \sum_{k=1}^{\infty} |x_k|^2. \tag{11}$$

The norm is positive for every vector except the zero vector, which has norm zero. Two vectors $\mathbf{u}$ and $\mathbf{v}$ in H are said to be *orthogonal* (to each other) if their scalar product is zero, i.e., if $(\mathbf{u}, \mathbf{v}) = 0$, or equivalently, $(\mathbf{v}, \mathbf{u}) = 0$. The scalar product obeys the same basic rules as in finite-dimensional spaces (Secs. 13 and 30). In particular, the inequality

$$|(\mathbf{x}, \mathbf{y})| \le \|\mathbf{x}\|\,\|\mathbf{y}\| \tag{12}$$

holds, and the triangle inequality

$$\|\mathbf{x} + \mathbf{y}\| \le \|\mathbf{x}\| + \|\mathbf{y}\| \tag{13}$$

can be derived just as in Sec. 30. If the vectors $\mathbf{x}^{(k)}$ $(k = 1, 2, \ldots, m)$ are (pairwise) orthogonal, i.e., if $(\mathbf{x}^{(i)}, \mathbf{x}^{(j)}) = 0$ for $i \neq j$, then we obviously have

$$(\mathbf{x}^{(1)} + \cdots + \mathbf{x}^{(m)}, \mathbf{x}^{(1)} + \cdots + \mathbf{x}^{(m)}) = (\mathbf{x}^{(1)}, \mathbf{x}^{(1)}) + \cdots + (\mathbf{x}^{(m)}, \mathbf{x}^{(m)}),$$

or equivalently,

$$\|\mathbf{x}^{(1)} + \cdots + \mathbf{x}^{(m)}\|^2 = \|\mathbf{x}^{(1)}\|^2 + \cdots + \|\mathbf{x}^{(m)}\|^2, \tag{14}$$

i.e., the square of the norm of a sum of orthogonal vectors equals the sum of the squares of the norms of the summands, a fact which can be called the *Pythagorean theorem.* It is an immediate consequence of the definition of the norm that

$$\|c\mathbf{x}\| = |c|\,\|\mathbf{x}\|,$$

where c is any complex number. If the vectors $\mathbf{z}^{(1)}, \mathbf{z}^{(2)}, \ldots, \mathbf{z}^{(m)}$ are orthogonal and each vector has unit norm, i.e., if

$$(\mathbf{z}^{(p)}, \mathbf{z}^{(q)}) = 0 \qquad \text{for } p \neq q,$$
$$(\mathbf{z}^{(p)}, \mathbf{z}^{(p)}) = 1,$$

then (14) gives

$$\|c_1\mathbf{z}^{(1)} + \cdots + c_m\mathbf{z}^{(m)}\|^2 = |c_1|^2 + \cdots + |c_m|^2,$$

where the c_s are arbitrary complex numbers.

The fundamental basis vectors in the space H are the unit orthogonal (*orthonormal*) vectors

$$\mathbf{a}^{(1)} = (1, 0, 0, \ldots), \mathbf{a}^{(2)} = (0, 1, 0, \ldots), \ldots$$

The components x_k of the vector $\mathbf{x}$ can be expressed as the scalar products

$$x_k = (\mathbf{x}, \mathbf{a}^{(k)}).$$

Finally, consider m arbitrary orthonormal vectors $\mathbf{z}^{(k)}$ $(k = 1, 2, \ldots, m)$. The quantity $(\mathbf{x}, \mathbf{z}^{(k)})\mathbf{z}^{(k)}$ is called the *projection of the vector* $\mathbf{x}$ *onto the vector* $\mathbf{z}^{(k)}$. These vectors do not determine a complete system of basis vectors for the space H, and therefore the sum

$$\sum_{k=1}^{m} (\mathbf{x}, \mathbf{z}^{(k)})\mathbf{z}^{(k)}$$

is in general different from $\mathbf{x}$, i.e., $\mathbf{x}$ can be written in the form

$$\mathbf{x} = \sum_{k=1}^{m} (\mathbf{x}, \mathbf{z}^{(k)})\mathbf{z}^{(k)} + \mathbf{u}. \tag{15}$$

Taking the scalar product of this equation with $\mathbf{z}^{(i)}$ and bearing in mind that the $\mathbf{z}^{(k)}$ are orthonormal, we obtain

$$(\mathbf{x}, \mathbf{z}^{(i)}) = (\mathbf{x}, \mathbf{z}^{(i)}) + (\mathbf{u}, \mathbf{z}^{(i)}),$$

so that $(\mathbf{u}, \mathbf{z}^{(i)}) = 0$, i.e., the vector $\mathbf{u}$ is orthogonal to all the vectors

$\mathbf{z}^{(k)}$. Thus, we can apply the Pythagorean theorem to the sum (15), obtaining

$$\|\mathbf{x}\|^2 = \sum_{k=1}^{m} |(\mathbf{x}, \mathbf{z}^{(k)})|^2 + \|\mathbf{u}\|^2,$$

which implies at once the inequality

$$\|\mathbf{x}\|^2 \geq \sum_{k=1}^{m} |(\mathbf{x}, \mathbf{z}^{(k)})|^2, \tag{16}$$

called *Bessel's inequality*. This inequality can be stated as follows: *The sum of the squares of the absolute values of the projections of a vector* $\mathbf{x}$ *onto any m orthonormal vectors does not exceed the square of the length (norm) of* $\mathbf{x}$. The equality sign holds in (16) if and only if the vector $\mathbf{u}$ in (15) is zero, i.e., has all its components equal to zero.

46. Convergence of Vectors. We now introduce the concept of the *limit of a sequence of vectors in H*. Suppose that we have a sequence of vectors $\mathbf{v}^{(k)}$, where k runs through the values 1, 2, 3, . . . ; denote the components of the vector $v^{(k)}$ by $v_1^{(k)}, v_2^{(k)}, \ldots$. We say that the vectors $\mathbf{v}^{(k)}$ approach the vector $\mathbf{v}$ as a limit and write $\mathbf{v}^{(k)} \Rightarrow \mathbf{v}$, if

$$\|\mathbf{v} - \mathbf{v}^{(k)}\| \to 0, \text{ i.e., } \|\mathbf{v} - \mathbf{v}^{(k)}\|^2 \to 0 \tag{17}$$

as $k \to \infty$. Denoting the components of $\mathbf{v}$ by $v_1, v_2, \ldots$, we can write (17) in expanded form as

$$\lim_{k\to\infty} \{|v_1 - v_1^{(k)}|^2 + |v_2 - v_2^{(k)}|^2 + \cdots\} = 0. \tag{18}$$

If a sum of positive terms converges to zero, then so does each term, i.e., it is an immediate consequence of (18) that

$$|v_m - v_m^{(k)}| \to 0 \text{ as } k \to \infty \qquad (m = 1, 2, \ldots). \tag{19}$$

Thus, every component $v_m^{(k)}$ must converge to the corresponding component v_m; more precisely, the real and imaginary parts of $v_m^{(k)}$ must converge to the real and imaginary parts of v_m. We note that the converse is not true, i.e., (19) does not imply (18). To see this, consider the case where $\mathbf{v}^{(k)}$ is the vector $(0, \ldots, 0, 1, 0, \ldots)$ where the 1 appears in the kth position. As $k \to \infty$, each component converges to zero, i.e., $v_m^{(k)} \to 0$ for any integer m, so that $v_m = 0$ $(m = 1, 2, \ldots)$, but nevertheless, the sum (18) always remains equal to unity. As an example of where there *is* convergence, take a vector $\mathbf{v} = (v_1, v_2, \ldots)$ and define the vectors $\mathbf{v}^{(k)}$ as follows: The vector $\mathbf{v}^{(k)}$ has the same first k components as $\mathbf{v}$, while the rest of its components equal zero, i.e.,

$$\mathbf{v}^{(k)} = (v_1, v_2, \ldots, v_k, 0, 0, \ldots).$$

It is not hard to prove that $\mathbf{v}^{(k)} \Rightarrow \mathbf{v}$. In fact,

$$\|\mathbf{v} - \mathbf{v}^{(k)}\|^2 = \sum_{n=k+1}^{\infty} |v_n|^2,$$

and this sum converges to zero as $k \to \infty$, since the series with the general term $|v_n|^2$ converges.

We now note some simple rules obeyed by limits. If $\mathbf{u}^{(k)} \Rightarrow \mathbf{u}$ and $\mathbf{v}^{(k)} \Rightarrow \mathbf{v}$, then

$$\mathbf{u}^{(k)} + \mathbf{v}^{(k)} \Rightarrow \mathbf{u} + \mathbf{v} \tag{20}$$

and

$$(\mathbf{u}^{(k)}, \mathbf{v}^{(k)}) \to (\mathbf{u}, \mathbf{v}). \tag{21}$$

Since the scalar product is a complex number, we write $\to$ instead of $\Rightarrow$ in (21). To prove (20), we use (13) to write

$$\begin{aligned}\|(\mathbf{u} + \mathbf{v}) - (\mathbf{u}^{(k)} + \mathbf{v}^{(k)})\| &= \|(\mathbf{u} - \mathbf{u}^{(k)}) + (\mathbf{v} - \mathbf{v}^{(k)})\| \\ &\le \|\mathbf{u} - \mathbf{u}^{(k)}\| + \|\mathbf{v} - \mathbf{v}^{(k)}\|,\end{aligned}$$

where by hypothesis $\|\mathbf{u} - \mathbf{u}^{(k)}\| \to 0$, $\|\mathbf{v} - \mathbf{v}^{(k)}\| \to 0$. It follows immediately that

$$\|(\mathbf{u} + \mathbf{v}) - (\mathbf{u}^{(k)} + \mathbf{v}^{(k)})\| \to 0,$$

i.e., $\mathbf{u}^{(k)} + \mathbf{v}^{(k)} \Rightarrow \mathbf{u} + \mathbf{v}$, as asserted. To prove (21), we first write

$$\mathbf{u}^{(k)} = \mathbf{u} + \mathbf{s}^{(k)}, \quad \mathbf{v}^{(k)} = \mathbf{v} + \mathbf{t}^{(k)},$$

where $\|\mathbf{s}^{(k)}\| \to 0$ and $\|\mathbf{t}^{(k)}\| \to 0$. Then the scalar product of $\mathbf{u}^{(k)}$ and $\mathbf{v}^{(k)}$ is

$$\begin{aligned}(\mathbf{u}^{(k)}, \mathbf{v}^{(k)}) = (\mathbf{u} + \mathbf{s}^{(k)}, \mathbf{v} + \mathbf{t}^{(k)}) = (\mathbf{u}, \mathbf{v}) + (\mathbf{u}, \mathbf{t}^{(k)}) + (\mathbf{s}^{(k)}, \mathbf{v}) \\ + (\mathbf{s}^{(k)}, \mathbf{t}^{(k)});\end{aligned}$$

from this it follows that

$$|(\mathbf{u}, \mathbf{v}) - (\mathbf{u}^{(k)}, \mathbf{v}^{(k)})| \le |(\mathbf{u}, \mathbf{t}^{(k)})| + |(\mathbf{s}^{(k)}, \mathbf{v})| + |(\mathbf{s}^{(k)}, \mathbf{t}^{(k)})|,$$

or, by (12),

$$|(\mathbf{u}, \mathbf{v}) - (\mathbf{u}^{(k)}, \mathbf{v}^{(k)})| \le \|\mathbf{u}\|\,\|\mathbf{t}^{(k)}\| + \|\mathbf{s}^{(k)}\|\,\|\mathbf{v}\| + \|\mathbf{s}^{(k)}\|\,\|\mathbf{t}^{(k)}\|.$$

Since the right-hand side goes to zero as $k \to \infty$, we have

$$|(\mathbf{u}, \mathbf{v}) - (\mathbf{u}^{(k)}, \mathbf{v}^{(k)})| \to 0,$$

i.e.,

$$(\mathbf{u}^{(k)}, \mathbf{v}^{(k)}) \to (\mathbf{u}, \mathbf{v}).$$

In particular, $(\mathbf{u}^{(k)}, \mathbf{u}^{(k)}) \to (\mathbf{u}, \mathbf{u})$, or equivalently $\|\mathbf{u}^{(k)}\|^2 \to \|\mathbf{u}\|^2$ and $\|\mathbf{u}^{(k)}\| \to \|\mathbf{u}\|$. The relation (21) can be regarded as expressing the continuity of the scalar product.

Another result involving limits is the following: If $\mathbf{u}^{(k)} \Rightarrow \mathbf{u}$ and the numbers $c_k \to c$, then $c_k\mathbf{u}^{(k)} \Rightarrow c\mathbf{u}$. We omit the proof, which is quite simple.

We can formulate a necessary and sufficient condition for the existence of a limit in terms of the familiar Cauchy criterion as follows: A necessary and sufficient condition for the sequence of vectors

$$\mathbf{v}^{(k)} = (\mathbf{v}_1^{(k)}, \mathbf{v}_2^{(k)}, \ldots) \qquad (k = 1, 2, \ldots) \tag{22}$$

to have a limit is that given any $\epsilon > 0$, however small, there exists an N such that

$$\|\mathbf{v}^{(n)} - \mathbf{v}^{(m)}\| < \epsilon \tag{23}$$

provided that n and $m > N$. We first prove the necessity of the condition. Suppose the sequence (22) has the limit $\mathbf{v}$. Then we can write

$$\mathbf{v}^{(n)} - \mathbf{v}^{(m)} = (\mathbf{v}^{(n)} - \mathbf{v}) + (\mathbf{v} - \mathbf{v}^{(m)}),$$

from which it follows by the triangle inequality that

$$\|\mathbf{v}^{(n)} - \mathbf{v}^{(m)}\| \leq \|\mathbf{v}^{(n)} - \mathbf{v}\| + \|\mathbf{v} - \mathbf{v}^{(m)}\|.$$

It is an immediate consequence of the definition of the limit that both terms in the right-hand side of this inequality converge to zero as n and m increase, and therefore the same is also true of the left-hand side, i.e., the condition (23) must be met.

To prove the sufficiency of the condition (23), we assume that (23) is met and show that then the sequence (22) converges to a limit. In expanded form, the condition (23) can be written as

$$\sum_{s=1}^{\infty} |v_s^{(n)} - v_s^{(m)}|^2 < \epsilon^2 \qquad \text{for } n \text{ and } m > N, \tag{24}$$

where the $v_s^{(j)}$ are the components of the $\mathbf{v}^{(j)}$. This implies at once that

$$|v_s^{(n)} - v_s^{(m)}| < \epsilon \qquad \text{for } n \text{ and } m > N,$$

or $\quad |\alpha_s^{(n)} - \alpha_s^{(m)}| < \epsilon, \qquad |\beta_s^{(n)} - \beta_s^{(m)}| < \epsilon \qquad \text{for } m \text{ and } n > N,$

where we have written $v_s^{(n)}$ in terms of its real and imaginary parts:

$$v_s^{(n)} = \alpha_s^{(n)} + i\beta_s^{(n)}.$$

Applying the usual Cauchy criterion, we can state that $\alpha_s^{(n)}$ and $\beta_s^{(n)}$ have limits α_s and β_s, and therefore $v_s^{(n)}$ has as its limit the complex number $\alpha_s + i\beta_s$. Denoting this limit by v_s, we now show that $\sum_{s=1}^{\infty} |v_s|^2$ converges, i.e., that the v_s are the components of some vector in H. Taking the first M terms in the sum (24) and then passing to the limit as $n \to \infty$, we obtain

$$\sum_{s=1}^{M} |v_s - v_s^{(m)}|^2 \le \epsilon^2, \tag{25}$$

where M is any integer; then, passing to the limit $M \to \infty$ in (25), we obtain

$$\sum_{s=1}^{\infty} |v_s - v_s^{(m)}|^2 \le \epsilon^2. \tag{26}$$

This immediately implies that the numbers $v_s - v_s^{(m)}$ are the components of a vector in H. Since we already know that the numbers $v_s^{(m)}$ are the components of a vector in H, it follows that the sum of $v_s - v_s^{(m)}$ and $v_s^{(m)}$, i.e., the numbers v_s, are components of a vector $\mathbf{v}$ in H. Thus, the inequality (26) can be written in the form

$$\|\mathbf{v} - \mathbf{v}^{(m)}\| \le \epsilon \qquad \text{for } m > N,$$

i.e., $\mathbf{v}^{(m)} \Rightarrow \mathbf{v}$; therefore, the sequence (21) does in fact have a limit, QED. Each component v_s of the vector $\mathbf{v}$ is obviously defined as the limit of $v_s^{(m)}$, from which it follows at once that this limit is unique.

We now consider an infinite sum of vectors

$$\mathbf{u}^{(1)} + \mathbf{u}^{(2)} + \cdots; \tag{27}$$

this sum is said to be convergent if the sum of the first n terms

$$\mathbf{s}^{(n)} = \mathbf{u}^{(1)} + \mathbf{u}^{(2)} + \cdots + \mathbf{u}^{(n)}$$

has a limit (in the sense indicated above) as $n \to \infty$. By the Cauchy criterion, a necessary and sufficient condition for the convergence of (27) is that the inequality

$$\|\mathbf{s}^{(n+p)} - \mathbf{s}^{(n)}\| = \|\mathbf{u}^{(n+1)} + \mathbf{u}^{(n+2)} + \cdots + \mathbf{u}^{(n+p)}\| < \epsilon \tag{28}$$

should hold for any p if $n > N$. Bearing in mind that the scalar product is continuous, we have

$$(\mathbf{x}, \mathbf{u}^{(1)} + \mathbf{u}^{(2)} + \cdots) = (\mathbf{x}, \mathbf{u}^{(1)}) + (\mathbf{x}, \mathbf{u}^{(2)}) + \cdots$$
$$(\mathbf{u}^{(1)} + \mathbf{u}^{(2)} + \cdots, \mathbf{x}) = (\mathbf{u}^{(1)}, \mathbf{x}) + (\mathbf{u}^{(2)}, \mathbf{x}) + \cdots$$

Applying these relations to the case where the vectors $\mathbf{u}^{(k)}$ are (pairwise) orthogonal, we have

$$(\mathbf{u}^{(1)} + \mathbf{u}^{(2)} + \cdots, \mathbf{u}^{(1)} + \mathbf{u}^{(2)} + \cdots) = (\mathbf{u}^{(1)}, \mathbf{u}^{(1)}) + (\mathbf{u}^{(2)}, \mathbf{u}^{(2)}) + \cdots$$

or
$$\|\mathbf{u}^{(1)} + \mathbf{u}^{(2)} + \cdots\|^2 = \|\mathbf{u}^{(1)}\|^2 + \|\mathbf{u}^{(2)}\|^2 + \cdots,$$

i.e., the Pythagorean theorem is also valid for the sum of an infinite set of orthogonal vectors.

Finally, we find a necessary and sufficient condition for the convergence of a series of the form (27) consisting of (pairwise) orthogonal vectors. According to the Cauchy criterion, we have to examine the expression (28), which by the Pythagorean theorem, is equal to

$$\|\mathbf{u}^{(n+1)}\|^2 + \|\mathbf{u}^{(n+2)}\|^2 + \cdots + \|\mathbf{u}^{(n+p)}\|^2.$$

It follows immediately that a necessary and sufficient condition for the convergence of the series (27) is that the series consisting of the squares of the norms of the vectors $\mathbf{u}^{(k)}$ converge. This result can be expressed somewhat differently as follows: Let $\mathbf{x}^{(k)}$ $(k = 1, 2, \ldots)$ be an infinite set of orthonormal vectors and form the series

$$\sum_{k=1}^{\infty} C_k \mathbf{x}^{(k)}, \tag{29}$$

where the C_k are certain numbers. In view of what has just been proved, a necessary and sufficient condition for the series (29) to be convergent is that the series

$$\sum_{k=1}^{\infty} |C_k|^2$$

be convergent. Incidentally, this shows that rearranging terms in (25) cannot destroy the convergence of (29). In fact, it is easy to show that the sum of the series (29) does not change if the terms in (29) are rearranged.

47. Complete Systems of Orthonormal Vectors. In this section, we study the important concept of a *complete system of (pairwise) orthogonal vectors in the space H*. Just as in the case of finite-dimensional spaces, it can be shown that every finite set of orthogonal vectors is linearly independent. In the n-dimensional case, we have seen that every set of n linearly independent vectors forms a complete system, in the sense that any vector can be expressed as a linear combination of n linearly independent vectors. In the case of the space H, we no longer have such a simple criterion for completeness, since H is infinite-dimensional.

Suppose we have an infinite system $\mathbf{x}^{(k)}$ $(k = 1, 2, \ldots)$ of orthonormal vectors in H. (Henceforth, we shall be concerned only with systems of *orthonormal* vectors.) Let $\mathbf{y}$ be an arbitrary vector in H. Just as in the case of a finite number of vectors, we form the sum

$$\sum_{k=1}^{\infty} (\mathbf{y}, \mathbf{x}^{(k)}) \mathbf{x}^{(k)} \tag{30}$$

of the projections of $\mathbf{y}$ onto the vectors $\mathbf{x}^{(k)}$. As we have seen above, the inequality

$$\sum_{k=1}^{m} |(\mathbf{y}, \mathbf{x}^{(k)})|^2 \leq \|\mathbf{y}\|^2$$

holds for any m; therefore, passing to the limit $m \to \infty$, we have

$$\sum_{k=1}^{\infty} |(\mathbf{y}, \mathbf{x}^{(k)})|^2 \leq \|\mathbf{y}\|^2, \tag{31}$$

so that the series in (31) must be convergent. It is an immediate consequence of the result given at the end of the preceding section that the series (30) is also convergent. If we set

$$\mathbf{y} = \sum_{k=1}^{\infty} (\mathbf{y}, \mathbf{x}^{(k)})\mathbf{x}^{(k)} + \mathbf{u}, \tag{32}$$

then, just as in Sec. 45, it is easily shown that the vector $\mathbf{u}$ is orthogonal to all the vectors $\mathbf{x}^{(k)}$, so that

$$\|\mathbf{y}\|^2 = \sum_{k=1}^{\infty} |(\mathbf{y}, \mathbf{x}^{(k)})|^2 + \|\mathbf{u}\|^2, \tag{33}$$

by the Pythagorean theorem. It follows that if the vector $\mathbf{u}$ in (32) is different from zero, the inequality sign holds in (31), while if $\mathbf{u}$ is zero (i.e., all its components are zero), the equality sign holds in (31). The system of vectors $\mathbf{x}^{(k)}$ is said to be *complete* if the equality sign holds in the formula (31) for any vector $\mathbf{y}$ of the space H. Thus, in terms of a complete system of vectors $\mathbf{x}^{(k)}$, we have the following expansion for the vector $\mathbf{y}$:

$$\mathbf{y} = \sum_{k=1}^{\infty} (\mathbf{x}^{(k)}, \mathbf{y})\mathbf{x}^{(k)}. \tag{34}$$

A complete system in H is sometimes said to be *closed,* and the equation

$$\sum_{k=1}^{\infty} |(\mathbf{y}, \mathbf{x}^{(k)})|^2 = \|\mathbf{y}\|^2 \tag{35}$$

is called the *closure relation.* The following consequence of (35) is called the *generalized closure relation.* Let $\mathbf{x}^{(k)}$ ($k = 1, 2, \ldots$) be a complete system in H, and let $\mathbf{y}$ and $\mathbf{z}$ be any two vectors in H, so that

$$\mathbf{y} = \sum_{k=1}^{\infty} (\mathbf{y}, \mathbf{x}^{(k)})\mathbf{x}^{(k)}, \qquad \mathbf{z} = \sum_{k=1}^{\infty} (\mathbf{z}, \mathbf{x}^{(k)})\mathbf{x}^{(k)}.$$

Applying (35) to the vectors $\mathbf{y} + \mathbf{z}$ and $\mathbf{y} + i\mathbf{z}$, we obtain

$$\sum_{k=1}^{\infty} [(\mathbf{y}, \mathbf{x}^{(k)}) + (\mathbf{z}, \mathbf{x}^{(k)})][\overline{(\mathbf{y}, \mathbf{x}^{(k)})} + \overline{(\mathbf{z}, \mathbf{x}^{(k)})}] = (\mathbf{y} + \mathbf{z}, \mathbf{y} + \mathbf{z}),$$

$$\sum_{k=1}^{\infty} [(\mathbf{y}, \mathbf{x}^{(k)}) + i(\mathbf{z}, \mathbf{x}^{(k)})][\overline{(\mathbf{y}, \mathbf{x}^{(k)})} - i\overline{(\mathbf{z}, \mathbf{x}^{(k)})}] = (\mathbf{y} + i\mathbf{z}, \mathbf{y} + i\mathbf{z}).$$

Using the closure equations for y and z, we find

$$\sum_{k=1}^{\infty} (\mathbf{y}, \mathbf{x}^{(k)})\overline{(\mathbf{z}, \mathbf{x}^{(k)})} + \sum_{k=1}^{\infty} (\mathbf{z}, \mathbf{x}^{(k)})\overline{(\mathbf{y}, \mathbf{x}^{(k)})} = (\mathbf{y}, \mathbf{z}) + (\mathbf{z}, \mathbf{y}),$$

$$\sum_{k=1}^{\infty} (\mathbf{y}, \mathbf{x}^{(k)})\overline{(\mathbf{z}, \mathbf{x}^{(k)})} - \sum_{k=1}^{\infty} (\mathbf{z}, \mathbf{x}^{(k)})\overline{(\mathbf{y}, \mathbf{x}^{(k)})} = (\mathbf{y}, \mathbf{z}) - (\mathbf{z}, \mathbf{y}),$$

which together imply the *generalized closure relation*†

$$\sum_{k=1}^{\infty} (\mathbf{y}, \mathbf{x}^{(k)})\overline{(\mathbf{z}, \mathbf{x}^{(k)})} = (\mathbf{y}, \mathbf{z}). \tag{36}$$

If $\mathbf{z}$ coincides with $\mathbf{y}$, then (36) reduces to (35).

To examine the situation in more detail, let $\mathbf{x}^{(k)}$ $(k = 1, 2, \ldots)$ be a system of orthonormal vectors, and denote the components of $\mathbf{x}^{(k)}$ by $x_s^{(k)}$ $(s = 1, 2, \ldots)$. Since the $\mathbf{x}^{(k)}$ are orthonormal, these components satisfy the relations

$$\sum_{s=1}^{\infty} x_s^{(p)}\bar{x}_s^{(q)} = \delta_{pq}, \tag{37}$$

where $\delta_{pq} = 0$ for $p \neq q$ and $\delta_{pp} = 1$. To find the conditions under which the system of vectors $\mathbf{x}^{(k)}$ is complete (or closed), let $\mathbf{y}^{(l)}$ be a vector whose lth component is 1 while the other components are 0. Then we have

$$(\mathbf{y}^{(l)}, \mathbf{x}^{(k)}) = x_l^{(k)},$$

and the closure relation (35) applied to $\mathbf{y}^{(l)}$ gives

$$\sum_{k=1}^{\infty} |x_l^{(k)}|^2 = 1 \qquad (l = 1, 2, 3, \ldots).$$

† Often called *Parseval's theorem.*

Similarly, applying the generalized closure relation (36) to the vectors $\mathbf{y}^{(p)}$ and $\mathbf{y}^{(q)}$, where $p \neq q$, and bearing in mind that $\mathbf{y}^{(p)}$ and $\mathbf{y}^{(q)}$ are orthogonal, we have

$$\sum_{k=1}^{\infty} x_p^{(k)} \bar{x}_q^{(k)} = 0 \qquad (p \neq q).$$

Together, the last two formulas give the single formula

$$\sum_{k=1}^{\infty} x_p^{(k)} \bar{x}_q^{(k)} = \delta_{pq}. \tag{38}$$

We now write the components of our vectors $\mathbf{x}^{(k)}$ in the form of an infinite matrix:

$$\left\| \begin{matrix} x_1^{(1)} & x_1^{(2)} & \cdots \\ x_2^{(1)} & x_2^{(2)} & \cdots \\ \cdots & \cdots & \cdots \end{matrix} \right\|. \tag{39}$$

The relations (37) which express the orthonormality of the vectors $\mathbf{x}^{(k)}$ are equivalent to the orthonormality of the *columns* of the matrix (39), while the relations (38) show that for the system of vectors $\mathbf{x}^{(k)}$ to be complete, it is also necessary that the *rows* of the matrix (39) be orthonormal.

We now show that a sufficient condition for the completeness of the $\mathbf{x}^{(k)}$ is that the relations (38) hold for $p = q$. Thus, suppose that these relations hold for $p = q$. Then the closure relation is satisfied by the vectors

$$\mathbf{y}^{(l)} = (0, \ldots, 0, 1, 0, \ldots) \qquad (l = 1, 2, \ldots),$$

where the 1 appears in the lth position. Therefore, all the $\mathbf{y}^{(l)}$ can be expressed as linear combinations of the $\mathbf{x}^{(k)}$:

$$\mathbf{y}^{(l)} = \sum_{k=1}^{\infty} c_k^{(l)} \mathbf{x}^{(k)}.$$

We must now show that the same is true of any vector $\mathbf{z}$. To see this, denote by $\mathbf{z}^{(l)}$ the vector whose first l components $z_1, z_2, \ldots, z_l$ are the same as those of $\mathbf{z}$, while the rest are zero. We obviously have

$$\mathbf{z}^{(l)} = z_1\mathbf{y}^{(1)} + z_2\mathbf{y}^{(2)} + \cdots + z_l\mathbf{y}^{(l)},$$

and since the $\mathbf{y}^{(m)}$ are linear combinations of the $\mathbf{x}^{(k)}$, so are the $\mathbf{z}^{(l)}$, i.e.,

$$\mathbf{z}^{(l)} = \sum_{k=1}^{\infty} d_k^{(l)} \mathbf{x}^{(k)}.$$

Taking the scalar product of both sides of this equation with $\mathbf{x}^{(k)}$, we obtain the following expression for the coefficients $d_k^{(l)}$:

$$d_k^{(l)} = (\mathbf{z}^{(l)}, \mathbf{x}^{(k)}).$$

On the other hand, we have already seen that

$$\mathbf{z} = \sum_{k=1}^{\infty} d_k \mathbf{x}^{(k)} + \mathbf{u}, \tag{40}$$

where $\mathbf{u}$ is orthogonal to all the $\mathbf{x}^{(k)}$. Consider the difference

$$\mathbf{z} - \mathbf{z}^{(l)} = \sum_{k=1}^{\infty} (d_k - d_k^{(l)})\mathbf{x}^{(k)} + \mathbf{u}.$$

By the Pythagorean theorem, we have

$$\|\mathbf{z} - \mathbf{z}^{(l)}\|^2 = \sum_{k=1}^{\infty} \|d_k - d_k^{(l)}\|^2 + \|\mathbf{u}\|^2,$$

and therefore

$$\|\mathbf{u}\|^2 \leq \|\mathbf{z} - \mathbf{z}^{(l)}\|^2. \tag{41}$$

The vector u does not depend on l, and as we know from Sec. 46, the right-hand side of (41) goes to zero as $l \to \infty$. It follows at once that $\mathbf{u} = 0$, so that (40) gives the following expansion for the arbitrary vector $\mathbf{z}$ in terms of the vectors $\mathbf{x}^{(k)}$:

$$\mathbf{z} = \sum_{k=1}^{\infty} d_k \mathbf{x}^{(k)}, \tag{42}$$

where $d_k = (\mathbf{x}^{(k)}, \mathbf{z})$. Therefore, the closure relation also holds for any vector in H. The final result can be stated as follows: *A necessary and sufficient condition for the orthonormal vectors* $\mathbf{x}^{(k)}$ *to form a complete* (*or closed*) *system is that the sum of the squares of the absolute values of the elements of any row of the matrix* (39) *be equal to unity.* If this condition is satisfied, then the rows of (39) are automatically orthogonal.

48. Linear Transformations in Infinitely Many Variables. In this section, we shall examine briefly the subject of linear transformations of the form

$$\begin{aligned} x_1' &= a_{11}x_1 + a_{12}x_2 + \cdots \\ x_2' &= a_{21}x_1 + a_{22}x_2 + \cdots \\ &\cdots\cdots\cdots\cdots\cdots \end{aligned} \tag{43}$$

or

$$\mathbf{x}' = A\mathbf{x}, \tag{44}$$

in infinitely many variables $x_1, x_2, \ldots$; here A is an infinite matrix with elements a_{ik}. First of all, we give the condition for the infinite series appearing in the right-hand side of (43) to be convergent for any vector **x** of the space H. As we know, a sufficient condition for this to be the case is that the series

$$\sum_{k=1}^{\infty} |a_{ik}|^2 \qquad (i = 1, 2, \ldots)$$

converge for every i. It can be shown that this condition is not only sufficient but also necessary. Thus, if this condition is not met, the series appearing in the right-hand side of (43) do not converge for the whole space H but only for a part of H. Another natural condition to impose on A is that the numbers x'_k obtained as a result of the transformation (43) be the components of a vector in H whenever the numbers x_k are the components of a vector in H, i.e., that the series $\sum_{k=1}^{\infty} |x'_k|^2$ converges whenever the series $\sum_{k=1}^{\infty} |x_k|^2$ converges. If the matrix A satisfies both of the conditions just stated, then the corresponding transformation, defined by (43) or (44), is called a *bounded transformation*, a designation which stems from the fact that for such transformations, we can find a positive number M such that

$$\|\mathbf{x}'\|^2 \leq M \|\mathbf{x}\|^2, \tag{45}$$

or in expanded form,

$$\sum_{k=1}^{\infty} |x'_k|^2 \leq M \sum_{k=1}^{\infty} |x_k|^2. \tag{46}$$

We now consider a special kind of linear transformation in infinitely many variables. Let the linear transformation be

$$\begin{aligned} x'_1 &= u_{11}x_1 + u_{12}x_2 + \cdots \\ x'_2 &= u_{21}x_1 + u_{22}x_2 + \cdots \\ &\cdots\cdots\cdots\cdots\cdots \end{aligned} \tag{47}$$

where, as always, we assume that the series

$$\sum_{k=1}^{\infty} |u_{ik}|^2$$

converges for every i. Introduce the vectors $\mathbf{u}^{(k)} = (\bar{u}_{k1}, \bar{u}_{k2}, \ldots)$, where the u_{ik} are the coefficients of (47), and assume that the u_{ik} are such that the vectors $\mathbf{u}^{(k)}$ form a complete system of orthonormal vectors. As we have seen in the preceding section, this is equivalent to the rows and columns of the infinite matrix $\|u_{ik}\|$ being orthonormal:

$$\sum_{s=1}^{\infty} u_{sp}\bar{u}_{sq} = \delta_{pq}, \qquad \sum_{s=1}^{\infty} u_{ps}\bar{u}_{qs} = \delta_{pq}. \tag{48}$$

With this choice of the coefficients u_{ik}, the transformation (47) is said to be *unitary*. The equations (47) can be written in the form

$$\begin{aligned} (\mathbf{x}, \mathbf{u}^{(1)}) &= x_1', \\ (\mathbf{x}, \mathbf{u}^{(2)}) &= x_2', \\ &\cdots\cdots\cdots \end{aligned} \tag{49}$$

and the closure relation gives

$$\sum_{k=1}^{\infty} |x_k'|^2 = \|\mathbf{x}\|^2 = \sum_{k=1}^{\infty} |x_k|^2,$$

i.e., just as in the finite-dimensional case, a unitary transformation does not change the length of a vector, and we can take M to be 1 in the formula (45).

The equations (47) can easily be solved for the x_k; this gives the inverse of the transformation (47). Using the completeness of the system of vectors $\mathbf{u}^{(k)}$, we obtain the following unique expression for the vector $\mathbf{x}$:

$$\mathbf{x} = x_1'\mathbf{u}^{(1)} + x_2'\mathbf{u}^{(2)} + \cdots \tag{50}$$

or, in expanded form,

$$\begin{aligned} x_1 &= \bar{u}_{11}x_1' + \bar{u}_{21}x_2' + \cdots \\ x_2 &= \bar{u}_{12}x_1' + \bar{u}_{22}x_2' + \cdots \\ &\cdots\cdots\cdots\cdots\cdots\cdots \end{aligned} \tag{51}$$

where, of course, we have in mind only solutions x_k such that $\sum_{k=1}^{\infty} |x_k|^2$ converges. To see that (50) or (51) gives the unique solution of the system (47), we argue as follows: Since, by hypothesis, the given numbers x_k' are such that $\sum_{k=1}^{\infty} |x_k'|^2$ converges, and since the $\mathbf{u}^{(k)}$ constitute an orthonormal system, we know that the series (50) converges (Sec. 46). Therefore, $\mathbf{x}$ satisfies the relations

$$(\mathbf{x}, \mathbf{u}^{(k)}) = (\mathbf{x}_1' \mathbf{u}^{(1)} + \mathbf{x}_2' \mathbf{u}^{(2)} + \cdots, \mathbf{u}^{(k)}) = x_k' \qquad (k = 1, 2, \ldots),$$

i.e., $\mathbf{x}$ satisfies (49) and hence (47), as asserted. Equation (51) shows that the inverse of a unitary transformation is obtained by interchanging rows and columns in the infinite matrix $\|u_{ik}\|$ and then taking complex conjugates; thus, the situation is completely analogous to the finite-dimensional case.

Even in the case of bounded operators, the problem of finding inverses and reducing matrices to diagonal form is in general quite difficult and leads to results which have no analogs (in the strict sense of the word) in finite-dimensional spaces. A detailed analysis of the linear transformations corresponding to infinite matrices will be given in a later volume; here, we shall cite only some of the results. First, we note that a necessary and sufficient condition for the transformation (43) to be bounded is that there should exist a positive number N such that the coefficients a_{ik} in (43) satisfy the inequality

$$\left| \sum_{n,m=1}^{k} a_{nm}x_m\bar{x}_n \right| \leq N \sum_{m=1}^{k} |x_m|^2$$

for any positive integer k and any complex numbers x_s $(s = 1, 2, \ldots)$. We also note that a sufficient condition for the transformation (43) to be bounded is that there should exist a positive number M (independent of m and n) such that the inequalities

$$\sum_{m=1}^{\infty} |a_{nm}| \leq M \qquad (n = 1, 2, \ldots), \quad \sum_{n=1}^{\infty} |a_{nm}| \leq M \qquad (m = 1, 2, \ldots)$$

hold. If the matrix A defines a bounded transformation, then there exists a unique matrix $\tilde{A}$ such that the relation

$$(A\mathbf{x}, \mathbf{y}) = (\mathbf{x}, \tilde{A}\mathbf{y})$$

holds for any $\mathbf{x}$ and $\mathbf{y}$ in H, and the elements $\tilde{a}_{ik}$ of the matrix $\tilde{A}$ are given by the formula $\tilde{a}_{ik} = \bar{a}_{ki}$. If A coincides with $\tilde{A}$, i.e., if $a_{ik} = \bar{a}_{ki}$, then the bounded transformation (43) is called *Hermitian*, or *self-adjoint*. For bounded transformations, the formula

$$(A\mathbf{x}, \mathbf{y}) = \sum_{n=1}^{\infty} \left(\sum_{m=1}^{\infty} a_{nm}x_m \right) \bar{y}_n = \sum_{m=1}^{\infty} x_m \left(\sum_{n=1}^{\infty} a_{nm}\bar{y}_n \right)$$
$$= \lim_{\substack{k \to \infty \\ l \to \infty}} \sum_{m=1}^{k} \sum_{n=1}^{l} a_{nm}x_m\bar{y}_n$$

holds.

Finally, we note another important special case of bounded transformations, i.e., the case where the double series

$$\sum_{n,m=1}^{\infty} |a_{nm}|^2 \tag{52}$$

converges. In this case, the double series

$$\sum_{n,m=1}^{\infty} a_{nm}x_m\bar{y}_n$$

is absolutely convergent for any choice of the vectors $\mathbf{x} = (x_1, x_2, \ldots)$ and $\mathbf{y} = (y_1, y_2, \ldots)$. If, in addition to the convergence of (52), we have the condition $a_{ik} = \bar{a}_{ki}$, then the corresponding Hermitian form can be reduced to a sum of squares by using a unitary transformation; thus, we have

$$\sum_{n,m=1}^{\infty} a_{nm} x_m \bar{y}_n = \sum_{k=1}^{\infty} \lambda_k z_k \bar{z}_k,$$

where the vector $\mathbf{z} = (z_1, z_2, \ldots)$ is obtained by applying a unitary transformation U to the vector $\mathbf{x} = (x_1, x_2, \ldots)$, i.e., $\mathbf{z} = U\mathbf{x}$. Moreover, $\lambda_k \to 0$ as $k \to \infty$. If A and B are two infinite matrices corresponding to bounded transformations, then successive application of A and B also gives a bounded transformation, whose coefficients are calculated by the usual formula

$$\{BA\}_{ik} = \sum_{s=1}^{\infty} \{B\}_{is}\{A\}_{sk}.$$

We note also that if A is the matrix of a bounded transformation and if $\mathbf{x}^{(k)} \Rightarrow \mathbf{x}$, then $A\mathbf{x}^{(k)} \Rightarrow A\mathbf{x}$. Unbounded linear transformations also play a very important role in mathematical physics.

49. Function Space. So far, we have considered the space H, where a vector is defined as an infinite set of numbers (the components), indexed by the positive integers. Thus, x_1 denotes the first component of the vector $\mathbf{x}$, x_2 the second component, etc. We now consider a function space F, where continuously varying functions of one or several variables play the role of vectors. Let $f(x)$ be a function defined on the interval $a \leq x \leq b$. We can regard $f(x)$ as a vector, where the value $f(x_0)$ is associated with the point x_0 in the interval $a \leq x \leq b$ and gives the "component of $f(x)$ with index x_0." Thus, in this case, the independent, variable x, which varies continuously through all values in the interval $a \leq x \leq b$, plays the role of an index for the components, i.e., $f(x)$ has a "continuous set of components." We shall always assume that the independent variable x ranges over a finite interval of the real axis and that the function $f(x)$ is in general complex-valued. To be explicit, for the time being, we shall study *continuous* complex functions $f(x) = f_1(x) + if_2(x)$, defined on the finite interval $a \leq x \leq b$.

The functions $f(x)$ can be added and multiplied by complex numbers, just like vectors of the space H. To define the norm and the scalar product, we need only replace summation by integration everywhere in our previous formulas. We define the *scalar product* by

$$(\varphi(x), \psi(x)) = \int_a^b \varphi(x)\overline{\psi(x)}\, dx \tag{53}$$

and the square of the *norm* by

$$\|f(x)\|^2 = (f(x), f(x)) = \int_a^b |f(x)|^2\, dx. \tag{54}$$

Now let the system of functions $\varphi_k(x)$ ($k = 1, 2, \ldots$) be a system of (pairwise) orthogonal "unit vectors" in F, i.e.,

$$\int_a^b \varphi_p(x)\overline{\varphi_q(x)}\, dx = \delta_{pq}. \tag{55}$$

We have already studied such systems of *orthonormal functions* in an earlier volume, so that here we merely recall some results which have direct bearing on our present considerations. The only departure from the previous discussion is that we are now dealing with functions which can take complex values.

Thus, let the $\varphi_k(x)$ be an orthonormal system, and let $f(x)$ be any function (vector) in F. We introduce the *Fourier coefficients*

$$a_k = (f(x), \varphi_k(x)) = \int_a^b f(x)\overline{\varphi_k(x)}\, dx \tag{56}$$

of the function $f(x)$, or with our present terminology, the magnitudes of the projections of $f(x)$ onto the vectors $\varphi_k(x)$ in F. Consider the integral

$$I_n = \int_a^b \left| f(x) - \sum_{k=1}^{n} a_k\varphi_k(x) \right|^2 dx \tag{57}$$

or

$$I_n = \int_a^b \left[f(x) - \sum_{k=1}^{n} a_k\varphi_k(x)\right]\left[\overline{f(x)} - \sum_{k=1}^{n} \overline{a_k\varphi_k(x)}\right] dx.$$

In view of (55) and (56), this integral becomes

$$I_n = \int_a^b |f(x)|^2\, dx - \sum_{k=1}^{n} |a_k|^2;$$

since $I_n \geq 0$, we have

$$\sum_{k=1}^{n} |a_k|^2 \leq \int_a^b |f(x)|^2\, dx, \tag{58}$$

which becomes

$$\sum_{k=1}^{\infty} |a_k|^2 \leq \int_a^b |f(x)|^2\, dx \tag{59}$$

in the limit as $n \to \infty$. This inequality is called *Bessel's inequality*. If the equality sign holds in (59), then the integral I_n goes to zero as $n \to \infty$, and conversely.

If the equality sign holds in (59), i.e., if

$$\sum_{k=1}^{\infty} |a_k|^2 = \int_a^b |f(x)|^2\,dx \tag{60}$$

for any continuous function $f(x)$, then the system of functions $\varphi_k(x)$ is said to be a *complete*, or *closed*, system, and (60) is called the *closure relation.* Then I_n goes to zero as $n \to \infty$ for any continuous function $f(x)$:

$$\lim_{n\to\infty} \int_a^b \left| f(x) - \sum_{k=1}^{n} a_k\varphi_k(x) \right| dx = 0. \tag{61}$$

Thus, any continuous function can be approximated arbitrarily closely by a finite linear combination of the functions $\varphi_k(x)$; however, here the phrase "arbitrarily closely" is not to be understood as meaning that the difference

$$\left| f(x) - \sum_{k=1}^{n} a_k\varphi_k(x) \right|$$

is itself arbitrarily small, but rather that the integral (57) is arbitrarily small for sufficiently large n. If we wish to be more precise, we speak of approximating $f(x)$ by a finite linear combination of the $\varphi_k(x)$ with an arbitrarily small *mean square error.*

Let a_k and b_k be the Fourier coefficients of the functions $f(x)$ and $g(x)$:

$$a_k = \int_a^b f(x)\overline{\varphi_k(x)}\,dx, \qquad b_k = \int_a^b g(x)\overline{\varphi_k(x)}\,dx. \tag{62}$$

Then, just as in Sec. 47, we can derive the *generalized closure relation*

$$\sum_{k=1}^{\infty} a_k\bar{b}_k = \int_a^b f(x)\overline{g(x)}\,dx. \tag{63}$$

Suppose that we form the Fourier series

$$\sum_{k=1}^{\infty} a_k\varphi_k(x)$$

of the function $f(x)$, where the a_k are the appropriate Fourier coefficients. We cannot even assert that this series converges, much less that its sum is $f(x)$. Thus, one often writes

$$f(x) \sim \sum_{k=1}^{\infty} a_k\varphi_k(x), \tag{64}$$

where the symbol $\sim$ means only that the infinite series in (64) is the Fourier series of the function $f(x)$. Therefore, (64) is not an equation in the usual sense of the word. However, it is well known that if the $\varphi_k(x)$ form a complete system, (64) becomes an exact equality when integrated term by term, i.e.,

$$\int_{x_1}^{x_2} f(x)\,dx = \sum_{k=1}^{\infty} a_k \int_{x_1}^{x_2} \varphi_k(x)\,dx \qquad (a \le x_1 < x_2 \le b).$$

If we multiply (64) by a continuous function $\psi(x)$ and then integrate, we obtain

$$\int_{x_1}^{x_2} f(x)\psi(x)\,dx = \sum_{k=1}^{\infty} a_k \int_{x_1}^{x_2} \varphi_k(x)\psi(x)\,dx.$$

If we integrate over the whole interval, this becomes

$$\int_a^b f(x)\psi(x)\,dx = \sum_{k=1}^{\infty} a_k \int_a^b \varphi_k(x)\psi(x)\,dx. \tag{65}$$

It is easy to see that (65) is just the generalized closure relation for the functions $f(x)$ and $\overline{\psi(x)}$.

50. Relation between the Spaces *F* and *H*. We now establish the relation between the function space F discussed in the preceding section and the space H studied earlier; this relation is of great importance in theoretical physics. Let

$$\varphi_k(x) \qquad (k = 1, 2, \ldots) \tag{66}$$

be a complete orthonormal system of functions in F, so that every continuous function $f(x)$ satisfies the closure relation (60), where the Fourier coefficients a_k are defined by (56). Let $g(x)$ be another continuous function, with Fourier coefficients

$$b_k = \int_a^b g(x)\overline{\varphi_k(x)}\,dx.$$

Applying the closure relation to the difference $f(x) - g(x)$, we obtain

$$\sum_{k=1}^{\infty} |a_k - b_k|^2 = \int_a^b |f(x) - g(x)|^2\,dx. \tag{67}$$

If the continuous functions $f(x)$ and $g(x)$ are different, the right-hand side of (67) is certainly greater than zero, and therefore the coefficients b_k are different from the a_k, i.e., different continuous functions have different Fourier coefficients with respect to the system (66). Therefore, every continuous function is completely characterized by its Fourier coefficients, and moreover, the sum of the squares of the absolute values of these Fourier coefficients is a convergent series. Thus, every continuous function corresponds to a definite vector in H, and different continuous functions correspond to different vectors in H. The converse is not true, i.e., there are vectors in H which do not correspond to functions in F. In other words, the vectors in H which correspond to continuous functions form only a part of H. For the converse to be true, F would have to consist of a wider class of functions than the continuous functions; we defer further discussion of this matter until a later volume.

Let $f_n(x)$ $(n = 1, 2, \ldots)$ be a sequence of continuous functions with Fourier coefficients

$$a_k^{(n)} = \int f_n(x)\overline{\varphi_k(x)}\,dx. \tag{68}$$

Applied to $f(x) - f_n(x)$, the closure relation gives

$$\sum_{k=1}^{\infty} |a_k - a_k^{(n)}|^2 = \int_a^b |f(x) - f_n(x)|^2\,dx,$$

from which it follows at once that the convergence of the vector in H with components $a_k^{(n)}$ $(k = 1, 2, \ldots)$ to the vector with components a_k is equivalent to the relation

$$\lim_{n\to\infty} \int_a^b |f(x) - f_n(x)|^2\,dx = 0,$$

i.e., to the convergence of $f_n(x)$ to $f(x)$ in the space F. In particular, if $\mathbf{z} = (a_1, a_2, \ldots)$ is the vector in H which corresponds to $f(x)$, and if we form the vector $\mathbf{z}^{(n)} = (a_1, a_2, \ldots, a_n, 0, 0, \ldots)$, then the finite Fourier series

$$\sum_{k=1}^{n} a_k\varphi_k(x)$$

is the function in F corresponding to $\mathbf{z}^{(n)}$. Then, as we have seen in Sec. 46, $\mathbf{z}^{(n)} \Rightarrow \mathbf{z}$, which corresponds to the relation

$$\int_a^b \left| f(x) - \sum_{k=1}^{n} a_k\varphi_k(x) \right|^2 dx \to 0.$$

In establishing the correspondence between the space of continuous functions and the space H, we started with the complete system of orthonormal functions (66). If instead of (66), we used another complete system of orthonormal functions

$$\psi_k(x) \qquad (k = 1, 2, \ldots), \tag{69}$$

then, of course, we would have a different correspondence between F and H. We now show that the vectors in H obtained using the new system (69) are related by a unitary transformation to those obtained using the old system (66). Every function $\psi_m(x)$ of the system (69) has definite

Fourier coefficients with respect to the system (66) and therefore corresponds to a definite Fourier series

$$\psi_m(x) \sim \sum_{k=1}^{\infty} u_{km}\varphi_k(x),$$

where, as before, the sign $\sim$ means merely that the Fourier series on the right corresponds to the function on the left. Using the closure relation (60) and the fact that the $\psi_k(x)$ are normalized, we obtain

$$\sum_{k=1}^{\infty} |u_{km}|^2 = 1. \tag{70}$$

Moreover, the generalized closure relation

$$\sum_{k=1}^{\infty} u_{kp}\bar{u}_{kq} = \int_a^b \psi_p(x)\overline{\psi_q(x)}\,dx$$

holds, which implies

$$\sum_{k=1}^{\infty} u_{kp}\bar{u}_{kq} = \delta_{pq}, \tag{71}$$

in view of (70) and the orthogonality of the functions $\psi_k(x)$. This formula shows that the columns of the matrix U with elements u_{ik} are orthonormal. Using the results of Sec. 47, we can show that a necessary and sufficient condition for the system (69) to be complete (which it is, by hypothesis) is that the sum of the squares of the absolute values of the elements of every row of U be unity:

$$\sum_{k=1}^{\infty} |u_{ik}|^2 = 1 \qquad (i = 1, 2, \ldots). \tag{72}$$

As in Sec. 47, it follows from (71) and (72) that the rows of U are also orthogonal:

$$\sum_{k=1}^{\infty} u_{pk}\bar{u}_{qk} = \delta_{pq}.$$

Thus, U is unitary, as asserted.

As an example of a complete orthonormal system of functions in F, consider the system

$$\varphi_k(x) = \frac{1}{\sqrt{2\pi}}\, e^{ikx} \qquad (k = 0, \pm 1, \pm 2, \ldots), \tag{73}$$

where the basic interval is $[-\pi, \pi]$. It is easy to see that the functions (71) form an orthonormal system. In fact,

$$\int_{-\pi}^{\pi} \varphi_p(x)\overline{\varphi_q(x)}\, dx = \frac{1}{2\pi}\int_{-\pi}^{\pi} e^{i(p-q)x}\, dx = \frac{1}{2\pi i(p-q)}\, e^{i(p-q)x}\Big|_{x=-\pi}^{x=\pi} = 0$$

for $p \neq q$, and

$$\int_{-\pi}^{\pi} |\varphi_p(x)|^2 = \frac{1}{2\pi}\int_{-\pi}^{\pi} dx = 1$$

for $p = q$. The proof that the system (73) is complete on the interval $[-\pi, \pi]$ was given in an earlier volume.

So far, we have considered the case where the function space consists of functions of one independent variable. We can also consider functions of several independent variables, defined on some region in a multi-dimensional space. Then, all the above considerations remain valid, the only difference being that we must everywhere replace single integrals by multiple integrals taken over the region of definition of the functions.

51. Linear Operators. For a function space F, we can introduce a concept analogous to the concept of a linear transformation in the space H. This leads us to the notion of a *linear operator defined on a function space*. Suppose that we have a definite rule which associates with every function $f(x)$ (with certain properties) another function $F(x)$; symbolically we denote the rule which establishes this correspondence by

$$F(x) = Lf(x).$$

This is a kind of generalization of the function concept, where the argument is not a variable number but a function $f(x)$, arbitrarily chosen from some class, and the value of the function is not a number but another function $F(x)$. This generalized functional dependence is usually called a *(functional) operator*. The operator concept is implicit in many problems of mathematical physics. For example, consider the problem of a vibrating string with fixed end points. The configuration of the string at any instant of time t is determined by two functions, i.e., the function which gives the initial displacement at every point of the string and the function which gives the initial velocity at every point. Thus, we are dealing here with a functional operator. A similar situation exists in many other problems of mathematical physics. Sometimes the argument is not a function specifying the initial conditions, but rather the boundary of some relevant region.

The operator L is said to be *linear* if the conditions

$$L[f_1(x) + f_2(x)] = Lf_1(x) + Lf_2(x)$$

and

$$L[cf(x)] = cLf(x),$$

where c is a constant, are satisfied. The condition for boundedness of the operator L is

$$\|L[f(x)]\| \leq M\|f(x)\|,$$

where M is a positive constant, and $f(x)$ is any function of the function space under consideration.

At this point, we shall not study the general theory of linear operators; instead, we confine ourselves to an examination of some examples. These examples illustrate the essential features of the concept of linear operators and also show the relation between linear operators and linear transformations in the space H. (It will be recalled that we have already established a correspondence between the function space F and the space H.)

Sometimes, we can write a linear operator in the form

$$F(x) = \int_a^b K(x, t)f(t)\,dt, \tag{74}$$

where $K(x, t)$ is a given function of two variables, called the *kernel* of the operator. In this case, the kernel is completely analogous to the matrix $\|a_{ik}\|$ of a linear transformation in the space H, but instead of the indices i and k, we have two variables x and t which vary continuously. The formula (74) is the analog of the formula (43). Operators of the form (74) are studied in the theory of integral equations.

It is not hard to extend the definitions of unitary and Hermitian transformations to the present context. A linear operator L is said to be *unitary* if the condition

$$(Lf(x), Lg(x)) = (f(x), g(x)) \tag{75}$$

is met for any two functions $f(x)$ and $g(x)$ of some class. Similarly, a *Hermitian* operator L_1 is defined by the relation

$$(L_1f(x), g(x)) = (f(x), L_1g(x)). \tag{76}$$

Suppose that this Hermitian operator has the form (74), i.e.,

$$L_1f(x) = \int_a^b K(x, t)f(t)\,dt.$$

Forming the scalar products appearing in (76), we obtain

$$(f(x), L_1g(x)) = \int_a^b \int_a^b \overline{K(x, t)}f(x)\overline{g(t)}\,dt\,dx,$$
$$(L_1f(x), g(x)) = \int_a^b \int_a^b K(x, t)f(t)\overline{g(x)}\,dt\,dx.$$

Changing the variables of integration in the last integral, we can write (76) in the form

$$\int_a^b \int_a^b [\overline{K(x, t)} - K(t, x)]f(x)\overline{g(t)}\,dt\,dx = 0. \tag{77}$$

If the kernel of the operator L_1 satisfies the condition

$$K(x, t) = \overline{K(t, x)}, \tag{78}$$

then the condition (77) is satisfied for any choice of $f(x)$ and $g(x)$; thus, in this case, the operator L_1 is indeed Hermitian. Since the functions $f(x)$ and $g(x)$ are arbitrary, (78) is not only a sufficient condition but also a necessary condition for (77) to hold, i.e., for the operator L_1 to be Hermitian. If the kernel $K(x, t)$ is real, then the condition (78) becomes

$$K(x, t) = K(t, x); \tag{79}$$

thus, in this case, the kernel has to be a symmetric function of its arguments.

Next, we consider two other examples of linear operators. As our first example, we take the operator

$$Lf(x) = \frac{1}{i}\frac{df(x)}{dx} = \frac{1}{i}f'(x), \tag{80}$$

corresponding to differentiation followed by multiplication by $1/i$, and we choose $[-\pi, \pi]$ as the basic interval. Forming the scalar product

$$(f(x), Lg(x)) = \left(f(x), \frac{1}{i}g'(x)\right) = -\frac{1}{i}\int_{-\pi}^{\pi} f(x)\overline{g'(x)}\,dx,$$

then integrating by parts and assuming that the functions $f(x)$ and $g(x)$ are periodic with period 2π, we have

$$\left(f(x), \frac{1}{i}g'(x)\right) = -\frac{1}{i}f(x)\overline{g(x)}\Bigg|_{x=-\pi}^{x=\pi} + \frac{1}{i}\int_{-\pi}^{\pi} f'(x)\overline{g(x)}\,dx.$$

This implies at once that

$$\left(f(x), \frac{1}{i}g'(x)\right) = \left(\frac{1}{i}f'(x), g(x)\right),$$

i.e., the operator (80) is Hermitian with respect to the class of periodic functions with continuous derivatives.

If we introduce the system of functions (73) as a coordinate system in our function space, $f(x)$ is characterized by its Fourier coefficients a_k, which are defined by the formula

$$a_k = \frac{1}{\sqrt{2\pi}}\int_{-\pi}^{\pi} e^{-ikx}f(x)\,dx.$$

The function $(1/i)\,f'(x)$ has different Fourier coefficients a'_k. It is easy to find the linear transformation which expresses the a'_k in terms of the a_k. This linear transformation represents the operator (80) in the form of an infinite matrix, where it must be borne in mind that the representation pertains to a definite choice of a coordinate system in our function space, namely, that corresponding to the functions (73). Thus, we have

$$a'_k = \frac{1}{\sqrt{2\pi}\, i} \int_{-\pi}^{\pi} e^{-ikx} f'(x)\, dx.$$

Integrating by parts and assuming that $f(x)$ is periodic, we obtain

$$a'_k = \frac{k}{\sqrt{2\pi}} \int_{-\pi}^{\pi} e^{-ikx} f(x)\, dx,$$

i.e.,

$$a'_k = k a_k \qquad (k = 0, \pm 1, \pm 2, \ldots). \tag{81}$$

This formula gives the desired linear transformation, which has a diagonal matrix of the form

$$\left\| \begin{matrix} \cdot & \cdot & \cdot & \cdot & \cdot & \cdot & \cdot & \cdot & \cdot \\ \cdots & -2 & 0 & 0 & 0 & 0 & \cdots \\ \cdots & 0 & -1 & 0 & 0 & 0 & \cdots \\ \cdots & 0 & 0 & 0 & 0 & 0 & \cdots \\ \cdots & 0 & 0 & 0 & 1 & 0 & \cdots \\ \cdots & 0 & 0 & 0 & 0 & 2 & \cdots \\ \cdot & \cdot & \cdot & \cdot & \cdot & \cdot & \cdot & \cdot & \cdot \end{matrix} \right\|. \tag{82}$$

A new but unimportant feature of the matrix (82) is that the enumeration of rows and columns goes from $-\infty$ to $+\infty$ rather than from 1 to ∞; this corresponds merely to the way we enumerated the functions (73). We note that the functions (73) obviously satisfy the relation

$$\frac{1}{i}\, \varphi'_k(x) = k\varphi_k(x),$$

or in terms of the operator (80),

$$L\varphi_k(x) = k\varphi_k(x).$$

In analogy with Sec. 32, we call the functions $\varphi_k(x)$ the *eigenfunctions* of the operator L, and the numbers k the corresponding *eigenvalues*. The diagonal form of the matrix (82) is directly related to the fact that the $\varphi_k(x)$ are eigenfunctions of the operator (80).

As a last example, we consider the operator

$$L_1 f(x) = x f(x), \tag{83}$$

corresponding to multiplication by the independent variable. Choosing the functions (73) as a coordinate system in our function space, we construct the linear transformation which represents the operator L_1 in the space H. As above, let the a_m be the Fourier coefficients of the function $f(x)$, and let a'_m be the Fourier coefficients of the function $xf(x)$, i.e.,

$$a'_m = \frac{1}{\sqrt{2\pi}} \int_{-\pi}^{\pi} e^{-imx} x f(x)\, dx \qquad (m = 0, 1, 2, \ldots). \tag{84}$$

We must now construct the linear transformation which expresses the a'_m in terms of the a_m. To evaluate the integral (84), we first determine the Fourier coefficients

$$c_k = \frac{1}{2\pi} \int_{-\pi}^{\pi} e^{i(m-k)x} x\, dx$$

of the function $(1/\sqrt{2\pi})e^{imx}x$. Integrating by parts, we find

$$c_k = \frac{1}{i(m-k)} e^{i(m-k)\pi} = \frac{(-1)^{m-k}}{i(m-k)} \tag{85}$$

for $m - k \neq 0$, while the coefficient c_m is just

$$c_m = \frac{1}{2\pi} \int_{-\pi}^{\pi} x\, dx = 0. \tag{86}$$

Then, we rewrite (84) in the form

$$a'_m = \frac{1}{\sqrt{2\pi}} \int_{-\pi}^{\pi} f(x) \overline{(e^{imx}x)}\, dx$$

and apply the generalized closure relation (63), where the Fourier coefficients of $f(x)$ are a_k and those of $(1/\sqrt{2\pi})e^{imx}x$ are given by (85) and (86). As a result, we obtain the expression

$$a'_m = i \sum_{k=-\infty}^{\infty}{}' \frac{(-1)^{m-k}}{m-k} a_k, \tag{87}$$

where the prime on the summation sign means that the term corresponding to $k = m$ is to be omitted. The formula (87) gives the linear transformation in the space H which corresponds to the operator (83), if the functions (73) are chosen as a coordinate system in the function space.

The device used to derive the transformation (87) is a special case of the following general procedure for representing an operator L in terms of a coordinate system corresponding to a complete orthonormal system. Let

L be a linear operator, and let

$$\varphi_1(x),\ \varphi_2(x),\ \ldots$$

be a complete orthonormal system. Write the Fourier series

$$L\varphi_i(x) \sim \sum_{k=1}^{\infty} a_{ki}\varphi_k(x);$$

if L is Hermitian, then $a_{ik} = \bar{a}_{ki}$. Let $\psi(x)$ be a function with the Fourier series

$$\psi(x) \sim \sum_{k=1}^{\infty} c_k\varphi_k(x).$$

Then the function $L\psi(x)$ has the Fourier series

$$L\psi(x) \sim \sum_{k=1}^{\infty} c_k'\varphi_k(x),$$

where, as is easily seen, the c_k' are given by the linear transformation

$$c_k' = \sum_{j=1}^{\infty} a_{kj}c_j \qquad (k = 1, 2, \ldots). \tag{88}$$

Returning to the differential operator (80), we note that this operator cannot be applied to every function in a space consisting only of *continuous* functions, since there exist nondifferentiable continuous functions; in fact, there exist continuous functions $f(x)$ which are not differentiable for any value of x. This can be seen in another way: The operator (80) corresponds to the linear transformation (81) in the space H. Even if the series

$$\sum_{-\infty}^{\infty} |a_k|^2$$

converges, the series

$$\sum_{-\infty}^{\infty} |a_k'|^2 = \sum_{-\infty}^{\infty} k^2|a_k|^2$$

may turn out to be divergent. This shows that the transformation (81) is not applicable to the whole space H.

A more detailed and rigorous discussion of this material will be given in a later volume.

PROBLEMS†

1. Let $\{x_n\}$ be an infinite sequence such that

$$\sum_{n=1}^{\infty} |x_n|^2 = \infty.$$

Show that there exists another infinite sequence $\{a_n\}$, with

$$\sum_{n=1}^{\infty} |a_n|^2 < \infty,$$

such that

$$\sum_{n=1}^{\infty} a_n x_n$$

does not converge.

2. In connection with (12), show that if

$$|(\mathbf{x}, \mathbf{y})| = \|\mathbf{x}\| \, \|\mathbf{y}\|,$$

then $\mathbf{x} = c\mathbf{y}$ for some constant c; in connection with (13), show that if

$$\|\mathbf{x} + \mathbf{y}\| = \|\mathbf{x}\| + \|\mathbf{y}\|,$$

then $\mathbf{x} = c\mathbf{y}$ for some constant $c \geq 0$.

3. Show that

$$\|\mathbf{x}\| = \sup_{\mathbf{y}} |(\mathbf{x}, \mathbf{y})|,$$

where the supremum (least upper bound) is taken over all vectors $\mathbf{y}$ for which $\|\mathbf{y}\| \leq 1$.

4. Show that if $\mathbf{u}^{(k)} \Rightarrow \mathbf{u}$ and $c_k \to c$, then $c_k\mathbf{u}^{(k)} \Rightarrow c\mathbf{u}$ (cf. Sec. 46).

5. Let $\{\mathbf{x}_n\}$ be an infinite sequence for which

$$\sum_{n=1}^{\infty} |x_n|^2 < \infty.$$

Show that

$$\sum_{n=1}^{\infty} |x_n x_{n+1}| \leq \sum_{n=1}^{\infty} |x_n|^2.$$

Show that the equality sign cannot hold unless all the x_n are zero.

6. The infinite sequence of vectors $\mathbf{x}^{(k)}$ is said to converge *weakly* to the vector $\mathbf{x}$ if the scalar products $(\mathbf{x}^{(k)}, \mathbf{y})$ converge to the scalar product $(\mathbf{x}, \mathbf{y})$ for every vector $\mathbf{y}$. Show that if $\mathbf{x}^{(k)} \Rightarrow \mathbf{x}$, then $\mathbf{x}^{(k)}$ converges weakly to $\mathbf{x}$.

7. The converse of Prob. 6 is not true. Let $\mathbf{e}^{(n)}$ be the vector in H with 1 in the nth position and 0 in all other positions. Show that $\mathbf{e}^{(n)}$ converges weakly to the zero vector, but that $\mathbf{e}^{(n)}$ does not $\Rightarrow 0$.

8. Show that if $\mathbf{x}^{(k)}$ converges weakly to $\mathbf{x}$, and if in addition $\|\mathbf{x}^{(k)}\|$ converges to $\|\mathbf{x}\|$, then $\mathbf{x}^{(k)} \Rightarrow \mathbf{x}$.

† All equation numbers refer to the corresponding equations of Chap. 5. For hints and answers, see p. 443. These problems are by A. L. Shields.

9. Show that a system of orthonormal vectors $\{\mathbf{x}^{(k)}\}$ is complete if and only if there is no nonzero vector orthogonal to all the vectors $\mathbf{x}^{(k)}$.

10. Let $\mathbf{e}^{(n)}$ be defined as in Prob. 7, and let $\mathbf{x}^{(n)} = \mathbf{e}^{(n)} + \mathbf{e}^{(n+1)}$. Show that there is no nonzero vector orthogonal to all the vectors $\mathbf{x}^{(n)}$.

11. Consider the linear transformation

$$x_k' = x_k + ax_{k+1} \qquad (a > 0),$$

and set $\mathbf{x} = (x_1', x_2', \ldots), \mathbf{x} = (x_1, x_2, \ldots)$. Find the smallest number M such that

$$\|\mathbf{x}'\|^2 \le M\|\mathbf{x}\|^2$$

(cf. (45)).

12. Let $A = \|a_{nm}\|$ be an infinite matrix for which

$$\|A\mathbf{x}\| \le N\|\mathbf{x}\|$$

for all $\mathbf{x}$ in H Prove that

$$\left| \sum_{n,m=1}^{\infty} a_{nm}x_m\bar{x}_n \right| \le N\|\mathbf{x}\|^2.$$

13. Let A be a bounded operator, and let $\mathbf{x}^{(k)} \Rightarrow \mathbf{x}$. Prove that $A\mathbf{x}^{(k)} \Rightarrow A\mathbf{x}$.

14. Show that if the infinite matrix $A = \|a_{nm}\|$ satisfies the two conditions

$$\sum_{m=1}^{\infty} |a_{nm}| \le M \qquad (n = 1, 2, \ldots),$$

$$\sum_{n=1}^{\infty} |a_{nm}| \le N \qquad (m = 1, 2, \ldots),$$

then

$$\|A\mathbf{x}\|^2 \le MN\|\mathbf{x}\|^2$$

for all $\mathbf{x}$ in H.

Comment: This test for the boundedness of a matrix transformation is due to I. Schur, *J. Reine Angew. Math.*, **140:** 1–128 (1911).

15. Show that if both A and B satisfy the conditions of the preceding problem, then so does BA.

16. Show that the infinite matrix

$$A = \left\| \frac{1}{(n+m)^\alpha} \right\|$$

is a bounded operator if $\alpha > 1$.

Comment: A is still bounded if $\alpha = 1$ (the *Hilbert transform*), but the proof is more delicate. For $\alpha < 1$, A is not bounded.

17. Let $A = \|a_{nm}\|$ have the property that

$$\sum_{n,m=1}^{\infty} |a_{nm}|^2 = M.$$

(Transformations whose matrices have this property are called *Hilbert-Schmidt transformations.*) Show that

$$\|A\mathbf{x}\|^2 \le M\|\mathbf{x}\|^2$$

for all $\mathbf{x}$ in H.

18. Show that if both A and B satisfy the condition of the preceding problem, then so does BA.

19. Let $A = \|a_{nm}\|$ be an infinite matrix, and let B be the matrix whose elements are $|a_{nm}|$. Show that if B is bounded, then A is also bounded. (In this case, A is said to be *absolutely bounded.*)

Comment: The converse is not true: It can be shown that the matrix $\|a_{nm}\|$ with elements

$$a_{nm} = \frac{1}{n-m} \qquad (n \neq m),$$
$$a_{nn} = 0$$

is bounded, but not absolutely bounded. See G. H. Hardy, J. E. Littlewood, and G. Pólya, "Inequalities," 2d ed., p. 214, Cambridge University Press, 1952.

20. Let $f_n(x)$, $n = 1, 2, \ldots$, be a sequence of continuous functions on the interval $[a, b]$, which converge uniformly to a function $f(x)$. Show that

$$\|f_n(x) - f(x)\| \to 0$$

(cf. (54)). Give an example to show that the converse statement is not true in general.

21. Let $f(x)$ be a square integrable function defined on the interval $[a, b]$, and let $\varphi_k(x)$, $k = 1, 2, \ldots$, be a complete orthonormal system on $[a, b]$. Show that the choice of the constants a_k minimizing the *mean square error*

$$\left\| f(x) - \sum_{k=1}^{n} a_k\varphi_k(x) \right\|$$

(cf. (54)) is given by the Fourier coefficients (56), *regardless of the value of n.*

22. The functions

$$\frac{1}{\sqrt{2\pi}} e^{ikx} \qquad (k = 0, \pm1, \pm2, \ldots)$$

form a complete orthonormal system on $[-\pi, \pi]$. Show that the *discontinuous* function $f(x)$, equal to 1 for $-\pi/2 \leq x \leq \pi/2$ and equal to 0 elsewhere, has Fourier coefficients a_k with respect to this system (cf. (56)), for which

$$\sum_{k=1}^{\infty} |a_k|^2 < \infty.$$

Comment: This shows that there are vectors in H which do not correspond to functions in F, as remarked in Sec. 50.

23. Which of the following operators L are linear $(0 \leq x \leq 1)$?

(a) $Lf(x) = x^2f(x)$; (b) $Lf(x) = f(1 - x)$; (c) $Lf(x) = \int_0^1 \sin(xf(t))\,dt$;
(d) $Lf(x) = |f(x)|$.

24. In equation (74), let $K(x, t) = 1/(x + t)$, $f(t) = t$, $a = 0$, $b = 1$, and find $F(x)$. Is it true that

$$\int_0^1 |F(x)|^2\,dx < \infty?$$

25. Let L be the linear operator of equation (80), let S be defined by

$$Sf(x) = i \int_0^x f(t)\,dt,$$

and let I be the identity transformation. Show that $LS = I$, but $SL \neq I$.

26. Find the matrix corresponding to the operator $Lf(x) = f(-x)$, with respect to the coordinate system

$$\varphi_k(x) = \frac{1}{\sqrt{2\pi}} e^{ikx} \qquad (k = 0, \pm 1, \pm 2, \ldots)$$

on $[-\pi, \pi]$.

27. A set K of vectors is said to be *convex* if for each pair of vectors $\mathbf{x}$ and $\mathbf{y}$ in K, the quantity

$$t\mathbf{x} + (1 - t)\mathbf{y} \qquad 0 \leq t \leq 1,$$

the "line segment joining $\mathbf{x}$ and $\mathbf{y}$," is also in K.

Show that

(a) The *unit ball*, i.e., the set of all vectors whose norms are ≤ 1, is convex;
(b) If K is convex and if L is a linear operator, then $L(K)$, the image of K under the transformation L, is convex.

Chapter 6

Reduction of Matrices to Canonical Form

52. Preliminary Considerations. This chapter is devoted to a proof of the following proposition, which was stated without proof in Sec. 27: Given a matrix A, we can always find a matrix V with a nonvanishing determinant, such that the matrix VAV^{-1}, which is similar to A, has the quasi-diagonal (or perhaps diagonal) form

$$VAV^{-1} = [I_{\rho_1}(\lambda_1), I_{\rho_2}(\lambda_2), \ldots, I_{\rho_p}(\lambda_p)], \tag{1}$$

where the matrix $I_\rho(\lambda)$ has the form

$$I_\rho(\lambda) = \begin{Vmatrix} \lambda & 0 & 0 & \cdots & 0 & 0 \\ 1 & \lambda & 0 & \cdots & 0 & 0 \\ 0 & 1 & \lambda & \cdots & 0 & 0 \\ \cdot & \cdot & \cdot & \cdot & \cdot & \cdot \\ 0 & 0 & 0 & \cdots & \lambda & 0 \\ 0 & 0 & 0 & \cdots & 1 & \lambda \end{Vmatrix}. \tag{2}$$

The subscript ρ indicates the order of the matrix $I_\rho(\lambda)$, and the argument λ gives the number appearing in the principal diagonal. If $\rho = 1$, the matrix $I_1(\lambda)$ reduces to the number λ. In the process of proving this assertion, we shall supplement it in a few important respects.

We begin by recalling the geometrical meaning of the transition to a similar matrix. The matrix A of order n is an operator in n-dimensional space in the sense that it brings about a linear transformation of the space. As we have seen (Sec. 21), the form of the matrix A depends on the choice of coordinate system, i.e., on the choice of basis vectors. If the matrix A represents a linear transformation relative to a given set of basis vectors, and if we transform to new coordinates in such a way that the new components of every vector are expressed in terms of the old components by using the transformation V, then in the new coordinate system our linear transformation is represented by the matrix VAV^{-1}. Thus, the problem just formulated is essentially the problem of finding new basis vectors that are in a certain sense more natural for the linear transformation which is represented by the matrix A relative to the old basis vectors; in fact, the problem reduces to finding new basis vectors relative to which our linear transformation is represented by a matrix of the canonical form $[I_{p_1}(\lambda_1), I_{p_2}(\lambda_2), \ldots, I_{p_s}(\lambda_s)]$.

Before solving this problem, we first present some auxiliary material, which will be needed later. Even though most of this material has been discussed previously, we gather it together here in the interest of completeness. We begin by discussing the concept of a subspace, a concept we have already encountered frequently. If $\mathbf{x}_1, \mathbf{x}_2, \ldots, \mathbf{x}_k$ are k linearly independent vectors in an n-dimensional space ($k \leq n$), then by the subspace of dimension k spanned by these vectors is meant the set of all vectors of the form

$$c_1\mathbf{x}_1 + c_2\mathbf{x}_2 + \cdots + c_k\mathbf{x}_k, \tag{3}$$

where the c_s are arbitrary numbers. In the case $k = n$, the subspace coincides with the whole n-dimensional space. As we have seen (Sec. 12), there is another definition of a subspace equivalent to the one just given, i.e., a subspace is a set of vectors M with the following two properties: (*a*) if a vector $\mathbf{x}$ belongs to M, then the vector $c\mathbf{x}$ also belongs to M, where c is an arbitrary number; and (*b*) if two vectors $\mathbf{x}_1$ and $\mathbf{x}_2$ belong to M, then their sum $\mathbf{x}_1 + \mathbf{x}_2$ also belongs to M. In other words, subspaces are sets of vectors which are *closed* under the operations of addition and multiplication by numbers (scalars).

In what follows, we shall be concerned with the following two methods of defining subspaces.† Let A be a matrix of order n, and let $\mathbf{x}$ be an arbitrary vector in an n-dimensional space. The set of vectors $\mathfrak{R}$ defined by the formula

$$\boldsymbol{\xi} = A\mathbf{x} \tag{4}$$

is clearly a subspace, which may coincide with the whole space. In fact, if $\boldsymbol{\xi} = A\mathbf{x} \in \mathfrak{R}$, where the symbol $\in$ means "is an element of," then $c\boldsymbol{\xi} = A(c\mathbf{x}) \in \mathfrak{R}$, and if $\boldsymbol{\xi}_1 = A\mathbf{x}_1 \in \mathfrak{R}$, $\boldsymbol{\xi}_2 = A\mathbf{x}_2 \in \mathfrak{R}$, then

$$\boldsymbol{\xi}_1 + \boldsymbol{\xi}_2 = A(\mathbf{x}_1 + \mathbf{x}_2) \in \mathfrak{R};$$

thus, the formula (4), where $\mathbf{x}$ is a variable vector, does define a subspace. As we have already noted (Sec. 15), the dimension of this subspace equals the rank of the matrix A.

We now indicate a second way of defining a subspace. Let A be a matrix of order n, and consider the set of vectors satisfying the equation

$$A\mathbf{x} = 0, \tag{5}$$

where 0 stands for the zero vector. As before, we can show that this set of vectors forms a subspace, and, as we have already seen (Sec. 14), the

† Cf. Problem 15.

dimension of this subspace is k, where $n - k$ is the rank of the matrix A. Of course, when we speak of a subspace, we always assume that it is not the subspace Θ whose only element is the zero vector, i.e., that it actually contains nonzero vectors. The formula (4) leads to the subspace Θ (i.e., yields the zero vector for any choice of the vector $\mathbf{x}$) if and only if the matrix A is zero (i.e., all the elements of A are zero). This is a direct consequence of the form of the linear transformation corresponding to A.

We say that the subspaces $E_1, E_2, \ldots, E_m$ form a *complete system of subspaces* if every vector $\mathbf{x}$ of the original space can be uniquely represented as a sum of vectors

$$\mathbf{x} = \boldsymbol{\xi}_1 + \boldsymbol{\xi}_2 + \cdots + \boldsymbol{\xi}_n, \tag{6}$$

where $\boldsymbol{\xi}_s$ belongs to E_s.† It follows from the assumption that the representation (6) is unique that the zero vector cannot be represented as a sum (6) unless all the summands are zero; this is equivalent to the fact that there can be no linear dependence between the vectors of the subspaces E_s. As an example, consider the ordinary three-dimensional space spanned by the vectors emanating from an origin O. For a complete system of subspaces, we can take a plane L passing through O and a line l passing through O which does not lie in L. The first subspace is two-dimensional, and can be spanned by any two noncollinear vectors lying in L. The second subspace is one-dimensional, and can be spanned by any vector lying on the line l. Every vector of our three-dimensional space can be uniquely represented as a sum of two vectors, one lying in L and the other in l.

Let A be a matrix corresponding to a linear transformation in an n-dimensional space, and suppose that we have succeeded in finding a complete system of subspaces $E_1, E_2, \ldots, E_m$ of dimensions $\rho_1, \rho_2, \ldots, \rho_m$, respectively, such that each subspace is *invariant* with respect to the linear transformation corresponding to A. In other words, as a result of the linear transformation produced by the matrix A, any vector of the subspace E_s ($s = 1, 2, \ldots, m$) goes into a vector of the same subspace. In this case, we have the following natural choice of basis vectors, relative to which the matrix A has the quasi-diagonal form of structure $\{\rho_1, \rho_2, \ldots, \rho_m\}$ (Sec. 26): For the first ρ_1 basis vectors, take any ρ_1 linearly independent vectors spanning the subspace E_1; for the next ρ_2 basis vectors, take any ρ_2 linearly independent basis vectors spanning the subspace E_2; etc. Since the E_s are a complete system of subspaces, we obviously have $\rho_1 + \rho_2 + \cdots + \rho_m = n$. It is easily seen that with this choice of basis vectors, our matrix A actually has quasi-diagonal form. To keep the notation simple, we show this in detail only for the case $m = 2$.

† Cf. Probs. 1 to 3.

Let $\mathbf{x} = (x_1, x_2, \ldots, x_n)$ be any vector, and let $\mathbf{x}' = (x_1', x_2', \ldots, x_n')$ be the new vector obtained from $\mathbf{x}$ by applying the linear transformation A, i.e., $\mathbf{x}' = A\mathbf{x}$. Because of our choice of basis vectors and the invariance of the subspace E_1,

$$x_{p_1+1} = x_{p_1+2} = \cdots = x_n = 0$$

must imply

$$x'_{p_1+1} = x'_{p_1+2} = \cdots = x'_n = 0;$$

similarly, since E_2 is invariant,

$$x_1 = x_2 = \cdots = x_{p_1} = 0$$

implies

$$x_1' = x_2' = \cdots = x'_{p_1} = 0.$$

It follows at once that with the choice of basis vectors just made, our linear transformation corresponds to a quasi-diagonal matrix of the form

$$\left\| \begin{matrix} a_{11} & a_{12} & \cdots & a_{1p_1} & 0 & 0 & \cdots & 0 \\ a_{21} & a_{22} & \cdots & a_{2p_1} & 0 & 0 & \cdots & 0 \\ \cdot & \cdot & \cdot & \cdot & \cdot & \cdot & \cdot & \cdot \\ a_{p_11} & a_{p_12} & \cdots & a_{p_1p_1} & 0 & 0 & \cdots & 0 \\ 0 & 0 & \cdots & 0 & b_{11} & b_{12} & \cdots & b_{1p_2} \\ 0 & 0 & \cdots & 0 & b_{21} & b_{22} & \cdots & b_{2p_2} \\ \cdot & \cdot & \cdot & \cdot & \cdot & \cdot & \cdot & \cdot \\ 0 & 0 & \cdots & 0 & b_{p_21} & b_{p_22} & \cdots & b_{p_2p_2} \end{matrix} \right\| = [A', B']. \tag{7}$$

It should be noted that the choice of basis vectors in each subspace is completely arbitrary; subsequently, we shall use this arbitrariness to reduce each of the submatrices A', B' appearing in the quasi-diagonal matrix (7) to an especially simple form.

We now consider polynomials with matrix arguments. Let $f(z)$ be a polynomial

$$f(z) = a_0z^p + a_1z^{p-1} + \cdots + a_{p-1}z + a_p,$$

where the c_s are arbitrary complex numbers. Replacing z by a matrix A, we obtain a *matrix polynomial*

$$f(A) = a_0A^p + a_1A^{p-1} + \cdots + a_{p-1}A + a_p, \tag{8}$$

i.e., the operations indicated in the right-hand side of (8) yield a new matrix $f(A)$. Since positive integral powers of the same matrix commute with each other (and with arbitrary complex numbers), polynomials in a matrix A can be added and multiplied by numbers according to the usual rules of algebra, just as in the case of polynomials with a numerical argument. Thus, if we have an identity which relates several polynomials with a numerical argument z and which involves the operations of addition and multiplication, the same identity is valid if we replace the argument z by any matrix A.

A basic role in the problem of reducing a matrix to canonical form is played by the characteristic equation

$$\varphi(\lambda) = \begin{vmatrix} a_{11} - \lambda & a_{12} & \cdots & a_{1n} \\ a_{21} & a_{22} - \lambda & \cdots & a_{2n} \\ \cdots & \cdots & \cdots & \cdots \\ a_{n1} & a_{n2} & \cdots & a_{nn} - \lambda \end{vmatrix}$$
$$= (-1)^n\lambda^n + \Delta_1\lambda^{n-1} + \cdots + \Delta_{n-1}\lambda + \Delta_n = 0, \quad (9)$$

where the a_{ik} are the elements of the matrix A. This equation can be written in the form

$$\det (A - \lambda) = 0. \quad (10)$$

For simplicity, we write $A - \lambda$ instead of $A - \lambda I$, where I is the unit matrix; a similar convention prevails throughout this chapter. We now prove the *Hamilton-Cayley theorem*, according to which

$$\varphi(A) = 0, \quad (11)$$

i.e., if the matrix A is substituted for the argument of the *characteristic polynomial* $\varphi(\lambda)$, the result is the matrix all of whose elements are zero.

The proof is as follows: If $\varphi(\lambda) \neq 0$, then

$$(A - \lambda)^{-1} = \frac{B(\lambda)}{\varphi(\lambda)}, \quad (12)$$

where each element of the matrix $B(\lambda)$ is the cofactor of the corresponding element of the matrix $A - \lambda$ (Sec. 3). It follows that $B(\lambda)$ is a polynomial in λ of degree $n - 1$:

$$B(\lambda) = B_{n-1}\lambda^{n-1} + B_{n-2}\lambda^{n-2} + \cdots + B_0, \quad (13)$$

where the matrices $B_0, B_1, \ldots, B_{n-1}$ are numerical matrices independent of the parameter λ. Equation (12) is equivalent to

$$(A - \lambda)B(\lambda) = \varphi(\lambda). \quad (14)$$

Substituting (9) and (13) into (14), and equating corresponding powers of λ, we obtain

$$\begin{aligned} -B_{n-1} &= (-1)^n, \\ B_{n-1}A - B_{n-2} &= \Delta_1, \\ \cdots\cdots & \cdots\cdots \\ B_1A - B_0 &= \Delta_{n-1}, \\ B_0A &= \Delta_n. \end{aligned} \quad (15)$$

Multiplying the first of the equations (15) by A^n, the second by A^{n-1}, etc., and then adding the resulting equations, we obtain

$$(-1)^nA^n + \Delta_1A^{n-1} + \cdots + \Delta_{n-1}A + \Delta_n = \varphi(A) = 0,$$

which proves the Hamilton-Cayley theorem.

We now prove the following theorem about the roots of the characteristic equation (9) (i.e., the eigenvalues of the matrix A): *If* $\lambda_1, \lambda_2, \ldots, \lambda_n$ *are the eigenvalues of the matrix* A, *then the eigenvalues of the matrix* A^s, *where* s *is a positive integer, are* $\lambda_1^s, \lambda_2^s, \ldots, \lambda_n^s$.

PROOF. Bearing in mind that the leading term of the polynomial $\varphi(\lambda)$ is $(-\lambda)^n$, we can write the following identity in λ:

$$\det(A - \lambda) = \prod_{k=1}^{n} (\lambda_k - \lambda). \tag{16}$$

Let $\epsilon = \exp(2\pi i/s)$ be an sth root of unity. Then we have the identity

$$(z - \lambda)(z - \epsilon\lambda) \cdots (z - \epsilon^{s-1}\lambda) = z^s - \lambda^s. \tag{17}$$

Using (16) and (17) and the fact that the determinant of a product of matrices equals the product of the determinants of the factors, we can write

$$\det(A^s - \lambda^s) = \prod_{k=1}^{n} (\lambda_k - \lambda) \prod_{k=1}^{n} (\lambda_k - \epsilon\lambda) \cdots \prod_{k=1}^{n} (\lambda_k - \epsilon^{s-1}\lambda)$$

or
$$\det(A^s - \lambda^s) = \prod_{k=1}^{n} [(\lambda_k - \lambda)(\lambda_k - \epsilon\lambda) \cdots (\lambda_k - \epsilon^{s-1}\lambda)].$$

Then, using the identity (17) again, we obtain

$$\det(A^s - \lambda^s) = \prod_{k=1}^{n} (\lambda_k^s - \lambda^s),$$

or

$$\det(A^s - \mu) = \prod_{k=1}^{n} (\lambda_k^s - \mu),$$

which completes the proof of the theorem.

Later, we shall have to calculate determinants of matrices of quasi-diagonal form

$$A = [A_1, A_2, \ldots, A_k].$$

It is not hard to see that the determinant of A equals the product of the determinants of the submatrices A_s:

$$\det A = \det A_1 \det A_2 \cdots \det A_k. \tag{18}$$

For simplicity, we verify this only for the case $k = 2$. By the multiplication theorem for matrices, we have

$$[A_1, A_2] = [A_1, I][I, A_2],$$

which implies that

$$\det A = \det [A_1, I] \det [I, A_2].$$

Applying the rules for expanding determinants with respect to elements of a row or column, we obtain

$$\det [A_1, I] = \det A_1,$$
$$\det [I, A_2] = \det A_2,$$

from which the formula (18) follows at once, for the case $k = 2$.

Finally, we note that similar matrices have identical eigenvalues (Sec. 27).

53. The Case of Distinct Roots. A complete discussion of the reduction of a matrix to canonical form in the case where the matrix has distinct eigenvalues was given in Sec. 27. We now solve this problem somewhat differently, using an approach which can be generalized later to the case where there are identical eigenvalues.

Let the matrix A have distinct eigenvalues $\lambda_1, \lambda_2, \ldots, \lambda_n$. In this case, as we have seen (Sec. 27), there exist n linearly independent vectors $\mathbf{v}_1, \mathbf{v}_2, \ldots, \mathbf{v}_n$ which satisfy the equations

$$A\mathbf{v}_k = \lambda_k\mathbf{v}_k \qquad (k = 1, 2, \ldots, n)$$

or

$$(A - \lambda_k)\mathbf{v}_k = 0 \qquad (k = 1, 2, \ldots, n). \tag{19}$$

Each of these vectors $\mathbf{v}_k$ spans a one-dimensional subspace E_k, and the subspaces $E_1, E_2, \ldots, E_n$ constitute a complete system of subspaces. Every vector of the form $c_k\mathbf{v}_k$, where c_k is any number, obviously satisfies the equation $Ac_k\mathbf{v}_k = \lambda_k c_k\mathbf{v}_k$, i.e., $c_k\mathbf{v}_k$ is just multiplied by the number λ_k when acted upon by the transformation A. Thus, each of the subspaces E_k is invariant under the linear transformation corresponding to the matrix A. Choosing the vectors $\mathbf{v}_k$ as basis vectors, we reduce A not only to quasi-diagonal form, but actually to diagonal form, since each of the subspaces E_k is one-dimensional.

Now we write the equation

$$(A - \lambda_k)\mathbf{x} = 0 \qquad (k = 1, 2, \ldots, n), \tag{20}$$

which is obviously satisfied by the vectors of the subspace E_k. It is easy to see that (20) has no solutions other than these vectors, i.e., that the subspace defined by (20) (cf. equation (5) of Sec. 52) is one-dimensional. In fact, suppose that (20) defines a subspace of dimension greater than 1, e.g., a two-dimensional subspace. Then, as we proved in Sec. 27, each vector of this subspace is linearly independent of the vectors of the other subspaces E_s ($s \neq k$), and we thereby obtain $n + 1$ linearly independent vectors in an n-dimensional space, which is impossible. Thus, in this case, (20) defines the subspaces $E_1, E_2, \ldots, E_n$.

These subspaces can be defined somewhat differently as follows: Consider the characteristic polynomial $\varphi(z)$, and expand $1/\varphi(z)$ in partial fractions, obtaining

$$\frac{1}{\varphi(z)} = \sum_{k=1}^{n} \frac{a_k}{z - \lambda_k},$$

or

$$\sum_{k=1}^{n} a_k \frac{\varphi(z)}{z - \lambda_k} = 1,$$

where the a_k are numbers different from zero. Replacing z by the matrix A, we obtain

$$\sum_{k=1}^{n} a_k \frac{\varphi(A)}{A - \lambda_k} = 1. \tag{21}$$

Consider the subspaces $E_1', E_2', \ldots, E_n'$ defined by the formulas

$$\xi = a_k \frac{\varphi(A)}{A - \lambda_k} \mathbf{x} \qquad (k = 1, 2, \ldots, n), \tag{22}$$

where $\mathbf{x}$ is a variable vector ranging over the whole space (cf. equation (4) of Sec. 52). The constant factor a_k obviously plays no role in (22). It follows from (21) that any vector $\mathbf{x}$ can be written in the form

$$\mathbf{x} = \sum_{k=1}^{n} a_k \frac{\varphi(A)}{A - \lambda_k} \mathbf{x}, \tag{23}$$

where each of the summands belongs to one of the subspaces E_k'.

We now show that the subspaces E_k' defined by (22) are the same as the subspaces E_k defined by (20). The proof is as follows: If ξ is a vector in the subspace E_k' defined by (22), then, by the Hamilton-Cayley theorem (Sec. 52), we have

$$(A - \lambda_k)\xi = a_k\varphi(A)\mathbf{x} = 0,$$

i.e., every vector of E_k' belongs to E_k. Next we show that the converse is true, i.e., that any vector η_k in E_k can be obtained from the formula (22) for a suitable choice of $\mathbf{x}$. Since each polynomial

$$\frac{\varphi(A)}{A - \lambda_s} \qquad (s \neq k)$$

contains the factor $A - \lambda_k$, then, by the formula (20) defining E_k, we have

$$\frac{\varphi(A)}{A - \lambda_s} \eta_k = 0 \qquad (s \neq k).$$

Thus, replacing $\mathbf{x}$ in (23) by η_k, we obtain

$$\eta_k = a_k \frac{\varphi(A)}{A - \lambda_k} \eta_k,$$

i.e., η_k can actually be obtained from (22) by suitably choosing $\mathbf{x}$, namely, by choosing $\mathbf{x}$ to be η_k itself, QED.

We shall use considerations completely analogous to those just given, when we discuss the case where the characteristic equation has multiple roots. This will allow us to decompose the whole n-dimensional space into a complete system of subspaces (each corresponding to a multiple root of the characteristic equation) which are invariant under the linear transformation corresponding to the matrix A. Then, the next step will be to choose basis vectors in every subspace such that A is reduced to the quasi-diagonal form (1).

54. The Case of Multiple Roots. First Step in the Reduction. Suppose that the characteristic equation (9) has a root α_1 of multiplicity r_1, a root α_2 of multiplicity r_2, etc., and finally a root α_s of multiplicity r_s. Expanding $1/\varphi(z)$ in partial fractions, we obtain an expression of the form

$$\frac{1}{\varphi(z)} = \sum_{k=1}^{s} \frac{g_k(z)}{(z - \alpha_k)^{r_k}},$$

where $g_k(z)$ is a polynomial in z of degree no greater than $r_k - 1$, for which $g_k(\alpha_k) \neq 0$. Consider the polynomials

$$f_k(z) = g_k(z) \frac{\varphi(z)}{(z - \alpha_k)^{r_k}} \qquad (k = 1, 2, \ldots, s). \tag{24}$$

We have the obvious identity

$$1 = \sum_{k=1}^{s} f_k(z);$$

if we replace z by the matrix A, this becomes

$$1 = \sum_{k=1}^{s} f_k(A).$$

Thus, we can represent any vector $\mathbf{x}$ in the form of a sum

$$\mathbf{x} = \sum_{k=1}^{s} f_k(A)\mathbf{x} \tag{25}$$

of s vectors.

Now let E_k ($k = 1, 2, \ldots, s$) be the subspace defined by the formula

$$\boldsymbol{\xi} = f_k(A)\mathbf{x} \qquad (k = 1, 2, \ldots, s). \tag{26}$$

We shall see later that none of these subspaces $E_1, E_2, \ldots, E_s$ is the set Θ consisting only of the zero vector. Let $\mathbf{x}_k$ be any vector of the subspace E_k; then we have

$$f_p(A)\mathbf{x}_q = 0 \qquad \text{if } p \neq q \tag{27}$$

and

$$f_p(A)\mathbf{x}_p = \mathbf{x}_p. \tag{28}$$

To see this, we note that, by definition

$$\mathbf{x}_q = f_q(A)\mathbf{x}$$

for some $\mathbf{x}$ in the original n-dimensional space. Therefore, by (24), we have

$$f_p(A)\mathbf{x}_q = g_p(A)g_q(A)\frac{[\varphi(A)]^2}{(A - \alpha_p)^{r_p}(A - \alpha_q)^{r_q}}\mathbf{x}. \tag{29}$$

If $p \neq q$, then the fraction in the right-hand side of (29) is a matrix polynomial containing the factor $\varphi(A)$, and therefore, by the Hamilton-Cayley theorem, this polynomial vanishes, which proves (27). To prove (28), it is sufficient to set $\mathbf{x} = \mathbf{x}_p$ in (25) and then apply (27); this immediately yields (28).

Next, we show that the subspaces $E_1, E_2, \ldots, E_k$ form a complete system of subspaces. Since (25) shows that every vector can be represented as a sum of vectors from the subspaces E_k, it remains only to show that there can be no linear dependence between the vectors of these subspaces. Suppose that there is a linear relation

$$C_1\mathbf{x}_1 + C_2\mathbf{x}_2 + \cdots + C_s\mathbf{x}_s = 0, \tag{30}$$

where the vector $\mathbf{x}_k$ belongs to the subspace E_k. We have to show that the coefficient C_k vanishes if $\mathbf{x}_k$ is nonzero. Applying the linear transformation $f_k(A)$ to both sides of (30) and using (27) and (28), we obtain

$$C_k\mathbf{x}_k = 0,$$

so that $C_k = 0$, if $\mathbf{x}_k \neq 0$, as required. Thus, the subspaces $E_1, E_2, \ldots, E_s$ actually form a complete system of subspaces, and the sum of the dimensions of these subspaces must equal n, the dimension of the whole space.

Another definition of the subspaces E_k ($k = 1, 2, \ldots, s$) is possible, i.e., the subspace E_k is the set of vectors satisfying the equation

$$(A - \alpha_k)^{r_k}\mathbf{x} = 0. \tag{31}$$

To see this, we first show that any vector satisfying (26) also satisfies (31). In fact, substituting $\boldsymbol{\xi} = f_k(A)\mathbf{x}$ for $\mathbf{x}$ in (31), we obtain

$$(A - \alpha_k)^{r_k} f_k(A)\mathbf{x} = g_k(A)\varphi(A)\mathbf{x} = 0,$$

since $\varphi(A) = 0$, by the Hamilton-Cayley theorem. It remains to prove that the converse is true, i.e., that any solution $\boldsymbol{\eta}$ of (31) can be obtained from (26) by suitably choosing $\mathbf{x}$. In fact, the equation

$$(A - \alpha_k)^{r_k}\boldsymbol{\eta} = 0 \tag{32}$$

implies

$$\boldsymbol{\eta} = f_k(A)\boldsymbol{\eta}. \tag{33}$$

To see this, we note that

$$\boldsymbol{\eta} = \sum_{p=1}^{s} f_p(A)\boldsymbol{\eta}, \tag{34}$$

according to (25). Since each polynomial $f_p(A)$ contains the factor $(A - \alpha_k)^{r_k}$ if $p \neq k$, (32) implies $f_p(A)\boldsymbol{\eta} = 0$ if $p \neq k$, so that (34) reduces to (33), as required.

If α_k is a root of the characteristic equation (9), then the equation

$$(A - \alpha_k)\mathbf{x} = 0$$

has a nontrivial solution $\mathbf{x} = \mathbf{v}_k$, since $\det (A - \alpha_k) = 0$ (cf. Sec. 27); hence, *a fortiori*, (31) is satisfied for $\mathbf{x} = \mathbf{v}_k$, so that $\mathbf{v}_k$ is an element of the subspace E_k. Thus, E_k contains vectors other than the zero vector, i.e., $E_k \neq \Theta$. Moreover, it follows from (31) that

$$(A - \alpha_k)^{r_k} A\mathbf{x} = A(A - \alpha_k)^{r_k}\mathbf{x},$$

i.e., if $\mathbf{x}$ satisfies (31), so does $A\mathbf{x}$. Thus, each subspace E_k is invariant with respect to the transformation corresponding to the matrix A.

Suppose now that the subspaces $E_1, E_2, \ldots, E_s$ have dimensions $q_1, q_2, \ldots, q_s$, respectively. Choosing basis vectors in the subspaces in the way indicated in the preceding section, we go from the matrix A to the similar matrix

$$S_1 A S_1^{-1} = [A_1, A_2, \ldots, A_s] \tag{35}$$

with quasi-diagonal form, where the submatrix A_k is of order q_k ($k = 1, 2, \ldots, s$). We shall show that the numbers q_k coincide with the multiplicities r_k of the roots of the characteristic equation and that each submatrix A_k has a single eigenvalue α_k of multiplicity r_k. The proof is as follows: Any vector ξ in the subspace E_k must satisfy (31). Relative to the new basis vectors, (31) becomes

$$S_1(A - \alpha_k)^{r_k}S_1^{-1}\xi = 0. \tag{36}$$

Since

$$S_1(A - \alpha_k)^2S_1^{-1} = S_1(A - \alpha_k)S_1^{-1}S_1(A - \alpha_k)S_1^{-1} = (S_1AS_1^{-1} - \alpha_k)^2,$$

and similarly for powers higher than two, (36) becomes

$$(S_1AS_1^{-1} - \alpha_k)^{r_k}\xi = 0,$$

or, in view of (35),

$$[A_1 - \alpha_k, A_2 - \alpha_k, \ldots, A_s - \alpha_k]^{r_k}\xi = 0. \tag{37}$$

Now let $k = 1$, say. Then all but the first q_1 components of ξ vanish, and instead of (37), we can write

$$(A_1 - \alpha_1)^{r_1}\xi' = 0, \tag{38}$$

where ξ' denotes an arbitrary vector in a q_1-dimensional space, and the matrix $(A_1 - \alpha_1)^{r_1}$ is of order q_1. Since (30) holds for any ξ', we must have

$$(A_1 - \alpha_1)^{r_1} = 0,$$

which implies at once that all the eigenvalues of the matrix $(A_1 - \alpha_1)^{r_1}$ vanish. Since these eigenvalues are obtained from the eigenvalues of the matrix $A_1 - \alpha_1$ by raising the latter to the power r_1 (Sec. 52), all the eigenvalues of $A_1 - \alpha_1$ are zero, so that all q_1 eigenvalues of the matrix A_1 are equal to α_1. In just the same way, we can prove more generally that all q_k eigenvalues of the matrix A_k (which has order q_k) are equal to α_k. But the matrix (35) is similar to A and hence has the same eigenvalues as A. Moreover, the characteristic equation of (35) is

$$\det [A_1 - \lambda, A_2 - \lambda, \ldots, A_s - \lambda] = 0,$$

or, by Sec. 52,

$$\det (A_1 - \lambda) \det (A_2 - \lambda) \cdots \det (A_s - \lambda) = 0.$$

It follows at once that the q_k must coincide with the r_k and that the matrix A_k has a single eigenvalue α_k of multiplicity r_k.

55. Reduction to Canonical Form. We have seen that each submatrix A has a single eigenvalue α_k of multiplicity r_k. To reduce A_k to the canonical form given in Sec. 52, we have to make a suitable choice of basis vectors in the subspace E_k. Thus, our problem has been reduced to studying the special case of a matrix D, say, with a single eigenvalue α of multiplicity r.

Let B be the matrix $D - \alpha$ of order r; clearly, B has a single eigenvalue 0 of multiplicity r. Therefore, B has the characteristic equation $(-1)^r\lambda^r = 0$, and the Hamilton-Cayley equation is just $B^r = 0$. However, it may happen that $B^s = 0$, where s is a positive integer less than r. Let l be the smallest such integer s, i.e.,

$$B^l = 0, \tag{39}$$

but $B^{l-1} \neq 0$. For example, if the matrix B itself is zero, then $l = 1$. If B is the matrix,

$$\begin{Vmatrix} 0 & 0 & 0 & 0 \\ 0 & 0 & 0 & 0 \\ 0 & 0 & 0 & 0 \\ 1 & 0 & 0 & 0 \end{Vmatrix},$$

it is easy to see that $l = 2$, since $B^2 = 0$. If the matrix B is zero, then $D = B + \alpha$ is the diagonal matrix

$$D = [\alpha, \alpha, \ldots, \alpha],$$

i.e., D is already in canonical form. Thus, we need only investigate the case $l > 1$.

In view of (39), the equation

$$B^l\mathbf{x} = 0$$

holds for every vector $\mathbf{x}$ in the r-dimensional space; this space will henceforth be denoted by ω. Consider next the equation

$$B^{l-1}\mathbf{x} = 0.$$

Since B^{l-1} is nonzero, this equation defines a subspace of ω of dimension less than r. In general, we define a whole series of subspaces by the equations

$$B^l\mathbf{x} = 0, B^{l-1}\mathbf{x} = 0, \ldots, B\mathbf{x} = 0. \tag{40}$$

Let F_m be the subspace defined by the equation $B^m\mathbf{x} = 0$, and let τ_m be the dimension of F_m. As already remarked, F_l coincides with the whole space ω and $\tau_l = r$, while $\tau_{l-1} < \tau_l$. If ξ is an element of F_m, i.e., if $B^m\xi = 0$, then $B^{m-1}(B\xi) = 0$, i.e., $B\xi$ is an element of F_{m-1}. Moreover, it is obvious that every element of F_m is also an element of F_{m+1}, i.e., F_m is a subset of F_{m+1}. In fact, we shall see below that the dimension of F_m is always less than the dimension of F_{m+1}, i.e., F_m is a *proper* subset of F_{m+1} and cannot coincide with F_{m+1}. Thus, we have (for the time being) the series of inequalities

$$\tau_l > \tau_{l-1} \geq \tau_{l-2} \geq \cdots \geq \tau_1. \tag{41}$$

(As noted, we shall show later that the sign > actually holds in all the inequalities of (41) as well as in the first inequality.)

Let $r_l = \tau_l - \tau_{l-1}$, where r_l is some positive integer. We can construct r_l linearly independent vectors $\xi_1, \ldots, \xi_{r_l}$ in the subspace F_l (i.e., in the whole space ω) such that no linear combination of these vectors belongs to F_{l-1}. Then, any vector in F_l can be represented as a linear combination of the vectors $\xi_1, \ldots, \xi_{r_l}$ and a vector in F_{l-1}; in particular, we can choose any τ_{l-1} linearly independent vectors in the subspace F_{l-1} and then supplement these vectors with the vectors $\xi_1, \ldots, \xi_{r_l}$ to make a complete system of linearly independent vectors in the space ω. In just the same way, let $r_{l-1} = \tau_{l-1} - \tau_{l-2}$, where r_{l-1} is some nonnegative integer, and construct r_{l-1} linearly independent vectors in the subspace F_{l-1} so that no linear combination of these vectors belongs to F_{l-2}. Now consider the vectors

$$B\xi_1, \ldots, B\xi_{r_l}. \tag{42}$$

All these vectors belong to F_{l-1}, but no linear combination of them can belong to F_{l-2}, for otherwise we would have

$$B^{l-2}(c_1 B\xi_1 + \cdots + c_{r_l} B\xi_{r_l}) = 0$$

or

$$B^{l-1}(c_1\xi_1 + \cdots + c_{r_l}\xi_{r_l}) = 0,$$

i.e., a linear combination of the vectors $\xi_1, \ldots, \xi_{r_l}$ would belong to the subspace F_{l-1}, which contradicts the definition of these vectors. Thus, we see that the vectors (42) are linearly independent vectors of F_{l-1} such that no linear combination of them belongs to F_{l-2}. Since, as we have just seen, we can construct r_{l-1} linearly independent vectors such that no linear combination of them belongs to F_{l-2}, it follows at once that $r_{l-1} \geq r_l$. In just the same way, writing $r_{l-2} = \tau_{l-2} - \tau_{l-3}$, we find that $r_{l-2} \geq r_{l-1}$, and more generally, writing $r_m = \tau_m - \tau_{m-1}$, we have

$$0 < r_l \leq r_{l-1} \leq r_{l-2} \leq \cdots \leq r_1 \qquad (r_1 = \tau_1). \tag{43}$$

This immediately implies, among other things, that we have the sign > everywhere in (41):

$$\tau_l > \tau_{l-1} > \cdots > \tau_1.$$

The number r_l can be called the dimension of the subspace F_l relative to the subspace F_{l-1} (which is a proper subset of F_l). More precisely, r_l is the number of linearly independent vectors in F_l such that no linear combination of them belongs to F_{l-1}. These vectors span a subspace G_l which is contained in F_l. Similarly, r_{l-1} is the dimension of F_{l-1} relative to F_{l-2}, and we again obtain a subspace G_{l-1} which this time is contained in F_{l-1}. More generally, r_m is the dimension of F_m relative to F_{m-1}, and any r_m linearly independent vectors in F_m with the property that no linear combination of them belongs to F_{m-1} span a subspace G_m contained in

F_m. The subspace G_1 coincides with F_1 itself. If ξ is a vector in G_m and hence in F_m, then $B\xi$ belongs to F_{m-1} but not to F_{m-2}, for otherwise we would have $B^{m-2}(B\xi) = 0$, so that ξ would belong not only to F_m but also to F_{m-1}, which contradicts the definition of G_m. Thus, if the linear transformation B acts on the subspace G_m, we obtain a part (or all) of the subspace G_{m-1}, and linearly independent vectors in G_m go into linearly independent vectors in G_{m-1}. It follows immediately from

$$r_m = \tau_m - \tau_{m-1} \qquad (r_1 = \tau_1)$$

that

$$r_l + r_{l-1} + \cdots + r_1 = r,$$

since $\tau_l = r$; therefore, the subspaces $G_l, \ldots, G_1$ form a complete system of subspaces.

We now discuss the last step, which consists of constructing the final subspaces, which are invariant under the linear transformation corresponding to the matrix B. We choose a vector ξ_1 in G_l as the first basis vector and then construct $l - 1$ more basis vectors by the following rule:

$$\xi_2 = B\xi_1,\ \xi_3 = B\xi_2,\ \ldots,\ \xi_l = B\xi_{l-1} \qquad (B\xi_l = B^l\xi_1 = 0). \quad (44)$$

It is an immediate consequence of the considerations given above that these basis vectors are linearly independent and that they belong in turn to the subspaces $G_l, G_{l-1}, \ldots, G_1$, respectively. It is not hard to see that they span a subspace which is invariant under the linear transformation. In fact, it follows from (44) that

$$B(c_1\xi_1 + c_2\xi_2 + \cdots + c_l\xi_l) = c_1\xi_2 + c_2\xi_3 + \cdots + c_{l-1}\xi_l,$$

for any choice of the constants c_k. This implies that the matrix of the linear transformation B relative to the basis vectors ξ_k of this subspace is a matrix of order l with the canonical form

$$I_l(0) = \left\| \begin{matrix} 0 & 0 & 0 & \cdots & 0 & 0 \\ 1 & 0 & 0 & \cdots & 0 & 0 \\ 0 & 1 & 0 & \cdots & 0 & 0 \\ \cdot & \cdot & \cdot & \cdots & \cdot & \cdot \\ 0 & 0 & 0 & \cdots & 1 & 0 \end{matrix} \right\|.$$

Having made this use of the vector ξ_1 from G_l, we choose another vector η_1 from G_l, which is linearly independent of ξ_1, and use η_1 to generate $l - 1$ more vectors,

$$\eta_2 = B\eta_1,\ \eta_3 = B\eta_2,\ \ldots,\ \eta_l = B\eta_{l-1}.$$

The l vectors $\eta_1, \eta_2, \ldots, \eta_l$ constructed in this way are linearly independent not only with respect to each other, but also with respect to the vectors ξ_k. This is an immediate consequence of the fact that linearly independent vectors in G_m go into linearly independent vectors in G_{m-1} under the transformation B. Choosing the vectors η_k as basis vectors, we obtain an invariant subspace in which our linear transformation is accomplished by using the matrix $I_l(0)$. Proceeding in this way, we use all r_l vectors in G_l to construct r_l invariant subspaces of dimension l, in each of which our linear transformation corresponds to a matrix of the form $I_l(0)$.

Now, we turn to the next subspace G_{l-1}, with dimension $r_{l-1} \geq r_l$; r_l of the vectors in G_{l-1} have just been used above in constructing basis vectors. Treating the $r_{l-1} - r_l$ remaining basis vectors in just the same way as before, we construct $r_{l-1} - r_l$ invariant subspaces, in each of which our linear transformation corresponds to a matrix of order $l - 1$ with the canonical form $I_{l-1}(0)$. More generally, when we arrive at the subspace G_m, we find that it contains $r_m - r_{m+1}$ linearly independent vectors which have not yet been used. Selecting these vectors in any way and applying the transformation B to each of them in turn, we obtain $m - 1$ more vectors from each of them; choosing these new vectors as basis vectors, we obtain $r_m - r_{m+1}$ sets of basis vectors, where each set contains m basis vectors and defines an invariant subspace of dimension m, in which our linear transformation is accomplished by using a matrix of order m with the canonical form $I_m(0)$. Finally, we arrive at the last subspace G_1, which contains $r_1 - r_2$ unused linearly independent vectors ξ, which obey the formula $B\xi = 0$. Choosing each of these vectors as a basis vector, we obtain $r_1 - r_2$ invariant one-dimensional subspaces, in each of which our linear transformation corresponds to the vanishing matrix $I_1(0)$ of order 1.

The net effect of this new choice of basis vectors is expressed by a linear transformation σ, which acts on the vectors of the space ω. As a result of applying the transformation σ, the linear transformation expressed by the matrix B relative to the old basis vectors is now expressed by the following quasi-diagonal matrix (which is similar to B) relative to the new basis vectors:

$$\sigma B \sigma^{-1} = [I_{\beta_1}(0), I_{\beta_2}(0), \ldots, I_{\beta_r}(0)]; \tag{45}$$

here, r_l of the subscripts appearing in the right-hand side are equal to l, $r_{l-1} - r_l$ are equal to $l - 1$, etc., and finally $r_1 - r_2$ are equal to 1. For the matrix $D = B + \alpha$, we obviously have

$$\sigma B \sigma^{-1} = \sigma B \sigma^{-1} + \sigma \alpha \sigma^{-1} = \sigma B \sigma^{-1} + \alpha;$$

this means that the number α has to be added to all the diagonal elements of (45), which leads to the formula

$$\sigma D \sigma^{-1} = [I_{\beta_1}(\alpha), I_{\beta_2}(\alpha), \ldots, I_{\beta_r}(\alpha)]. \tag{46}$$

We now return to our original matrix A. The formula (35) of the preceding section represents A in the form of a quasi-diagonal matrix $[A_1, A_2, \ldots, A_s]$, where each submatrix A_k has a single eigenvalue α_k of multiplicity r_k. According to the construction just given, each of these submatrices A_k can be reduced to the canonical form (46) by using a matrix σ_k of order r_k. Introducing the matrix

$$S_2 = [\sigma_1, \sigma_2, \ldots, \sigma_s],$$

we have

$$S_2[A_1, A_2, \ldots, A_s]S_2^{-1} = [\sigma_1 A_1 \sigma_1^{-1}, \sigma_2 A_2 \sigma_2^{-1}, \ldots, \sigma_s A_s \sigma_s^{-1}],$$

so that

$$(S_2S_1)A(S_2S_1)^{-1} = [I_{\rho_1}(\lambda_1), I_{\rho_2}(\lambda_2), \ldots, I_{\rho_p}(\lambda_p)], \tag{47}$$

i.e., *our matrix A has finally been reduced to canonical form.* The numbers λ_k must coincide with the numbers α_k, and the sum of the subscripts of the submatrices $I_{\rho_j}(\lambda_j)$ for which $\lambda_j = \alpha_k$ must equal r_k. Equation (47) completely solves the problem of reducing a given matrix A to *canonical form.*†

The problem now arises of proving the uniqueness of the representation (47), i.e., of proving that regardless of how A is reduced to canonical form, there appears in the right-hand side of (47) a perfectly definite number of matrices $I_{\rho_j}(\lambda_j)$ with given values of ρ_j and λ_j. Thus, suppose that the reduction of the matrix A to the canonical form

$$VAV^{-1} = [I_{\rho_1}(\lambda_1), I_{\rho_2}(\lambda_2), \ldots, I_{\rho_p}(\lambda_p)]$$

has been accomplished in any way. Bearing in mind that similar matrices must have the same characteristic equation, we can write the characteristic equation of the matrix A in the form

$$\det([I_{\rho_1}(\lambda_1), I_{\rho_2}(\lambda_2), \ldots, I_{\rho_p}(\lambda_p)] - \lambda) = 0,$$

or

$$\det [I_{\rho_1}(\lambda_1 - \lambda), I_{\rho_2}(\lambda_2 - \lambda), \ldots \quad I_{\rho_p}(\lambda_p - \lambda)] = 0.$$

It follows from the form of the matrix $I_\rho(a)$ that

$$\det I_\rho(a - \lambda) = (a - \lambda)^\rho.$$

Therefore, the numbers λ_j must coincide with the eigenvalues α_k of the matrix A, and the sum of indices ρ_j for which $\lambda_j = \alpha_k$ must equal the multiplicity r_k of the eigenvalue α_k. It remains to show that the numbers ρ_j can take only certain definite values. By studying the invariant subspaces, this fact can be proved with the help of the geometrical argument used to reduce A to canonical form. We do not carry out this proof,

† Synonymously, *Jordan form.*

since in the next section, we shall establish an algebraic criterion which completely specifies the indices ρ_j in terms of the given matrix A and therefore provides us with a proof of the uniqueness of the representation of A in canonical form. This criterion, which was given without proof in Sec. 27, is based on an examination of the greatest common divisors of the minors of a given order of the matrix $A - \lambda$.

56. Determination of the Structure of the Canonical Form. We begin by proving some lemmas.

LEMMA 1. *If A and B are two square matrices of order n and $C = AB$ is their product, then every minor of order t of the matrix C, where $t \leq n$, can be represented as a sum of products of certain minors of order t of the matrix A with certain minors of order t of the matrix B.*

This lemma is an immediate consequence of the corollary to the theorem proved in Sec. 7.

COROLLARY. Let the elements of the matrix $A(\lambda)$ be polynomials in λ; let the elements of the matrix B be independent of λ, and let $\det B \neq 0$. Let $d_t(\lambda)$ denote the greatest common divisor of all the minors of order t of the matrix $A(\lambda)$, and let $d_t'(\lambda)$ denote the greatest common divisor of all the minors of order t of the matrix $A(\lambda)B$. Then it is an immediate consequence of Lemma 1 that $d_t(\lambda)$ must divide $d_t'(t)$. But since we can write

$$A(\lambda) = [A(\lambda)B]B^{-1},$$

Lemma 1 shows in just the same way that $d_t'(\lambda)$ divides $d_t(\lambda)$, i.e.,

$$d_t(\lambda) = d_t'(\lambda).$$

We can obtain the same result by forming the matrix $BA(\lambda)$ instead of the matrix $A(\lambda)B$. It also follows from Lemma 1 that the matrices $A(\lambda)$ and the similar matrix $BA(\lambda)B^{-1}$ have the same greatest common divisor $d_t(\lambda)$ for their minors of order t.

We now show another important property of the greatest common divisor $d_t(\lambda)$. First, we introduce a new definition:

DEFINITION. By an *elementary transformation*† of the matrix $A(\lambda)$ whose elements are polynomials in λ, we mean a transformation of $A(\lambda)$ consisting of a finite number of the following three kinds of operations: (a) interchange of two rows (or columns); (b) multiplication of all the elements of a row (or column) by the same nonzero number; (c) addition of the elements of a row (or column), multiplied by the same number or by the same polynomial in λ, to the corresponding elements of another row (or column).

† Synonymously, *elementary operation* on the matrix $A(\lambda)$.

If the matrix $A_1(\lambda)$ is obtained from $A(\lambda)$ by performing elementary transformations, then obviously the converse is true, i.e., $A(\lambda)$ can be obtained from $A_1(\lambda)$ by performing elementary transformations. Two matrices which can be obtained from each other by using elementary transformations are said to be *equivalent.*

LEMMA 2. *Equivalent matrices have identical greatest common divisors* $d_t(\lambda)$ $(t = 1, 2, \ldots, n)$.

PROOF. It is sufficient to prove that if all the minors of order t of the matrix $A(\lambda)$ have a polynomial $\varphi(\lambda)$ as a common factor, then all the minors of order t of the equivalent matrix $A_1(\lambda)$ have the same common factor. The lemma is obviously true for the first and second kinds of operations listed above, since these operations just multiply the minors of order t of $A(\lambda)$ by a numerical factor. It remains to prove that the common factor $\varphi(\lambda)$ remains the same when the third type of transformation is carried out. For example, consider the transformation consisting of adding to the elements of the pth row the corresponding elements of the qth row $(q \neq p)$ multiplied by a polynomial $\psi(\lambda)$. According to Property 6 of determinants (Sec. 3), all the minors of order t which do not contain the pth row, or which contain the pth row and the qth row, remain unchanged as a result of this transformation. However, as a result of the transformation, the minors of order t which contain the pth row but do not contain the qth row have the form $A'(\lambda) \pm \psi(\lambda)A''(\lambda)$, where $A'(\lambda)$ and $A''(\lambda)$ are minors of order t of the matrix $A(\lambda)$. It follows at once that the common factor $\varphi(\lambda)$ of the minors of order t of the matrix $A(\lambda)$ divides all the minors of order t of the matrix $A_1(\lambda)$.

LEMMA 3. *The matrix*

$$I_\rho(a-\lambda) = \left\| \begin{matrix} a-\lambda & 0 & 0 & \cdots & 0 & 0 \\ 1 & a-\lambda & 0 & \cdots & 0 & 0 \\ 0 & 1 & a-\lambda & \cdots & 0 & 0 \\ \cdots & \cdots & \cdots & \cdots & \cdots & \cdots \\ 0 & 0 & 0 & \cdots & 1 & a-\lambda \end{matrix} \right\|$$

of order ρ can be reduced to the diagonal matrix

$$[1, \ldots, 1, (a-\lambda)^\rho]$$

by means of elementary transformations.

PROOF. In the case $\rho = 1$, the lemma is obvious. In the case $\rho = 2$, we perform the following elementary transformations in order: Interchange the first and second rows, multiply the first column by $-(a-\lambda)$ and add it to the second column, multiply the first row by $-(a-\lambda)$ and add it to the second row, and finally multiply the second column by -1. The separate steps of this process are

$$\begin{Vmatrix} a-\lambda & 0 \\ 1 & a-\lambda \end{Vmatrix} \to \begin{Vmatrix} 1 & a-\lambda \\ a-\lambda & 0 \end{Vmatrix} \to \begin{Vmatrix} 1 & 0 \\ a-\lambda & -(a-\lambda)^2 \end{Vmatrix} \to$$
$$\begin{Vmatrix} 1 & 0 \\ 0 & -(a-\lambda)^2 \end{Vmatrix} \to [1, (a-\lambda)^2].$$

In the case $\rho = 3$, we begin by performing the same elementary transformations as in the case $\rho = 2$, obtaining

$$\begin{Vmatrix} a-\lambda & 0 & 0 \\ 1 & a-\lambda & 0 \\ 0 & 1 & a-\lambda \end{Vmatrix} \to \begin{Vmatrix} 1 & 0 & 0 \\ 0 & (a-\lambda)^2 & 0 \\ 0 & -1 & a-\lambda \end{Vmatrix}.$$

We then interchange the second and third rows and carry out the following elementary transformations:

$$\begin{Vmatrix} 1 & 0 & 0 \\ 0 & (a-\lambda)^2 & 0 \\ 0 & -1 & a-\lambda \end{Vmatrix} \to \begin{Vmatrix} 1 & 0 & 0 \\ 0 & -1 & a-\lambda \\ 0 & (a-\lambda)^2 & 0 \end{Vmatrix} \to$$
$$\begin{Vmatrix} 1 & 0 & 0 \\ 0 & -1 & 0 \\ 0 & (a-\lambda)^2 & (a-\lambda)^3 \end{Vmatrix} \to \begin{Vmatrix} 1 & 0 & 0 \\ 0 & -1 & 0 \\ 0 & 0 & (a-\lambda)^3 \end{Vmatrix}.$$

Multiplying the second column by -1, we finally obtain

$$[1, 1, (a-\lambda)^3].$$

Thus, proceeding step by step, we prove the lemma for matrices of any order.

We now come to the proof of the algebraic criterion for determining the structure of the canonical form of a matrix A (cited in Sec. 27). According to Lemma 1, we can reduce the problem of finding the greatest common divisors $d_i(\lambda)$ of the matrix $A - \lambda$ to the problem of finding the greatest common divisors of the similar matrix

$$\begin{aligned} V(A-\lambda)V^{-1} &= VAV^{-1} - \lambda \\ &= [I_{\rho_1}(\lambda_1 - \lambda), I_{\rho_2}(\lambda_2 - \lambda), \ldots, I_{\rho_p}(\lambda_p - \lambda)]. \quad (48) \end{aligned}$$

Applying Lemma 3 to each of the submatrices appearing in this quasi-diagonal matrix, we can replace the matrix (48) by a purely diagonal matrix, insofar as finding greatest common divisors is concerned, i.e., the diagonal matrix which has $\rho_1 - 1$ ones along the principal diagonal, then $(\lambda_1 - \lambda)^{\rho_1}$, followed by $\rho_2 - 1$ ones, then $(\lambda_2 - \lambda)^{\rho_2}$, etc. Moreover, we note that if we construct a minor of order t from this matrix by deleting a set of rows and a set of columns which are not the same, then the minor so obtained vanishes, since at least one row and one column consist entirely of zeros. Thus, in constructing minors of our diagonal matrix, we must always delete the same set of rows and columns; this reduces to simply deleting some of the diagonal elements, whose product is just the value of the determinant.

To be explicit, let $\lambda = \alpha$ be a root of multiplicity k of the characteristic equation, so that the nth order determinant obviously contains the factor $(\lambda - \alpha)^k$. Suppose that the greatest common divisor of the minors of order $n - 1$ contains $(\lambda - \alpha)^{k_1}$ but not $(\lambda - \alpha)^{k_1+1}$; this means that the largest power of $\lambda - \alpha$ appearing in the diagonal matrix just constructed is $k - k_1$, i.e., the matrix A has the matrix $I_{k-k_1}(\alpha)$ in its canonical representation. In just the same way, suppose that the greatest common divisor of the minors of order $n - 2$ contains $(\lambda - \alpha)^{k_2}$ but not $(\lambda - \alpha)^{k_2+1}$; this means that after $(\lambda - \alpha)^{k-k_1}$, the next largest power of $\lambda - \alpha$ appearing in the diagonal matrix is $(\lambda - \alpha)^{k_1-k_2}$, i.e., the canonical form of A contains $I_{k_1-k_2}(\alpha)$ as well as $I_{k-k_1}(\alpha)$. Continuing in this way, suppose that we finally arrive at minors of order $n - m - 1$ such that at least one of them does not contain the factor $\lambda - \alpha$ at all; this means that we have exhausted all the constituents of the canonical form of A corresponding to $\lambda = \alpha$. Applying the same argument to the other roots of the characteristic equation, we finally complete the proof of the algebraic criterion (given in Sec. 27) for determining the structure of the canonical form of a matrix. We note that our considerations imply not only that

$$k > k_1 > k_2 > \cdots > k_m,$$

but also that

$$l_1 \geq l_2 \geq \cdots \geq l_m \geq l_{m+1},$$

where

$$l_1 = k - k_1, \qquad l_2 = k_1 - k_2, \ \ldots, \qquad l_m = k_{m-1} - k_m, \qquad l_{m+1} = k_m$$

(see Sec. 27).

57. An Example. If we can solve the characteristic equation of a given matrix A, then the canonical form of A can be determined by using the algebraic criterion proved in the preceding section. However, there still remains the problem of constructing the matrix V, with $\det V \neq 0$, which reduces A to canonical form. It will be recalled that our approach to this problem is successively to choose new basis vectors until A is ultimately reduced to canonical form; let T be the resulting linear transformation which carries the old basis vectors into the new basis vectors. Using the result of Sec. 21, we see that the matrix V which reduces A to the canonical form VAV^{-1} is just $(T^*)^{-1}$, i.e., to obtain V from T we must interchange the rows and columns of T and then take the inverse.

Thus, in more detail, our plan is the following: First we choose new basis vectors which reduce A to the quasi-diagonal form corresponding to the various eigenvalues of A. This is done in the way indicated in Sec. 52, i.e., the problem of choosing new basis vectors is reduced to the problem of solving systems of linear equations of the form

$$(A - \alpha_k)^{r_k}\mathbf{x} = 0, \tag{49}$$

which leads to a certain transformation of basis vectors and a corresponding similarity transformation $S_1AS_1^{-1}$. The next step consists of reducing to canonical form each of the matrices B which have the single eigenvalue zero (see Sec. 55). To do this, we first find the least number l for which $B^l = 0$, and then form the system of equations

$$B^{l-1}\mathbf{x} = 0. \tag{50}$$

Determining the rank of the system (50), we take the vectors which do *not* satisfy (50) and subject them repeatedly to the transformation B, thereby obtaining new basis vectors. Then, we take the remaining vectors (if any) in the subspace determined by (50) which do not satisfy $B^{l-2}\mathbf{x} = 0$ and subject them repeatedly to the transformation B, thereby obtaining more basis vectors, etc. We continue in this way with each of the matrices B in turn, until we have constructed n new basis vectors; this leads to a second transformation of basis vectors and a corresponding similarity transformation $S_2AS_2^{-1}$. Finally, the similarity transformation $(S_2S_1)A(S_2S_1)^{-1}$ reduces A to canonical form.

We now illustrate these general considerations by a numerical example. Consider the following matrix of order 5:

$$A = \begin{Vmatrix} -2 & -1 & -1 & 3 & 2 \\ -4 & 1 & -1 & 3 & 2 \\ 1 & 1 & 0 & -3 & -2 \\ 4 & -2 & -1 & 5 & 1 \\ 4 & 1 & 1 & -3 & 0 \end{Vmatrix}. \tag{51}$$

The characteristic equation of (51), formed in the usual way, turns out to be

$$(\lambda - 2)^3(\lambda + 1)^2 = 0, \tag{52}$$

i.e., (52) has a root $\lambda = 2$ of multiplicity 3 and a root $\lambda = -1$ of multiplicity 2. Next, we form the matrices $(A - 2)^3$ and $(A + 1)^2$. The equation $(A - 2)^3\mathbf{x} = 0$ must define a subspace of dimension 3, i.e., the matrix $(A - 2)^3$ must have rank 2. Similarly, the matrix $(A + 1)^2$

must have rank 3. Elementary calculations lead to the result

$$(A - 2)^3 = \left\| \begin{array}{rrrrr} -54 & 0 & -27 & 27 & 27 \\ -54 & 0 & -27 & 27 & 27 \\ 27 & 0 & 0 & -27 & -27 \\ -54 & 0 & -27 & 27 & 27 \\ 54 & 0 & 27 & -27 & -27 \end{array} \right\|,$$

so that the system $(A - 2)^3\mathbf{x} = 0$ reduces to the following two equations in the components x_1, x_2, x_3, x_4, x_5 of the vector $\mathbf{x}$:

$$\begin{aligned} -54x_1 - 27x_3 + 27x_4 + 27x_5 &= 0, \\ 27x_1 \qquad\quad - 27x_4 - 27x_5 &= 0. \end{aligned}$$

Thus, we have

$$x_1 = x_4 + x_5, \qquad x_3 = -x_4 - x_5,$$

where x_2, x_4, and x_5 are still arbitrary. Setting each of the parameters x_2, x_4, x_5 in turn equal to 1 and the others equal to 0, we obtain three new basis vectors, which have the form

$$(0, 1, 0, 0, 0), \qquad (1, 0, -1, 1, 0), \qquad (1, 0, -1, 0, 1) \tag{53}$$

relative to the old coordinate system.

In just the same way, elementary calculations lead to the result

$$(A + 1)^2 = \left\| \begin{array}{rrrrr} 0 & -6 & 0 & 9 & 3 \\ -9 & 3 & 0 & 9 & 3 \\ 0 & 6 & 0 & -9 & -3 \\ -9 & -12 & 0 & 18 & -3 \\ 9 & 6 & 0 & -9 & 6 \end{array} \right\|,$$

and the equation $(A + 1)^2\mathbf{x} = 0$ leads to a system of three equations,

$$\begin{aligned} - 2x_2 + 3x_4 + x_5 &= 0, \\ -3x_1 + x_2 + 3x_4 + x_5 &= 0, \\ 3x_1 + 2x_2 - 3x_4 + 2x_5 &= 0, \end{aligned}$$

or

$$x_2 = x_1, \qquad x_4 = x_1, \qquad x_5 = -x_1,$$

where x_1 and x_3 are still arbitrary. This gives us two new basis vectors

$$(1, 1, 0, 1 - 1), \qquad (0, 0, 1, 0, 0). \tag{54}$$

The new basis vectors (53) and (54) are expressed in terms of the old basis vectors by the formulas

$$\begin{aligned} \mathbf{e}_1' &= \mathbf{e}_2, \\ \mathbf{e}_2' &= \mathbf{e}_1 - \mathbf{e}_3 + \mathbf{e}_4, \\ \mathbf{e}_3' &= \mathbf{e}_1 - \mathbf{e}_3 + \mathbf{e}_5, \\ \mathbf{e}_4' &= \mathbf{e}_1 + \mathbf{e}_2 + \mathbf{e}_4 - \mathbf{e}_5, \\ \mathbf{e}_5' &= \mathbf{e}_3. \end{aligned}$$

The matrix of this linear transformation has the form

$$T = \begin{Vmatrix} 0 & 1 & 0 & 0 & 0 \\ 1 & 0 & -1 & 1 & 0 \\ 1 & 0 & -1 & 0 & 1 \\ 1 & 1 & 0 & 1 & -1 \\ 0 & 0 & 1 & 0 & 0 \end{Vmatrix}.$$

Interchanging rows and columns, and then taking the inverse, we obtain

$$S_1^{-1} = T^* = \begin{Vmatrix} 0 & 1 & 1 & 1 & 0 \\ 1 & 0 & 0 & 1 & 0 \\ 0 & -1 & -1 & 0 & 1 \\ 0 & 1 & 0 & 1 & 0 \\ 0 & 0 & 1 & -1 & 0 \end{Vmatrix},$$

$$S_1 = (T^*)^{-1} = \begin{Vmatrix} -1 & 1 & 0 & 1 & 1 \\ -1 & 0 & 0 & 2 & 1 \\ 1 & 0 & 0 & -1 & 0 \\ 1 & 0 & 0 & -1 & -1 \\ 0 & 0 & 1 & 1 & 1 \end{Vmatrix}.$$

Then, forming the matrix product $S_1AS_1^{-1}$ in the usual way, we reduce A to a quasi-diagonal form consisting of a third order matrix and a second order matrix:

$$S_1AS_1^{-1} = \begin{Vmatrix} 1 & 0 & -1 & 0 & 0 \\ -2 & 2 & -2 & 0 & 0 \\ 1 & 0 & 3 & 0 & 0 \\ 0 & 0 & 0 & -2 & -1 \\ 0 & 0 & 0 & 1 & 0 \end{Vmatrix}. \qquad (55)$$

The third order matrix

$$D_1 = \begin{Vmatrix} 1 & 0 & -1 \\ -2 & 2 & -2 \\ 1 & 0 & 3 \end{Vmatrix}$$

must have the eigenvalue $\lambda = 2$ of multiplicity 3. We form the new matrix

$$B_1 = D_1 - 2 = \begin{Vmatrix} -1 & 0 & -1 \\ -2 & 0 & -2 \\ 1 & 0 & 1 \end{Vmatrix},$$

which has the eigenvalue $\lambda = 0$ of multiplicity 3. Squaring B_1, we find that $B_1^2 = 0$. Thus, in this case, we have $l = 2$, and the single system $B_1\mathbf{x} = 0$ leads to just one equation,

$$x_1 + x_3 = 0.$$

The space F_2 (in the notation of Sec. 55) coincides with the whole three-dimensional space, and the subspace F_1 can be spanned by the vectors $(1, 0, -1)$ and $(0, 1, 0)$. Take the vector $(1, 0, 0)$, which is not an element of F_1; this vector spans a subspace G_2. Applying the operator B_1 to $(1, 0, 0)$, we obtain the vector $(-1, -2, 1)$. The vectors $(1, 0, 0)$ and $(-1, -2, 1)$ form the first set of new basis vectors, corresponding to the canonical matrix

$$\begin{Vmatrix} 0 & 0 \\ 1 & 0 \end{Vmatrix}.$$

For the third basis vector, we have to take any vector in F_1 which is linearly independent of the vector $(-1, -2, 1)$. Choose the vector $(0, 1, 0)$, which corresponds to the first order canonical matrix consisting of the single element 0. The new basis vectors are expressed in terms of the old basis vectors by the formulas

$$\begin{aligned} \mathbf{e}_1'' &= \mathbf{e}_1', \\ \mathbf{e}_2'' &= -\mathbf{e}_1' - 2\mathbf{e}_2' + \mathbf{e}_3', \\ \mathbf{e}_3'' &= \mathbf{e}_2'. \end{aligned} \tag{56}$$

Next, we consider the second order matrix

$$D_2 = \begin{Vmatrix} -2 & -1 \\ 1 & 0 \end{Vmatrix},$$

appearing in the quasi-diagonal matrix (55). D_2 must have the eigenvalue $\lambda = -1$ of multiplicity 2. We form the matrix

$$B_2 = D_2 + 1 = \begin{Vmatrix} -1 & -1 \\ 1 & 1 \end{Vmatrix},$$

with the eigenvalue $\lambda = 0$ of multiplicity 2. Clearly, $B_2^2 = 0$, as is to be expected from the Hamilton-Cayley theorem. The equation $B_2\mathbf{x} = 0$ is equivalent to $x_4 + x_5 = 0$, where x_4 and x_5 are the components of $\mathbf{x}$ in the two-dimensional subspace under consideration. As the first basis vector, we take the vector $(1, 0)$, i.e., $x_4 = 1$, $x_5 = 0$, which does not satisfy the equation $B_2\mathbf{x} = 0$; then, applying the operator B_2 to $(1, 0)$, we obtain the vector $(-1, 1)$. Thus, we obtain two new basis vectors $(1, 0)$ and $(-1, 1)$, which are expressed in terms of the old basis vectors by the formulas

$$\mathbf{e}_4'' = \mathbf{e}_4', \qquad \mathbf{e}_5'' = -\mathbf{e}_4' + \mathbf{e}_5'. \tag{57}$$

Together, (56) and (57) give

$$
\begin{aligned}
\mathbf{e}_1'' &= \mathbf{e}_1', \\
\mathbf{e}_2'' &= -\mathbf{e}_1' - 2\mathbf{e}_2' + \mathbf{e}_3', \\
\mathbf{e}_3'' &= \mathbf{e}_2', \\
\mathbf{e}_4'' &= \mathbf{e}_4', \\
\mathbf{e}_5'' &= -\mathbf{e}_4' + \mathbf{e}_5'.
\end{aligned}
\tag{58}
$$

The last two basis vectors correspond to the canonical matrix $I_2(0)$. We have to add the number 2 to the principal diagonals of the first two canonical matrices and the number -1 to the principal diagonal of the last canonical matrix. The final canonical form of the matrix A is then

$$
\begin{Vmatrix}
2 & 0 & 0 & 0 & 0 \\
1 & 2 & 0 & 0 & 0 \\
0 & 0 & 2 & 0 & 0 \\
0 & 0 & 0 & -1 & 0 \\
0 & 0 & 0 & 1 & -1
\end{Vmatrix} = [I_2(2),\ I_1(2),\ I_2(-1)]. \tag{59}
$$

Finally, we find the matrix V, which transforms A into the form (59). As we have seen, V is the product S_2S_1, where S_1 was obtained above, and S_2 is given by the formula

$$S_2 = (T_1^*)^{-1},$$

where T_1 is the matrix of the linear transformation (58). Thus, we have

$$
T_1^* = \begin{Vmatrix}
1 & -1 & 0 & 0 & 0 \\
0 & -2 & 1 & 0 & 0 \\
0 & 1 & 0 & 0 & 0 \\
0 & 0 & 0 & 1 & -1 \\
0 & 0 & 0 & 0 & 1
\end{Vmatrix}
$$

and

$$
S_2 = (T_1^*)^{-1} = \begin{Vmatrix}
1 & 0 & 1 & 0 & 0 \\
0 & 0 & 1 & 0 & 0 \\
0 & 1 & 2 & 0 & 0 \\
0 & 0 & 0 & 1 & 1 \\
0 & 0 & 0 & 0 & 1
\end{Vmatrix}.
$$

Carrying out the multiplication, we have

$$
V = S_2S_1 = \begin{Vmatrix}
0 & 1 & 0 & 0 & 1 \\
1 & 0 & 0 & -1 & 0 \\
1 & 0 & 0 & 0 & 1 \\
1 & 0 & 1 & 0 & 0 \\
0 & 0 & 1 & 1 & 1
\end{Vmatrix}.
$$

The final result is

$$VAV^{-1} = [I_2(2),\ I_1(2),\ I_2(-1)].$$

PROBLEMS†

1. Let L_1 and L_2 be two subspaces of the n-dimensional complex space R_n whose sum $L_1 + L_2$ equals R_n and whose intersection $L_1 \cap L_2$ contains only the zero vector. (Cf. Prob. 18 of Chap. 2.) Then R_n is said to be the *direct sum* of L_1 and L_2, written $R_n = L_1 \oplus L_2$. Show that

(**a**) Every vector $\mathbf{x} \in R_n$ has a unique representation $\mathbf{x} = \mathbf{x}_1 + \mathbf{x}_2$, where $\mathbf{x} \in L_1$, $\mathbf{x}_2 \in L_2$;

(**b**) Conversely, if L_1 and L_2 are any two subspaces of R_n and if every vector $\mathbf{x}$ in R_n has a unique representation $\mathbf{x} = \mathbf{x}_1 + \mathbf{x}_2$ where $\mathbf{x}_1 \in L_1$, $\mathbf{x}_2 \in L_2$, then $R_n = L_1 \oplus L_2$.

2. Let $R_n = L_1 \oplus L_2$, so that every $\mathbf{x}$ in R_n has a unique representation $\mathbf{x} = \mathbf{x}_1 + \mathbf{x}_2$, where $\mathbf{x}_1 \in L_1$, $\mathbf{x}_2 \in L_2$. Define a linear transformation P by the formula $P\mathbf{x} = \mathbf{x}_1$. Show that

(**a**) $P^2 = P$;

(**b**) P is Hermitian if and only if L_1 and L_2 are orthogonal complements of each other (Sec. 14).

3. Let P be a linear transformation such that $P^2 = P$ (i.e., an *idempotent* transformation). Let L_1 be the set of vectors $\mathbf{x} \in R_n$ for which $P\mathbf{x} = \mathbf{x}$, and let L_2 be the set of vectors $\mathbf{x} \in R_n$ for which $P\mathbf{x} = 0$. Show that L_1 and L_2 are subspaces and that $R_n = L_1 \oplus L_2$.

Comment: The projection matrices of Sec. 43 are Hermitian as well as idempotent.

4. Show that if a linear transformation T satisfies any two of the following conditions, then it also satisfies the other:

(**a**) T is Hermitian;

(**b**) T is unitary;

(**c**) T is an *involution*, i.e., $T^2 = I$ (the identity transformation).

5. Find all linear transformations T on R_n with the three properties in Prob. 4.

6. Let A be a normal matrix (cf. Prob. 29 of Chap. 4), let $f(x)$ be any function defined for all x, and let U be a unitary matrix diagonalizing A, i.e., such that $U^{-1}AU = [\lambda_1, \ldots, \lambda_n]$. *Define* $f(A)$ to be $U[f(\lambda_1), \ldots, f(\lambda_n)]U^{-1}$. Show that if V is any unitary matrix, then $f(V^{-1}AV) = V^{-1}f(A)V$.

7. Show that

(**a**) The matrix A is unitary if and only if there is a Hermitian matrix B such that $A = e^{iB}$;

(**b**) The matrix A is Hermitian and positive definite if and only if there is a Hermitian matrix B such that $A = e^B$.

8. Show that $\tilde{A}$ is a polynomial in A if and only if A is normal.

9. Find the invariant subspaces of the linear transformation with matrix $\begin{Vmatrix} 0 & 0 \\ 1 & 1 \end{Vmatrix}$.

† For hints and answers see page 443. These problems are by A. L. Shields and the translator, making considerable use of the book by Proskuryakov (cited in the Preface).

10. Let T be a linear transformation on R_n, and let M be an invariant subspace of T. Show that M contains an eigenvector of T.

11. Let P_n be the $(n + 1)$-dimensional space of all polynomials of degree $\leq n$, and let D be the differentiation operator. Find the matrix of D relative to the basis $1, x, x^2, \ldots, x^n$. Find the eigenvalues and eigenvectors of D. Is P_n the direct sum of two invariant subspaces?

12. Let T be a linear transformation, and let M be an invariant subspace of T. Show that the orthogonal complement of M is an invariant subspace of $\tilde{T}$.

13. Write the equation of the plane which is invariant under the transformation with matrix

$$\begin{Vmatrix} 4 & -23 & 17 \\ 11 & -43 & 30 \\ 15 & -54 & 37 \end{Vmatrix}.$$

14. Show that AB and BA are similar if either A or B has an inverse, but that otherwise AB and BA need not be similar.

15. If T is a linear transformation defined on R_n, the subspace consisting of all vectors $\mathbf{y} \in R_n$ such that $\mathbf{y} = T\mathbf{x}$ for some $\mathbf{x} \in R_n$ is called the *range* of T, while the subspace consisting of all vectors $\mathbf{x} \in R_n$ such that $T\mathbf{x} = 0$ is called the *null space* of T (cf. Sec. 52). Give an example of a transformation whose range and null space have a nonzero vector in common.

16. Show that two idempotent transformations P and Q are similar if and only if they have the same rank.

Comment: The *rank* of a linear transformation is the dimension of its range.

17. Show that if T is any linear transformation on R_n, then T has an invariant subspace of dimension $n - 1$.

18. Let T be a linear transformation on R_n. Show that T has a family of invariant subspaces $M_1, M_2, \ldots, M_n = R_n$, each contained in the next, with the dimension of M_k equal to k. Hence, show that a basis can be chosen relative to which T has a triangular matrix (cf. Prob. 41 of Chap. 3) whose diagonal elements are just the eigenvalues of T.

19. Let A be a matrix with eigenvalues $\lambda_1, \ldots, \lambda_n$, and let $f(A)$ be a polynomial in A. Show that $f(A)$ has the eigenvalues $f(\lambda_1), \ldots, f(\lambda_n)$.

20. Show that if two linear transformations commute, then they have a common eigenvector.

21. Show that if A and B commute, then they can be simultaneously reduced to triangular form.

22. Show that a triangular matrix is normal if and only if it is diagonal.

23. Suppose that the matrix A has the elementary divisors

$$(\lambda - \lambda_1)^{\rho_1}, (\lambda - \lambda_1)^{\rho_2}, \ldots, (\lambda - \lambda_1)^{\rho_j}, \ldots, (\lambda - \lambda_m)^{\tau_1}, (\lambda - \lambda_m)^{\tau_2}, \ldots, (\lambda - \lambda_m)^{\tau_k},$$

where $\lambda_1, \ldots, \lambda_m$ are the distinct eigenvalues of A, and $\rho_1 \geq \rho_2 \geq \cdots \geq \rho_j, \ldots, \tau_1 \geq \tau_2 \geq \cdots \geq \tau_k$. By the *minimal polynomial* of A is meant the product

$$\mu(\lambda) = (\lambda - \lambda_1)^{\rho_1} \cdots (\lambda - \lambda_m)^{\tau_1}.$$

Show that

(a) $\mu(\lambda)$ *annihilates* A, i.e., $\mu(A) = 0$;

(b) $\mu(\lambda)$ is the polynomial of smallest degree (with leading coefficient 1) that annihilates A;
(c) $\mu(\lambda)$ divides the characteristic polynomial of A.

24. Find the minimal polynomial of

(a) The unit matrix;
(b) The zero matrix (i.e., the matrix with all elements equal to 0);
(c) The matrix $I_p(\lambda - \alpha)$ given by equation (2) of Sec. 52.

25. For what matrices does the minimal polynomial have the form $\lambda - \alpha$, where α is a constant?

26. Prove that some power of the minimal polynomial of a matrix A is divisible by the characteristic polynomial of A.

27. Find the minimal polynomials of the following matrices:

$$\text{(a)} \begin{Vmatrix} 3 & 1 & -1 \\ 0 & 2 & 0 \\ 1 & 1 & 1 \end{Vmatrix}; \quad \text{(b)} \begin{Vmatrix} 4 & -2 & 2 \\ -5 & 7 & -5 \\ -6 & 6 & -4 \end{Vmatrix}.$$

28. Prove that two matrices are similar if and only if

(a) They have the same canonical form;
(b) They have the same elementary divisors.

29. Prove that a necessary (but not sufficient) condition for two matrices to be similar is that they have the same characteristic polynomial and the same minimal polynomial. Give an example of two matrices which are not similar that have the same characteristic polynomial $\varphi(\lambda)$ and the same minimal polynomial $\mu(\lambda)$.

30. Prove that every matrix is similar to its own transpose.

31. Prove that any square matrix can be represented as the product of two symmetric matrices, one of which is nonsingular.

32. In Prob. 32 of Chap. 2 it was shown that a matrix A can be brought into diagonal form by making suitable elementary row and column transformations. In the same way, show that the $n \times n$ matrix $A - \lambda$ can be reduced to a (non-unique) diagonal form $[E_1(\lambda), E_2(\lambda), \ldots, E_n(\lambda)]$ by making suitable elementary row and column transformations, where if each polynomial $E_i(\lambda)$ is factored into a product of powers of distinct linear terms $\lambda - \lambda_i$, then these factors are the elementary divisors of A. (It is possible to obtain polynomials $E_i(\lambda)$ such that each polynomial divides those with higher indices.)

33. Which of the following matrices are similar:

(a)
$$A = \begin{Vmatrix} 3 & 2 & -5 \\ 2 & 6 & -10 \\ 1 & 2 & -3 \end{Vmatrix}, \quad B = \begin{Vmatrix} 6 & 20 & -34 \\ 6 & 32 & -51 \\ 4 & 20 & -32 \end{Vmatrix};$$

(b)
$$A = \begin{Vmatrix} 6 & 6 & -15 \\ 1 & 5 & -5 \\ 1 & 2 & -2 \end{Vmatrix}, \quad B = \begin{Vmatrix} 37 & -20 & -4 \\ 34 & -17 & -4 \\ 119 & -70 & -11 \end{Vmatrix};$$

(c)
$$A = \begin{Vmatrix} 4 & 6 & -15 \\ 1 & 3 & -5 \\ 1 & 2 & -4 \end{Vmatrix}, \quad B = \begin{Vmatrix} 1 & -3 & 3 \\ -2 & -6 & 13 \\ -1 & -4 & 8 \end{Vmatrix},$$

$$C = \begin{Vmatrix} -13 & -70 & 119 \\ -4 & -19 & 34 \\ -4 & -20 & 35 \end{Vmatrix};$$

(d)
$$A = \begin{Vmatrix} 14 & -2 & -7 & -1 \\ 20 & -2 & -11 & -2 \\ 19 & -3 & -9 & -1 \\ -6 & 1 & 3 & 1 \end{Vmatrix}, \quad B = \begin{Vmatrix} 4 & 10 & -19 & 4 \\ 1 & 6 & -8 & 3 \\ 1 & 4 & -6 & 2 \\ 0 & -1 & 1 & 0 \end{Vmatrix},$$

$$C = \begin{Vmatrix} 41 & -4 & -26 & -7 \\ 14 & -13 & -91 & -18 \\ 40 & -4 & -25 & -8 \\ 0 & 0 & 0 & 1 \end{Vmatrix}?$$

34. Show that the arrangement of the submatrices in the canonical form of a matrix A is not unique.

35. Prove that the number of linearly independent eigenvectors of a matrix A which belong to the same eigenvalue λ_0 equals the number of submatrices with diagonal element λ_0 in the canonical form of A.

36. Find the elementary divisors and canonical form of each of the following matrices:

(a) $\begin{Vmatrix} 0 & -4 & -2 \\ 1 & 4 & 1 \\ 0 & 0 & 2 \end{Vmatrix}$; (b) $\begin{Vmatrix} 2 & 1 & 1 \\ 6 & 1 & 2 \\ -15 & -5 & -6 \end{Vmatrix}$; (c) $\begin{Vmatrix} 9 & 18 & 18 \\ -6 & -12 & -9 \\ -2 & -3 & -6 \end{Vmatrix}$;

(d) $\begin{Vmatrix} 4 & 1 & 1 \\ 6 & 3 & 2 \\ -15 & -5 & -4 \end{Vmatrix}$; (e) $\begin{Vmatrix} 0 & 1 & 1 \\ -4 & -4 & -2 \\ 0 & 0 & -2 \end{Vmatrix}$; (f) $\begin{Vmatrix} \alpha & 0 & \alpha \\ 0 & \alpha & 0 \\ 0 & 0 & \alpha \end{Vmatrix}$, where $\alpha \neq 0$;

(g) $\begin{Vmatrix} 1 & -2 & 0 & -1 \\ -3 & -6 & -3 & -4 \\ 0 & 0 & 1 & 0 \\ 3 & 13 & 3 & 8 \end{Vmatrix}$; (h) $\begin{Vmatrix} 3 & 9 & 0 & 0 \\ -1 & -3 & 0 & 0 \\ 1 & -7 & 4 & 2 \\ -7 & -1 & -8 & -4 \end{Vmatrix}$;

(i) $\begin{Vmatrix} 3 & 4 & 0 & 0 \\ -4 & -5 & 0 & 0 \\ 0 & -2 & 3 & 2 \\ 2 & 4 & -2 & -1 \end{Vmatrix}$; (j) $\begin{Vmatrix} n & 0 & 0 & \cdots & 0 \\ n-1 & n & 0 & \cdots & 0 \\ n-2 & n-1 & n & \cdots & 0 \\ \cdots & \cdots & \cdots & \cdots & \cdots \\ 1 & 2 & 3 & \cdots & n \end{Vmatrix}$.

37. For each of the following matrices A, find a matrix S such that SAS^{-1} is in canonical form, and write the corresponding canonical form:

(a) $A = \begin{Vmatrix} 3 & 4 & 3 \\ 2 & 10 & 6 \\ -3 & -12 & -7 \end{Vmatrix}$; (b) $A = \begin{Vmatrix} 1 & -3 & -2 \\ 1 & -3 & -2 \\ -1 & 3 & 2 \end{Vmatrix}$;

(c) $A = \begin{Vmatrix} 0 & -1 & 2 \\ 3 & 8 & -14 \\ 3 & 6 & -10 \end{Vmatrix}$; (d) $A = \begin{Vmatrix} 6 & 1 & 1 \\ 6 & 5 & 2 \\ -15 & -5 & -2 \end{Vmatrix}$;

(e)
$$A = \begin{Vmatrix} 6 & 7 & 8 & 1 \\ -9 & -13 & -17 & -2 \\ 5 & 8 & 11 & 1 \\ 4 & 7 & 8 & 3 \end{Vmatrix}.$$

38. Find A^{100}, where

$$A = \left\| \begin{matrix} 0 & -1 & -1 & -1 \\ 1 & 2 & -1 & 1 \\ -1 & 1 & 1 & 0 \\ -1 & 1 & 0 & 1 \end{matrix} \right\|.$$

39.† Let T be a finite-dimensional linear transformation of norm ≤ 1 (see Prob. 24 of Chap. 4). Show that

(**a**) All the powers of T have norm ≤ 1;
(**b**) All the eigenvalues of T are ≤ 1;
(**c**) Every eigenvalue of T of absolute value 1 is associated with a diagonal submatrix in the canonical representation of T, i.e., if $|\lambda_1| = 1$, T has no elementary divisors of the form $(\lambda - \lambda_1)^m$, where $m > 1$.

40. Let T be a finite-dimensional linear transformation of norm ≤ 1. Prove that

$$T_\infty = \lim_{M \to \infty} \frac{1}{M} \sum_{m=0}^{M-1} T^{m-1}$$

exists. Prove that T_∞ is an idempotent transformation.

Comment: This is the finite-dimensional version of the *ergodic theorem.*

† Problems 39 and 40 were suggested by J. T. Schwartz.

PART III

Group Theory

Chapter 7

Elements of the General Theory of Groups

58. Groups of Linear Transformations. Consider the set of all unitary transformations in n-dimensional space. Regarded as matrices, these transformations all have nonvanishing determinants, so that every unitary transformation U has a well-defined inverse transformation U^{-1}, which is also unitary (Sec. 28). Moreover, if U_1 and U_2 are two unitary transformations, then their product U_2U_1 is also a unitary transformation. These properties of unitary transformations can be summarized by saying that *the set of all unitary transformations forms a group.*

More generally, a set of linear transformations with nonvanishing determinants forms a group if the following two conditions are satisfied: (*a*) *The inverse of a transformation belonging to the set also belongs to the set;* (*b*) *The product (in either order) of two transformations belonging to the set also belongs to the set (the factors can be identical).* Bearing in mind that the product of every transformation with its inverse is the identity transformation, we can assert that *a group must contain the identity transformation (i.e., the unit matrix).* Since a linear transformation is always completely specified by its matrix, we can speak either of a *group of linear transformations* or of a *group of matrices.* (This applies both to the definition just given and to all subsequent considerations.)

We now give some other examples of groups of linear transformations. It is easy to see that the set of all real orthogonal transformations forms a group. As we know (Sec. 28), real orthogonal transformations have determinants equal to $+1$ or -1. The set of real orthogonal transformations with determinants equal to $+1$ also forms a group. However, the set of real orthogonal matrices with determinants equal to -1 does not form a group, since the product of two matrices with determinants equal to -1 gives a matrix with determinant equal to $+1$. In fact, the group of all real orthogonal transformations in three variables is the group consisting of pure spatial rotations about the origin and of transformations which are the combined result of rotations and symmetry operations with respect to the origin. On the other hand, the group of linear orthogonal transformations in three variables with determinants equal to +1 is the group consisting only of spatial rotations about the origin. In all these examples, the groups contain an infinite number of transformations; in particular, the rotation group in three-dimensional space depends on three arbitrary real parameters, the Eulerian angles, which we discussed previously (Sec. 20).

As another example of a group, consider spatial rotations about the z-axis through the angle φ; the corresponding transformation formulas have the form

$$\begin{aligned} x' &= x \cos \varphi - y \sin \varphi, \\ y' &= x \sin \varphi + y \cos \varphi. \end{aligned} \tag{1}$$

As the real parameter φ runs through values in the interval $[0, 2\pi]$, we obviously obtain a group containing an infinite number of transformations and depending on one real parameter φ. We introduce the following notation for the transformation matrix:

$$Z_\varphi = \begin{Vmatrix} \cos \varphi & -\sin \varphi \\ \sin \varphi & \cos \varphi \end{Vmatrix}. \tag{2}$$

It is immediately obvious that the product of two rotations through the angles φ_1 and φ_2, respectively, gives a rotation through the angle $\varphi_1 + \varphi_2$, i.e.,

$$Z_{\varphi_2} Z_{\varphi_1} = Z_{\varphi_2+\varphi_1}; \tag{3}$$

in just the same way, we have

$$Z_{\varphi_1} Z_{\varphi_2} = Z_{\varphi_1+\varphi_2}.$$

Thus, we see that in this case all the transformations of the group, or as one says, all the *elements* of the group, commute in pairs. Such a group is called an *Abelian group*. Moreover, in the example just given, multiplication of two elements of the group reduces simply to addition of the parameter values φ of the corresponding matrices.

We can enlarge this last group somewhat by considering not only the operation of rotation of the xy-plane about the origin, but also reflection (i.e., symmetry) with respect to the y-axis. Clearly, from the standpoint of the total set of elements in the group, it does not matter in which order the operations are performed, i.e., whether we first perform a rotation about the origin and then a reflection in the y-axis, or vice versa. (Although the particular group element obtained depends on the order of the operations, the total set of transformations obtained does not.) The transformations obtained in this way are the orthogonal transformations in two variables, whose matrices have the general form

$$\{\varphi, d\} = \begin{Vmatrix} d\cos\varphi & -d\sin\varphi \\ \sin\varphi & \cos\varphi \end{Vmatrix}, \tag{4}$$

where φ is the previous parameter, and d is a number equal to ± 1. For $d = 1$, we obtain a simple rotation of the xy-plane about the origin, while for $d = -1$, we obtain a rotation followed by the symmetry operation just described. It is not hard to verify that the product of matrices of the form (4) obeys the rule

$$\{\varphi_2, d_2\}\{\varphi_1, d_1\} = \{\varphi_1 + d_1\varphi_2, d_1d_2\}. \tag{5}$$

In this case, the product does in fact depend on the order of the factors, i.e., the group is no longer Abelian. In the same way, it is clear that the group of real orthogonal transformations in three-dimensional space, or for that matter, the group of three-dimensional rotations about the origin, is not Abelian.

So far, we have cited examples of groups containing an infinite number of transformations (elements), where the corresponding matrices contain arbitrary real parameters. We now give some examples of groups containing a finite number of elements. Thus, let m be any positive integer, and consider the set of rotations of the xy-plane about the origin through the angles

$$0, \frac{2\pi}{m}, \frac{4\pi}{m}, \ldots, \frac{2(m-1)\pi}{m}.$$

Then we have a total of m transformations, whose matrices are

$$Z_{2k\pi/m} = \begin{Vmatrix} \cos(2k\pi/m) & -\sin(2k\pi/m) \\ \sin(2k\pi/m) & \cos(2k\pi/m) \end{Vmatrix} \qquad (k = 0, 1, \ldots, m-1).$$

These transformations obviously form a group, and the elements of this group are the positive integral powers of a single transformation:

$$Z_{2k\pi/m} = Z^k_{2\pi/m} \qquad (k = 0, 1, \ldots, m-1). \tag{6}$$

A finite group of this type, consisting of the powers of a certain transformation, is called a *cyclic group*.

If we choose an angle φ_0 which is incommensurate with π, then the transformations (matrices)

$$Z^k_{\varphi_0} = Z_{k\varphi_0} \qquad (k = 0, 1, 2, \ldots) \tag{7}$$

obviously form a group also. However, this group consists of an infinite number of elements, since the matrix $Z^k_{\varphi_0}$ does not coincide with the unit matrix $Z^0_{\varphi_0} = I$ for any positive k. Thus, the group (7) is an infinite group, although its matrices do not contain a continuously varying parameter. In this case, one says that the number of elements in the group is *countable*, i.e., we can enumerate all the elements of the group by supplying each element of the group with an integral index, such that different elements correspond to different indices and every integer is the index of some element. This cannot be done in the case of a group containing a continuously varying parameter.

59. The Polyhedral Groups. We now give further examples of finite groups, which this time are formed from three-dimensional spatial rotations about the origin. As we know, relative to a given coordinate system, such rotations are expressed by linear coordinate transformations. It should be noted that in referring to a spatial rotation about the origin,

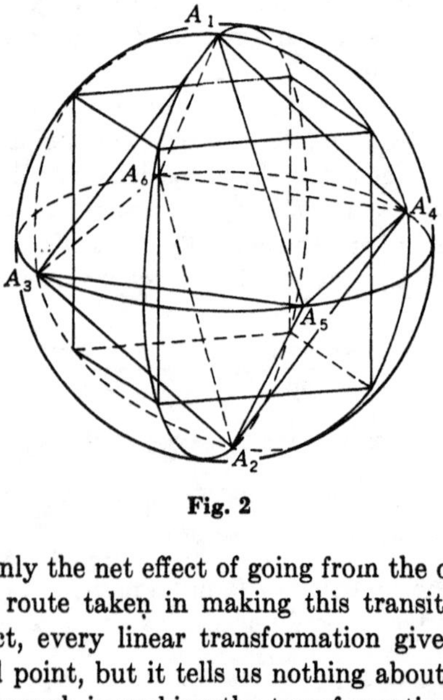

Fig. 2

we have in mind only the net effect of going from the old position to the new position; the route taken in making this transition is completely irrelevant. In fact, every linear transformation gives the coordinates of the transformed point, but it tells us nothing about the intermediate positions passed through in making the transformation, a matter which does not concern us at all.

Consider a sphere of unit radius, with its center at the origin, and inscribe in this sphere a regular polyhedron, e.g., an octahedron (Fig. 2). As is well known, the surface of this polyhedron consists of eight equilateral triangles. Consider next those three-dimensional spatial rotations about the origin which carry the octahedron into itself. It is not hard to see that the set of these rotations forms a group which contains a finite number of elements. To calculate the number of elements in this group, we take any axis of the octahedron, and observe that the octahedron is transformed into itself if we perform spatial rotations about this axis through the angles 0, $\pi/2$, π, $3\pi/2$. The rotation through the angle 0 obviously corresponds to the identity transformation, i.e., the unit matrix. We denote these four rotations about the given axis by

$$S_0 = I,\ S_1,\ S_2,\ S_3. \tag{8}$$

Now let A be one of the vertices of the octahedron lying on the axis just chosen, and consider the five linear transformations

$$T_1,\ T_2,\ T_3,\ T_4,\ T_5$$

which carry the octahedron into itself, while moving the vertex A into the position of one of the other five vertices. Then, in addition to the four rotations (8), we construct 20 more spatial rotations about the origin, of the form

$$T_kS_0,\ T_kS_1,\ T_kS_2,\ T_kS_3 \qquad (k = 1, 2, 3, 4, 5). \tag{9}$$

It is easily verified that the 24 rotations (8) and (9) are different. This is quite obvious geometrically, but can also be proved as follows: Suppose that

$$T_pS_q = T_{p_1}S_{q_1}. \tag{10}$$

The transformations S_i correspond to rotations about an axis passing through the vertex A, which leave the position of A unchanged. Since the transformations T_p and T_{p_1} carry the vertex A into different vertices when the indices p and p_1 are different, it follows from (10) that the indices p and p_1 must be the same. But then, by multiplying (10) from the left by $T_p^{-1} = T_{p_1}^{-1}$, it is an obvious consequence of (10) that the indices q and q_1 must also be the same, i.e., (10) can hold if and only if the left-hand and right-hand sides consist of the same factors.

Thus, the expressions (8) and (9) give 24 different rotations carrying the octahedron into itself. We now show that there are no more rotations with this property. To see this, let V be a rotation carrying the octahedron into itself, and suppose that the vertex A is thereby carried into another vertex A_j; let T_j be the transformation from the set T_1, T_2, T_3, T_4, T_5 which transforms A into A_j, and form the transformation $T_j^{-1}V$. This transformation carries the octahedron into itself while leaving the position of the vertex A unchanged. Therefore, the position of the vertex opposite A also remains unchanged, and the transformation $T_j^{-1}V$ is one of the rotations S_i about an axis passing through the vertex A, i.e., $T_j^{-1}V = S_i$, so that $V = T_jS_i$. In other words, every rotation carrying the octahedron into itself must be one of the 24 rotations already given. Thus, finally, *the rotation group carrying the octahedron into itself contains* 24 *elements.*

Incidentally, it is clear that we can inscribe a cube in the unit sphere in such a way that the radii drawn to the centers of the faces of the octahedron terminate on the vertices of the cube (Fig. 2). This implies at once that the cube has the same rotation group as the octahedron.

Suppose now that the position of the octahedron is chosen differently and that the new position of the octahedron is obtained from the old position by means of the rotation accomplished by using the matrix U. Then, if V is a rotation which carries the old position of the octahedron into itself, UVU^{-1} clearly gives a rotation which carries the new octahedron into itself, and conversely. Thus, if the rotation group of the old octahedron consists of the matrices V_k ($k = 1, 2, \ldots, 24$), the rotation group of the new octahedron consists of the *similar* matrices UV_kU^{-1}, i.e., we obtain a *similar group*. In general, *if a set of matrices V_k forms a group, the set of similar matrices UV_kU^{-1} (where U is any fixed matrix) also forms a group.*

Consider now the tetrahedron, which has four vertices and whose surface consists of four equilateral triangles. Take any axis of the tetrahedron joining a vertex A with the center of the opposite face. The tetrahedron is carried into itself if we perform spatial rotations (in some direction) about this axis through angles 0, $2\pi/3$, and $4\pi/3$. Let S_0, S_1, and S_2 denote these rotations, and introduce three more linear transformations T_1, T_2, and T_3, which carry the tetrahedron into itself, while carrying the vertex A into one of the other three vertices. Then, in addition to the rotations S_0, S_1, and S_2, form the nine rotations T_kS_0, T_kS_1, T_kS_2 ($k = 1, 2, 3$). The 12 rotations so obtained are different, and moreover these are *all* the rotations carrying the tetrahedron into itself.

Next consider the icosahedron, which has 12 vertices and whose surface consists of 20 equilateral triangles. As before, we take any axis of the icosahedron joining a vertex A with an opposite vertex. The icosahedron is carried into itself if we perform a spatial rotation through the angle $2k\pi/5$ ($k = 0, 1, 2, 3, 4$); let S_k denote this rotation. We also have 11 rotations T_l ($l = 1, 2, \ldots, 11$) which carry the icosahedron into itself, while carrying the vertex A into one of the other vertices. The complete group of rotations carrying the icosahedron into itself consists of the five rotations S_k and the 55 rotations T_lS_k, i.e., contains 60 rotations. The dodecahedron, which has 20 vertices and whose surface consists of 12 regular pentagons, has the same rotation group as the icosahedron. To see this, we have to position the dodecahedron with respect to the icosahedron in just the same way as we previously positioned the cube with respect to the octahedron.

Finally, we consider one more group consisting of three-dimensional rotations. Suppose that we have a regular n-gon in the xy-plane, with center at the origin of coordinates. Take any axis of the n-gon joining a vertex with the opposite vertex, if n is even, or with the mid-point of the opposite side, if n is odd. When the xy-plane is rotated about this axis through the angles 0 and π, the n-gon is carried into itself. The first rotation is the identity transformation I, and the second will be denoted by S. In addition to these transformations, we also have each rotation T_k about the z-axis through the angle $2k\pi/n$ ($k = 1, 2, \ldots, n - 1$), which transforms the n-gon into itself, while carrying the vertex A into one of the other vertices. (For $k = 0$, we obtain the identity transformation $T_0 = I$.) The complete group of transformations carrying the n-gon into itself consists of the $2n$ transformations T_k and T_kS ($k = 0, 1, 2, \ldots, n - 1$). This n-gon, whose surface has been taken into account twice (both the top and the bottom), is usually called a *dihedron*, and the group just constructed is called the *dihedral group*.

60. Lorentz Transformations. In all the examples given above, the groups of linear transformations consisted of unitary transformations or of three-dimensional spatial rotations (a special case of unitary transformations). We now consider a group whose elements are not unitary transformations. This group plays an important role in relativity theory, electrodynamics, and relativistic quantum mechanics.

Consider the four variables x_1, x_2, x_3, x_4, where the variables x_1, x_2, x_3 are the spatial coordinates of a point and the variable x_4 is the time. The basic requirement of the special theory of relativity, i.e., that a certain velocity c (the velocity of light) be invariant with respect to relative motion, raises the following question. Under which linear transformations of the variables x_1, x_2, x_3, x_4 is the expression

$$x_1^2 + x_2^2 + x_3^2 - c^2x_4^2$$

invariant? More explicitly, we have to find linear transformations which express the new variables x_k' in terms of the old variables x_k and satisfy the identity

$$x_1'^2 + x_2'^2 + x_3'^2 - c^2x_4'^2 = x_1^2 + x_2^2 + x_3^2 - c^2x_4^2.$$

We first consider the case where the coordinates x_2 and x_3 remain unchanged and only the variables x_1 and x_4 participate in the linear transformation. Then, we must find linear transformations

$$x_1' = a_{11}x_1 + a_{14}x_4, \qquad x_4' = a_{41}x_1 + a_{44}x_4 \tag{11}$$

such that

$$x_1'^2 - c^2x_4'^2 = x_1^2 - c^2x_4^2. \tag{12}$$

If, instead of x_4, we introduce a new purely imaginary variable

$$y_1 = icx_4,$$

then the required linear transformations take the form

$$x_1' = \alpha_{11}x_1 + \alpha_{12}y_1, \qquad y_1' = \alpha_{21}x_1 + \alpha_{22}y_1, \tag{13}$$

where $\alpha_{11} = a_{11}, \qquad \alpha_{12} = \dfrac{a_{14}}{ic}, \qquad \alpha_{21} = ica_{41}, \qquad \alpha_{22} = a_{44},$

and the condition (12) can be rewritten as

$$x_1'^2 + y_1'^2 = x_1^2 + y_1^2. \tag{14}$$

The coefficients α_{11}, α_{22} have to be real, and the coefficients α_{12}, α_{21} have to be purely imaginary; therefore, we write $\alpha_{12} = i\beta_{12}$ and $\alpha_{21} = i\beta_{21}$. Clearly, the condition (14) is equivalent to the transformation (13) being orthogonal. Thus, the sum of the squares of the elements of each row and column must equal 1. This immediately gives

$$\beta_{12}^2 = \beta_{21}^2 = \alpha_{11}^2 - 1 = \alpha_{22}^2 - 1, \qquad \alpha_{11}^2 = \alpha_{22}^2.$$

We set $\alpha_{22} = \alpha$ and $\beta_{12} = \alpha\beta$, and assume that the coefficients α_{11} and α_{22} are positive, which corresponds to no change in the directions in which x_1 and x_4 are measured. Then, instead of (13), we obtain

$$x_1' = \alpha x_1 + i\alpha\beta y_1, \qquad y_1' = \alpha_{21}x_1 + \alpha y_1.$$

The row orthogonality condition

$$\alpha\alpha_{21} + i\alpha^2\beta = 0$$

gives us $\alpha_{21} = -i\alpha\beta$, i.e., β_{12} and β_{21} must have opposite signs. Then the condition

$$\alpha_{11}^2 + \alpha_{12}^2 = 1$$

gives

$$\alpha^2 - \alpha^2\beta^2 = 1,$$

or

$$\alpha = \frac{1}{\sqrt{1-\beta^2}} \qquad (\beta^2 < 1).$$

Thus, we finally arrive at the formulas

$$x_1' = \frac{x_1 + i\beta y_1}{\sqrt{1-\beta^2}}, \qquad y_1' = \frac{-i\beta x_1 + y_1}{\sqrt{1-\beta^2}},$$

or, going back from the new variable $y_1 = icx_4$ to the old variable x_4,

$$x_1' = \frac{x_1 - \beta c x_4}{\sqrt{1-\beta^2}}, \qquad x_4' = \frac{-(\beta/c)x_1 + x_4}{\sqrt{1-\beta^2}}. \tag{15}$$

It follows at once from these formulas that the coordinate system corresponding to the primed variables moves relative to the original coordinate system with the velocity

$$v = \beta c \tag{16}$$

along the x_1-axis. In fact, taking x_1' to be constant, we obtain

$$dx_1 - \beta c\, dx_4 = 0,$$

or

$$\frac{dx_1}{dx_4} = \beta c.$$

Using (16) to introduce the velocity v instead of β, and replacing x_1 by x and x_4 by t, we obtain the usual form of the *Lorentz transformation* in two variables:

$$x_1' = \frac{x - vt}{\sqrt{1 - v^2/c^2}}, \qquad t' = \frac{-(v/c^2)x + t}{\sqrt{1 - v^2/c^2}}. \tag{17}$$

In the limiting case $c \to \infty$, we obtain the familiar relative motion formulas of classical mechanics:

$$x' = x - vt, \qquad t' = t.$$

It is not hard to verify that the Lorentz transformations (17), which depend on the single real parameter v, form a group. In fact, the transformation which is the inverse of (17) is obtained from (17) by replacing v by $-v$. To see this, we solve (17) for x and t, obtaining

$$(1 - v^2/c^2)x = \sqrt{1 - v^2/c^2}\,(x' + vt'),$$
$$(1 - v^2/c^2)t = \sqrt{1 - v^2/c^2}\left(\frac{v}{c^2}x' + t'\right),$$

from which it follows immediately that

$$x = \frac{x' + vt'}{\sqrt{1 - v^2/c^2}}, \qquad t = \frac{(v/c^2)x' + t'}{\sqrt{1 - v^2/c^2}}.$$

Next, we consider two linear transformations L_1 and L_2, corresponding to the parameter values $v = v_1$ and $v = v_2$. We form the product L_2L_1 and show that L_2L_1 is also a Lorentz transformation. Thus, we have to form the matrix product

$$\left\|\begin{matrix} \dfrac{1}{\sqrt{1-\beta_2^2}} & -\dfrac{\beta_2 c}{\sqrt{1-\beta_2^2}} \\ -\dfrac{\beta_2/c}{\sqrt{1-\beta_2^2}} & \dfrac{1}{\sqrt{1-\beta_2^2}} \end{matrix}\right\| \left\|\begin{matrix} \dfrac{1}{\sqrt{1-\beta_1^2}} & -\dfrac{\beta_1 c}{\sqrt{1-\beta_1^2}} \\ -\dfrac{\beta_1/c}{\sqrt{1-\beta_1^2}} & \dfrac{1}{\sqrt{1-\beta_1^2}} \end{matrix}\right\|$$

where

$$\beta_1 = v_1/c, \qquad \beta_2 = v_2/c.$$

Applying the usual rules of matrix multiplication, we find that the product is the following matrix:

$$\frac{1 + \beta_1\beta_2}{\sqrt{1-\beta_2^2}\,\sqrt{1-\beta_1^2}} \left\|\begin{matrix} 1 & -\dfrac{\beta_1 c + \beta_2 c}{1 + \beta_1\beta_2} \\ -\dfrac{\beta_1/c + \beta_2/c}{1 + \beta_1\beta_2} & 1 \end{matrix}\right\|. \tag{18}$$

If we introduce the new quantity

$$v_3 = \frac{v_1 + v_2}{1 + v_1 v_2/c^2}, \tag{19}$$

it is not hard to verify the formula

$$\frac{1 + v_1 v_2/c^2}{\sqrt{1 - v_2^2/c^2}\sqrt{1 - v_1^2/c^2}} = \frac{1}{\sqrt{1 - v_3^2/c^2}}.$$

As a result, the matrix (18) can be written in the following form:

$$\left\| \begin{array}{cc} \dfrac{1}{\sqrt{1-\beta_3^2}} & -\dfrac{\beta_3 c}{\sqrt{1-\beta_3^2}} \\ -\dfrac{\beta_3/c}{\sqrt{1-\beta_3^2}} & \dfrac{1}{\sqrt{1-\beta_3^2}} \end{array} \right\| \qquad (\beta_3 = v_3/c).$$

This matrix corresponds to a Lorentz transformation with the parameter $v = v_3$. Thus, the formula (19) gives the *addition rule for velocities in the special theory of relativity.* If we set $v_1 = c$ in (19), it is easily seen that $v_3 = c$ also, i.e., *the velocity c is in fact invariant with respect to relative motion.*

In deriving the formula (15), we made a definite choice of sign in the coefficients of the linear transformation (11); we assumed that the coefficients a_{11} and a_{44} are positive. This requirement can be replaced by the requirement that a_{44} be positive and that the determinant

$$a_{11}a_{44} - a_{14}a_{41} \tag{20}$$

be $+1$. In fact, if $a_{11} > 0$, $a_{44} > 0$, then it follows from (17) that (20) equals $+1$, while if we had taken $a_{11} = -\alpha$, $a_{44} = \alpha$, where $\alpha > 0$, we would have obtained a transformation with determinant -1. Lorentz transformations for which the determinant (20) equals $+1$ are said to be *proper.* The condition that a_{44} be positive is equivalent to requiring that $x_4' \to \infty$ as $x_4 \to \infty$, for fixed x_1, which corresponds to invariance of the direction in which time is measured. Lorentz transformations for which $a_{44} > 0$ are said to be *orthochronous.* Thus, our formulas do not give all linear transformations satisfying the condition (12), but only those for which the determinant (20) is $+1$ and the time direction is preserved.

We now consider the Lorentz transformation in *four* variables x_k ($k = 1, 2, 3, 4$), where the condition

$$x_1'^2 + x_2'^2 + x_3'^2 - c^2 x_4'^2 = x_1^2 + x_2^2 + x_3^2 - c^2 x_4^2 \tag{21}$$

must be met. We shall show that by regarding x_k and x_k' ($k = 1, 2, 3$) as Cartesian coordinates in two different three-dimensional spaces R and R', and then suitably choosing coordinate axes in R and R', a Lorentz transformation in four variables can be reduced to the special case considered above. Thus, let T denote a Lorentz transformation in four variables, and let S denote the special Lorentz transformation in two variables of the type studied above. Our assertion is equivalent to stating that T can be represented in the form

$$T = VSU, \tag{22}$$

where U and V are real orthogonal transformations corresponding to appropriate coordinate transformations in the spaces R and R'.

As before, we introduce four new variables

$$y_1 = x_1, \qquad y_2 = x_2, \qquad y_3 = x_3, \qquad y_4 = icx_4,$$

and similarly,

$$y_1' = x_1', \qquad y_2' = x_2', \qquad y_3' = x_3', \qquad y_4' = icx_4'.$$

For the new variables, we obtain the usual condition for an orthogonal transformation

$$y_1'^2 + y_2'^2 + y_3'^2 + y_4'^2 = y_1^2 + y_2^2 + y_3^2 + y_4^2, \tag{23}$$

instead of the condition (21). The linear transformation we are looking for has the form

$$y_k' = \alpha_{k1}y_1 + \alpha_{k2}y_2 + \alpha_{k3}y_3 + \alpha_{k4}y_4 \qquad (k = 1, 2, 3, 4). \tag{24}$$

Bearing in mind that y_4 and y_4' have to be purely imaginary, we can assert that the coefficients α_{k1}, α_{k2}, α_{k3} $(k = 1, 2, 3)$ and also α_{44} must be real, while the coefficients α_{4k}, α_{k4} $(k = 1, 2, 3)$ must be purely imaginary.

A change of coordinate axes in the space R' is equivalent to a real orthogonal transformation of the variables y_1', y_2', y_3'. Consider the coefficients

$$\alpha_{14} = i\beta_{14}, \qquad \alpha_{24} = i\beta_{24}, \qquad \alpha_{34} = i\beta_{34}.$$

The three real numbers β_{14}, β_{24}, β_{34} define a vector; if we take the direction of this vector to be the new x_1-axis in the space R', then, as a result of the corresponding orthogonal transformation, the coefficients α_{24} and α_{34} vanish. To see this, it is enough to observe that according to (24), an orthogonal transformation of the variables y_1', y_2', y_3' reduces to the same transformation of β_{14}, β_{24}, β_{34}. Thus, we shall assume that this coordinate transformation in the space R' has already been performed, so that we have $\alpha_{24} = \alpha_{34} = 0$. It follows from (23) that the coefficients of the transformation (24) must satisfy the usual conditions for an orthogonal transformation. Bearing in mind that $\alpha_{24} = \alpha_{34} = 0$, and considering the second and third rows, we obtain the following conditions:

$$\alpha_{k1}^2 + \alpha_{k2}^2 + \alpha_{k3}^2 = 1 \qquad (k = 2, 3),$$
$$\alpha_{21}\alpha_{31} + \alpha_{22}\alpha_{32} + \alpha_{23}\alpha_{33} = 0$$

involving only real coefficients. Because of these conditions, the two vectors $(\alpha_{21}, \alpha_{22}, \alpha_{23})$ and $(\alpha_{31}, \alpha_{32}, \alpha_{33})$ have unit length and are orthogonal to each other. If we choose these two vectors as basis vectors in the space R, directed along the x_2-axis and the x_3-axis, respectively, then the two sums

$$\alpha_{k1}y_1 + \alpha_{k2}y_2 + \alpha_{k3}y_3 \qquad (k = 2, 3),$$

which express the scalar products of these two vectors with the variable vector (y_1, y_2, y_3), become simply y_2 and y_3, i.e., for this choice of axes we have

$$\alpha_{22} = \alpha_{33} = 1, \qquad \alpha_{21} = \alpha_{23} = \alpha_{31} = \alpha_{32} = 0.$$

Thus, finally, if we make this choice of axes in the two spaces R and R', the transformation matrix (24) takes the form

$$\begin{Vmatrix} \alpha_{11} & \alpha_{12} & \alpha_{13} & \alpha_{14} \\ 0 & 1 & 0 & 0 \\ 0 & 0 & 1 & 0 \\ \alpha_{41} & \alpha_{42} & \alpha_{43} & \alpha_{44} \end{Vmatrix}. \tag{25}$$

This matrix was obtained by multiplying the original matrix by the matrices corresponding to the two orthogonal transformations in R and R'; although these transformations affect only the first three variables, they can of course be regarded as orthogonal transformations in four variables, with the fourth variable remaining unchanged. Bearing in mind that the product of two orthogonal transformations is also orthogonal, we can assert that the elements of the matrix (25) must also satisfy the orthogonality conditions. Writing the conditions for the first row to be orthogonal to the second and third rows, we obtain

$$\alpha_{12} = \alpha_{13} = 0;$$

similarly, writing the conditions for the fourth row to be orthogonal to the second and third rows, we obtain

$$\alpha_{42} = \alpha_{43} = 0.$$

As a result, we obtain the matrix

$$\begin{Vmatrix} \alpha_{11} & 0 & 0 & \alpha_{14} \\ 0 & 1 & 0 & 0 \\ 0 & 0 & 1 & 0 \\ \alpha_{41} & 0 & 0 & \alpha_{44} \end{Vmatrix},$$

i.e., in this case, we have the linear transformation

$$y_1' = \alpha_{11}y_1 + \alpha_{14}y_4,$$
$$y_4' = \alpha_{41}y_1 + \alpha_{44}y_4,$$

which must satisfy the condition

$$y_1'^2 + y_4'^2 = y_1^2 + y_4^2.$$

This is a Lorentz transformation in two variables, of the type dealt with previously (which, for $\alpha_{11} > 0$, $\alpha_{44} > 0$ led to a transformation of the form (17)). Thus, the matrix T of a Lorentz transformation in four variables, as defined only by the condition (21), can be represented by the formula (22), where U and V are rotations, and S is a Lorentz transformation in two variables. (The matrices U, V, and S are to be regarded as fourth order matrices.) Since the determinant of S can equal ± 1 while the determinants of U and V equal $+1$, the determinant of T can equal ± 1. However, if we require that T be a *proper* Lorentz transformation (i.e., have determinant $+1$), then det S must also equal $+1$. If, in addition, we require that T be an *orthochronous* Lorentz transformation (i.e., preserve the time direction), then, since U and V do not involve the fourth variable, S must also be orthochronous. Thus, proper orthochronous Lorentz transformations have matrices given by the formula (22), where S is a Lorentz transformation of the special form (17), and U and V are spatial rotations. It can be shown that proper orthochronous Lorentz transformations in four variables form a group, just like the transformations (17).

61. Permutations. So far, we have considered examples of groups whose elements are linear transformations. However, there is no need to relate the group concept to linear transformations, and groups can be constructed by using other kinds of operations. Thus, we now consider the operations already encountered in Sec. 2, namely *permutations*. We begin by explaining certain basic facts and concepts pertaining to permutations.

Suppose that we have n objects of any kind, which we number as in Sec. 2, i.e., we simply assume that the objects are the integers 1, 2, . . . , n. As is well known, $n!$ permutations can be formed from these numbers. We take any of these permutations

$$p_1, p_2, \ldots, p_n, \tag{26}$$

where the numbers p_k are the integers 1 to n, arranged in some order. We then compare the permutation (26) with the basic permutation 1, 2, . . . , n and write

$$P = \begin{pmatrix} 1 & 2 & \cdots & n \\ p_1 & p_2 & \cdots & p_n \end{pmatrix}, \tag{27}$$

which means that the transition from the basic permutation to the permutation (26) is accomplished by replacing 1 by p_1, 2 by p_2, etc. We denote this operation by a single letter P, henceforth referred to as a permutation.

Next, we define the *inverse permutation* P^{-1} as the operation which takes (26) into the basic permutation, i.e., which replaces p_1 by 1, p_2 by 2, etc. The following example illustrates this concept. Let $n = 5$, and let

$$P = \begin{pmatrix} 1 & 2 & 3 & 4 & 5 \\ 3 & 2 & 5 & 1 & 4 \end{pmatrix}.$$

Then, the inverse permutation has the form

$$P^{-1} = \begin{pmatrix} 1 & 2 & 3 & 4 & 5 \\ 4 & 2 & 1 & 5 & 3 \end{pmatrix}.$$

It is easy to see that

$$(P^{-1})^{-1} = P, \tag{28}$$

quite generally.

We now introduce the concept of the *product of two permutations*. Let P_1 and P_2 be any two permutations. By the product P_2P_1 of these two permutations, we mean the permutation obtained as a result of applying first the permutation P_1 and then the permutation P_2. For example, if we have two permutations,

$$P_2 = \begin{pmatrix} 1 & 2 & 3 & 4 & 5 \\ 5 & 1 & 4 & 3 & 2 \end{pmatrix}, \qquad P_1 = \begin{pmatrix} 1 & 2 & 3 & 4 & 5 \\ 3 & 1 & 5 & 2 & 4 \end{pmatrix},$$

then the product P_2P_1 is the permutation

$$P_2P_1 = \begin{pmatrix} 1 & 2 & 3 & 4 & 5 \\ 4 & 5 & 2 & 1 & 3 \end{pmatrix}.$$

It is immediately clear that the inverse permutation P^{-1} is completely specified by the condition

$$P^{-1}P = PP^{-1} = I, \tag{29}$$

where I denotes the *unit* (or *identity*) *permutation*, i.e., the permutation which replaces each element by itself.

Applying three permutations P_1, P_2, P_3 in succession, we obtain the product $P_3P_2P_1$. It is not hard to see that this product obeys the associative law

$$P_3(P_2P_1) = (P_3P_2)P_1. \tag{30}$$

Thus, we can first apply the permutation P_1 and then apply the single permutation P_3P_2 corresponding to consecutive application of P_2 and P_3, or we can first apply the single permutation corresponding to consecutive application of P_1 and P_2 and then apply the permutation P_3. Finally, we note that the unit permutation obviously satisfies the relation

$$IP = PI = P, \tag{31}$$

where P is any permutation. However, a product of permutations does not in general satisfy the commutative law, i.e., the products P_2P_1 and P_1P_2 are usually different permutations. The reader can verify this, using the example given above.

In this way, we have established the fundamental concepts of the unit permutation, the inverse permutation, and the product of permutations, just as was done in the case of linear transformations (matrices). We now pursue this analogy further and introduce the *group* concept. Thus, *a set of permutations forms a group if the following two conditions are satisfied: (a) The inverse of a permutation belonging to the set also belongs to the set; (b) the product (in either order) of two permutations belonging to the set also belongs to the set.* Just as in the case of linear transformations, the unit permutation must belong to the group.

Obviously, the set of all $n!$ permutations forms a group. We now exhibit another group, which is only a part of the whole group of $n!$ permutations. To do this, we observe that every permutation can be accomplished by means of transpositions, where, although a given permutation can be obtained by using different numbers of transpositions, the number of transpositions must always be either even or odd for a given permutation, as was proved above (Sec. 2). Permutations consisting of an even number of transpositions form a group by themselves. The group formed by all $n!$ permutations is called the *symmetric group of degree n*, while the group consisting of all $n!/2$ *even* permutations, i.e., permutations which reduce to an even number of transpositions, is called the *alternating group of degree n*.

We now consider permutations of a special type. Let $l_1, l_2, \ldots, l_m$ be any m different numbers chosen from the numbers $1, 2, \ldots, n$. Suppose that our permutation consists of changing l_1 to l_2, l_2 to l_3, . . . , l_{m-1} to l_m, and finally l_m to l_1. A permutation of this type is called a *cycle* (of *length m*) and is denoted by the symbol $(l_1, l_2, \ldots, l_m)$. By cyclic permutation of the numbers inside the parentheses, we obtain the cycles

$$(l_2, l_3, \ldots, l_m, l_1), (l_3, l_4, \ldots, l_m, l_1, l_2), \ldots,$$

which obviously give the same permutation as $(l_1, l_2, \ldots, l_m)$. If $m = 1$, we have the cycle (l_1), which is equivalent to replacing a number by itself; there is no point in considering such cycles. A cycle (l_1, l_2), consisting of two numbers, is equivalent to a transposition of the elements l_1 and l_2. The product of two cycles with no elements in common is independent of the order of the factors. For example, let $n = 5$, and consider the following products of two cycles without common elements

$$(1, 3)(2, 4, 5), \qquad (2, 4, 5)(1, 3).$$

Both products give the same permutation

$$\begin{pmatrix} 1 & 2 & 3 & 4 & 5 \\ 3 & 4 & 1 & 5 & 2 \end{pmatrix}.$$

We can represent every permutation P as a *product of cycles*. To do this, we choose the element 1 as the first element of a cycle. For the second element of the cycle, we take the element which is obtained from 1 by applying P; let this element be l_2. For the third element, we take the element obtained from l_2 by applying P, and so forth, until finally we arrive at an element which goes into 1 as a result of applying P; this will be the last element of the cycle. It is easy to see that this cycle cannot have identical elements. The cycle formed in this way will in general not use up all n elements. Then, we take one of the remaining elements as the first element of a new cycle, and proceed as before to form a second cycle, and so on. As an example, let $n = 6$, and consider the permutation

$$\begin{pmatrix} 1 & 2 & 3 & 4 & 5 & 6 \\ 3 & 6 & 4 & 1 & 2 & 5 \end{pmatrix}.$$

Applying the process just indicated, we can represent this permutation as the following product of cycles:

$$\begin{pmatrix} 1 & 2 & 3 & 4 & 5 & 6 \\ 3 & 6 & 4 & 1 & 2 & 5 \end{pmatrix} = (1, 3, 4)(2, 6, 5),$$

where the order of the factors on the right plays no role.

It is not hard to see that a product of two transpositions can be represented as a product of two cycles of length 3, i.e., given two cycles of length 2 with no common elements, it can easily be verified that

$$(l_3, l_4)(l_1, l_2) = (l_1, l_3, l_4)(l_1, l_2, l_4),$$

whereas

$$(l_1, l_3)(l_1, l_2) = (l_1, l_2, l_3)$$

in the case where the transpositions have common elements. Thus, every permutation in the alternating group can be represented as a product of cycles of length 3.

It should be noted that we can write the numbers in the first row of a permutation in any order, instead of in the natural order $1, 2, \ldots, n$. The only thing that matters is that the number appearing below a given number should be the number obtained from the given number as a result of the permutation in question. For example, we have the following two ways of writing the same permutation:

$$\begin{pmatrix} 1 & 2 & 3 & 4 & 5 \\ 3 & 2 & 5 & 1 & 4 \end{pmatrix} = \begin{pmatrix} 3 & 1 & 5 & 4 & 2 \\ 5 & 3 & 4 & 1 & 2 \end{pmatrix}.$$

Thus, suppose that we have a permutation

$$P = \begin{pmatrix} a_1 & a_2 & \cdots & a_n \\ b_1 & b_2 & \cdots & b_n \end{pmatrix}.$$

Then, the inverse permutation can obviously be written in the form

$$P^{-1} = \begin{pmatrix} b_1 & b_2 & \cdots & b_n \\ a_1 & a_2 & \cdots & a_n \end{pmatrix}.$$

Now suppose that we have two permutations

$$P = \begin{pmatrix} 1 & 2 & \cdots & n \\ c_1 & c_2 & \cdots & c_n \end{pmatrix}, \quad Q = \begin{pmatrix} 1 & 2 & \cdots & n \\ d_1 & d_2 & \cdots & d_n \end{pmatrix} = \begin{pmatrix} c_1 & c_2 & \cdots & c_n \\ f_1 & f_2 & \cdots & f_n \end{pmatrix},$$

where we have written the second permutation in two different ways. Then we have

$$PQ^{-1} = \begin{pmatrix} 1 & 2 & \cdots & n \\ c_1 & c_2 & \cdots & c_n \end{pmatrix} \begin{pmatrix} d_1 & d_2 & \cdots & d_n \\ 1 & 2 & \cdots & n \end{pmatrix} = \begin{pmatrix} d_1 & d_2 & \cdots & d_n \\ c_1 & c_2 & \cdots & c_n \end{pmatrix},$$

and therefore

$$QPQ^{-1} = \begin{pmatrix} c_1 & c_2 & \cdots & c_n \\ f_1 & f_2 & \cdots & f_n \end{pmatrix} \begin{pmatrix} d_1 & d_2 & \cdots & d_n \\ c_1 & c_2 & \cdots & c_n \end{pmatrix} = \begin{pmatrix} d_1 & d_2 & \cdots & d_n \\ f_1 & f_2 & \cdots & f_n \end{pmatrix}.$$

This leads to the following rule: *To obtain the permutation QPQ^{-1}, we have to apply the permutation Q to both rows of the permutation*

$$P = \begin{pmatrix} 1 & 2 & \cdots & n \\ c_1 & c_2 & \cdots & c_n \end{pmatrix}.$$

62. Abstract Groups. To define a group, we can depart completely from the concrete meaning of the operations forming the group, which so far have been linear transformations or permutations. This leads us to the concept of an *abstract group*.

An abstract group is a set of symbols (elements) for which multiplication is defined in the sense that a rule is specified which associates with every two elements P and Q (different or identical) belonging to the set, a third element, also belonging to the set, which is called the *product* of P and Q and is denoted by PQ.† Moreover, the following conditions must be satisfied:

† Or by QP, as in the case of linear transformations or permutations. This is merely a matter of convention.

(a) *Multiplication must obey the associative law*, i.e., $(PQ)R = P(QR)$. More generally, this implies that we can group together any factors in a product (without changing the order of the factors, of course).

(b) *The set must contain a unique element E which when multiplied by any other element (from the left or from the right) gives back the element itself:*

$$EP = PE = P. \tag{32}$$

This element is called the *unit element* (or *identity*).

(c) *For any element P of the set, there exists another element Q of the set, which satisfies the relation*

$$QP = PQ = E. \tag{33}$$

(We shall sometimes write I instead of E.) The element Q defined by (33) is called the *inverse* of P and is denoted by P^{-1}. The relation (28) obviously holds, since it follows from (33) that P is the inverse of Q. Setting $P = E$ in (32), we obtain $EE = E$, i.e., according to the definition of the inverse, E is its own inverse ($E^{-1} = E$).

It is possible to give a less redundant version of the postulates defining an abstract group, from which the present postulates follow as formal consequences; however, this will not be done here. Rather, we shall consider only the most elementary and basic facts connected with the notion of an abstract group. A more detailed study of group theory would lead to an amount of material which could easily fill a whole book by itself. Our aim is merely to acquaint the reader with the fundamental concepts of group theory, and thereby facilitate his reading of the literature of physics, where the group concept and the basic properties of groups are often applied.

Having established the concept of an abstract group, we now introduce some new concepts and prove some new properties of abstract groups. First, we note that the number of elements in a group can be either finite or infinite (as we saw in Sec. 58). Next, consider a product PQR of three elements of a group; PQR is itself an element of the group. The inverse element is obtained in just the same way as in the case of the group of linear transformations (Sec. 26), i.e.,

$$(PQR)^{-1} = R^{-1}Q^{-1}P^{-1};$$

this is easily verified by performing the multiplication and using the associative law.

Let P be an element of a group. The integral powers of P:

$$P^0 = I, P^1, P^2, \ldots,$$

are also elements of a group. If there exists a positive integer m such that $P^m = I$, then we say that the element P is of *finite order*, and the *order of the element* is the least positive integer m for which $P^m = I$. Moreover, no two of the elements

$$I, P, P^2, \ldots, P^{m-1}$$

can be identical, since the relation $P^k = P^l$ $(k < l)$ immediately implies $P^{l-k} = I$. Clearly, all the elements of a finite group are of finite order.

Suppose that the elements of a group are denoted by P_α. If the group is finite, then we can assume that the index α runs through a finite number of positive integral values. If the group is infinite, α can take all integral values (Sec. 52), it may vary continuously, or it may even be equivalent to several continuously varying indices. Now let U be a fixed element of our group, and form all possible products UP_α. It is not hard to see that as the index α takes all possible values, the product UP_α runs through all the elements of the group just once. In fact, multiplying the relation

$$UP_{\alpha_1} = UP_{\alpha_2}$$

by U^{-1} from the left, we obtain $P_{\alpha_1} = P_{\alpha_2}$, i.e., the products UP_α are different for different α. Moreover, UP_α gives any element of the group, since the relation $UP_\alpha = UP_{\alpha_0}$ is equivalent to $P_\alpha = U^{-1}P_{\alpha_0}$, so that the product UP_α gives the element P_{α_0} if the factor P_α is the element $U^{-1}P_{\alpha_0}$. The same conclusion would have been reached if we had written the fixed element U on the right instead of on the left. Thus, we obtain the following result: *Let U be a fixed element of a group, and let P_α run through all the elements of the group. Then the product UP_α (or $P_\alpha U$) runs through all the elements of the group just once.*

Consider the special case of a group consisting of six elements (a group of order 6), and denote these elements by

$$E, A, B, C, D, F.$$

We define multiplication by the following table:

	E	A	B	C	D	F
E	E	A	B	C	D	F
A	A	E	D	F	B	C
B	B	F	E	D	C	A
C	C	D	F	E	A	B
D	D	C	A	B	F	E
F	F	B	C	A	E	D

(34)

For example, the product DB is the element A, located at the intersection of the row marked D and the column marked B. It is easily verified that all the conditions given in the definition of an abstract group are satisfied and that the element E plays the role of the unit element.

In preceding sections, we encountered concrete realizations of the abstract concept of a group. In one case, the elements were linear transformations (matrices), and multiplication of two elements consisted in consecutive application of two linear transformations, i.e., multiplication of the corresponding matrices. In another case, the elements were permutations, and multiplication of two elements consisted in performing two permutations in succession. Consider now two more concrete realizations of a group. Let the elements be all possible complex numbers, and let multiplication of two elements consist in addition of the corresponding complex numbers. In this case, the number zero plays the role of the unit element, and $-\alpha$ is the inverse of the complex number α. Instead of complex numbers, we could have taken the elements to be all possible vectors

$$\mathbf{x} = (x_1, x_2, \ldots, x_n)$$

of the complex n-dimensional space R_n, and multiplication of elements could have been defined as addition of the corresponding vectors, with the zero vector playing the role of the unit element, i.e., the elements of the group are the vectors of R_n, and the group operation is vector addition. We note that in the last two examples, the result of multiplying two elements of the group does not depend on the order of the factors, a property which is expressed by saying that any two elements of the group *commute*. Groups of this type are called *Abelian groups*. The simplest example of an Abelian group is the so-called *cyclic group*, which consists of the unit element E and the powers of some element P. If m is the least positive integer for which $P^m = E$, the cyclic group contains the m elements $E, P, P^2, \ldots, P^{m-1}$. If there is no such positive integer, the group becomes $E, P, P^2, \ldots$, and is infinite.

63. Subgroups. Given a group $\mathcal{G}$, suppose that the set of elements $\mathcal{H}$, consisting of only some of the elements of $\mathcal{G}$, is also a group (with the original definition of multiplication). In this case, the group $\mathcal{H}$ is called a *subgroup* of the group $\mathcal{G}$. It is easy to see that the set containing only the unit element of $\mathcal{G}$ is always a subgroup, and henceforth, in discussing subgroups, we shall rule out the possibility of this trivial subgroup.

Let H_α denote a generic element of $\mathcal{H}$, and let G_1 be an element of the original group $\mathcal{G}$ which does not belong to $\mathcal{H}$. As we saw above (Sec. 56), the products G_1H_α give us group elements which are distinct for different α. Moreover, these elements do not belong to $\mathcal{H}$, for otherwise we would have $G_1H_{\alpha_1} = H_{\alpha_2}$ for the parameter values α_1 and α_2, which would imply $G_1 = H_{\alpha_2}H_{\alpha_1}^{-1}$, i.e., G_1 belongs to $\mathcal{H}$, contrary to hypothesis. Suppose now that G_1 and G_2 are two different elements of the group $\mathcal{G}$ which do not belong to the subgroup $\mathcal{H}$. Then, the sets of elements $G_1\mathcal{H}$ and $G_2\mathcal{H}$ either have no elements in common, or else they coincide, i.e., consist of exactly the same elements. (By $G_1\mathcal{H}$ is meant the set of elements of the form G_1H_α, where the index α runs through the values corresponding to the subgroup $\mathcal{H}$, etc.) In fact, if we have $G_2H_{\alpha_2} = G_1H_{\alpha_1}$ for certain values of the parameter α, then $G_2 = G_1H_{\alpha_1}H_{\alpha_2}^{-1} = G_1H_{\alpha_3}$, i.e., G_2 belongs to the set $G_1\mathcal{H}$; in just the same way, G_1 belongs to the set $G_2\mathcal{H}$. From this it follows that $G_1\mathcal{H}$ and $G_2\mathcal{H}$ define the same set of elements.

Since the elements H_α of the subgroup $\mathcal{H}$ do not exhaust all the elements of the group $\mathcal{G}$, we can take an element G_1 which does not belong to $\mathcal{H}$ and form all possible products G_1H_α; as we have just seen, these products are all different from each other and from the elements H_α. It may happen that the elements H_α and G_1H_α still do not exhaust the whole group. We then take another element G_2 which does not belong to the sets $\mathcal{H}$ and $G_1\mathcal{H}$ and form all possible products G_2H_α. Again, as already noted, the elements G_2H_α are all different from each other and from the elements H_α and G_1H_α. If the sets $\mathcal{H}$, $G_1\mathcal{H}$, and $G_2\mathcal{H}$ do not exhaust all the elements of the group $\mathcal{G}$, we take a third element G_3 which does not belong to any of the three sets so far considered and form the products G_3H_α. In this way, we obtain new elements of the group, and so on. Suppose that after a finite number of such operations, we finally use up all the elements of the group $\mathcal{G}$, and let $m - 1$ be the number of elements G_k needed to do this. Then every element of the group $\mathcal{G}$ belongs to one of the sets

$$\mathcal{H}, G_1\mathcal{H}, G_2\mathcal{H}, \ldots, G_{m-1}\mathcal{H}. \tag{35}$$

If we set $G_k' = G_kH_{\alpha_0}$, where α_0 is any fixed index, then, as proved above, the set $G_k'\mathcal{H}$ coincides with the set $G_k\mathcal{H}$. In other words, any element of a given set $G_k\mathcal{H}$ $(G_0 = I)$ can "represent" G_k. It immediately follows that for a given subgroup $\mathcal{H}$, the decomposition of the whole group $\mathcal{G}$ into sets of the form (35) is unique. The sets $G_k\mathcal{H}$ are called *cosets of the subgroup* $\mathcal{H}$. When we have a decomposition like (35), the subgroup $\mathcal{H}$ is called a *subgroup of finite index*, i.e., a *subgroup of index m*.

If the group $\mathcal{G}$ is finite, the index of the subgroup $\mathcal{H}$ obviously equals the ratio of the order of the whole group $\mathcal{G}$ to the order of the subgroup $\mathcal{H}$, where by the *order* of a finite group we mean the number of elements in the group. We note that only the first of the cosets (35) is a subgroup, since none of the other cosets $G_k\mathcal{H}$ contains the unit element, so that none of the other cosets can be a subgroup.

In constructing the cosets (35), we multiplied the elements of $\mathcal{H}$ from the left by the elements G_k of the group $\mathcal{G}$. We could just as well have performed this multiplication from the right. Thus, writing G_k' instead of G_k, we could have decomposed the group $\mathcal{G}$ into cosets of the form

$$\mathcal{H},\ \mathcal{H}G'_1,\ \mathcal{H}G'_2,\ \ldots,\ \mathcal{H}G'_{m-1}, \tag{36}$$

instead of (35), where, as we shall show next, the index of the subgroup $\mathcal{H}$ remains unchanged. (As before, $\mathcal{H}G'_1$ means the set of products $H_\alpha G'_1$, where H_α is in $\mathcal{H}$, etc.) The sets $G_k\mathcal{H}$ are called *left cosets*, while the sets $\mathcal{H}G'_k$ are called *right cosets.*

To prove that the number of right cosets equals the number of left cosets, we first observe that as α runs through all the values indexing elements H_α of the subgroup $\mathcal{H}$, H_α^{-1} runs through all the elements of $\mathcal{H}$; this follows at once from the fact that the inverse of an element belonging to $\mathcal{H}$ also belongs to $\mathcal{H}$. Now take any two distinct cosets of the form (35), consisting of the elements G_pH_α and G_qH_α ($p \neq q$), where the possibility $G_p = E$ is not excluded. Taking inverses, we obtain

$$(G_pH_\alpha)^{-1} = H_\alpha^{-1}G_p^{-1}, \qquad (G_qH_\alpha)^{-1} = H_\alpha^{-1}G_q^{-1}.$$

In view of the observation just made, we can write the corresponding cosets as $\mathcal{H}G_p^{-1}$, $\mathcal{H}G_q^{-1}$. These cosets have no elements in common, since if they did, e.g., if

$$H_{\alpha_1}G_p^{-1} = H_{\alpha_2}G_q^{-1},$$

this would imply that

$$G_p^{-1}G_q = H_{\alpha_1}^{-1}H_{\alpha_2} = H_{\alpha_3},$$

or

$$G_q = G_pH_{\alpha_3},$$

so that G_q would belong to the coset $G_p\mathcal{H}$, contrary to assumption. Therefore, the sets

$$\mathcal{H},\ \mathcal{H}G_1^{-1},\ \mathcal{H}G_2^{-1},\ \ldots,\ \mathcal{H}G_{m-1}^{-1}$$

are right cosets, i.e., we can take $G'_s = G_s^{-1}$ in (36).

We now consider some examples of subgroups. Let $\mathcal{G}$ be the set of real orthogonal transformations in three variables, and let $\mathcal{H}$ be the set of real orthogonal transformations in three variables with determinant $+1$. Every real orthogonal transformation is either a rotation, i.e., belongs to $\mathcal{H}$, or a product of a rotation with the reflection S in the origin, given by the formulas

$$x' = -x, \qquad y' = -y, \qquad z' = -z. \tag{37}$$

Thus, this group $\mathcal{G}$ can be represented as

$$\mathcal{H},\ S\mathcal{H} \tag{38}$$

or

$$\mathcal{H},\ \mathcal{H}S. \tag{39}$$

In this case, $\mathcal{H}$ is a subgroup of index 2.

Next, let $\mathcal{G}$ be the symmetric group of permutations on n elements, and let $\mathcal{H}$ be the alternating group consisting of even permutations. Moreover, let S be any odd permutation, e.g., the permutation consisting of the single cycle (1, 2), which transposes the elements 1 and 2. Then, we can obviously represent $\mathcal{G}$ by either (38) or (39), i.e., multiplication from the left leads to the same result as multiplication from the right. Thus, the alternating group is a subgroup of index 2 of the symmetric group.

Next, consider the finite group of the regular octahedron, discussed above (Sec. 59). Let A be any vertex of the octahedron, and let l be the axis going through A. Let S_0, S_1, S_2, and S_3 be rotations about l through the angles 0, $\pi/2$, π, and $3\pi/2$, respectively. These rotations form a subgroup of the whole group of rotations of the octahedron; we denote this subgroup by $\mathcal{S}$. Let T_k ($k = 1, 2, 3, 4, 5$) denote the rotations carrying the vertex A into the other five vertices of the octahedron. Then, the complete group of the octahedron can be written in the form

$$\mathcal{S},\ T_1\mathcal{S},\ T_2\mathcal{S},\ T_3\mathcal{S},\ T_4\mathcal{S},\ T_5\mathcal{S},$$

i.e., $\mathcal{S}$ is a subgroup of index 6.

Finally, let G_s, G_s^{-1} ($s = 1, 2, \ldots, k$) be any elements of the group $\mathcal{G}$, and consider the set of all elements of $\mathcal{G}$ which can be represented in the form of products of the elements G_s, G_s^{-1} ($s = 1, 2, \ldots, k$). This set of elements obviously forms a group, which either is a subgroup of $\mathcal{G}$ or coincides with $\mathcal{G}$. This subgroup is said to be *generated* by the elements G_s, G_s^{-1} ($s = 1, 2, \ldots, k$).

64. Classes and Normal Subgroups. Let U and V be any two elements of a group. The element $W = VUV^{-1}$ is said to be *conjugate* to U. It is easy to see that, conversely, U is conjugate to W, since $U = V^{-1}WV$. Two elements U_1 and U_2 which are conjugate to a third element W, i.e., such that

$$U_1 = V_1WV_1^{-1}, \qquad U_2 = V_2WV_2^{-1},$$

are also conjugate to each other, since

$$U_2 = V_2V_1^{-1}U_1(V_2V_1^{-1})^{-1}.$$

A set of group elements which are conjugate to each other is said to form a *class* of the group. A class is uniquely defined by giving one of its elements U. In fact, from U we obtain the whole class by using the formula $G_\alpha UG_\alpha^{-1}$, where G_α runs through all the elements of the group. Thus, we can decompose the whole group into classes.

Because of the basic property of the unit element (Sec. 56), we have

$$G_\alpha IG_\alpha^{-1} = I,$$

i.e., the unit element forms a class by itself. If the element U is of order m, i.e., if m is the least positive integer for which $U^m = I$, then every conjugate element $G_\alpha U G_\alpha^{-1}$ has the same order m; this is an immediate consequence of the relation

$$(G_\alpha U G_\alpha^{-1})^m = G_\alpha U^m G_\alpha^{-1} = I.$$

In other words, all the elements of a given class have the same order. We note that as G_α runs through all the elements of the group $\mathcal{G}$, the product $G_\alpha U G_\alpha^{-1}$ may run through the elements of a class several times. For example, if $U = I$, the product $G_\alpha U G_\alpha^{-1}$ is always I, as just noted.

As an illustration, we again consider the rotation group of the octahedron. Let U be a rotation through the angle $\pi/2$ about an axis A_pA_q of the octahedron, and let T_k be a rotation belonging to the octahedral group, which takes the axis $l = A_pA_q$ into another axis l_1, thereby carrying the vertex A_p into A_r and the vertex A_q into A_s. Then, the group element $T_kUT_k^{-1}$ represents a rotation through the angle $\pi/2$ about the axis A_rA_s. If, for example, T_k carries A_p into A_q, then $T_kUT_k^{-1}$ represents a rotation through the angle $\pi/2$ about the axis A_qA_p (a rotation through $3\pi/2$ about the axis A_pA_q). If T_k takes the axis A_pA_q into itself, i.e., if T_k is a rotation about A_pA_q, then $T_kUT_k^{-1}$ coincides with U. Thus, in this case, the class of elements conjugate to U is the set of rotations through the angle $\pi/2$ about the axes of the octahedron.

In just the same way, if $\mathcal{G}$ is the group of three-dimensional rotations about the origin, then, as we know from Sec. 42, every element U of $\mathcal{G}$ is a rotation through an angle φ about some axis. In this case, the class of elements conjugate to U is the class of rotations through the angle φ about all possible axes passing through the origin.

Closely related to the concept of a class is the concept of a *normal subgroup*, which we now introduce. Let G_1 be a fixed element of the group $\mathcal{G}$, and consider the set of elements of $\mathcal{G}$ which can be represented as products of the form

$$G_1 H_\alpha G_1^{-1}, \tag{40}$$

where H_α denotes a variable element of a subgroup $\mathcal{H}$ of $\mathcal{G}$, i.e., H_α runs through all the elements of $\mathcal{H}$. It is not hard to see that the products (40) also form a subgroup. In fact, the product of two elements of the form (40) is also of the form (40) since

$$(G_1 H_{\alpha_1} G_1^{-1})(G_1 H_{\alpha_2} G_1^{-1}) = G_1 H_{\alpha_1} H_{\alpha_2} G_1^{-1} = G_1 H_{\alpha_3} G_1^{-1},$$

and similarly, the other group properties are satisfied. The subgroup $G_1 \mathcal{H} G_1^{-1}$, consisting of elements of the form (40), is said to be *conjugate* to the subgroup $\mathcal{H}$. If G_1 belongs to $\mathcal{H}$, the subgroup $G_1 \mathcal{H} G_1^{-1}$ consists of elements belonging to $\mathcal{H}$, and in fact coincides with $\mathcal{H}$, since every element H_{α_0} of $\mathcal{H}$ can be represented in the form (40) if we take

$$H_\alpha = G_1^{-1} H_{\alpha_0} G_1.$$

If the element G_1 does not belong to $\mathcal{H}$, then the subgroup $G_1\mathcal{H}G_1^{-1}$ may be different from the subgroup $\mathcal{H}$. A subgroup $\mathcal{H}$ is called a *normal subgroup* (synonymously, *normal divisor* or *invariant subgroup*) of the group $\mathcal{G}$ if the subgroup $G_1\mathcal{H}G_1^{-1}$, consisting of all elements of the form (40), coincides with the subgroup $\mathcal{H}$ for any choice of the element G_1 in $\mathcal{G}$. Before giving examples of normal subgroups, we discuss some new ideas related to this concept.

Suppose that the subgroup $\mathcal{H}$ is a normal subgroup of the group $\mathcal{G}$, and for simplicity, suppose that $\mathcal{H}$ has finite index m. Then, the group $\mathcal{G}$ can be decomposed into cosets of the form

$$\mathcal{H},\ G_1\mathcal{H},\ G_2\mathcal{H},\ \ldots\ ,\ G_{m-1}\mathcal{H}, \tag{41}$$

where, as always, $G_k\mathcal{H}$ denotes the set of all products of the form G_kH_α. Since $\mathcal{H}$ is a normal subgroup, $G_k\mathcal{H}G_k^{-1}$ coincides with $\mathcal{H}$, i.e., $G_k\mathcal{H}$ coincides with $\mathcal{H}G_k$. Thus, if $\mathcal{H}$ is a normal subgroup, the decomposition of the whole group $\mathcal{G}$ into cosets of the form (41) is the same as the decomposition of $\mathcal{G}$ into cosets of the form

$$\mathcal{H},\ \mathcal{H}G_1,\ \mathcal{H}G_2,\ \ldots\ ,\ \mathcal{H}G_{m-1}, \tag{42}$$

i.e., in this case, *the left and right cosets of $\mathcal{H}$ coincide.*

If H is an element of a normal subgroup $\mathcal{H}$, then, for any element G in $\mathcal{G}$, the element GHG^{-1} also belongs to $\mathcal{H}$, i.e., if an element H belongs to a normal subgroup $\mathcal{H}$, then the whole class containing H in the basic group $\mathcal{G}$ also belongs to $\mathcal{H}$. Conversely, suppose that a subgroup has the property that if it contains an element, it also contains the whole class to which the element belongs in the basic group. Then, it is not hard to show that such a subgroup is a normal subgroup.

We now examine once more the cosets (41) and (42) of a normal subgroup $\mathcal{H}$. Consider the product $G_lH_\alpha G_kH_\beta$ obtained by multiplying an element of the coset $G_l\mathcal{H}$ with an element of the coset $G_k\mathcal{H}$. We can write this product as

$$G_l(H_\alpha G_k)H_\beta = G_lG_kH_\gamma H_\beta = G_lG_kH_\delta,$$

where $H_\alpha G_k = G_kH_\gamma$, since $\mathcal{H}$ is a normal subgroup, and $H_\gamma H_\beta = H_\delta$, since $\mathcal{H}$ is a subgroup. Thus, all products of the form $G_lH_\alpha G_kH_\beta$ are elements of the same coset $G_lG_k\mathcal{H}$, and it is also not hard to show that all elements of the coset $G_lG_k\mathcal{H}$ can be obtained in this way. More succinctly, if a subgroup is a normal subgroup, the product of two cosets is also a coset. We shall regard each of the cosets $G_k\mathcal{H}$ (or $\mathcal{H}G_k$) as a new kind of element, and the first of the cosets (41) (or (42)) as the unit element. Then, the result just given, concerning products of cosets, furnishes us with a rule for multiplying these new elements. The reader can easily verify that this multiplication rule satisfies all the conditions required to form a group, i.e., with this multiplication rule, our new elements constitute a group, with the first of the cosets (41) playing the role of the unit element. This new group, whose order equals the index of the normal subgroup $\mathcal{H}$, is called the *factor group of* $\mathcal{H}$ *in* $\mathcal{G}$ (or the *quotient group of* $\mathcal{G}$ *relative* to $\mathcal{H}$).

Every group $\mathcal{G}$ has two trivial normal subgroups, one consisting of the unit element alone, and the other consisting of the whole group. A group which has no normal subgroup except these two trivial ones is said to be *simple.*

65. Examples

1. Consider the group $\mathcal{G}$ of real orthogonal transformations in three-dimensional space. Let $\mathcal{H}$ be the subgroup consisting of the rotations, i.e., the orthogonal transformations with determinants equal to $+1$, and let S be the operation of reflection in the origin, defined by (37). Then the whole group $\mathcal{H}$ has the decomposition

$$\mathcal{H},\ S\mathcal{H} \qquad \text{or} \qquad \mathcal{H},\ \mathcal{H}S. \tag{43}$$

If G is any transformation in $\mathcal{G}$ and H_α is a variable element of $\mathcal{H}$, then $GH_\alpha G^{-1}$ has determinant $+1$, and hence belongs to $\mathcal{H}$; therefore, $\mathcal{H}$ is a normal subgroup of index 2.

We now consider the factor group of $\mathcal{H}$ in $\mathcal{G}$. The coset consisting of $\mathcal{H}$ itself is the unit element of the factor group. Moreover, the product of two elements of the coset $S\mathcal{H}$ (or $\mathcal{H}S$), i.e., two orthogonal transformations with determinant -1, gives an orthogonal transformation with determinant $+1$ belonging to $\mathcal{H}$, i.e., if K is the element corresponding to $S\mathcal{H}$ (or $\mathcal{H}S$), then $K^2 = E$. Thus, the factor group of $\mathcal{H}$ in G consists of two elements E and K, where $K^2 = E$, and hence is a cyclic group of order 2. This is always the situation for normal subgroups of index 2.

2. The alternating group is a normal subgroup (of index 2) of the symmetric group.† First, we write down the elements of the group of permutations on three elements, denoting each by a single letter and using the cycle notation of Sec. 61:

$$E,\ A = (2, 3),\ B = (1, 2),\ C = (1, 3),\ D = (1, 3, 2),\ F = (1, 2, 3).$$

It is easily verified that the multiplication rule for the elements of this symmetric group is the rule defined by the table (34) of Sec. 62.

In this case, the alternating group consists of the permutations E, D, and F ($F = D^2$, $D = F^2$) and is a cyclic group of order 3, i.e.,

$$D^3 = F^3 = E.$$

† Cf. Probs. 29 and 33.

The whole group consists of the three classes

$$\begin{aligned} \text{I}&: E, \\ \text{II}&: A, B, C, \\ \text{III}&: D, F. \end{aligned}$$

The alternating group also consists of three classes, each containing a single element:

$$\begin{aligned} \text{I}&: E, \\ \text{II}&: D, \\ \text{III}&: F. \end{aligned}$$

For $n = 4$, the alternating group contains 12 elements, which are divided into the following four classes:

$$\begin{aligned} \text{I}&: E, \\ \text{II}&: A_1 = (1, 2)(3, 4),\ A_2 = (1, 3)(2, 4),\ A_3 = (1, 4)(2, 3), \\ \text{III}&: B_1 = (1, 2, 3),\ B_2 = (2, 1, 4),\ B_3 = (3, 4, 1),\ B_4 = (4, 3, 2), \\ \text{IV}&: C_1 = (1, 2, 4),\ C_2 = (2, 1, 3),\ C_3 = (3, 4, 2),\ C_4 = (4, 3, 1). \end{aligned}$$

The class II contains three elements of order 2, while the classes III and IV each contain four elements of order 3. It is easily verified that the product of two elements of order 2 is again an element of order 2. Thus, since the class II contains all the elements of order 2, these three elements together with the unit element form a normal subgroup (of order 4 and index 3) of the alternating group. The normal subgroup itself corresponds to the unit element E of the factor group. Moreover, it is easily seen that a product of two elements of the class III is an element of the class IV, while a product of two elements of the class IV is an element of the class III. Thus, if A and B denote the other elements of the factor group, we have $A^2 = B$ and $B^2 = A$, so that the factor group consists of the elements E, A, and A^2, where $A^3 = E$, and is obviously a cyclic group of order 3. We observe that the elements E, (1, 2, 3) and (2, 1, 3) of the alternating group, also form a cyclic subgroup of order 3, but this subgroup is not a normal subgroup.

If we number the vertices of a tetrahedron in any order, then it can easily be verified that the alternating group (for $n = 4$) corresponds to the rotations which carry the tetrahedron into itself. Each permutation corresponds to the transformation of certain vertices into others. The permutations of the class III correspond to rotations through the angle $2\pi/3$ about one of the axes of the tetrahedron, while the permutations of the class IV correspond to rotations through the same angle about the same axes, but in the opposite direction. Thus, for example, the permutations (1, 2, 3) and (2, 1, 3) correspond to rotations about an axis going through the vertex numbered 4. The permutations of the class II correspond to rotations of the tetrahedron which change the position of all the vertices.

It can be shown that the alternating group is simple when $n > 4$.

3. Let $\mathcal{G}$ be an Abelian group, and let $\mathcal{H}$ be any subgroup of $\mathcal{G}$. Then for any choice of the elements H in $\mathcal{H}$ and G in $\mathcal{G}$, we have $GH = HG$, or $GHG^{-1} = H$, so that it is immediately apparent that $\mathcal{H}$ is a normal subgroup, i.e., every subgroup of an Abelian group is a normal subgroup. As an example, consider the group $\mathcal{G}$ corresponding to addition of vectors in R_n (discussed in Sec. 62). For the subgroup $\mathcal{H}$, we take the vectors belonging to a k-dimensional subspace L_k of R_n $(0 < k < n)$. The cosets are obtained by adding all the vectors of the subspace L_k to any vector $\mathbf{x}$ in R_n; if $\mathbf{x}$ belongs to L_k, the coset coincides with the subgroup $\mathcal{H}$.

We now introduce basis vectors $\mathbf{x}^{(1)}, \ldots, \mathbf{x}^{(k)}$ in L_k and basis vectors $\mathbf{x}^{(k+1)}, \ldots, \mathbf{x}^{(n)}$ in the subspace M_{n-k}, which is the orthogonal complement of L_k. In view of what has been said, every coset consists of the vectors

$$c_1\mathbf{x}^{(1)} + \cdots + c_k\mathbf{x}^{(k)} + c_{k+1}\mathbf{x}^{(k+1)} + \cdots + c_n\mathbf{x}^{(n)},$$

where $c_{k+1}, \ldots, c_n$ have fixed values and $c_1, \ldots, c_k$ can take any values. Thus, with every coset we can associate a definite vector in M_{n-k}, and conversely, a definite coset corresponds to every vector in M_{n-k}. Addition of two vectors from any two cosets corresponds to addition of the corresponding vectors from M_{n-k}, i.e., we can regard the vectors of M_{n-k} as the elements of the factor group, with the same group operation (vector addition) as before.

In this example, the order and index of the normal subgroup $\mathcal{H}$ are both infinite.

66. Isomorphic and Homomorphic Groups. Two groups $\mathcal{A}$ and $\mathcal{B}$ are said to be *isomorphic* if a correspondence can be established between the elements of $\mathcal{A}$ and $\mathcal{B}$ such that (1) a definite element of $\mathcal{B}$ corresponds to every element of $\mathcal{A}$, and conversely, a definite element of $\mathcal{A}$ corresponds to every element of $\mathcal{B}$ (a one-to-one correspondence), and (2) the product of any two elements of $\mathcal{B}$ corresponds to the product of the corresponding elements of $\mathcal{A}$, i.e., if the elements B_1 and B_2 of $\mathcal{B}$ are the images of the elements A_1 and A_2 of $\mathcal{A}$, then the element B_1B_2 of $\mathcal{B}$ is the image of the element A_1A_2 of $\mathcal{A}$. Two isomorphic groups $\mathcal{A}$ and $\mathcal{B}$ have identical structure, i.e., they do not differ in any essential way.

We now introduce a new concept, which is a generalization of the concept of isomorphic groups. The group $\mathcal{B}$ is said to be *homomorphic* to the group $\mathcal{A}$ if (1) a definite element of $\mathcal{B}$ corresponds to every element of $\mathcal{A}$ and at least one element of $\mathcal{A}$ corresponds to every element of $\mathcal{B}$, and (2) the product of any two elements of $\mathcal{B}$ corresponds to the product of the corresponding elements of $\mathcal{A}$, i.e., if the elements B_1 and B_2 are the images of the elements A_1 and A_2 of $\mathfrak{A}$, then the element B_1B_2 of $\mathfrak{B}$ is the image of the element A_1A_2 of $\mathfrak{A}$. Unlike the case of isomorphic groups, this correspondence does not have to be one-to-one, i.e., several different elements of $\mathfrak{A}$ can correspond to the same element of the group $\mathfrak{B}$. If the group $\mathfrak{B}$ is homomorphic to the group $\mathfrak{A}$ and if just one element of $\mathfrak{A}$ corresponds to every element of $\mathfrak{B}$, then the groups $\mathfrak{A}$ and $\mathfrak{B}$ are actually isomorphic.

Suppose that the groups $\mathfrak{A}$ and $\mathfrak{B}$ are homomorphic, and let A_0 be the unit element of $\mathfrak{A}$ and B_0 the corresponding element of $\mathfrak{B}$. Then, it is easy to see that B_0 is the unit element of $\mathfrak{B}$. In fact, for any A_1 in $\mathfrak{A}$, we have the relation

$$A_0A_1 = A_1A_0 = A_1,$$

which implies the relation

$$B_0B_1 = B_1B_0 = B_1$$

between the corresponding elements of $\mathfrak{B}$, where by the definition of the homomorphism, B_1 can be any element of $\mathfrak{B}$. This relation shows that B_0 is the unit element of the group $\mathfrak{B}$. Thus, in either an isomorphism or a homomorphism, the unit element of $\mathfrak{B}$ corresponds to the unit element of $\mathfrak{A}$.

Next, let A_1 and A_1^{-1} be two elements of $\mathfrak{A}$ which are inverses of each other, and let B_1 and B_2 be the corresponding elements of $\mathfrak{B}$. Then, by the definition of homomorphic groups, the relation

$$A_1A_1^{-1} = A_1^{-1}A_1 = A_0,$$

where A_0 is the unit element of $\mathfrak{A}$, implies

$$B_1B_2 = B_2B_1 = B_0,$$

where, as we have just seen, B_0 is the unit element of $\mathfrak{B}$. Therefore, $B_2 = B_1^{-1}$, i.e., inverse elements of $\mathfrak{B}$ correspond to inverse elements of $\mathfrak{A}$.

Let the groups $\mathfrak{A}$ and $\mathfrak{B}$ be homomorphic but not isomorphic, and consider the set of elements C_α in $\mathfrak{A}$ which have the unit element B_0 of $\mathfrak{B}$ as their image. If C_α has the image B_0, then, as we have just seen, C_α^{-1} has the image $B_0^{-1} = B_0$. Moreover, every product $C_{\alpha_1}C_{\alpha_2}$ has $B_0B_0 = B_0$ as its image. Thus, the set of elements C_α of the group $\mathfrak{A}$ which have the unit element of $\mathfrak{B}$ as their image form a subgroup $\mathfrak{C}$ of $\mathfrak{A}$. We now show that $\mathfrak{C}$ is a normal subgroup. To see this, let A be any element of the group $\mathfrak{A}$, and let B be the corresponding element of $\mathfrak{B}$. Then, every element of the form $AC_\alpha A^{-1}$ has the image $BB_0B^{-1} = B_0$, where B_0 is the unit element of $\mathfrak{B}$. Thus, every element $AC_\alpha A^{-1}$ is one of the elements C_α, i.e., belongs to the subgroup $\mathfrak{C}$. Therefore $\mathfrak{C}$ is a normal subgroup.

Next, we consider the decomposition of the group $\mathcal{A}$ into the cosets

$$\mathcal{C},\ A_1\mathcal{C},\ A_2\mathcal{C},\ \ldots\ . \tag{44}$$

Let B_k be the image of A_k, and take two elements $A_kC_{\alpha_1}$ and $A_kC_{\alpha_2}$ belonging to the same coset. Their images are B_kB_0 and B_kB_0, i.e., the same element B_k of $\mathcal{B}$. Moreover, let $A_kC_{\alpha_1}$ and $A_lC_{\alpha_2}$ be two elements from different cosets, with images B_k and B_l, respectively. Then, B_k and B_l are different elements, since if they were the same, the element $A_k^{-1}A_l$ would have as its image the unit element B_0 of $\mathcal{B}$, so that $A_k^{-1}A_l$ would have to be one of the elements C_α, i.e., $A_k^{-1}A_l = C_{\alpha_0}$ or $A_l = A_kC_{\alpha_0}$, contrary to the decomposition (44). Thus, *if the group $\mathcal{B}$ is homomorphic to the group $\mathcal{A}$, the set of elements $\mathcal{C}$ of $\mathcal{A}$ which have the unit element of $\mathcal{B}$ as their image, forms a normal subgroup, and every coset of this normal subgroup is the set of all elements of $\mathcal{A}$ which have the same element of $\mathcal{B}$ as their image.* Moreover, it follows at once from the definition of homomorphic groups that the product of any two elements from different (or identical) cosets, with images B_k and B_l, respectively, has as its image the product B_kB_l. More succinctly, every coset in $\mathcal{A}$ has as its image a definite element in $\mathcal{B}$, different cosets have different images, and this correspondence establishes an isomorphism between the group $\mathcal{B}$ and the factor group of $\mathcal{C}$ in $\mathcal{A}$.

If the group $\mathcal{B}$ is homomorphic (but not isomorphic) to the group $\mathcal{A}$, the set of elements in $\mathcal{A}$ which have as their image the unit element of $\mathcal{B}$ is called the *kernel of the homomorphism.* As we have just seen, the kernel of a homomorphism is a normal subgroup of the group $\mathcal{A}$.

67. Examples

1. Let $\mathcal{G}$ be the group of real orthogonal transformations in three-dimensional space; associate with each transformation T in $\mathcal{G}$ the number equal to the determinant of T, and define the group operation on these numbers to be ordinary multiplication. Then the group $\mathcal{G}'$, consisting of the numbers $+1$ and -1, is homomorphic to the group $\mathcal{G}$. The unit element $+1$ of the group $\mathcal{G}'$ corresponds to the three-dimensional spatial rotations in $\mathcal{G}$. These rotations form a normal subgroup, and the corresponding factor group is a cyclic group of order 2 (Sec. 65).

2. Consider the equilateral triangle in the xy-plane with the vertices

$$(1, 0),\ (\cos 120°, \sin 120°),\ (\cos 240°, \sin 240°),$$

and form the group consisting of rotations of the plane about the origin through the angles 0°, 120°, and 240° (operations denoted by E, F, and D, respectively) and of reflection of the plane in the x-axis, followed by rotations through the angles 0°, 120°, and 240° (operations denoted by A, B, and C, respectively). This is the dihedral group for $n = 3$ (Sec. 59).

The matrices corresponding to these group elements are

$$E = \begin{Vmatrix} 1 & 0 \\ 0 & 1 \end{Vmatrix}, \quad A = \begin{Vmatrix} 1 & 0 \\ 0 & -1 \end{Vmatrix}, \quad B = \begin{Vmatrix} -\tfrac{1}{2} & \tfrac{1}{2}\sqrt{3} \\ \tfrac{1}{2}\sqrt{3} & \tfrac{1}{2} \end{Vmatrix},$$

$$C = \begin{Vmatrix} -\tfrac{1}{2} & -\tfrac{1}{2}\sqrt{3} \\ -\tfrac{1}{2}\sqrt{3} & \tfrac{1}{2} \end{Vmatrix}, \quad D = \begin{Vmatrix} -\tfrac{1}{2} & \tfrac{1}{2}\sqrt{3} \\ -\tfrac{1}{2}\sqrt{3} & -\tfrac{1}{2} \end{Vmatrix},$$

$$F = \begin{Vmatrix} -\tfrac{1}{2} & -\tfrac{1}{2}\sqrt{3} \\ \tfrac{1}{2}\sqrt{3} & -\tfrac{1}{2} \end{Vmatrix}.$$

Referring to the multiplication table (34) of Sec. 62, we see that this table describes the multiplication law of the matrices in our group. Moreover, as we saw above (Sec. 65), this multiplication law also corresponds to the symmetric group of permutations of three elements:

$$E,\ A = (2, 3),\ B = (1, 2),\ C = (1, 3),\ D = (1, 3, 2),\ F = (1, 2, 3). \qquad (45)$$

Thus, we can set up an isomorphism between the elements of these two groups by having elements denoted by the same letter correspond to each other. The permutations of the group (45) correspond to permutations of the vertices of the equilateral triangle, if these vertices are appropriately numbered.

In just the same way, the tetrahedral group is isomorphic to the alternating group for $n = 4$, as already noted in Sec. 65.

3. We can give a general method for constructing permutation groups homomorphic to a given group $\mathcal{G}$. Let $\mathcal{H}$ be any subgroup of $\mathcal{G}$, with finite index n, and write the corresponding cosets

$$\mathcal{H},\ \mathcal{H}S_1,\ \mathcal{H}S_2,\ \ldots,\ \mathcal{H}S_{n-1},$$

where $S_1, S_2, \ldots, S_{n-1}$ denote elements of $\mathcal{G}$. Multiplying each of these cosets from the right by an element S of $\mathcal{G}$ produces a permutation of the order of the cosets, and this permutation can be regarded as corresponding to the given element S of $\mathcal{G}$. It is not hard to see that in this way we obtain a permutation group $\mathcal{G}'$ which is homomorphic to the group $\mathcal{G}$.

A necessary and sufficient condition for the unit element of $\mathcal{G}'$ to correspond to the element S in $\mathcal{G}$ is that every coset go into itself when multiplied from the right by S, i.e., that

$$H_\alpha S = H_\beta \quad \text{and} \quad H_\alpha S_k S = H_\gamma S_k \quad (k = 1, 2, \ldots, n-1),$$

where H_α is an arbitrary element of $\mathcal{H}$, and H_β, H_γ are also elements of $\mathcal{H}$. These relations can be written in the form

$$S = H_\alpha^{-1} H_\beta, \qquad S = (S_k^{-1} H_\alpha S_k)^{-1} (S_k^{-1} H_\gamma S_k),$$

from which it follows that a necessary and sufficient condition for the unit element of $\mathcal{G}'$ to correspond to the element S is that S should simultaneously belong to $\mathcal{H}$ and to all the conjugate subgroups $S_k^{-1}\mathcal{H}S_k$ ($k = 1, 2, \ldots, n - 1$). If $\mathcal{H}$ is a normal subgroup of $\mathcal{G}$, this requirement implies that S belongs to $\mathcal{H}$, and in this case, the group $\mathcal{G}'$ is isomorphic to the factor group of $\mathcal{H}$ in $\mathcal{G}$. If $\mathcal{H}$ is the group consisting only of the unit element, then the group $\mathcal{G}$ is isomorphic to the permutation group $\mathcal{G}'$ obtained by multiplying the elements of $\mathcal{G}$ from the right by each element S in $\mathcal{G}$ (which gives some permutation of the elements of $\mathcal{G}$).

Later, we shall consider in detail the problems of constructing groups of linear transformations isomorphic to a given group (see Chap. 8).

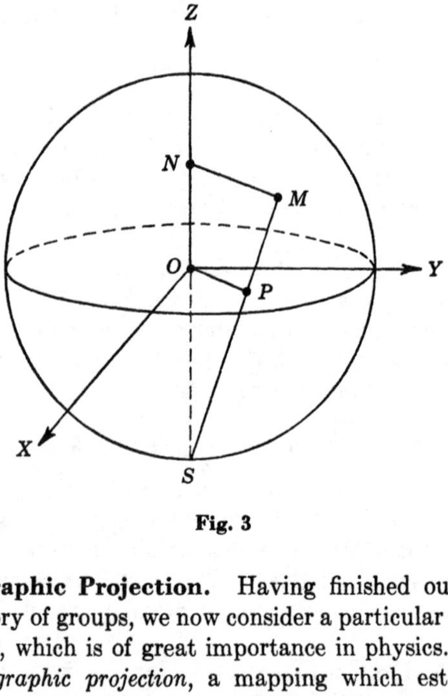

Fig. 3

68. Stereographic Projection. Having finished our discussion of the general theory of groups, we now consider a particular correspondence between groups, which is of great importance in physics. We begin by studying *stereographic projection*, a mapping which establishes a correspondence between the points of a sphere and the points of a plane.

Let C be a sphere of unit radius, which has its center at the origin of a three-dimensional rectangular coordinate system, with axes OX, OY, OZ. Let S be the point of the sphere with coordinates $(0, 0, -1)$, and let M be a variable point on the sphere (see Fig. 3). The line SM intersects the XY-plane at the point P, and thereby establishes a well-defined correspondence between the points of the sphere C and the points of the XY-plane; the point $S = (0, 0, -1)$ of the sphere corresponds to the point at infinity in the plane. This mapping constitutes a *stereographic projection* of the sphere onto the plane.

We now derive the formulas describing stereographic projection. Let MN be the perpendicular dropped from the point M onto the z-axis. Since $\overline{SO} = 1$ (where $\overline{SO}$ denotes the length of the segment SO), we find by similar triangles that

$$\overline{NM} = (1 + \overline{ON})\overline{OP}. \tag{46}$$

Denoting the coordinates of the point M by x, y, z and the coordinates of P by α, β, we can write

$$\overline{NM} = (1 + z)\overline{OP}$$

or

$$x = (1 + z)\alpha, \qquad y = (1 + z)\beta, \tag{47}$$

where we have projected the parallel segments OP and NM onto the x-axis and y-axis, respectively. Substituting these values into the equation $x^2 + y^2 + z^2 = 1$ of the unit sphere gives a quadratic equation in z:

$$(\alpha^2 + \beta^2)(1 + z)^2 + z^2 = 1.$$

Solving this equation, we obtain

$$z = \frac{\pm 1 - (\alpha^2 + \beta^2)}{1 + (\alpha^2 + \beta^2)}.$$

For all finite points (α, β) we must have $z > -1$; therefore, we must choose the plus sign in the formula for z. Then, using the formulas (47) again, we finally obtain the following expressions for x, y, and z in terms of α and β:

$$x = \frac{2\alpha}{1 + \alpha^2 + \beta^2}, \qquad y = \frac{2\beta}{1 + \alpha^2 + \beta^2}, \qquad z = \frac{1 - (\alpha^2 + \beta^2)}{1 + \alpha^2 + \beta^2}. \tag{48}$$

Instead of two real coordinates α and β in the plane, we can introduce a single complex coordinate $\zeta = \alpha + i\beta$. Letting $\bar{\zeta}$ denote the complex conjugate of ζ (as usual), we can rewrite (48) in the form

$$x + iy = \frac{2\zeta}{1 + \zeta\bar{\zeta}}, \qquad x - iy = \frac{2\bar{\zeta}}{1 + \zeta\bar{\zeta}}, \qquad z = \frac{1 - \zeta\bar{\zeta}}{1 + \zeta\bar{\zeta}}. \tag{49}$$

We now write the complex number ζ as a ratio of two other complex numbers ξ and η:

$$\zeta = \frac{\eta}{\xi}. \tag{50}$$

Pairs of values of ξ and η which differ by a common factor, e.g., pairs of the form $k\xi$, $k\eta$, and ξ, η, give the same value of ζ, and therefore the same point of the plane. Moreover, the values $\eta \neq 0$, $\xi = 0$ correspond to the point at infinity. The complex numbers ξ and η are called *homogeneous complex coordinates* in the plane. By using (50) and separating real and imaginary parts, we can rewrite (49) in the form

$$x = \frac{\xi\bar{\eta} + \bar{\xi}\eta}{\xi\bar{\xi} + \eta\bar{\eta}}, \qquad y = \frac{1}{i}\frac{\xi\bar{\eta} - \bar{\xi}\eta}{\xi\bar{\xi} + \eta\bar{\eta}}, \qquad z = \frac{\xi\bar{\xi} - \eta\bar{\eta}}{\xi\bar{\xi} + \eta\bar{\eta}}. \tag{51}$$

For any pair of complex numbers ξ and η, the formulas (51) give real numbers x, y, and z, satisfying the relation $x^2 + y^2 + z^2 = 1$, as is to be expected, since the point (x, y, z) lies on the unit sphere.

69. The Unitary Group and the Rotation Group. We now consider a unitary transformation

$$\xi' = a\xi + b\eta, \qquad \eta' = c\xi + d\eta \tag{52}$$

of the variables ξ and η; since (52) is unitary, we have

$$\xi'\bar{\xi}' + \eta'\bar{\eta}' = \xi\bar{\xi} + \eta\bar{\eta}. \tag{53}$$

The new variables ξ', η' give us a new point on the unit sphere:

$$x' = \frac{\xi'\bar{\eta}' + \bar{\xi}'\eta'}{\xi'\bar{\xi}' + \eta'\bar{\eta}'}, \qquad y' = \frac{1}{i}\frac{\xi'\bar{\eta}' - \bar{\xi}'\eta'}{\xi'\bar{\xi}' + \eta'\bar{\eta}'}, \qquad z' = \frac{\xi'\bar{\xi}' - \eta'\bar{\eta}'}{\xi'\bar{\xi}' + \eta'\bar{\eta}'}. \tag{54}$$

As we know (Sec. 28), the determinant of the unitary transformation (52) has absolute value 1, i.e., the determinant is a complex number of the form $e^{i\varphi}$. Multiplying all the coefficients of the transformation (52) by $e^{-i\varphi/2}$, we obtain a unitary transformation with determinant equal to 1. This also multiplies ξ' and η' by $e^{-i\varphi/2}$, but the extra factor has no effect on the quantity $\zeta' = \xi'/\eta'$. Thus, we need only consider unitary transformations (52), satisfying the condition that the determinant of the transformation equals 1, i.e.,

$$ad - bc = 1. \tag{55}$$

This constraint is compatible with two transformations whose coefficients differ in sign; however, these two transformations give pairs of values of ξ', η' which differ in sign, and therefore lead to the same point ζ'.

We now find the general form of unitary transformations of the form (52) with unit determinants. The general conditions for a unitary transformation (Sec. 28) give

$$a\bar{c} + b\bar{d} = 0, \qquad c\bar{c} + d\bar{d} = 1.$$

Multiplying (55) by $\bar{c}$, and using the condition $a\bar{c} + b\bar{d} = 0$, we obtain

$$-bd\bar{d} - bc\bar{c} = \bar{c}:$$

since $c\bar{c} + d\bar{d} = 1$, this gives $\bar{c} = -b$ or $c = -\bar{b}$. In just the same way, using the conditions

$$a\bar{c} + b\bar{d} = 0, \qquad a\bar{a} + b\bar{b} = 1,$$

we find that $d = \bar{a}$. Thus, we can write all the unitary transformations (52) with unit determinants in the form

$$\xi' = a\xi + b\eta, \qquad \eta' = -\bar{b}\xi + \bar{a}\eta, \tag{56}$$

where a and b are any complex numbers satisfying the condition

$$a\bar{a} + b\bar{b} = 1. \tag{57}$$

If we replace ξ' and η' in (54) by the expressions (56), using the condition (53) and the formulas (51), we see that the coordinates x', y', z' are linear combinations of the coordinates x, y, z. Since by (53) the denominators in (51) and (54) are the same, the variables x, y, z undergo the same linear transformation as the quantities

$$u = \bar{\xi}\eta + \xi\bar{\eta}, \qquad v = \frac{1}{i}(\bar{\xi}\eta - \xi\bar{\eta}), \qquad w = \xi\bar{\xi} - \eta\bar{\eta}, \tag{58}$$

when the unitary transformation (56) is performed. We now find the exact form of this linear transformation.

In the new variables, the quantities (58) become

$$u' + iv' = 2\bar{\xi}'\eta', \qquad u' - iv' = 2\xi'\bar{\eta}', \qquad w' = \xi'\bar{\xi}' - \eta'\bar{\eta}'.$$

Using (56), we obtain

$$\begin{aligned} u' + iv' &= \overline{a}^2 2\bar{\xi}\eta - \overline{b}^2 2\xi\bar{\eta} - 2\overline{ab}(\xi\bar{\xi} - \eta\bar{\eta}), \\ u' - iv' &= -b^2 2\bar{\xi}\eta + a^2 2\xi\bar{\eta} - 2ab(\xi\bar{\xi} - \eta\bar{\eta}), \\ w' &= \bar{a}b2\bar{\xi}\eta + a\bar{b}2\xi\bar{\eta} + (a\bar{a} - b\bar{b})(\xi\bar{\xi} - \eta\bar{\eta}). \end{aligned}$$

Making the substitutions,

$$2\bar{\xi}\eta = u + iv, \qquad 2\xi\bar{\eta} = u - iv, \qquad \xi\bar{\xi} - \eta\bar{\eta} = w,$$

and first adding and then subtracting the first two equations, we find expressions for u', v', w' in terms of u, v, w, or equivalently, expressions for x', y', z' in terms of x, y, z:

$$\begin{aligned} x' &= \frac{1}{2}(a^2 + \overline{a^2} - b^2 - \overline{b^2})x + \frac{i}{2}(\overline{a^2} + \overline{b^2} - a^2 - b^2)y - (ab + \overline{ab})z, \\ y' &= \frac{i}{2}(a^2 + \overline{b^2} - \overline{a^2} - b^2)x + \frac{1}{2}(a^2 + \overline{a^2} + b^2 + \overline{b^2})y + i(\overline{ab} - ab)z, \\ z' &= (\bar{a}b + a\bar{b})x + i(\bar{a}b - a\bar{b})y + (a\bar{a} - b\bar{b})z. \end{aligned} \tag{59}$$

Every unitary transformation (56) corresponds to a transformation of the XY-plane into itself, and then stereographic projection gives in turn a transformation of the unit sphere into itself. Thus, (59) is a real transformation which takes the equation

$$x^2 + y^2 + z^2 = 1$$

into the equation

$$x'^2 + y'^2 + z'^2 = 1.$$

Since the linear homogeneous transformation (59) does not change the constant term 1, it follows that the left-hand sides of the equations just written must be the same:

$$x'^2 + y'^2 + z'^2 = x^2 + y^2 + z^2.$$

(All these results can also be obtained directly from the form of the transformation (59).) Thus, (59) gives a real orthogonal transformation in three variables. Moreover, the determinant of (59) equals $+1$. To see this, note that the determinant of the transformation is a continuous function of the real and imaginary parts of the complex parameters a and b (where $a\bar{a} + b\bar{b} = 1$) and therefore is either always $+1$ or always -1, since the determinant of an orthogonal transformation can only be $+1$ or -1. But for $a = 1$, $b = 0$, (59) gives the identity transformation, with determinant $+1$. Thus, the determinant of (59) is in fact always $+1$, so that the transformation (59) represents a spatial rotation about the origin.

We now prove that every spatial rotation can be represented in the form (59). If we set

$$a = e^{-i\varphi/2}, \qquad \bar{a} = e^{i\varphi/2}, \qquad b = \bar{b} = 0,$$

which corresponds to the unitary transformation with the matrix

$$A_\varphi = \begin{Vmatrix} e^{-i\varphi/2} & 0 \\ 0 & e^{i\varphi/2} \end{Vmatrix}, \tag{60}$$

then the formula (59) becomes

$$\begin{aligned} x' &= x \cos \varphi - y \sin \varphi, \\ y' &= x \sin \varphi + y \cos \varphi, \\ z' &= z, \end{aligned} \tag{61}$$

i.e., we obtain a rotation through the angle φ about the z-axis. Next, if we set

$$a = \bar{a} = \cos \frac{\psi}{2}, \qquad b = -i \sin \frac{\psi}{2}, \qquad \bar{b} = i \sin \frac{\psi}{2},$$

which corresponds to the unitary transformation with the matrix

$$B_\psi = \begin{Vmatrix} \cos\frac{\psi}{2} & -i\sin\frac{\psi}{2} \\ -i\sin\frac{\psi}{2} & \cos\frac{\psi}{2} \end{Vmatrix}, \tag{62}$$

then the formula (59) becomes

$$\begin{aligned} x' &= x, \\ y' &= y\cos\psi - z\sin\psi, \\ z' &= y\sin\psi + z\cos\psi, \end{aligned} \tag{63}$$

i.e., we obtain a rotation through the angle ψ about the x-axis.

As we saw in Sec. 20, the rotation $\{\alpha, \beta, \gamma\}$ with Eulerian angles α, β, γ can be obtained as a result of three rotations, first a rotation through the angle α about the z-axis, then a rotation through the angle β about the new x-axis, and finally a rotation through the angle γ about the new z-axis. Let Z_φ denote the (third order) matrix corresponding to the transformation (61), and let X_ψ be the matrix corresponding to the transformation (63). The matrix Z_α produces a rotation through the angle α about the z-axis, which carries the old x-axis into the new x-axis. Moreover, it is not hard to see that the matrix $Z_\alpha X_\beta Z_\alpha^{-1}$ produces a rotation through the angle β about the new x-axis, so that the first two rotations are accomplished by using the matrix

$$Z_\alpha X_\beta Z_\alpha^{-1} Z_\alpha = Z_\alpha X_\beta.$$

Similarly, the matrix

$$(Z_\alpha X_\beta) Z_\gamma (Z_\alpha X_\beta)^{-1}$$

produces a rotation through the angle γ about the new z-axis, so that finally the rotation $\{\alpha, \beta, \gamma\}$ is accomplished by using the matrix

$$(Z_\alpha X_\beta) Z_\gamma (Z_\alpha X_\beta)^{-1} (Z_\alpha X_\beta)$$

or

$$Z_\alpha X_\beta Z_\gamma. \tag{64}$$

(In this derivation, we have twice used the obvious fact that if X_φ is a matrix producing a rotation through the angle φ about an axis l passing through the origin, and if M carries l into a new axis l_1, then the similar matrix $MZ_\varphi M^{-1}$ produces a rotation through the angle φ about l_1.)

Suppose that by (56) and (59), the orthogonal transformations T_1 and T_2 correspond to the unitary transformations A_1 and A_2, respectively. Then the product T_2T_1 obviously corresponds to the product A_2A_1.

Therefore, by (64), the rotation $\{\alpha, \beta, \gamma\}$ corresponds to the product

$$\begin{Vmatrix} e^{-i\frac{\alpha}{2}} & 0 \\ 0 & e^{i\frac{\alpha}{2}} \end{Vmatrix} \begin{Vmatrix} \cos\frac{\beta}{2} & -i\sin\frac{\beta}{2} \\ -i\sin\frac{\beta}{2} & \cos\frac{\beta}{2} \end{Vmatrix} \begin{Vmatrix} e^{-i\frac{\gamma}{2}} & 0 \\ 0 & e^{i\frac{\gamma}{2}} \end{Vmatrix} \tag{65}$$

of three unitary matrices, which is itself a unitary matrix. Thus, not only does every unitary transformation (with determinant 1) define a three-dimensional rotation, but every rotation corresponds to a unitary transformation; moreover, the product of two rotations corresponds to the product of the corresponding unitary transformations. We can say that the formulas (59) define a homomorphism between the group of unitary transformations with determinant 1 and the three-dimensional rotation group.

We now find the unitary transformations corresponding to the unit element of the rotation group (i.e., the identity transformation). According to the third of the formulas (59), in the case of the identity transformation, we have

$$a\bar{b} = 0, \qquad a\bar{a} - b\bar{b} = 1,$$

so that $|a| = 1$ and $b = 0$. If we write $a = e^{i\vartheta}$, the first of the formulas (59) gives

$$\tfrac{1}{2}(e^{2i\vartheta} + e^{-2i\vartheta}) = 1,$$

from which it immediately follows that $\vartheta = 0$ or π, i.e., $a = \pm 1$. Thus, we find that the two unitary transformations with matrices

$$E = \begin{Vmatrix} 1 & 0 \\ 0 & 1 \end{Vmatrix}, \qquad S = \begin{Vmatrix} -1 & 0 \\ 0 & -1 \end{Vmatrix} = -E$$

both correspond to the unit element of the rotation group.

Suppose now that the two unitary matrices U and V give the same rotation. Then $V^{-1}U$ corresponds to the identity transformation of the rotation group, i.e., $V^{-1}U = E$ or $V^{-1}U = -E$, so that $U = V$ or $U = -V$. (We recall that the minus sign before a matrix means that the signs of all the matrix elements have to be changed.) Thus, this argument shows that a necessary condition for two unitary transformations of the form (56) to lead to the same rotation is that they differ only in sign. On the other hand, as we have already noted (and as follows from the form of (59)), a sufficient condition for two unitary transformations to give the same rotation is that they differ in sign. Thus, finally, we can assert that *the three-dimensional rotation group is homomorphic to the group of unitary transformations of the form* (56) *with unit determinants, and identical rotations are obtained if and only if the corresponding unitary matrices have different signs.* The matrices E and $-E$ form a normal subgroup $\mathcal{H}$ of the group $\mathcal{G}$ of unitary transformations of the form (56) with unit determinants. Every coset of this normal subgroup $\mathcal{H}$ consists of two elements G and $-G$, where G is any element of the group $\mathcal{G}$. It follows at once that *the rotation group is isomorphic to the factor group of $\mathcal{H}$ in $\mathcal{G}$.*

The formulas (59) contain two complex parameters a and b, which must satisfy the condition (57). Since each complex parameter contains two real parameters, i.e.,

$$a = a_1 + ia_2, \qquad b = b_1 + ib_2,$$

the condition (57) is equivalent to

$$a_1^2 + a_2^2 + b_1^2 + b_2^2 = 1.$$

Thus, the formulas (59) contain four real parameters, which have to satisfy one condition, i.e., the formulas (59) contain three independent real parameters, as must be the case for the rotation group. The parameters a and b are usually called the *Cayley-Klein parameters.* It is not hard to express these parameters in terms of the Eulerian angles. In fact, as already noted, the product of the three unitary matrices (65) is the unitary matrix corresponding to the rotation $\{\alpha,\beta,\gamma\}$, with Eulerian angles α, β, γ; carrying out the multiplication, we see that this matrix has the parameters

$$a = e^{-i\frac{\gamma+\alpha}{2}} \cos\frac{\beta}{2}, \qquad b = -ie^{i\frac{\gamma-\alpha}{2}} \sin\frac{\beta}{2}. \tag{66}$$

If we add 2π to α or γ, then a and b change sign, and the rotation remains unchanged. This fact has already been noted.

70. The Unimodular Group and the Lorentz Group. We have just established a close relation between the three-dimensional rotation group and the group of two-dimensional unitary transformations with determinant 1. In the same way, we can establish a relation between the Lorentz group and the group of general two-dimensional linear transformations with determinant 1, the so-called (two-dimensional) *unimodular group.* (The group of unitary transformations with determinant 1 is often called the *unitary unimodular group,* and is a subgroup of the unimodular group.)

We introduce the four new variables

$$x_1,\ x_2,\ x_3,\ x_0,$$

and write

$$x = \frac{x_1}{x_0}, \qquad y = \frac{x_2}{x_0}, \qquad z = \frac{x_3}{x_0} \tag{67}$$

in the formulas (51) defining stereographic projection. As a result, we obtain

$$\frac{x_1}{x_0} = \frac{\bar{\xi}\eta + \xi\bar{\eta}}{\xi\bar{\xi} + \eta\bar{\eta}}, \qquad \frac{x_2}{x_0} = \frac{1}{i}\frac{\bar{\xi}\eta - \xi\bar{\eta}}{\xi\bar{\xi} + \eta\bar{\eta}}, \qquad \frac{x_3}{x_0} = \frac{\xi\bar{\xi} - \eta\bar{\eta}}{\xi\bar{\xi} + \eta\bar{\eta}}.$$

These formulas define the x_k to within an arbitrary common factor, so that we can write

$$\begin{aligned} x_0 &= \xi\bar{\xi} + \eta\bar{\eta}, \qquad x_1 = \bar{\xi}\eta + \xi\bar{\eta}, \\ x_2 &= \frac{1}{i}(\bar{\xi}\eta - \xi\bar{\eta}), \; x_3 = \xi\bar{\xi} - \eta\bar{\eta}. \end{aligned} \tag{68}$$

Since the old variables x, y, and z satisfy the relation $x^2 + y^2 + z^2 - 1 = 0$ (Sec. 68), the new variables defined by (68) satisfy the relation

$$x_1^2 + x_2^2 + x_3^2 - x_0^2 = 0, \tag{69}$$

for any complex numbers ξ and η.

The expression $\xi\bar{\xi} + \eta\bar{\eta}$, and therefore by (68), the variable x_0 (which here represents the time), remain unchanged under unitary transformations on ξ and η, so that in this case we obtain a three-dimensional rotation. We now drop the requirement that the transformation be unitary, and consider the general group of linear transformations of the form

$$\xi' = a\xi + b\eta, \qquad \eta' = c\xi + d\eta. \tag{70}$$

Adopting an approach similar to that used in the case of unitary transformations, we form the expressions

$$\begin{aligned} x_1 + ix_2 &= 2\bar{\xi}\eta, \qquad x_1 - ix_2 = 2\xi\bar{\eta}, \\ x_0 + x_3 &= 2\xi\bar{\xi}, \qquad x_0 - x_3 = 2\eta\bar{\eta}. \end{aligned} \tag{71}$$

For new values of the variables ξ', η', we obtain new values of the x_k':

$$\begin{aligned} x_1' + ix_2' &= 2\bar{\xi}'\eta', \qquad x_1' - ix_2' = 2\xi'\bar{\eta}', \\ x_0' + x_3' &= 2\xi'\bar{\xi}', \qquad x_0' - x_3' = 2\eta'\bar{\eta}'. \end{aligned}$$

Substituting (70) into these formulas and using (71), we obtain

$$\begin{aligned} x_1' + ix_2' &= \bar{a}d(x_1 + ix_2) + \bar{b}c(x_1 - ix_2) + \bar{a}c(x_0 + x_3) + \bar{b}d(x_0 - x_3), \\ x_1' - ix_2' &= b\bar{c}(x_1 + ix_2) + a\bar{d}(x_1 - ix_2) + a\bar{c}(x_0 + x_3) + b\bar{d}(x_0 - x_3), \\ x_0' + x_3' &= \bar{a}b(x_1 + ix_2) + a\bar{b}(x_1 - ix_2) + a\bar{a}(x_0 + x_3) + b\bar{b}(x_0 - x_3), \\ x_0' - x_3' &= \bar{c}d(x_1 + ix_2) + c\bar{d}(x_1 - ix_2) + c\bar{c}(x_0 + x_3) + d\bar{d}(x_0 - x_3), \end{aligned} \tag{72}$$

from which we can immediately obtain linear expressions, with real coefficients, for the x_k' in terms of the x_k. We shall not give these expressions, but merely note that if the last two equations of (72) are added, then the coefficient of x_0 in the expression for x'_0 turns out to be

$$\tfrac{1}{2}(a\bar{a} + b\bar{b} + c\bar{c} + d\bar{d})$$

and is therefore positive.

The new variables x'_k satisfy the same relation as the old variables x_k, i.e.,

$$x_1'^2 + x_2'^2 + x_3'^2 - x_0'^2 = 0. \tag{73}$$

If we replace the x'_k in (73) by their expressions in terms of the x_k, we have to obtain the relation (69). However, the left-hand sides of (69) and (73) may still differ by a common factor, i.e., we may have

$$x_1'^2 + x_2'^2 + x_3'^2 - x_0'^2 = k(x_1^2 + x_2^2 + x_3^2 - x_0^2),$$

where k is a constant. Using the relation

$$x_1'^2 + x_2'^2 + x_3'^2 - x_0'^2 = (x_1' + ix_2')(x_1' - ix_2') - (x_0' + x_3')(x_0' - x_3')$$

and the formulas (72), we find that

$$k = (ad - bc)(\overline{ad} - \overline{bc}) = |ad - bc|^2.$$

If the formulas (72) are to describe a Lorentz transformation, which must satisfy the relation

$$x_1'^2 + x_2'^2 + x_3'^2 - x_0'^2 = x_1^2 + x_2^2 + x_3^2 - x_0^2, \tag{74}$$

then k must equal 1. This means that we have to choose transformations (70) whose determinants have absolute value 1, i.e., are numbers of the form $e^{i\varphi}$. If we multiply all the coefficients of the transformation (70) by $e^{-i\varphi/2}$ (as was done in Sec. 69), we change the value of the determinant to 1; however, this does not change the quantities x_1', x_2', x_3', x_0' defined by the formulas (68) (with ξ, η replaced by ξ', η'), since these formulas contain only products of one of the quantities ξ', η' multiplied by one of the quantities $\bar{\xi}'$, $\bar{\eta}'$. Thus, we need only consider transformations of the form (70) with determinants equal to 1, i.e., *unimodular* transformations, for which

$$ad - bc = 1. \tag{75}$$

As in Sec. 69, we can show that the linear transformations expressing the x'_k in terms of the x_k all have determinant +1. Moreover, as already noted, the coefficient of x_0 in the expression for x'_0 is positive, so that the transformations (72) not only have determinant +1, but also preserve the time direction. Thus, finally, we find that the linear transformations (72), subject to the condition (75), are just the *proper orthochronous* Lorentz transformations defined in Sec. 60.

Just as in Sec. 69, we now ask whether every proper orthochronous Lorentz transformation can be obtained from the formulas (72). First, using (70) and (72), we note that if the Lorentz transformations T_1 and T_2 correspond to the unimodular transformations A_1 and A_2, respectively, then the product T_2T_1 corresponds to the product A_2A_1, i.e., a product of two Lorentz transformations corresponds to the product of the corresponding unimodular transformations of the form (70). As we saw in Sec. 60, every proper orthochronous Lorentz transformation can be represented in the form

$$T = VSU,$$

where U and V are three-dimensional rotations, and S is a proper orthochronous Lorentz transformation in two variables. Moreover, according to Sec. 69, any rotation can be obtained by using a *unitary* transformation of the form (70) with determinant 1. Thus, it remains only to show that any proper orthochronous Lorentz transformation in two variables can be obtained by suitably choosing a unimodular transformation of the form (70). Comparing the relations (74) and (21) (Sec. 60), we see that we have assumed here that $c = 1$. As a result, the formulas (17) of Sec. 60, which define proper orthochronous Lorentz transformations in two variables, can be rewritten in the form

$$x_3' = \frac{-vx_0 + x_3}{\sqrt{1 - v^2}}, \qquad x_0' = \frac{x_0 - vx_3}{\sqrt{1 - v^2}}, \qquad x_1' = x_1, \qquad x_2' = x_2. \tag{76}$$

We now introduce the quantity

$$u = \frac{1}{\sqrt{1 - v^2}} > 1$$

and consider the special linear transformation

$$\xi' = l\xi, \qquad \eta' = \frac{1}{l}\eta$$

of the form (70), where l is a real constant. In this case, we have $a = l$, $d = 1/l$, $b = c = 0$, and the determinant of the transformation obviously equals 1. Substituting these values in (72), we find just the transformation (76), provided that l satisfies the conditions

$$\frac{l^2}{2} + \frac{1}{2l^2} = u, \qquad \frac{l^2}{2} - \frac{1}{2l^2} = -vu.$$

The first condition immediately gives

$$l^2 = u \pm \sqrt{u^2 - 1}. \tag{77}$$

The second condition shows that if $v > 0$, the minus sign has to be chosen in (77) ($l^2 < 1$), while if $v < 0$, the plus sign has to be chosen ($l^2 > 1$).

Then, taking the square root of (77), we obtain two values of l, which differ only in sign. Thus, finally, we can assert that *the group of proper orthochronous Lorentz transformations is homomorphic to the group of linear transformations of the form* (70) *with unit determinants, where the formula* (72) *establishes the homomorphism.* Just as in Sec. 69, this homomorphism is not an isomorphism, i.e., different transformations of the form (70) can lead to the same Lorentz transformation. In fact, it follows at once from (72) that the identity transformation of the Lorentz group is obtained from the two linear transformations with the matrices

$$E = \begin{Vmatrix} 1 & 0 \\ 0 & 1 \end{Vmatrix}, \qquad S = \begin{Vmatrix} -1 & 0 \\ 0 & -1 \end{Vmatrix} = -E,$$

and just as in Sec. 69, we can show that *every transformation in the Lorentz group can be obtained from just two linear transformations of the form* (70), *whose coefficients differ only in sign.* Moreover, just as before, the two elements E and $-E$ form a normal subgroup $\mathcal{H}$ of the group of linear transformations with unit determinants, and *the group of proper orthochronous Lorentz transformations is isomorphic to the factor group of* $\mathcal{H}$.

The linear transformations of the form (70) contain four complex coefficients related by the condition (75). Thus, the formulas (72) contain three arbitrary complex parameters, or equivalently, six arbitrary real parameters.

PROBLEMS†

1. Which of the following sets of matrices of order n are groups (with respect to multiplication):

   (a) The nonsingular matrices with integral elements ($0, \pm 1, \pm 2, \ldots$);
   (b) The matrices with integral elements and determinant ± 1;
   (c) The matrices with nonnegative elements and determinant $+1$;
   (d) The positive definite matrices;
   (e) The nonsingular triangular matrices (see Prob. 41 of Chap. 3)?

2. Prove that every rotation R is the product of a rotation about the x-axis, a rotation about the y-axis, and a rotation about the z-axis.
3. Prove that every 3×3 rotation matrix R can be expressed in the form $R = e^A$, where A is a real skew-symmetric matrix.
4. Find the group which transforms a rectangular lamina (with unequal sides) into itself. Write the group multiplication table.
5. Find the group which transforms a rectangular parallelepiped (with unequal sides) into itself. Write the group multiplication table.

† For hints and answers, see page 445. These problems are by J. S. Lomont and the translator; some use was made of Proskuryakov's book (cited in the Preface).

6. Show that the set of all real symmetric matrices of the form

$$L(\theta) = \begin{Vmatrix} \cosh\theta & \sinh\theta \\ \sinh\theta & \cosh\theta \end{Vmatrix} \qquad (-\infty < \theta < \infty)$$

is an Abelian group, and show that $L(\theta_1)L(\theta_2) = L(\theta_1 + \theta_2)$. If

$$G = \begin{Vmatrix} 1 & 0 \\ 0 & -1 \end{Vmatrix},$$

show that $L^*(\theta)GL(\theta) = G$, where $L^*(\theta)$ denotes the transpose of $L(\theta)$. Also show that

$$L(\theta) = \exp \begin{Vmatrix} 0 & \theta \\ \theta & 0 \end{Vmatrix}.$$

7. Let G be defined as in Prob. 6, and let $\mathfrak{L}$ be the set of all real 2×2 matrices L for which $L^*GL = L$. Show that

(a) $\mathfrak{L}$ is a group;
(b) Every element of $\mathfrak{L}$ can be written in one of the forms

$$\pm \begin{Vmatrix} \cosh\theta & \sinh\theta \\ \sinh\theta & \cosh\theta \end{Vmatrix},$$

$$\pm \begin{Vmatrix} \cosh\theta & -\sinh\theta \\ \sinh\theta & -\cosh\theta \end{Vmatrix} \qquad (-\infty < \theta < \infty);$$

(c) Every matrix of the above form satisfies $L^*GL = G$;
(d) The matrices of Prob. 6 form a subgroup of index 4 (i.e., with four distinct cosets) of $\mathfrak{L}$.

8. Find the eigenvalues and the eigenvectors of the matrix $L(\theta)$ of Prob. 6. Show that in units such that c (the velocity of light) is 1, we can choose θ so that $L(\theta)$ is the Lorentz transformation (17) of Sec. 60.

Comment: The condition $L^*GL = G$ expresses the invariance of the quadratic form $x^2 - t^2$ under the Lorentz transformation L (cf. Sec. 32).

9. Let

$$L = \begin{Vmatrix} a & b \\ c & d \end{Vmatrix}$$

be a real 2×2 *Lorentz matrix*, i.e., a matrix satisfying $L^*GL = G$, where G is defined as in Prob. 6. (We exclude the trivial case where L is the unit matrix.) Show that

(a) $d^2 > 1$;
(b) If L is *orthochronous* (i.e., $d > 0$ as in Sec. 60), then $d > 1$;
(c) The product of two real orthochronous 2×2 Lorentz matrices is itself orthochronous.

10. Choose units in which c (the velocity of light) equals 1. Let $\mathfrak{S}$ and $\mathfrak{S}'$ be two coordinate systems *with parallel axes*, and let the velocity of $\mathfrak{S}'$ relative to $\mathfrak{S}$ be given by the vector $\mathbf{v}$, of magnitude $v < 1$. Show that the proper orthochronous

Lorentz transformation from $\mathfrak{S}$ to $\mathfrak{S}'$ is described by the formulas

$$\mathbf{x}' = \mathbf{x} + \alpha\left[\frac{\alpha}{\alpha+1}(\mathbf{v}, \mathbf{x}) - t\right]\mathbf{v},$$
$$t' = \alpha[t - (\mathbf{v}, \mathbf{x})],$$

where $\alpha = (1 - v^2)^{-1/2}$.

11. Show that any (real) 4×4 Lorentz matrix L satisfies the relation $L^*GL = G$, where

$$G = \begin{Vmatrix} 1 & 0 & 0 & 0 \\ 0 & 1 & 0 & 0 \\ 0 & 0 & 1 & 0 \\ 0 & 0 & 0 & -1 \end{Vmatrix}.$$

Show that

$$L_0 = \begin{Vmatrix} 1 & -1 & 0 & 1 \\ 1 & 0 & 1 & 1 \\ 0 & -1 & 1 & 1 \\ 1 & -1 & 1 & 2 \end{Vmatrix}$$

is a proper orthochronous Lorentz matrix, but that it is not a *parallel-axis Lorentz* transformation, i.e., a transformation of the type given in Prob. 10. Show that L_0 cannot be diagonalized.

12. Find the parallel-axis Lorentz transformation L_1 whose matrix has the same last row as the matrix L_0 of Prob. 11. Find the rotation matrix

$$R = \begin{Vmatrix} a_{11} & a_{12} & a_{13} & 0 \\ a_{21} & a_{22} & a_{23} & 0 \\ a_{31} & a_{32} & a_{33} & 0 \\ 0 & 0 & 0 & 1 \end{Vmatrix}$$

such that $L_0 = RL_1$.

Comment: It can in fact be shown that every proper orthochronous 4×4 Lorentz matrix can be expressed as the product of a spatial rotation R and a parallel-axis Lorentz transformation.

13. Show that any cycle $(i_1, i_2, \ldots, i_m)$ of length m of the symmetric group of degree n $(m \leq n)$ is of order m, i.e., $(i_1, i_2, \ldots, i_m)^m = I$, where I is the unit permutation. Write the powers of the cycle (1, 2, 3, 4, 5, 6) as products of cycles.

14. Suppose that a permutation has been written as a product of cycles. Prove that the order of the permutation is the least common multiple of the lengths of its cycles.

15. Find A^{100}, where

$$A = \begin{pmatrix} 1 & 2 & 3 & 4 & 5 & 6 & 7 & 8 & 9 & 10 \\ 3 & 5 & 4 & 1 & 7 & 10 & 2 & 6 & 9 & 8 \end{pmatrix}.$$

Find B^{150}, where

$$B = \begin{pmatrix} 1 & 2 & 3 & 4 & 5 & 6 & 7 & 8 & 9 & 10 \\ 3 & 5 & 4 & 6 & 9 & 7 & 1 & 10 & 8 & 2 \end{pmatrix}.$$

16. Find the permutation X satisfying the equation $AXB = C$, where

$$A = \begin{pmatrix} 1 & 2 & 3 & 4 & 5 & 6 & 7 \\ 7 & 3 & 2 & 1 & 6 & 5 & 4 \end{pmatrix}, \quad B = \begin{pmatrix} 1 & 2 & 3 & 4 & 5 & 6 & 7 \\ 3 & 1 & 2 & 7 & 4 & 5 & 6 \end{pmatrix},$$
$$C = \begin{pmatrix} 1 & 2 & 3 & 4 & 5 & 6 & 7 \\ 5 & 1 & 3 & 6 & 4 & 7 & 2 \end{pmatrix}.$$

17. By the *decrement* of a permutation is meant the difference between the number of elements actually changed by the permutation and the number of cycles (of length ≥ 2) in the permutation. Prove that a permutation is even (odd) if and only if its decrement is even (odd).

18. Find all permutations of the integers 1, 2, 3, 4 which commute with the permutation

$$\begin{pmatrix} 1 & 2 & 3 & 4 \\ 2 & 1 & 4 & 3 \end{pmatrix}.$$

19. Is the following the multiplication table of a group?

	E	*A*	*B*	*C*	*D*	*F*
E	*E*	*A*	*B*	*C*	*D*	*F*
A	*A*	*E*	*C*	*D*	*F*	*B*
B	*B*	*C*	*E*	*F*	*A*	*D*
C	*C*	*D*	*F*	*E*	*B*	*A*
D	*D*	*F*	*A*	*B*	*E*	*C*
F	*F*	*B*	*D*	*A*	*C*	*E*

20. Prove that the order of any element of a finite group must divide the order of the group.

21. Prove that if A, B, and C are any three elements of a group, then the products ABC, BCA, and CAB have the same order.

22. Let A be an element of order m of a finite group. Prove that the order of any power of A cannot exceed m. Prove that if m and n are relatively prime, then A^n and A are of the same order.

23. Let C be an element of a finite group $\mathcal{G}$, and let the order of C be mn, where m and n are relatively prime. Prove that C can be expressed uniquely as a product of two elements A and B of $\mathcal{G}$, of orders m and n, respectively, where A and B commute.

24. Let $\mathcal{G}$ be the Abelian group consisting of the integers 1, 2, . . . , 6, with the group operation (denoted by $\odot$) being multiplication *modulo* 7. For example, $2 \odot 4 = 1$, $3 \odot 3 = 2$, etc. Show that $\mathcal{G}$ is a cyclic group of order 6, and show that the only elements which *generate* $\mathcal{G}$ (i.e., of whose powers $\mathcal{G}$ consists) are the integers 3 and 5. Write both 3 and 5 as the product of two elements of orders 2 and 3, respectively (cf. preceding problem).

25. Show that the symmetric group of degree n ($n > 1$) is generated by the set of all transpositions of the form (1, 2), (1, 3), . . . , (1, n). (A group $\mathcal{G}$ is said to be *generated* by the elements $G_1, G_2, \ldots, G_k$ of $\mathcal{G}$ if $\mathcal{G}$ is the smallest subgroup containing $G_1, G_2, \ldots, G_k$.)

26. Show that the alternating group of degree n ($n \geq 3$) is generated by the cycles of the form (1, 2, 3), (1, 2, 4), . . . , (1, 2, n).

27. Let p be a positive prime number. Show that there is only one group of order p (if isomorphic groups are regarded as identical), and show that this group is cyclic.

28. What is the smallest integer n such that there is a non-Abelian group of order n? Describe the group.

29. Prove that the alternating group is the only subgroup of index 2 of the symmetric group of degree $n > 1$.

30. Let P be a permutation of the integers 1, 2, . . . , n, and write P as a product of cycles $C_1C_2 \cdots C_r$ ($r \leq n$), where it is assumed that the lengths of the cycles do not decrease as we read $C_1C_2 \cdots C_r$ from left to right. Then, if l_i is the length of the cycle C_i ($i = 1, 2, \ldots, r$), we have

$$1 \le l_1 \le l_2 \le \cdots \le l_r \le n,$$

where

$$l_1 + l_2 + \cdots + l_r = n.$$

Thus, the lengths of the cycles written in this order constitute a *partition* of n into positive integers. Show that two permutations P and Q correspond to the same partition of n if and only if they are conjugate, i.e., if and only if there exists a permutation T such that $TPT^{-1} = Q$.

Comment: Thus, for the symmetric group of degree n, each class is just the set of permutations associated with a given partition of n.

31. Suppose that each permutation of a given class consists of α_i cycles of length i ($i = 1, 2, \ldots, n$), so that $\alpha_1 + 2\alpha_2 + \cdots + n\alpha_n = n$. How many different permutations does the class contain?

32. By the *normalizer* $\mathfrak{N}_A$ of an element A of a group $\mathcal{G}$, we mean the set of elements of $\mathcal{G}$ which commute with A. Prove that

(a) $\mathfrak{N}_A$ is a subgroup of $\mathcal{G}$;
(b) The index of $\mathfrak{N}_A$ equals the number of *distinct* conjugates of A;
(c) The number of elements in each class of $\mathcal{G}$ divides the order of $\mathcal{G}$.

33. Prove that every subgroup of index 2 is a normal subgroup.

34. A class of a group is said to be *ambivalent* if the inverse of every element in the class is also in the class. Let $\mathcal{G}$ be a finite group of order g, with α ambivalent classes. Let A be an element of $\mathcal{G}$, and let $\rho(A)$ be the number of elements of $\mathcal{G}$ whose squares equal A. Prove that

$$\frac{1}{g} \sum_A \rho^2(A) = \alpha,$$

where the sum extends over all A in $\mathcal{G}$.

35. Let A be any element of a finite group $\mathcal{G}$, let $\rho(A)$ be the number of elements in $\mathcal{G}$ whose squares equal A, and let $n(A)$ be the number of elements in $\mathcal{G}$ which commute with A, i.e., the order of $\mathfrak{N}_A$, the normalizer of A. Prove that every class of $\mathcal{G}$ is ambivalent if and only if

$$\sum_A \rho^2(A) = \sum_A n(A),$$

where the sum extends over all A in $\mathcal{G}$.

36. By the *center* $\mathcal{Z}$ of a group $\mathcal{G}$ is meant the set of elements in $\mathcal{G}$ which commute with *all* elements of $\mathcal{G}$. Show that

(a) $\mathcal{Z}$ is the set of elements of $\mathcal{G}$ which are their own conjugates;
(b) $\mathcal{Z}$ is an Abelian subgroup of $\mathcal{G}$;
(c) $\mathcal{Z}$ is a normal subgroup of $\mathcal{G}$;
(d) If $\mathcal{G}$ is of order p^m, where p is prime, then the order of $\mathcal{Z}$ is a multiple of p.

37. Prove that the group of real numbers under addition is isomorphic to the group of positive real numbers under multiplication.

38. Let $\mathcal{H}$ and $\mathcal{K}$ be two normal subgroups of a group $\mathcal{G}$, and let $\mathcal{H}$ be a subgroup of $\mathcal{K}$. Prove that $\mathcal{G}/\mathcal{H}$ (the quotient group of $\mathcal{G}$ relative to $\mathcal{H}$) is homomorphic to $\mathcal{G}/\mathcal{K}$.

39. An isomorphism of a group with itself is called an *automorphism.* Let φ and ψ be two automorphisms of a group $\mathcal{G}$, with φ taking the element A into $\varphi(A)$ and ψ taking A into $\psi(A)$. Define the product $\varphi\psi(A)$ as $\varphi(\psi(A))$. Show that with this product, the set of all automorphisms of $\mathcal{G}$ is itself a group. Prove that the operation of conjugation, which takes A into BAB^{-1} (B fixed) is an automorphism; such an automorphism is called an *inner automorphism.* Automorphisms which are not inner automorphisms are called *outer automorphisms.* Prove that the set of all inner automorphisms of $\mathcal{G}$ is a subgroup of the group of all automorphisms of $\mathcal{G}$.

40. Show that the symmetric group of degree 3 has six inner automorphisms and no outer automorphisms. Show that the group of Prob. 4 has one inner automorphism and five outer automorphisms.

41. Show that the quotient group of the additive group of integers relative to numbers that are multiples of the integer $n > 0$ is isomorphic to the group of rotations of a regular n-gon about its center.

42. Show that the quotient group of the additive group of real numbers relative to the subgroup of integers is isomorphic to the multiplicative group of complex numbers of absolute value 1.

43. Prove that the quotient group of a non-Abelian group $\mathcal{G}$ relative to its center $\mathcal{Z}$ cannot be cyclic.

44. Let A and B be elements of a group $\mathcal{G}$. The element $ABA^{-1}B^{-1}$ is called the *commutator* of A and B and equals the unit element if and only if A and B commute (cf. Sec. 95). Show that

(a) The subgroup of $\mathcal{G}$ generated by all the commutators is a normal subgroup (the *commutator subgroup*);
(b) The factor group of the commutator group is Abelian;
(c) The commutator subgroup is contained in every normal subgroup which has an Abelian factor group.

45. Prove that the alternating group of degree n is the commutator subgroup of the symmetric group of degree n.

46. Prove that if $\mathcal{H}$ and $\mathcal{K}$ are two normal subgroups which have only the unit element in common, then every element of $\mathcal{H}$ commutes with every element of $\mathcal{K}$.

47. Prove that the stereographic projection of Sec. 68 takes circles on the sphere into circles in the XY-plane, and conversely. (Straight lines are regarded as special cases of circles.)

Chapter 8

Representations of Groups

71. Representation of Groups by Linear Transformations. Let $\mathcal{G}$ be a group with elements G_α, and suppose that a definite matrix A_α corresponds to every element G_α, where all the matrices are nonsingular (i.e., have nonvanishing determinants) and are of the same order. Moreover, suppose that the correspondence is such that the matrix product $A_{\alpha_2}A_{\alpha_1}$ corresponds to the product $G_{\alpha_2}G_{\alpha_1}$, for every α_1 and α_2. Then, we say that *the matrices A_α form a (linear) representation of the group* $\mathcal{G}$. Let G_0 be the unit element of $\mathcal{G}$, and let A_0 be the corresponding matrix. Since $G_0G_\alpha = G_\alpha$, we must have $A_0A_\alpha = A_\alpha$; multiplying this relation from the right by A_α^{-1}, we obtain $A_0 = I$, i.e., the image of the unit element G_0 of $\mathcal{G}$ must be the unit matrix. Moreover, let G_{α_1} and G_{α_2} be two elements of $\mathcal{G}$ which are inverses of each other, and let A_{α_1} and A_{α_2} be the corresponding matrices. Then, the relation $G_{\alpha_2}G_{\alpha_1} = G_0$ implies that $A_{\alpha_2}A_{\alpha_1} = I$, i.e., the images of inverse elements are inverse matrices.

From all this, it follows at once that the matrices A_α (or the corresponding linear transformations) *form a group $\mathcal{A}$ homomorphic to the group* $\mathcal{G}$. If different matrices of $\mathcal{A}$ correspond to different elements of $\mathcal{G}$, then $\mathcal{A}$ is not only homomorphic but also isomorphic to $\mathcal{G}$. In this case, we say that $\mathcal{A}$ is a *faithful representation of the group* $\mathcal{G}$. If this is not the case, the set of elements of $\mathcal{G}$ which have the unit matrix of $\mathcal{A}$ as their image form a normal subgroup $\mathcal{H}$ of the group $\mathcal{G}$, and the group $\mathcal{A}$ is isomorphic to the factor group of $\mathcal{H}$ in $\mathcal{G}$ (Sec. 63). If the basic group $\mathcal{G}$ is itself a group of linear transformations, then obviously, it is one of the possible representations of itself.

We now make an observation concerning the above definition of a representation of a group $\mathcal{G}$. Suppose that we know that every element G_α of $\mathcal{G}$ has a definite matrix A_α as its image and that products of matrices A_α correspond to products of elements G_α, but that we do not know whether the matrices A_α are nonsingular. We now show that if one of the A_α, say A_{α_0}, has a vanishing determinant, then all the A_α

have vanishing determinants. To see this, we note that for variable α, the set of matrix products $A_{\alpha_0}A_\alpha$ contains all matrices corresponding to elements of the group $\mathcal{G}$ (Sec. 62). But

$$\det A_{\alpha_0}A_\alpha = \det A_{\alpha_0} \det A_\alpha = 0,$$

since $\det A_{\alpha_0} = 0$, by hypothesis. Thus, a correspondence between the elements of $\mathcal{G}$ and a set $\mathcal{A}$ of matrices A_α of the same order, which preserves products, is a representation of $\mathcal{G}$ if at least one of the $\det A_\alpha$ is nonvanishing. For example, it is sufficient to verify that the unit element of $\mathcal{G}$ has the unit matrix as its image.

Let X be a nonsingular matrix of the same order as the matrices A_α. Since

$$(XA_{\alpha_2}X^{-1})(XA_{\alpha_1}X^{-1}) = XA_{\alpha_2}A_{\alpha_1}X^{-1},$$

the matrices $XA_\alpha X^{-1}$ also give a representation of the group $\mathcal{G}$. Representations of this kind, which use *similar* matrices, are called *equivalent representations.*

If the order of the matrices A_α is n, we can regard the group $\mathcal{A}$ as a group of linear transformations defined on an n-dimensional vector space, often called the *carrier space* (of the A_α). Let

$$\mathbf{x} = (x_1, x_2, \ldots, x_n)$$

be a vector in this space, and introduce new coordinate axes such that the new components $y_1, y_2, \ldots, y_n$ of $\mathbf{x}$ are related to the old components $x_1, x_2, \ldots, x_n$ by the formula

$$\mathbf{y} = X\mathbf{x}, \tag{1}$$

where $\mathbf{y} = (y_1, y_2, \ldots, y_n)$. Then, as we know from Sec. 21, if A_α carries $\mathbf{x}$ into $\mathbf{x}'$, i.e., if

$$\mathbf{x}' = A_\alpha\mathbf{x}, \tag{2}$$

the similar matrix $XA_\alpha X^{-1}$ carries $\mathbf{y}$ into $\mathbf{y}'$, i.e.,

$$\mathbf{y}' = XA_\alpha X^{-1}\mathbf{y}, \tag{3}$$

where $\mathbf{y}$ and $\mathbf{y}'$ are related to $\mathbf{x}$ and $\mathbf{x}'$ by the formula (1). Thus, the equivalent representation $XA_\alpha X^{-1}$ is obtained by changing coordinate axes in the carrier space in accordance with (1). In other words, the transition to an equivalent representation corresponds to replacing the components of vectors in the carrier space of the matrices A_α by new components in accordance with (1), where X is a nonsingular matrix.

Suppose that the matrices A_α of order n and the matrices B_α of order m are two representations of the same group $\mathcal{G}$, and form the quasi-diagonal matrices

$$[A_\alpha, B_\alpha] = \begin{Vmatrix} A_\alpha & 0 \\ 0 & B_\alpha \end{Vmatrix} \tag{4}$$

of order $m + n$. According to the multiplication rule for quasi-diagonal matrices (Sec. 26), we have

$$[A_{\alpha_2}, B_{\alpha_2}][A_{\alpha_1}, B_{\alpha_1}] = [A_{\alpha_2}A_{\alpha_1}, B_{\alpha_2}B_{\alpha_1}].$$

Thus, the matrices (4) are also a representation of the group $\mathcal{G}$. More generally, if we have three representations of a group $\mathcal{G}$, which use matrices A_α, B_α, and C_α, we can form a new representation which uses the quasi-diagonal matrices

$$D_\alpha = [A_\alpha, B_\alpha, C_\alpha] = \begin{Vmatrix} A_\alpha & 0 & 0 \\ 0 & B_\alpha & 0 \\ 0 & 0 & C_\alpha \end{Vmatrix}, \tag{5}$$

and so forth. It should be observed that if we go over to an equivalent representation which uses the matrices $XD_\alpha X^{-1}$, then the quasi-diagonal character of the matrices (5) is in general destroyed. Then, it will no longer be immediately apparent from the form of the new representation that, to within a similarity transformation, it is made up of representations of lower dimension, in the way described by (5). If our representation D_α is already in the quasi-diagonal form (5), then it obviously decomposes into the representations A_α, B_α, and C_α of lower dimension, i.e., into matrices of lower order; in this case the representation is said to be *reduced.* If a representation E_α does not have quasi-diagonal form, but some equivalent representation $XE_\alpha X^{-1}$ does have quasi-diagonal form, then E_α is said to be *reducible.* Finally, if neither the representation E_α nor any equivalent representation $XE_\alpha X^{-1}$ has quasi-diagonal form, then E_α is said to be an *irreducible representation.*

We now give some conditions which are sufficient for a representation to be reducible. First, consider a representation consisting of matrices A_α of order n, corresponding to linear transformations on the variables $x_1, x_2, \ldots, x_n$. Suppose that all the matrices A_α are unitary, and suppose that the subspace R' spanned by the first k basis vectors is carried into itself by the transformations A_α, so that if

$$x_{k+1} = x_{k+2} = \cdots = x_n = 0,$$

then $x'_{k+1} = x'_{k+2} = \cdots = x'_n = 0$. Then, all the matrices A_α have the form

$$\begin{Vmatrix} A'_\alpha & N_\alpha \\ 0 & A''_\alpha \end{Vmatrix}, \tag{6}$$

where the A'_α are matrices of order k, the A''_α are matrices of order $n - k$, and zeros appear everywhere in the lower left-hand corner (which has $n - k$ rows and k columns). Consider the subspace R'' spanned by the last $n - k$ basis vectors; R'' consists of the vectors orthogonal to all the vectors of the subspace R'. Since every transformation A_α carries R' into itself, and since orthogonality is preserved (A_α is unitary), the transformation A_α also carries every vector of R'' into a vector of R''. In other words, if $x_1 = x_2 = \cdots = x_k = 0$, then $x'_1 = x'_2 = \cdots x'_k = 0$. It follows at once that all the elements appearing in the upper right-hand corner of (6) (which has k rows and $n - k$ columns) must also be zero, i.e., the matrices of our representation must have the form

$$\begin{Vmatrix} A'_\alpha & 0 \\ 0 & A''_\alpha \end{Vmatrix} = [A'_\alpha, A''_\alpha],$$

so that the representation is in reduced form.

More generally, suppose that we have a representation consisting of unitary matrices A_α of order n, and suppose that a subspace R_1 of dimension k ($k < n$) is invariant under all the transformation A_α. Then, we transform the coordinate axes in such a way that R_1 is spanned by the first k basis vectors. This corresponds to going over to an equivalent representation and can be accomplished by using a unitary transformation. After this transformation, the representation is in reduced form, by the argument just given. Thus, we have proved the following theorem:

THEOREM 1. *If a representation of a group consists of unitary matrices A_α, and if every A_α transforms a certain subspace into itself, then the representation is reducible.*

As we have seen, the problem of reducing a representation $\mathcal{A}$ of a group $\mathcal{G}$ is intimately related to the problem of making appropriate similarity transformations $XA_\alpha X^{-1}$ on the matrices A_α in $\mathcal{A}$. We examine next the equivalent representations obtained by using certain special forms of the matrix X. Consider the matrix

$$X = \begin{Vmatrix} 0 & 1 & 0 & 0 & \cdots & 0 \\ 1 & 0 & 0 & 0 & \cdots & 0 \\ 0 & 0 & 1 & 0 & \cdots & 0 \\ 0 & 0 & 0 & 1 & \cdots & 0 \\ \cdot & \cdot & \cdot & \cdot & \cdots & \cdot \\ 0 & 0 & 0 & 0 & \cdots & 1 \end{Vmatrix},$$

obtained by interchanging the first and second rows of the unit matrix; obviously, $\det X = -1$. It is an immediate consequence of the usual rules of matrix multiplication that XYX^{-1} is the matrix obtained from Y by interchanging the first and second rows and the first and second columns. Similarly, any interchange of two rows of Y, accompanied by an interchange of the same two columns of Y, can be accomplished by making a similarity transformation XYX^{-1} with a (unitary) matrix X, which is obviously independent of Y. Since every permutation is a product of transpositions (Sec. 2), we have the following theorem:

THEOREM 2. *If a representation of a group $\mathcal{G}$ consists of matrices A_α, and if we perform the same permutation on both the rows and the columns of every A_α, we obtain an equivalent representation of $\mathcal{G}$.*

In certain cases, this result can be used to reduce a representation of $\mathcal{G}$. Thus, suppose that the matrices A_α, of order n, represent $\mathcal{G}$, and suppose that the integers $1, 2, \ldots, n$ can be divided into two disjoint classes $i_1, \ldots, i_k$ and $i_{k+1}, \ldots, i_n$, say, such that only zeros appear at the intersections of the rows numbered $i_1, \ldots, i_k$ with the columns numbered $i_{k+1}, \ldots, i_n$ and at the intersections of the rows numbered $i_{k+1}, \ldots, i_n$ with the columns numbered $i_1, \ldots, i_k$. Then, the matrices A_α can be reduced by simultaneously subjecting the rows and columns of every A_α to the permutation

$$P = \begin{pmatrix} i_1 & \cdots & i_k & i_{k+1} & \cdots & i_n \\ 1 & \cdots & k & k+1 & \cdots & n \end{pmatrix}.$$

Finally, suppose that we have a one-dimensional representation of a group $\mathcal{G}$, i.e., suppose that all the matrices A_α are ordinary numbers. Then a number m_α (more precisely, a transformation $x' = m_\alpha x$) corresponds to every element G_α of $\mathcal{G}$, and the product $m_{\alpha_2}m_{\alpha_1}$ corresponds to every product $G_{\alpha_2}G_{\alpha_1}$.

72. Basic Theorems. Let $\mathcal{G}$ be a finite group, containing m elements $G_1, G_2, \ldots, G_m$, and let $\mathcal{A}$ be a representation of $\mathcal{G}$, containing m matrices $A_1, A_2, \ldots, A_m$, of order n. Let $\mathbf{x} = (x_1, x_2, \ldots, x_m)$ be a vector in the carrier space of the matrices A_α, and consider the expression

$$\varphi(x_1, x_2, \ldots, x_n) = \sum_{s=1}^{m} \|A_s\mathbf{x}\|^2. \tag{7}$$

In expanded form, (7) becomes

$$\varphi = \sum_{s=1}^{m} \sum_{i=1}^{n} (a_{i1}^{(s)}x_1 + \cdots + a_{in}^{(s)}x_n)(\bar{a}_{i1}^{(s)}\bar{x}_1 + \cdots + \bar{a}_{1n}^{(s)}\bar{x}_n), \tag{8}$$

where the $a_{ik}^{(s)}$ $(i, k = 1, \ldots, n)$ denote the elements of the matrix A_s. It is easy to see that (8) is a Hermitian form (Sec. 30), i.e., that the coefficients of $\bar{x}_p x_q$ and $x_p\bar{x}_q$ are complex conjugates. Moreover, since (7) is the sum of the squares of certain vectors, the Hermitian form φ is positive definite (Sec. 40). Thus, if we perform a unitary transformation

$$\mathbf{y} = U\mathbf{x}$$

which reduces (8) to a sum of squares,

$$\varphi = \sum_{j=1}^{n} \lambda_j \bar{y}_j y_j,$$

we find that all the coefficients λ_j are positive. If we perform the further transformation $z_j = \sqrt{\lambda_j}\, y_j$, then in the new variables, the Hermitian form φ becomes

$$\varphi = \sum_{j=1}^{n} |z_j|^2, \tag{9}$$

i.e., φ reduces to a sum of squares of absolute values, with unit coefficients.

Now let A_k be one of the matrices representing the group $\mathcal{G}$, and apply A_k to the vector $\mathbf{x}$, obtaining

$$\mathbf{x}' = A_k\mathbf{x}. \tag{10}$$

It is easy to see that the Hermitian form

$$\varphi(x_1', x_2', \ldots, x_n') = \sum_{s=1}^{m} \|A_s A_k \mathbf{x}\|^2$$

does not change, since, as we know from Sec. 62, the set of transformations (matrices) $A_1A_k, A_2A_k, \ldots, A_mA_k$ coincides with the set of matrices $A_1, A_2, \ldots, A_m$ (in some order). Suppose now that we express the transformation (10) in the new variables $z_1, z_2, \ldots, z_n$ appearing in (9); these new variables are related to the old variables $x_1, x_2, \ldots, x_n$ by the formula

$$\mathbf{z} = B\mathbf{x},$$

where $\mathbf{x} = (x_1, x_2, \ldots, x_n)$, $\mathbf{z} = (z_1, z_2, \ldots, z_n)$, and B is a certain matrix. Then, instead of the group $A_1, A_2, \ldots, A_m$, we get the conjugate group (Sec. 64) $BA_1B^{-1}, BA_2B^{-1}, \ldots, BA_mB^{-1}$, and as before, all the transformations of this conjugate group leave the sum (9) unchanged, i.e., these transformations are all *unitary*. Thus we have just proved the following theorem:

THEOREM 1. *Every representation of a finite group has an equivalent unitary representation.*

If certain additional restrictions are imposed (which will not be discussed here), this result remains true even in the case of infinite groups, whose elements depend on a parameter.† Henceforth, when we talk about a representation of a group, unless the contrary is explicitly stated (cf. Sec. 86), we shall always mean a unitary representation.

† See however Prob. 5.

Next, we derive a necessary and sufficient condition for a representation to be reducible. First, we agree to call the diagonal matrix $[k, k, \ldots, k]$, with identical elements k on the principal diagonal, a *multiple of the unit matrix*, written kI. As already noted (Sec. 25), in matrix operations, kI behaves like the number k. Suppose now that we have a reducible representation of a group $\mathcal{G}$, e.g., a representation consisting of matrices of the form

$$D_\alpha = X[A_\alpha, B_\alpha, C_\alpha]X^{-1},$$

where X is some matrix and the inner matrix is quasi-diagonal. We form the matrix

$$Y = X[kI, lI, mI]X^{-1},$$

where the inner matrix is quasi-diagonal with the same structure as in the matrices D_α. It is not hard to see that the matrix Y commutes with all the matrices D_α, since

$$D_\alpha Y = X[A_\alpha k, B_\alpha l, C_\alpha m]X^{-1},$$

and similarly

$$YD_\alpha = X[kA_\alpha, lB_\alpha, mC_\alpha]X^{-1},$$

while the order of the factors is irrelevant in the product of a matrix with a number. Moreover, we shall assume that the numbers k, l, and m are all different, so that the matrix Y is not a multiple of the unit matrix; in fact, this implies that Y has the distinct eigenvalues k, l, and m. Thus, we arrive at the following theorem:

THEOREM 2. *If a representation consisting of the matrices D_α is reducible, then there exists a matrix Y which is not a multiple of the unit matrix and which commutes with all the matrices D_α.*

Next, we prove the converse of this theorem:

THEOREM 3. *If there exists a matrix Y which is not a multiple of the unit matrix and which commutes with all the matrices D_α of a representation, then the representation is reducible.*

PROOF. By hypothesis, we have

$$D_\alpha Y = YD_\alpha \tag{11}$$

for any index α. Let Z be a nonsingular matrix such that all the matrices $ZD_\alpha Z^{-1} = U_\alpha$ are unitary, and write (11) in the form

$$Z^{-1}U_\alpha ZY = YZ^{-1}U_\alpha Z.$$

Multiplying from the left by Z and from the right by Z^{-1}, we obtain

$$U_\alpha(ZYZ^{-1}) = (ZYZ^{-1})U_\alpha,$$

i.e., the matrix ZYZ^{-1} commutes with all the matrices of the unitary representation. This matrix is obviously not a multiple of the unit matrix, since $ZYZ^{-1} = kI$ implies $Y = kI$. Thus, it is sufficient to prove that the equivalent representation U_α is reducible, i.e., we can assume that the representation in the statement of Theorem 3 is unitary. For simplicity, we drop the notation U_α and assume that the matrices D_α are already unitary.

Let λ_1 be an eigenvalue of the matrix Y. Since the matrix $\lambda_1 I$ commutes with any matrix, the matrix $Y - \lambda_1 I$, as well as Y, satisfies the condition (11), i.e., commutes with all the matrices D_α. It is easy to see that at least one of the eigenvalues of the matrix $Y_1 = Y - \lambda_1 I$ is zero. In fact, the characteristic equation of the matrix Y_1 is

$$\det (Y_1 - \lambda I) = \det [Y - (\lambda + \lambda_1)I] = 0,$$

i.e., it is obtained from the characteristic equation for Y by replacing λ by $\lambda + \lambda_1$. Therefore, since one of the eigenvalues of Y is λ_1, at least one of the eigenvalues of Y_1 is zero. It follows that the determinant of Y_1, which equals the product of the eigenvalues of Y_1, is also zero. Thus, in proving our theorem, we can assume not only that all the matrices D_α are unitary, but also that the matrix Y appearing in (11) has a vanishing determinant.

Now consider the set of vectors with components

$$\begin{aligned} x_1 &= y_{11}u_1 + y_{12}u_2 + \cdots + y_{1n}u_n, \\ x_2 &= y_{21}u_1 + y_{22}u_2 + \cdots + y_{2n}u_n, \\ &\cdots\cdots\cdots\cdots\cdots\cdots \\ x_n &= y_{n1}u_1 + y_{n2}u_2 + \cdots + y_{nn}u_n, \end{aligned} \tag{12}$$

where the u_s take arbitrary values, and the y_{ik} are the elements of the matrix Y. Since the determinant of Y vanishes, the rank r of Y satisfies the inequality $0 < r < n$. Then, as we know from Sec. 15, the formulas (12) define an r-dimensional subspace R'. Consider the left-hand side of the equation

$$D_\alpha Y\mathbf{u} = YD_\alpha\mathbf{u}. \tag{13}$$

The vector $Y\mathbf{u}$ has just the components (12), and hence $D_\alpha Y\mathbf{u}$ is the result of applying the transformation D_α to an arbitrary vector of the space R'. The right-hand side of (13) is the result of applying Y to the vector $D_\alpha\mathbf{u}$. Thus, the components of the right-hand side of (13) are also given by (12), where instead of $u_1, u_2, \ldots, u_n$, we now have the components of the vector $D_\alpha\mathbf{u}$, i.e., the right-hand side of (13) represents a vector belonging to the subspace R'. Finally, comparing the left- and right-hand sides of (13), we see that applying the transformation D_α to any vector of the subspace R' gives a vector which also belongs to R'. But we already know (Theorem 1 of Sec. 71) that if the unitary transformations D_α carry a subspace into itself, then the D_α form a reducible representation, QED.

Together, Theorems 2 and 3 show that *a necessary and sufficient condition for a representation to be irreducible is that there should exist no matrix other than a matrix of the form kI which commutes with all the matrices of the representation.* It is an immediate consequence of Theorem 1 of this section that there is no need to stipulate in Theorem 1 of Sec. 71 that the representation be unitary. Thus, more generally, we can state that *if all the matrices of a representation transform a certain subspace into itself, the representation is reducible.* The converse statement is obvious.

73. Abelian Groups and One-Dimensional Representations. A group $\mathcal{G}$ is called *Abelian* if all its elements G_α commute, i.e., if

$$G_{\alpha_2}G_{\alpha_1} = G_{\alpha_1}G_{\alpha_2} \tag{14}$$

for any indices α_1 and α_2 (Sec. 62). Let A_{α_1} and A_{α_2} be the matrices corresponding to G_{α_1} and G_{α_2} in some representation. Since the products $G_{\alpha_2}G_{\alpha_1}$ and $G_{\alpha_1}G_{\alpha_2}$ have images $A_{\alpha_2}A_{\alpha_1}$ and $A_{\alpha_1}A_{\alpha_2}$, respectively, it follows from (14) that

$$A_{\alpha_2}A_{\alpha_1} = A_{\alpha_1}A_{\alpha_2};$$

thus, *the matrices forming a representation of an Abelian group commute.*

Suppose that the representation is unitary, i.e., that all the matrices A_α are unitary. Then, as we know (Sec. 42), there exists a unitary transformation U such that all the matrices $UA_\alpha U^{-1}$ have purely diagonal form, so that there exists an equivalent representation

$$UA_\alpha U^{-1} = [k_\alpha^{(1)}, k_\alpha^{(2)}, \ldots, k_\alpha^{(n)}],$$

consisting of diagonal matrices. Thus, in this case, the representation reduces to n one-dimensional representations

$$B_\alpha^{(s)} = k_\alpha^{(s} \qquad (s = 1, 2, \ldots, n).$$

In other words, by using a unitary matrix, every unitary representation of an Abelian group can be transformed to an equivalent set of one-dimensional representations.

We now consider various examples of representations of Abelian groups and also some examples of one-dimensional representations of non-Abelian groups.

EXAMPLE 1. Consider the cyclic (Abelian) group of order m, consisting of the elements

$$S^0 = I, S, S^2, \ldots, S^{m-1} \qquad (S^m = I). \tag{15}$$

If the number ω (more precisely, the linear transformation $x' = \omega x$) corresponds to the element S, then the elements

$$1, \omega, \omega^2, \ldots, \omega^{m-1}$$

correspond to the elements (15). Since $S^m = I$, we must have $\omega^m = 1$, i.e.,

$$\omega = e^{2\pi k i/m},$$

where k is an integer which can obviously take any of the values 0, 1, 2, $\ldots, m - 1$.

Let us examine the case $m = 2$ in more detail; in this case, we have

$$I, S, S^2 = I,$$

i.e., $S = S^{-1}$. For $k = 0$, the identity transformation (the number 1) corresponds to both elements I and S. For $k = 1$, the transformation $x' = x$ corresponds to the element I, while the transformation $x' = -x$ corresponds to S; more succinctly, the image of I is the number $+1$, and the image of S is the number -1. An important case in physical applications is where I is the identity transformation in three dimensions and the operation S corresponds to reflection in the origin, i.e.,

$$x' = -x, \qquad y' = -y, \qquad z' = -z.$$

In this case, $m = 2$, and the two representations just given may be called the *identity* representation and the *alternating* representation.

EXAMPLE 2. Consider the group of rotations about the z-axis. The matrices of this group have the form

$$Z_\varphi = \left\| \begin{matrix} \cos\varphi & -\sin\varphi \\ \sin\varphi & \cos\varphi \end{matrix} \right\|. \tag{16}$$

Moreover, as already noted (Sec. 58), we have the obvious relation

$$Z_{\varphi_2} Z_{\varphi_1} = Z_{\varphi_1} Z_{\varphi_2} = Z_{\varphi_1+\varphi_2},$$

which is also satisfied by the function $e^{l\varphi}$. However, it should be noted that if $\varphi = 2\pi$, the rotation Z_φ is equivalent to the identity transformation, and therefore we must have $e^{2\pi l} = 1$, i.e., l must equal mi, where m is any integer. Thus, we have an infinite set of representations of our rotation group, with the number $e^{\varphi m i}$ corresponding to the matrix Z_φ; the various representations are obtained by assigning the number m the values

$$m = 0, \pm 1, \pm 2, \ldots.$$

EXAMPLE 3. Consider the group consisting of $n!$ permutations on n elements. One possibility is to assign the number $+1$ to every permutation; this gives the so-called *identity* representation of the permutation group. Another possibility is to assign the number $+1$ to every permutation of the first class and the number -1 to every permutation of the second class (Sec. 2); this gives the so-called *alternating* representation of the permutation group. In this representation, the number $+1$ corresponds to every permutation of the alternating group, and -1 corresponds to every other permutation. It can be shown that these two cases exhaust all the possibilities of one-dimensional representations of the permutation group. However, the permutation group has other higher-dimensional representations.

EXAMPLE 4. Consider the group of all real orthogonal transformations in the plane, i.e., the group consisting of every rotation of the plane about the origin, together with reflection in the y-axis. As we have already seen (Sec. 60), the matrices of this group have the form

$$\{\varphi, d\} = \left\| \begin{matrix} d \cos \varphi & -d \sin \varphi \\ \sin \varphi & \cos \varphi \end{matrix} \right\|, \tag{17}$$

where $d = 1$ for a pure rotation, and $d = -1$ for a rotation plus a reflection. In addition to the obvious one-dimensional representation, which assigns the number $+1$ to every matrix (17), we can construct another one-dimensional representation which assigns the number $+1$ to the matrix (92) if $d = 1$, and the number -1 if $d = -1$. This actually gives a representation, since the product of two matrices of the form (92) corresponds to a pure rotation, if d has the same sign in both factors, and to a rotation plus a reflection, if d has different signs in the two factors.

74. Representations of the Two-Dimensional Unitary Group. We now consider the representations of the two-dimensional unitary group. As we know from Sec. 69, the transformations of this group have the form

$$\begin{aligned} x_1' &= ax_1 + bx_2, \\ x_2' &= -\bar{b}x_1 + \bar{a}x_2, \end{aligned} \tag{18}$$

where the complex numbers a and b satisfy the condition

$$a\bar{a} + b\bar{b} = 1. \tag{19}$$

We construct the $m + 1$ quantities

$$\xi_0 = x_1^m,\ \xi_1 = x_1^{m-1}x_2,\ \ldots,\ \xi_m = x_2^m. \tag{20}$$

If we take every $\xi_k' = x_1'^{m-k}x_2'^k$ and substitute the expressions (18) for x_1' and x_2', then obviously the ξ_k' become linear combinations of the ξ_k, i.e., corresponding to every transformation of the group (18), we obtain a linear transformation from the variables ξ_k to the variables ξ_k'. Moreover, it is clear that the correspondence between these transformations and the transformations (18) preserves products. Thus, we have an $(m+1)$-dimensional representation of the group (18).

However, it turns out that this representation is not unitary. To make the representation unitary, it is sufficient to equip each of the variables (20) with an extra constant factor, i.e., instead of using the variables (20), we define new variables η_k and η'_k by the formulas

$$\eta_k = \frac{x_1^{m-k}x_2^k}{\sqrt{(m-k)!k!}} \qquad (k = 0, 1, \ldots, m) \tag{21a}$$

and

$$\eta'_k = \frac{x_1'^{m-k}x_2'^k}{\sqrt{(m-k)!k!}} \qquad (k = 0, 1, \ldots, m) \tag{21b}$$

where, as always, we write $0! = 1$. With this new choice of variables, our transformation is unitary:†

$$\sum_{k=0}^{m} \eta'_k\bar{\eta}'_k = \sum_{k=0}^{m} \eta_k\bar{\eta}_k. \tag{22}$$

To see this, we apply the binomial theorem, obtaining

$$m!\sum_{k=0}^{m} \eta'_k\bar{\eta}'_k = m!\sum_{k=0}^{m} \frac{x_1'^{m-k}\bar{x}_1'^{m-k}x_2'^k\bar{x}_2'^k}{(m-k)!k!} = (x'_1\bar{x}'_1 + x'_2\bar{x}'_2)^m,$$

and similarly

$$m!\sum_{k=0}^{m} \eta_k\bar{\eta}_k = (x_1\bar{x}_1 + x_2\bar{x}_2)^m.$$

But since the transformation (18) is unitary, we have

$$x'_1\bar{x}'_1 + x_2\bar{x}'_2 = x_1\bar{x}_1 + x_2\bar{x}_2,$$

and therefore (22) also holds.

We now derive formulas which give explicit expressions for the coefficients of these unitary representations of the group (18). To do this, we first modify our notation somewhat, by setting

$$\eta_l = \frac{x_1^{j+l}x_2^{j-l}}{\sqrt{(j+l)!(j-l)!}} \qquad (l = -j, -j+1, \ldots, j-1, j). \tag{23}$$

In our previous notation, $m = 2j$, so that j is an integer if m is even and

† See footnote on p. 132.

an odd multiple of ½ if m is odd. For example, if $m = 5$, the formulas (23) give the six variables

$$\eta_{-5/2} = \frac{x_2^5}{\sqrt{5!}}, \quad \eta_{-3/2} = \frac{x_1 x_2^4}{\sqrt{1!4!}}, \quad \eta_{-1/2} = \frac{x_1^2 x_2^3}{\sqrt{2!3!}},$$
$$\eta_{1/2} = \frac{x_1^3 x_2^2}{\sqrt{3!2!}}, \quad \eta_{3/2} = \frac{x_1^4 x_2}{\sqrt{4!1!}}, \quad \eta_{5/2} = \frac{x_1^5}{\sqrt{5!}}.$$

In this case, instead of being indexed by the six integers 0, 1, . . . , 5, our variables are indexed by odd multiples of ½ ranging from $-5/2$ to $+5/2$. If $m = 4$, the formulas (23) give the five variables

$$\eta_{-2} = \frac{x_2^4}{\sqrt{4!}}, \quad \eta_{-1} = \frac{x_1 x_2^3}{\sqrt{1!3!}}, \quad \eta_0 = \frac{x_1^2 x_2^2}{\sqrt{2!2!}},$$
$$\eta_1 = \frac{x_1^3 x_2}{\sqrt{3!1!}}, \quad \eta_2 = \frac{x_1^4}{\sqrt{4!}}.$$

In this case, the variables are indexed by integers ranging from -2 to $+2$. Moreover, for every $m = 2j$, we obtain a similar labeling of the rows and columns in the matrices forming a $(2j + 1)$-dimensional representation of the group (18).

To determine the elements of these matrices, we proceed as follows. We have

$$\eta_l' = \frac{x_1'^{j+l} x_2'^{j-l}}{\sqrt{(j+l)!(j-l)!}} = \frac{(ax_1 + bx_2)^{j+l}(-\bar{b}x_1 + \bar{a}x_2)^{j-l}}{\sqrt{(j+l)!(j-l)!}}, \tag{24}$$

and we have to represent the right-hand side of (24) as a linear combination of the variables η_l. Applying the binomial theorem, we obtain

$$\eta_l' = \sum_{k=0}^{j+l} \sum_{k'=0}^{j-l} (-1)^{j-l-k'} \frac{\sqrt{(j+l)!(j-l)!}}{k!k'!(j+l-k)!(j-l-k')!}$$
$$\times \bar{a}^{k'} a^{j+l-k} \bar{b}^{j-l-k'} b^k x_1^{2j-k-k'} x_2^{k+k'}. \tag{25}$$

If we set $p! = \infty$ when p is a negative integer, then the summations in (25) can extend from $-\infty$ to $+\infty$, since the denominators of the additional terms will contain ∞ as a factor and will therefore vanish. Then, instead of k', we introduce a new summation variable $s = j - k - k'$, which we can again sum from $-\infty$ to $+\infty$, over integral values, if j is an integer, or over half-integral values, if j is a half integer. This gives

$$\eta_l' = \sum_k \sum_s (-1)^{k+s-l} \frac{\sqrt{(j+l)!(j-l)!}}{k!(j-k-s)!(j+l-k)!(k+s-l)!}$$
$$\times \bar{a}^{j-k-s} a^{j+l-k} \bar{b}^{k+s-l} b^k x_1^{j+s} x_2^{j-s}.$$

But according to (23) we have

$$x_1^{j+s}x_2^{j-s} = \sqrt{(j+s)!(j-s)!}\,\eta_s,$$

so that finally we obtain the required linear combination in the form

$$\eta_l' = \sum_k \sum_s (-1)^{k+s-l} \frac{\sqrt{(j+l)!(j-l)!(j+s)!(j-s)!}}{k!(j-k-s)!(j+l-k)!(k+s-l)!} \times \bar{a}^{j-k-s}a^{j+l-k}\bar{b}^{k+s-l}b^k\eta_s.$$

Thus, for a given fixed j, the matrix elements of the $(2j+1)$-dimensional linear transformation corresponding to the unitary transformation (18), with matrix

$$\begin{Vmatrix} a & b \\ -\bar{b} & \bar{a} \end{Vmatrix},$$

are given by

$$D_j\begin{Bmatrix} a & b \\ -\bar{b} & \bar{a} \end{Bmatrix}_{ls} = (-1)^{s-l} \sum_k (-1)^k \frac{\sqrt{(j+l)!(j-l)!(j+s)!(j-s)!}}{k!(j-k-s)!(j+l-k)!(k+s-l)!} \times \bar{a}^{j-k-s}a^{j+l-k}\bar{b}^{k+s-l}b^k. \quad (26)$$

Here, the indices . and s range over the values

$$l, s = -j, -j+1, \ldots, j-1, j,$$

and we note that if j is a half integer, the rows and columns of the matrix (26) are labeled with half integers. Bearing in mind that $p! = \infty$ if p is a negative integer, we find that the summation variable k in (26) lies within the range defined by the inequalities

$$k \geq 0, \qquad k \geq l - s, \qquad k \leq j - s, \qquad k \leq j + l. \quad (27)$$

The formula (26) can be simplified somewhat by transforming to an equivalent representation. Let A be a matrix with elements a_{ls}, and let $S = [\delta_1, \delta_2, \ldots, \delta_n]$ be a diagonal matrix. Applying the usual rule for matrix multiplication, it is easy to see that the matrix SAS^{-1} has the elements

$$\{SAS^{-1}\}_{ls} = \delta_l a_{ls} \delta_s^{-1}.$$

If we now apply this transformation to the matrices

$$D_j\begin{Bmatrix} a & b \\ -\bar{b} & \bar{a} \end{Bmatrix}_{ls},$$

choosing $\delta_l = (-1)^l$, the factor $(-1)^{s-l}$ disappears in the formula (26); thus, from now on, we shall omit the factor $(-1)^{s-l}$.

Next we prove that the representation (26) is irreducible. We begin by proving the following two preliminary lemmas.

LEMMA 1. *If a diagonal matrix, all of whose (diagonal) elements are different, commutes with a matrix A, then A is also a diagonal matrix.*

PROOF. By hypothesis, we have

$$A[\delta_1, \delta_2, \ldots, \delta_n] = [\delta_1, \delta_2, \ldots, \delta_n]A, \tag{28}$$

where all the δ_k are different. Denoting the elements of A by a_{pq}, and carrying out the multiplication in (28), we obtain

$$a_{pq}\delta_q = \delta_p a_{pq}$$

or

$$a_{pq}(\delta_q - \delta_p) = 0,$$

i.e., $a_{pq} = 0$ if $p \neq q$, and hence the matrix A is diagonal, QED.

LEMMA 2. *If a diagonal matrix $[\delta_1, \delta_2, \ldots, \delta_n]$ commutes with a matrix A which has at least one column containing no zeros, then*

$$\delta_1 = \delta_2 = \cdots = \delta_n.$$

PROOF. Suppose that the q_0th column of A contains no zeros, i.e.,

$$a_{pq_0} \neq 0 \qquad (p = 1, 2, \ldots, n).$$

Since $[\delta_1, \delta_2, \ldots, \delta_n]$ commutes with A, we have

$$a_{pq_0}(\delta_{q_0} - \delta_p) = 0 \qquad (p = 1, 2, \ldots, n),$$

as in the proof of Lemma 1. It follows that $\delta_1 = \delta_2 = \cdots = \delta_n$, QED.

We denote by

$$D_j \begin{Bmatrix} a & b \\ -\bar{b} & \bar{a} \end{Bmatrix} \tag{29}$$

the matrices with elements given by the formula (26). It will be recalled that the numbers a and b must satisfy the condition (19). We are now in a position to prove the following theorem:

THEOREM. *The representation of the unitary group* (18) *consisting of the matrices* (29) *is irreducible.*

PROOF. Let Y be a matrix of order $2j + 1$ which commutes with all the matrices (29). By Theorem 2 of Sec. 72, to prove the irreducibility of the representation consisting of the matrices (29), it suffices to show that Y must be a multiple of the unit matrix. Consider first the case where $a = e^{i\alpha}$ and $b = 0$; these numbers obviously satisfy the condition (19). Using (26), we find that in this case

$$D_j \begin{Bmatrix} e^{i\alpha} & 0 \\ 0 & e^{-i\alpha} \end{Bmatrix}_{ls} = 0 \qquad \text{for } l \neq s,$$

while the diagonal elements are

$$D_j\begin{Bmatrix} e^{i\alpha} & 0 \\ 0 & e^{-i\alpha}\end{Bmatrix}_{ll} = e^{2il\alpha} \qquad (l = -j, -j+1, \ldots, j-1, j).$$

Thus, when $a = e^{i\alpha}$ and $b = 0$, our matrix has the form

$$D_j\begin{Bmatrix} e^{i\alpha} & 0 \\ 0 & e^{-i\alpha}\end{Bmatrix} = \begin{Vmatrix} e^{-2ij\alpha} & 0 & 0 & \cdots & 0 \\ 0 & e^{-2i(j-1)\alpha} & 0 & \cdots & 0 \\ 0 & 0 & e^{-2i(j-2)\alpha} & \cdots & 0 \\ \cdots & \cdots & \cdots & \cdots & \cdots \\ 0 & 0 & 0 & \cdots & e^{2ij\alpha}\end{Vmatrix}, \tag{30}$$

which, for a suitable choice of α, is a diagonal matrix with distinct elements on the principal diagonal. It follows from Lemma 1 that the matrix Y, which must commute with the matrix (30), is also diagonal, i.e.,

$$Y = [\delta_1, \delta_2, \ldots, \delta_n]. \tag{31}$$

Next, suppose that the numbers a and b are both nonzero, and consider the first column of the matrix (29). Its elements are given by (26), if we set $s = -j$. Then, the inequalities (27) become

$$k \geq 0, \qquad k \geq l + j, \qquad k \leq 2j, \qquad k \leq j + l$$
$$(l = -j, -j+1, \ldots, j-1, j),$$

from which it is clear that the sum in (26) reduces to the single term which is obtained for $k = j + l$; this term is nonzero, since both a and b are nonzero. Thus, in this case, the first column of the matrix (29) contains no zeros. But since the diagonal matrix (31) commutes with this matrix, it follows from Lemma 2 that all the numbers δ_k are the same, i.e., Y is a multiple of the unit matrix. Thus, by Theorem 2 of Sec. 72, the matrices (29) are in fact an irreducible representation of the unitary group (18), QED.

By assigning j the values

$$j = 0, \tfrac{1}{2}, 1, \tfrac{3}{2}, 2, \ldots,$$

we obtain an infinite set of representations of the unitary group. For $j = 0$, we get the trivial identity representation, in which every element of the group (18) has the number $+1$ as its image. Next, we consider the case $j > 0$ and find the elements of the group (18) which have as their images the identity transformation of the group (29). The identity transformation is defined by the relations $\eta_l' = \eta_l$, or equivalently by the relations

$$(ax_1 + bx_2)^{j+l}(-\bar{b}x_1 + \bar{a}x_2)^{j-l} = x_1^{j+l}x_2^{j-l}$$
$$(l = -j, -j+1, \ldots, j-1, j). \tag{32}$$

Setting $j = l$, we obtain

$$(ax_1 + bx_2)^{2j} = x_1^{2j},$$

from which it follows that $b = 0$. Thus, (32) becomes

$$a^{j+l}\bar{a}^{j-l}x_1^{j+l}x_2^{j-l} = x_1^{j+l}x_2^{j-l} \qquad (l = -j, -j+1, \ldots, j-1, j),$$

whence

$$a^{j+l}\bar{a}^{j-l} = 1. \tag{33}$$

But $|a| = 1$ for $b = 0$, so that (33) can be written in the form

$$a^{2l} = 1 \qquad (l = -j, -j+1, \ldots, j-1, j). \tag{34}$$

If j is an odd multiple of $\frac{1}{2}$, then a possible value of l is $\frac{1}{2}$, and (34) gives $a = 1$. If j is an integer, then (34) reduces to $a^2 = 1$, i.e., $a = \pm 1$. Thus, finally, if j is an odd multiple of $\frac{1}{2}$, the identity transformation in the group of matrices (29) corresponds only to the identity transformation in the group (18), i.e., in this case, the matrices (29) constitute a *faithful* representation of the group (18). However, if j is an integer, the identity transformation in the group (29) corresponds to the two transformations

$$E = \begin{Vmatrix} 1 & 0 \\ 0 & 1 \end{Vmatrix}, \qquad S = \begin{Vmatrix} -1 & 0 \\ 0 & -1 \end{Vmatrix} = -E \tag{35}$$

in the group (18). The transformations (35) form a cyclic subgroup $\mathcal{H}$ of order 2, and the matrices (29) give a faithful representation of the factor group of $\mathcal{H}$ in the group $\mathcal{G}$ of all matrices of the form (18). In other words, for integral j, every matrix (29) corresponds to two transformations of the form (18), which differ only in the signs of the numbers a and b.

75. Representations of the Rotation Group. In view of the fact that the unitary group (18) is intimately connected with the three-dimensional rotation group, the results obtained in the preceding section are of particular importance, and in fact they lead to irreducible representations of the rotation group. It will be recalled (see Sec. 69) that every unitary transformation (18) has as its image a definite rotation and that changing the signs of both a and b gives another unitary transformation which has as its image the same rotation. The parameters a and b are related to the Eulerian angles of the corresponding rotation by the formulas

$$a = e^{-i\frac{\gamma+\alpha}{2}} \cos\frac{\beta}{2}, \qquad b = -ie^{i\frac{\gamma-\alpha}{2}} \sin\frac{\beta}{2} \tag{36}$$

(cf. equation (66) of Sec. 69).

We consider first the case where j is an integer. Then simultaneously changing the signs of a and b does not change the terms in the right-hand side of (26), since the sum of the exponents of a, $\bar{a}$, b, and $\bar{b}$ equals the even number $2j$. Thus, in this case, the same matrix in the representation (29) corresponds to the two unitary transformations which lead to the same rotation. In other words, when j is an integer, a unique matrix in the representation (29) corresponds to every rotation $\{\alpha, \beta, \gamma\}$, with Eulerian angles α, β, γ. This matrix will henceforth be written in the form

$$D_j\{\alpha, \beta, \gamma\}, \tag{37}$$

rather than in the form (29). However, if j is an odd multiple of $\frac{1}{2}$, then simultaneously changing the signs of a and b changes the sign of all the terms in (26), so that two different matrices, whose elements have different signs, correspond to the two different unitary transformations associated with the same rotation; thus, in this case, the two different matrices $\pm D_j\{\alpha, \beta, \gamma\}$ correspond to the rotation $\{\alpha, \beta, \gamma\}$. In other words, for integral j, the matrices $D_j\{\alpha, \beta, \gamma\}$ give a faithful representation of the rotation group, while for half integral j, we do not have, strictly speaking, a representation of the rotation group, but rather a so-called *two-valued representation*, which associates the two matrices $\pm D_j\{\alpha, \beta, \gamma\}$ with the same rotation $\{\alpha, \beta, \gamma\}$.

To find the elements of the matrices $D_j\{\alpha, \beta, \gamma\}$, we have to substitute the expressions (36) for a and b in the formula (26). Omitting the factor $(-1)^{s-l}$, for the reason given previously, we obtain

$$D_j\{\alpha, \beta, \gamma\}_{ls} = i^{s-l} \sum_k (-1)^k \frac{\sqrt{(j+l)!(j-l)!(j+s)!(j-s)!}}{k!(j-k-s)!(j+l-k)!(k+s-l)!} \times e^{-il\alpha - is\gamma} \left(\cos\frac{\beta}{2}\right)^{2j+l-2k-s} \left(\sin\frac{\beta}{2}\right)^{2k+s-l}. \tag{38}$$

Next, we transform to an equivalent representation, by using the matrix

$$X = \begin{Vmatrix} 0 & 0 & \cdots & 0 & 1 \\ 0 & 0 & \cdots & 1 & 0 \\ \cdot & \cdot & \cdots & \cdot & \cdot \\ 0 & 1 & \cdots & 0 & 0 \\ 1 & 0 & \cdots & 0 & 0 \end{Vmatrix},$$

i.e., we arrange the rows and columns in reverse order, which in this case corresponds to replacing l and s by $-l$ and $-s$, respectively. Then, instead of (38), we obtain new matrices with elements given by

$$D'_j\{\alpha, \beta, \gamma\}_{ls} = i^{l-s} \sum_k \frac{\sqrt{(j+l)!(j-l)!(j+s)!(j-s)!}}{k!(j-k+s)!(j-l-k)!(k-s+l)!} \times e^{il\alpha + is\gamma} \left(\cos\frac{\beta}{2}\right)^{2j-l-2k+s} \left(\sin\frac{\beta}{2}\right)^{2k-s+l}, \tag{39}$$

where, by the same argument as given in Sec. 74, we can omit the factor i^{l-s}.

We now study some special cases. For $j = 0$, we have the one-dimensional representation

$$\eta' = \eta,$$

i.e., the trivial identity representation. For $j = \frac{1}{2}$, we have $2j + 1 = 2$, and the variables $\eta_{-\frac{1}{2}}$, $\eta_{\frac{1}{2}}$ are just x_2 and x_1, respectively (cf. equation (23)); thus, in this case, apart from an interchange of rows and columns, the unitary group (18) is a representation of itself. The two-dimensional representation of the rotation group corresponding to $j = \frac{1}{2}$ is two-valued and consists of the matrices

$$D'_{\frac{1}{2}}\{\alpha, \beta, \gamma\} = \left\| \begin{matrix} e^{-\frac{1}{2}i(\gamma+\alpha)} \cos\dfrac{\beta}{2} & ie^{\frac{1}{2}i(\gamma-\alpha)} \sin\dfrac{\beta}{2} \\ ie^{-\frac{1}{2}i(\gamma-\alpha)} \sin\dfrac{\beta}{2} & e^{\frac{1}{2}i(\gamma+\alpha)} \cos\dfrac{\beta}{2} \end{matrix} \right\|,$$

taken with either sign. For $j = 1$, we have the three-dimensional representation

$$D'_1\{\alpha, \beta, \gamma\} = \left\| \begin{matrix} e^{-i(\gamma+\alpha)} \dfrac{1 + \cos\beta}{2} & -e^{-i\alpha} \dfrac{\sin\beta}{\sqrt{2}} & e^{i(\gamma+\alpha)} \dfrac{1 - \cos\beta}{2} \\ e^{-i\gamma} \dfrac{\sin\beta}{\sqrt{2}} & \cos\beta & -e^{i\gamma} \dfrac{\sin\beta}{\sqrt{2}} \\ e^{-i(\gamma-\alpha)} \dfrac{1 - \cos\beta}{2} & e^{i\alpha} \dfrac{\sin\beta}{\sqrt{2}} & e^{i(\gamma+\alpha)} \dfrac{1 + \cos\beta}{2} \end{matrix} \right\|.$$

The situation can be summarized as follows: For integral j, the matrices $D'_j\{\alpha, \beta, \gamma\}$ give a faithful representation of the rotation group; this is an immediate consequence of the fact that in this case, both a single matrix $D'_j\{\alpha, \beta, \gamma\}$ and a single rotation correspond to the two matrices of the group (18), which differ only in the signs of a and b. On the other hand, for half-integral j, the two matrices $\pm D'_j\{\alpha, \beta, \gamma\}$ both correspond to the same rotation $\{\alpha, \beta, \gamma\}$. In particular, the two matrices $\pm E$, where E is the unit matrix of order $2j + 1$, both correspond to the identity transformation of the rotation group. Thus, if we confine ourselves to transformations $D'_j\{\alpha, \beta, \gamma\}$ which are "sufficiently close" to the identity transformation E, we obtain a faithful representation of the rotation group. This corresponds to using only values of α, β, γ in the formula (39) which are "sufficiently close" to zero. However, if we add 2π to α or to γ, then since s and l are half integral, all the elements of the matrices $D'_j\{\alpha, \beta, \gamma\}$ change their signs, and we obtain a second representation of the same rotation. Later, we shall show that *to within an isomorphism, these are all the irreducible representations of the rotation group.*

Since the representations $D_j'\{\alpha, \beta, \gamma\}$ essentially give all irreducible representations of the rotation group, the matrix $D_1'\{\alpha, \beta, \gamma\}$ must be similar to the matrix $D\{\alpha, \beta, \gamma\}$ which describes the rotation $\{\alpha, \beta, \gamma\}$ with Eulerian angles α, β, γ. In Sec. 69, we saw that

$$D\{\alpha, \beta, \gamma\} = Z_\alpha X_\beta Z_\gamma.$$

Explicitly calculating this matrix product, we obtain

$$D\{\alpha, \beta, \gamma\} = \begin{Vmatrix} \cos\alpha\cos\gamma - \sin\alpha\cos\beta\sin\gamma & -\cos\alpha\sin\gamma - \sin\alpha\cos\beta\cos\gamma & \sin\alpha\sin\beta \\ \sin\alpha\cos\gamma + \cos\alpha\cos\beta\sin\gamma & -\sin\alpha\sin\gamma + \cos\alpha\cos\beta\cos\gamma & -\cos\alpha\cos\beta \\ \sin\beta\sin\gamma & \sin\beta\cos\gamma & \cos\beta \end{Vmatrix}.$$

It can easily be verified that

$$AD_1'\{\alpha, \beta, \gamma\}A^{-1} = D\{\alpha, \beta, \gamma\},$$

where

$$A = \begin{Vmatrix} 1 & 0 & 1 \\ i & 0 & -i \\ 0 & \sqrt{2}\,i & 0 \end{Vmatrix}.$$

76. Proof That the Rotation Group Is Simple. In this section, we prove that the rotation group is simple, i.e., that it has no nontrivial normal subgroups (Sec. 64). Let $\mathcal{G}$ be the group of (unitary) transformations of the form (18) with unit determinants. Then, according to Sec. 69, any nontrivial normal subgroup of the rotation group is the image of a normal subgroup of $\mathcal{G}$, which is different from the normal subgroup $\mathcal{H}$ consisting of the elements E and $-E$. Thus, it suffices to show that the group $\mathcal{G}$ has no nontrivial normal subgroups other than $\mathcal{H}$, i.e., if a normal subgroup $\mathcal{H}_1$ of $\mathcal{G}$ contains a matrix A different from E or $-E$, then $\mathcal{H}_1$ coincides with $\mathcal{G}$.

First, we note that if $\mathcal{H}_1$ contains a matrix B, then, by the definition of a normal subgroup, $\mathcal{H}_1$ contains all the matrices UBU^{-1}, where U is any matrix of the group $\mathcal{G}$. Moreover, by suitably choosing the matrix U, we can obtain any matrix of $\mathcal{G}$ which has the same eigenvalues as the matrix B. Therefore, to show that $\mathcal{H}_1$ coincides with $\mathcal{G}$, it suffices to show that $\mathcal{H}_1$ contains matrices with all admissible eigenvalues, i.e., all eigenvalues of the form $e^{i\omega}$ and $e^{-i\omega}$, where ω is a real number (since the matrices are unitary and have unit determinants). Since, by what has just been said, we can replace the matrix A by any matrix UAU^{-1}, it can be assumed that A is a diagonal matrix.

Thus, suppose that $\mathcal{H}_1$ contains the matrix $A = [e^{i\varphi}, e^{-i\varphi}]$, where φ is real and $e^{i\varphi} \neq 1$. Then we have $A^{-1} = [e^{-i\varphi}, e^{i\varphi}]$. Let

$$U = \begin{Vmatrix} x & y \\ -\bar{y} & \bar{x} \end{Vmatrix} \qquad (x\bar{x} + y\bar{y} = 1)$$

be an arbitrary matrix of $\mathcal{G}$, with inverse

$$U^{-1} = \begin{Vmatrix} \bar{x} & -y \\ \bar{y} & x \end{Vmatrix}.$$

Since $\mathcal{H}_1$ is a normal subgroup and contains A, it must also contain the matrix

$$Y = A(UA^{-1}U^{-1}).$$

Carrying out the multiplication and using the relation $x\bar{x} + y\bar{y} = 1$, we find that the trace of the matrix Y is

$$s = 2 - 4y\bar{y}\sin^2\varphi = 2 - 4\rho^2\sin^2\varphi,$$

where $\sin\varphi \neq 0$ and $\rho = |y|$ can take any value in the interval $0 \leq \rho \leq 1$. The eigenvalues $e^{i\alpha}$, $e^{-i\alpha}$ of the matrix Y are the roots of the equation (Sec. 27)

$$\lambda^2 - s\lambda + 1 = 0,$$

or

$$\lambda^2 + (4\rho^2\sin^2\varphi - 2)\lambda + 1 = 0.$$

Thus, as ρ varies from 0 to 1, the value of α ranges from 0 to 2φ. Therefore $\mathcal{H}_1$ contains all the matrices H_α for which $0 \leq \alpha \leq 2\varphi$ (since Y belongs to $\mathcal{H}_1$).

It is now easy to show that $\mathcal{H}_1$ contains any matrix U_β ($\beta > 0$). To prove this, choose a positive integer n such that the inequality

$$0 < \frac{\beta}{n} < 2\varphi$$

holds. Then $\mathcal{H}_1$ contains $U_{\beta/n}$ and therefore also contains

$$U_{\beta/n}^n = U_\beta,$$

i.e., $\mathcal{H}_1$ contains matrices with all admissible eigenvalues. Therefore, by the argument given above, $\mathcal{H}_1$ coincides with $\mathcal{G}$, and we have proved that *the rotation group is simple.* From this it follows at once that the rotation group cannot have homomorphic (unfaithful) representations. In fact, if such a homomorphic representation existed, then the transformations in the rotation group corresponding to the identity transformation in the representation group would be a normal subgroup (Sec. 66), and as just proved, no such normal subgroup can exist.

77. Laplace's Equation and Representations of the Rotation Group. We now examine the relation between differential equations and linear representations of groups, a relation which is the basis for the applications of representation theory to modern physics. We shall consider only the simplest case, i.e., Laplace's equation; although this case will lead to nothing new, it will serve to illustrate the general situation. We begin by proving some general results of great importance in the theory of group representation; we have already encountered special cases of these results in the examples given above.

Suppose we wish to construct a representation of a group $\mathcal{G}$, consisting of the n-dimensional linear transformations

$$x_k' = g_{k1}^{(\alpha)}x_1 + \cdots + g_{kn}^{(\alpha)}x_n \qquad (k = 1, 2, \ldots, n), \tag{40}$$

where the index α, specifying the element of $\mathcal{G}$, ranges over a finite or infinite set of values. Moreover, suppose that there exist m functions

$$\varphi_s(x_1, \ldots, x_n) \qquad (s = 1, 2, \ldots, m), \tag{41}$$

such that when the independent variables are changed in accordance with (40), the φ_s also undergo a linear transformation, of the form

$$\varphi_s(x_1', \ldots, x_n') = a_{s1}^{(\alpha)}\varphi_1(x_1, \ldots, x_n) + \cdots + a_{sm}^{(\alpha)}\varphi_m(x_1, \ldots, x_n) \qquad (s = 1, 2, \ldots, m). \tag{42}$$

Thus, corresponding to the transformation (40) of the group $\mathcal{G}$, we have a matrix A_α, with elements $A_{ik}^{(\alpha)}$. As an example, consider the construction given in Sec. 74 of the representations of the unitary group. In this case, the functions φ_s are defined by the formula (21a).

We now form the product $G_{\alpha_3} = G_{\alpha_2}G_{\alpha_1}$ of the two transformations G_{α_1}, taking $(x_1, \ldots, x_n)$ into $(x_1', \ldots, x_n')$, and G_{α_2}, taking $(x_1', \ldots, x_n')$ into $(x_1'', \ldots, x_n'')$. The corresponding transformations of the functions (41) are

$$\varphi_s(x_1', \ldots, x_n') = a_{s1}^{(\alpha_1)}\varphi_1(x_1, \ldots, x_n) + \cdots + a_{sm}^{(\alpha_1)}\varphi_m(x_1, \ldots, x_n) \tag{43a}$$

and

$$\varphi_s(x_1'', \ldots, x_n'') = a_{s1}^{(\alpha_2)}\varphi_1(x_1', \ldots, x_n') + \cdots + a_{sm}^{(\alpha_2)}\varphi_m(x_1, \ldots, x_n). \tag{43b}$$

Substituting (43a) into (43b), we find an expression for $\varphi_s(x_1'', \ldots, x_n'')$

in terms of $\varphi_s(x_1, \ldots, x_n)$, which defines a matrix A_{α_3} with elements

$$\{A_{\alpha_3}\}_{ik} = \sum_{s=1}^{m} a_{is}^{(\alpha_2)} a_{sk}^{(\alpha_1)},$$

i.e., $A_{\alpha_3} = A_{\alpha_2} A_{\alpha_1}$. Thus, the formulas (42) clearly define an m-dimensional representation of the group $\mathcal{G}$. In this derivation, we have assumed that the functions φ_s are linearly independent, which implies that the linear transformation (42) is uniquely defined and that $\det A_\alpha \neq 0$, for otherwise the $\varphi_s(x_1', \ldots, x_n')$ would be linearly dependent.

Now let $\mathcal{G}$ be the three-dimensional rotation group ($n = 3$), and suppose that the functions φ_s are orthonormal when integrated over a sphere K with center at the origin:

$$\iiint_K \varphi_p(x_1, x_2, x_3)\overline{\varphi_q(x_1, x_2, x_3)}\, dx_1\, dx_2\, dx_3 = \delta_{pq}, \tag{44}$$

where, as always, $\delta_{pq} = 0$ for $p \neq q$ and $\delta_{pp} = 1$. Then the representation (42) of $\mathcal{G}$ is unitary. To see this, we write (44) in the primed variables:

$$\iiint_K \varphi_p(x_1', x_2', x_3')\overline{\varphi_q(x_1', x_2', x_3')}\, dx_1'\, dx_2'\, dx_3' = \delta_{pq},$$

and then substitute (42), obtaining

$$\iiint_K \left[\sum_{i=1}^{m} a_{pi}^{(\alpha)}\varphi_i(x_1, x_2, x_3) \sum_{j=1}^{m} \overline{a_{qj}^{(\alpha)}\varphi_j(x_1, x_2, x_3)}\right] dx_1'\, dx_2'\, dx_3' = \delta_{pq}.$$

By the rule for changing variables in the triple integral, we can replace $dx_1'\, dx_2'\, dx_3'$ by $dx_1\, dx_2\, dx_3$ and integrate over the same sphere K, since any rotation G_α transforms K into itself and $\det G_\alpha = 1$. Then, using (44), we find that

$$\sum_{i=1}^{m} a_{pi}^{(\alpha)}\overline{a_{qi}^{(\alpha)}} = \delta_{pq} \qquad (p, q = 1, 2, \ldots, m),$$

i.e., every matrix A_α satisfies the row orthogonality conditions. Thus, the transpose of A_α satisfies the column orthogonality conditions, and hence also satisfies the row orthogonality conditions (see Sec. 28), i.e., the original matrix A_α satisfies both the row and column orthogonality conditions. It follows that A_α is in fact unitary for any value of α, as asserted.

We now consider Laplace's equation in two variables,

$$\frac{\partial^2 U}{\partial x^2} + \frac{\partial^2 U}{\partial y^2} = 0, \tag{45}$$

or in vector form,

$$\operatorname{div} \operatorname{grad} U = 0, \tag{46}$$

and we introduce the homogeneous polynomial

$$\phi_l(x, y) = a_0 x^l + a_1 x^{l-1} y + \cdots + a_k x^{l-k} y^k + \cdots + a_l y^l \tag{47}$$

of degree l in x and y. Then, *there exist two linearly independent polynomials of the form* (47) *which are solutions of* (45), *and every solution of* (45) *which is a homogeneous polynomial of degree* l *must be a linear combination (with constant coefficients) of these two polynomials.* The proof goes as follows: The coefficients of the polynomial (47) are given by

$$a_k = \frac{1}{(l-k)!k!} \frac{\partial^l \phi_l(x, y)}{\partial x^{l-k} \partial y^k}.$$

Moreover, if (47) satisfies the equation (45), we can replace two differentiations of ϕ_l with respect to y by two differentiations with respect to x and a change of sign. Thus, we obtain

$$a_k = \pm \frac{1}{(l-k)!k!} \frac{\partial^l \phi_l}{\partial x^l},$$

if k is even, and

$$a_k = \pm \frac{1}{(l-k)!k!} \frac{\partial^l \phi_l}{\partial x^{l-1} \partial y},$$

if k is odd, i.e., all the coefficients of the polynomial (47) can be expressed in terms of a_0 and a_1. This shows that there exist no more than two linearly independent homogeneous polynomials satisfying (45). To show that two such polynomials actually exist, we consider the homogeneous polynomial

$$\omega_l(x, y) = (x + iy)^l.$$

Expanding $(x + iy)^l$ and taking real and imaginary parts, we obtain

$$\omega_l(x, y) = \phi_l(x, y) + i\psi_l(x, y),$$

where the real homogeneous polynomials $\phi_l(x, y)$ and $\psi_l(x, y)$, of degree l, are linearly independent. Differentiating $\omega_l(x, y)$, we find that

$$\frac{\partial^2 \omega_l(x, y)}{\partial x^2} = l(l-1)(x + iy)^{l-2},$$

$$\frac{\partial^2 \omega_l(x, y)}{\partial y^2} = -l(l-1)(x + iy)^{l-2},$$

so that $\omega_l(x, y)$ satisfies the equation (45). The same is true of the real and imaginary parts of $\omega_l(x, y)$, i.e., the polynomials $\phi_l(x, y)$ and $\psi_l(x, y)$

give the required solutions of (45). In terms of the polar coordinates defined by

$$x = r\cos\varphi, \qquad y = r\sin\varphi,$$

we have

$$\omega_l(x, y) = r^l e^{il\varphi},$$

and the polynomials ϕ_l and ψ_l become simply

$$\phi_l(x, y) = r^l \cos l\varphi, \qquad \psi_l(x, y) = r^l \sin l\varphi.$$

It is easy to see that the equation (45) is invariant under any rotation

$$\begin{aligned} x' &= x\cos\vartheta - y\sin\vartheta, \\ y' &= x\sin\vartheta + y\cos\vartheta \end{aligned} \tag{48}$$

of the xy-plane through the angle ϑ about the origin; more precisely, (45) has exactly the same form in the new variables as in the old, i.e.,

$$\frac{\partial^2 U}{\partial x'^2} + \frac{\partial^2 U}{\partial y'^2} = 0. \tag{49}$$

This can be verified by using the formulas (48) and the rule for differentiation of composite functions, or alternatively, by observing that the left-hand side of equation (46) does not depend on the choice of axes, and hence has the same form for any choice of rectangular coordinates. Since the polynomials $\phi_l(x', y')$ and $\psi_l(x', y')$ must satisfy (49), they must also satisfy (45). It follows that $\phi_l(x', y')$ and $\psi_l(x', y')$ are linear combinations of $\phi_l(x, y)$ and $\psi_l(x, y)$; this leads to a representation of the two-dimensional rotation group (in the way explained above).

Instead of the polynomials ϕ_l and ψ_l, we can introduce two other polynomials which are linear combinations of ϕ_l and ψ_l, defined by

$$\phi_l'(x, y) = \phi_l(x, y) - i\psi_l(x, y), \qquad \psi_l'(x, y) = \phi_l(x, y) + i\psi_l(x, y),$$

or

$$\phi_l'(x, y) = (x - iy)^l = r^l e^{-il\varphi}, \qquad \psi_l'(x, y) = (x + iy)^l = r^l e^{il\varphi}.$$

These new polynomials lead to the transformation

$$\begin{aligned} \phi_l'(x', y') &= r^l e^{-il(\varphi+\vartheta)} = e^{-il\vartheta}\phi_l'(x, y), \\ \psi_l'(x', y') &= r^l e^{il(\varphi+\vartheta)} = e^{il\vartheta}\psi_l'(x, y), \end{aligned}$$

i.e., the matrix

$$\begin{Vmatrix} e^{-il\vartheta} & 0 \\ 0 & e^{il\vartheta} \end{Vmatrix} \tag{50}$$

represents the transformation (48), where the angle ϑ can take any value. It is obvious from (50) that this representation is in reduced form and gives two one-dimensional representations, characterized by the numbers $e^{-il\vartheta}$ and $e^{il\vartheta}$. Since, in the entire argument given above, l can take any integral value, we obtain the same representations of the two-dimensional rotation group as in Sec. 73.

Next, we consider Laplace's equation in three variables

$$\frac{\partial^2 U}{\partial x^2} + \frac{\partial^2 U}{\partial y^2} + \frac{\partial^2 U}{\partial z^2} = 0, \tag{51}$$

or

$$\text{div grad } U = 0.$$

We again introduce homogeneous polynomials of degree l, this time in three variables. Thus, we write

$$\phi_l(x, y, z) = a_0 z^l + X_1(x, y)z^{l-1} + \cdots + X_{l-1}(x, y)z + X_l(x, y), \tag{52}$$

where $X_k(x, y)$ is a homogeneous polynomial of degree k in x and y. Since each such polynomial $X_k(x, y)$ contains $k + 1$ arbitrary coefficients, the general homogeneous polynomial $\phi_l(x, y, z)$ contains

$$1 + 2 + 3 + \cdots + (l + 1) = \tfrac{1}{2}(l + 1)(l + 2)$$

arbitrary coefficients. Substituting (52) into (51), we find that the left-hand side is a homogeneous polynomial of degree $l - 2$, and equating its coefficients to zero, we obtain $\frac{1}{2}(l - 1)l$ homogeneous equations in the $\frac{1}{2}(l + 1)(l + 2)$ unknown coefficients of the polynomials $\phi_l(x, y, z)$. Since

$$\tfrac{1}{2}(l + 1)(l + 2) - \tfrac{1}{2}(l - 1)l = 2l + 1,$$

at least $2l + 1$ coefficients of the polynomials $\phi_l(x, y, z)$ are still arbitrary, i.e., there exist at least $2l + 1$ linearly independent homogeneous polynomials of degree l satisfying the equation (51). Using the same method as in the two-dimensional case, we can show that there are no more than $2l + 1$ such polynomials. Thus, there are exactly $2l + 1$ such polynomials, which we designate by

$$\psi_s^{(l)}(x, y, z) \qquad (s = 1, 2, \ldots, 2l + 1).$$

Then, since the equation (51) is invariant under any three-dimensional spatial rotation carrying the point (x, y, z) into (x', y', z'), the polynomials $\psi_s^{(l)}(x, y, z)$ generate a $(2l + 1)$-dimensional representation of the three-dimensional rotation group.

In a later volume, we shall give a detailed treatment of the theory of these so-called *spherical harmonics*, and we shall derive explicit formulas for them. We shall find that they can always be chosen to be orthonormal over any sphere with center at the origin. This means that the representation of the rotation group obtained by using these polynomials will be unitary. In fact, it can be shown that in this way we obtain a representation which is equivalent to the representation $D_l\{\alpha, \beta, \gamma\}$ constructed in Sec. 75.

78. The Direct Product of Two Matrices. Given two matrices

$$A = \begin{Vmatrix} a_{11} & a_{12} & \cdots & a_{1n} \\ a_{21} & a_{22} & \cdots & a_{2n} \\ \cdots & \cdots & \cdots & \cdots \\ a_{n1} & a_{n2} & \cdots & a_{nn} \end{Vmatrix}, \qquad B = \begin{Vmatrix} b_{11} & b_{12} & \cdots & b_{1m} \\ b_{21} & b_{22} & \cdots & b_{2m} \\ \cdots & \cdots & \cdots & \cdots \\ b_{m1} & b_{m2} & \cdots & b_{mm} \end{Vmatrix}, \tag{53}$$

the first of order n and the second of order m, we form a new matrix C with elements $c_{ij;kl}$ obtained by multiplying each element of A by each element of B in the following way:

$$\{C\}_{ij;kl} = c_{ij;kl} = a_{ik}b_{jl}. \tag{54}$$

Here, the pair of integers (i, j) acts as the first index, and the pair of integers (k, l) acts as the second index, where

$$i, k = 1, 2, \ldots, n,$$
$$j, l = 1, 2, \ldots, m.$$

Thus, in this special way of labeling the elements of C, every row and every column is labeled by a pair of integers, where the first number of each pair takes values from 1 to n, and the second number takes values from 1 to m. Of course, the rows and columns can also be labeled in the ordinary way, i.e., by single integers running from 1 to nm, where in this scheme a definite integer corresponds to each pair (i, j) or (k, l), and identical integers correspond to identical pairs. This labeling by integers can be done in various ways, and it does not matter which way it is done, since the transformation from one such scheme to another reduces to a simultaneous permutation of rows and columns, i.e., corresponds to going over to a similar matrix. The matrix C is called the *direct product* (or the *Kronecker product*) of the matrices A and B and is denoted by

$$C = A \times B. \tag{55}$$

This new kind of product is commutative.

As an example, let the two matrices (53) be second order matrices. Then, their direct product C is a fourth order matrix, which can be written as

$$C = \begin{Vmatrix} a_{11}b_{11} & a_{11}b_{12} & a_{12}b_{11} & a_{12}b_{12} \\ a_{11}b_{21} & a_{11}b_{22} & a_{12}b_{21} & a_{12}b_{22} \\ a_{21}b_{11} & a_{21}b_{12} & a_{22}b_{11} & a_{22}b_{12} \\ a_{21}b_{21} & a_{21}b_{22} & a_{22}b_{21} & a_{22}b_{22} \end{Vmatrix} = \begin{Vmatrix} c_{11;11} & c_{11;12} & c_{11;21} & c_{11;22} \\ c_{12;11} & c_{12;12} & c_{12;21} & c_{12;22} \\ c_{21;11} & c_{21;12} & c_{21;21} & c_{21;22} \\ c_{22;11} & c_{22;12} & c_{22;21} & c_{22;22} \end{Vmatrix},$$

or in other equivalent ways, by simultaneously permuting rows and columns.

Suppose that A and B are diagonal matrices

$$A = [\gamma_1, \ldots, \gamma_n], \qquad B = [\delta_1, \ldots, \delta_m].$$

Then $a_{ik} = 0$ and $b_{jl} = 0$ for $i \neq k$, $j \neq l$, and therefore, by (54), $c_{ij;kl}$ is nonzero only if the pair of numbers (i, j) coincides with the pair of numbers (k, l), i.e., only if C is also diagonal and its principal diagonal contains all possible products of the numbers γ_k with the numbers δ_l. If the γ_k and δ_l all equal 1, then C is also a unit matrix. Thus, we have proved the following theorem:

THEOREM 1. *The direct product of two diagonal matrices is a diagonal matrix, and the direct product of two unit matrices is a unit matrix.*

Next, we prove the following theorem:

THEOREM 2. *If $A^{(1)}$ and $A^{(2)}$ are two matrices of order n, and if $B^{(1)}$ and $B^{(2)}$ are two matrices of order m, then the formula*

$$(A^{(2)} \times B^{(2)})(A^{(1)} \times B^{(1)}) = A^{(2)}A^{(1)} \times B^{(2)}B^{(1)} \tag{56}$$

holds.

PROOF. As usual, the ordinary product of two matrices A and B of the same order is denoted by AB, and matrix elements are denoted by the corresponding small letters indexed by two subscripts. Then, by the definition of the direct product

$$\{A^{(t)} \times B^{(t)}\}_{ij;kl} = a_{ik}^{(t)} b_{jl}^{(t)} \qquad (t = 1, 2).$$

Using the rule for ordinary matrix multiplication, we find that the elements of the left-hand side of (56) are given by

$$d_{ij;kl} = \sum_{p=1}^{n} \sum_{q=1}^{m} a_{ip}^{(2)} b_{jq}^{(2)} a_{pk}^{(1)} b_{ql}^{(1)}. \tag{57}$$

However, since

$$\{A^{(2)}A^{(1)}\}_{ik} = \sum_{p=1}^{n} a_{ip}^{(2)} a_{pk}^{(1)}, \qquad \{B^{(2)}B^{(1)}\}_{jl} = \sum_{q=1}^{m} b_{jq}^{(2)} b_{ql}^{(1)},$$

we find, using the definition of the direct product, that the elements of the right-hand side of (56) are given by

$$d_{ij;kl} = \sum_{p=1}^{n} a_{ip}^{(2)} a_{pk}^{(1)} \sum_{q=1}^{m} b_{jq}^{(2)} b_{ql}^{(1)}.$$

Since this formula coincides with (57), the theorem is proved.

Finally, we prove another theorem on direct products:

THEOREM 3. *If the matrices A and B are unitary, their direct product* $C = A \times B$ *is also unitary.*

PROOF. By hypothesis, we have

$$\sum_{s=1}^{n} a_{sp}\bar{a}_{sq} = \delta_{pq}, \qquad \sum_{s=1}^{m} b_{sp}\bar{b}_{sq} = \delta_{pq}, \tag{58}$$

where $\delta_{pq} = 0$ if $p \neq q$ and $\delta_{pp} = 1$. We have to verify that C also satisfies the column orthonormality conditions, i.e., that

$$\sum_{i=1}^{n}\sum_{j=1}^{m} c_{ij;p_1q_1}\bar{c}_{ij;p_2q_2} = \delta_{p_1q_1;p_2q_2}, \tag{59}$$

where $\delta_{p_1q_1;p_2q_2} = 1$ if the pairs (p_1, q_1) and (p_2, q_2) are the same, and $\delta_{p_1q_1;p_2q_2} = 0$ if the pairs are different. By (54) we have

$$\begin{aligned}\sum_{i=1}^{n}\sum_{j=1}^{m} c_{ij;p_1q_1}\bar{c}_{ij;p_2q_2} &= \sum_{i=1}^{n}\sum_{j=1}^{m} a_{ip_1}b_{jq_1}\bar{a}_{ip_2}\bar{b}_{jq_2} \\ &= \sum_{i=1}^{n} a_{ip_1}\bar{a}_{ip_2}\sum_{j=1}^{m} b_{jq_1}\bar{b}_{jq_2}.\end{aligned} \tag{60}$$

If the two pairs (p_1, q_1) and (p_2, q_2) are different, then at least one of the factors in the right-hand side of (60) equals zero, while if the pairs are equal, both factors equal 1, by (58). Thus, the relation (59) is valid, QED.

If $A^{(1)} \times A^{(2)}$ is the direct product of two matrices, we can form the direct product of $A^{(1)} \times A^{(2)}$ with a third matrix $A^{(3)}$, thereby obtaining a direct product of three matrices, denoted by

$$C = A^{(1)} \times A^{(2)} \times A^{(3)}.$$

In our previous notation, the elements of this new matrix are

$$c_{ikl;i'k'l'} = a_{ii'}^{(1)}a_{kk'}^{(2)}a_{ll'}^{(3)}.$$

Similarly, we can define the direct product of any number of matrices, thereby obtaining a matrix whose order equals the product of the orders of the factors. The arrangement of the factors is immaterial, since this kind of multiplication is commutative.

79. The Direct Product of Two Representations of a Group. Let $\mathcal{G}$ be a group with elements G_α, and suppose that we have two linear representations

$$x_i' = a_{i1}^{(\alpha)}x_1 + \cdots + a_{in}^{(\alpha)}x_n \qquad (i = 1, 2, \ldots, n) \tag{61}$$

and

$$y_k' = b_{k1}^{(\alpha)}y_1 + \cdots + b_{km}^{(\alpha)}y_m \qquad (k = 1, 2, \ldots, m) \tag{62}$$

of $\mathcal{G}$, where the index α runs through a finite or infinite set of values. We denote the matrices of the transformations (61) and (62) by $A^{(\alpha)}$ and $B^{(\alpha)}$, respectively, and form the direct product

$$C^{(\alpha)} = A^{(\alpha)} \times B^{(\alpha)}. \tag{63}$$

The matrices $C^{(\alpha)}$ also form a representation of the group $\mathcal{G}$, i.e., if the matrix $C^{(\alpha)}$ corresponds to the element G_α of $\mathcal{G}$, then the matrix $C^{(\alpha_2)}C^{(\alpha_1)}$, which by (56) equals

$$C^{(\alpha_2)}C^{(\alpha_1)} = (A^{(\alpha_2)} \times B^{(\alpha_2)})(A^{(\alpha_1)} \times B^{(\alpha_1)}) = (A^{(\alpha_2)}A^{(\alpha_1)}) \times (B^{(\alpha_2)}B^{(\alpha_1)}), \tag{64}$$

corresponds to the product $G_{\alpha_2}G_{\alpha_1} = G_{\alpha_3}$. To see this, we note that

$$A^{(\alpha_2)}A^{(\alpha_1)} = A^{(\alpha_3)}, \qquad B^{(\alpha_2)}B^{(\alpha_1)} = B^{(\alpha_3)},$$

since the matrices $A^{(\alpha)}$ and $B^{(\alpha)}$ are representations of $\mathcal{G}$. Therefore, by (64),

$$C^{(\alpha_2)}C^{(\alpha_1)} = A^{(\alpha_3)} \times B^{(\alpha_3)},$$

or

$$C^{(\alpha_2)}C^{(\alpha_1)} = C^{(\alpha_3)},$$

by (63). Thus, the product of two elements G_α of $\mathcal{G}$ has as its image the product of the corresponding matrices $C^{(\alpha)}$, so that these matrices form a new representation of $\mathcal{G}$, as asserted. In this regard, we note that the direct product of the unit matrices of the representations $A^{(\alpha)}$, $B^{(\alpha)}$, i.e., the unit matrix of the representation $C^{(\alpha)}$, corresponds to the unit element of $\mathcal{G}$.

We now form nm products x_iy_k and subject each of the factors to the transformations (61) and (62). As a result, we obtain

$$x_i'y_k' = (a_{i1}^{(\alpha)}x_1 + \cdots + a_{in}^{(\alpha)}x_n)(b_{k1}^{(\alpha)}y_1 + \cdots + b_{km}^{(\alpha)}y_m),$$

so that

$$x_i'y_k' = \sum_{p=1}^{n}\sum_{q=1}^{m} c_{ik;pq}^{(\alpha)}x_py_q, \tag{65}$$

where $c_{ik;pq}^{(\alpha)} = a_{ip}^{(\alpha)}b_{kq}^{(\alpha)}$. *Thus, if the variables x_i and y_k are the components of vectors in the carrier spaces of the representations of $\mathcal{G}$ defined by the matrices $A^{(\alpha)}$ and $B^{(\alpha)}$, then the variables x_iy_k are the components of a vector in the carrier space of the representation of $\mathcal{G}$ defined by the matrices $C^{(\alpha)}$.* If the representations formed by the matrices $A^{(\alpha)}$ and $B^{(\alpha)}$ are irreducible, it does not follow that the representation formed by the matrices $C^{(\alpha)}$ is irreducible. In Sec. 81, we shall study in detail the case where the group $\mathcal{G}$ is the three-dimensional rotation group and the matrices $A^{(\alpha)}$

and $B^{(\alpha)}$ are different irreducible representations of $\mathcal{G}$ of the type constructed in Sec. 75. We shall show that the product

$$D_{j1}\{\alpha, \beta, \gamma\} \times D_{j2}\{\alpha, \beta, \gamma\}$$

is reducible, and we shall find the irreducible representations of which it is made up.

As an example, consider the Schrödinger equation for the case of two electrons in the field of a positively charged nucleus. This equation has the form

$$\left[-\frac{h^2}{8\pi^2 m}\sum_{s=1}^{2}\left(\frac{\partial^2}{\partial x_s^2}+\frac{\partial^2}{\partial y_s^2}+\frac{\partial^2}{\partial z_s^2}\right)+V\right]\psi = E\psi,$$

where

$$V = \sum_{s=1}^{2}\left(-\frac{ee_0}{\sqrt{x_s^2+y_s^2+z_s^2}}\right)+\frac{e^2}{\sqrt{(x_1-x_2)^2+(y_1-y_2)^2+(z_1-z_2)^2}}. \tag{66}$$

The constants h, m, e, and e_0 denote Planck's constant, the mass of the electron, the charge of the electron, and the charge of the nucleus, respectively. The first term in (66) corresponds to the attraction between the electrons and the nucleus, while the second term corresponds to the repulsive interaction between the two electrons. If, to a first approximation, we neglect this interaction, Schrödinger's equation becomes

$$(H_1 + H_2)\psi = E\psi, \tag{67}$$

where

$$H_s = -\frac{h^2}{8\pi^2 m}\left(\frac{\partial^2}{\partial x_s^2}+\frac{\partial^2}{\partial y_s^2}+\frac{\partial^2}{\partial z_s^2}\right)-\frac{ee_0}{\sqrt{x_s^2+y_s^2+z_s^2}} \qquad (s \doteq 1, 2).$$

Suppose that the individual equations $H_1\psi = E\psi$ and $H_2\psi = E\psi$ have eigenvalues E_1 and E_2, respectively, with corresponding eigenfunctions

$$\psi_1(x_1, y_1, z_1), \qquad \psi_2(x_2, y_2, z_2),$$

i.e., suppose that

$$H_1\psi_1 = E_1\psi_1, \qquad H_2\psi_2 = E_2\psi_2.$$

Then substituting the product

$$\psi = \psi_1(x_1, y_1, z_1)\psi_2(x_2, y_2, z_2)$$

into (67), we obtain, using (68),

$$(H_1 + H_2)\psi = \psi_2 H_1\psi_1 + \psi_1 H_2\psi_2 = (E_1 + E_2)\psi_1\psi_2 = (E_1 + E_2)\psi,$$

i.e., (67) has the eigenfunction $\psi_1\psi_2$, with corresponding eigenvalue $E_1 + E_2$. The equation $H_1\psi_1 = E_1\psi_1$ contains the Laplace operator and the distance of a point from the origin and is therefore invariant under rotations about the origin. It may happen that several eigenfunctions ψ_1 correspond to the same eigenvalue E_1 in the equation $H_1\psi_1 = E_1\psi_1$. Since all these functions are solutions of the equation $H_1\psi_1 = E_1\psi_1$, they generate a representation of the rotation group (just as the spherical harmonics of Sec. 77 led to a representation of the rotation group); we denote this representation by $D_{j1}\{\alpha, \beta, \gamma\}$. In exactly the same way, for a given eigenvalue E_2, the equation $H_2\psi_2 = E_2\psi_2$ leads to a representation $D_{j2}\{\alpha, \beta, \gamma\}$ of the rotation group. Then as we have just seen, the product $\psi_1\psi_2$ generates a representation of the rotation group given by the direct product $D_{j1} \times D_{j2}$. To discover the physical character of the corresponding eigenvalue $E_1 + E_2$, it is important to find the irreducible representations which make up the representation $D_{j1} \times D_{j2}$. This problem plays an essential role in the quantum-mechanical theory of perturbations.

80. The Direct Product of Two Groups and Its Representations. We now study another problem in which the concept of the direct product of matrices plays a role. Let $\mathcal{G}$ and $\mathcal{H}$ be two groups, whose elements are denoted by G_α and H_β, where α and β run independently through index sets, which are different in general. We define a new group $\mathcal{F}$, whose elements $F_{\alpha\beta}$ are defined as products of elements of $\mathcal{G}$ and $\mathcal{H}$, i.e.,

$$F_{\alpha\beta} = (G_\alpha, H_\beta),$$

where the first element of the pair is an element of $\mathcal{G}$ and the second is an element of $\mathcal{H}$. By the unit element of $\mathcal{F}$, we mean the pair for which G_α and H_β are the unit elements of $\mathcal{G}$ and $\mathcal{H}$; the inverse elements of $\mathcal{F}$ are defined similarly. We define multiplication of elements of $\mathcal{F}$ by the natural formula

$$F_{\alpha_2\beta_2}F_{\alpha_1\beta_1} = (G_{\alpha_2}G_{\alpha_1}, H_{\beta_2}H_{\beta_1}). \tag{68}$$

It is not hard to see that the set of elements $F_{\alpha\beta}$ actually does form a group. We call this group the *direct product of the groups* $\mathcal{G}$ *and* $\mathcal{H}$.

Let the matrices $A^{(\alpha)}$ form a representation of the group $\mathcal{G}$, and let the matrices $B^{(\beta)}$ form a representation of the group $\mathcal{H}$. By using the formula (56), as in Sec. 79, we can show that *the direct products*

$$C^{(\alpha,\beta)} = A^{(\alpha)} \times B^{(\beta)}$$

form a representation of the group $\mathcal{F}$. Moreover, if the representations $A^{(\alpha)}$ and $B^{(\beta)}$ are unitary, the representation $C^{(\alpha,\beta)}$ of $\mathcal{F}$ is also unitary (Sec. 78).

We now show that *if the representations $A^{(\alpha)}$ and $B^{(\beta)}$ are irreducible, then the representation $C^{(\alpha,\beta)}$ of the group $\mathfrak{F}$ is also irreducible.* The proof goes as follows: Let the matrices $A^{(\alpha)}$ be of order n and the matrices $B^{(\beta)}$ of order m; then the matrices $C^{(\alpha,\beta)}$ are of order nm. Suppose there exists a matrix X of order nm which commutes with all the matrices $C^{(\alpha,\beta)}$. Then, designating matrix elements by the appropriate small letters, for any indices i, j, p, q and any α, β, we have

$$\sum_{l=1}^{m} \sum_{k=1}^{n} x_{ij;kl} a_{kp}^{(\alpha)} b_{lq}^{(\beta)} = \sum_{l=1}^{m} \sum_{k=1}^{n} a_{ik}^{(\alpha)} b_{jl}^{(\beta)} x_{kl;pq}, \tag{69}$$

since
$$c_{kl;pq}^{(\alpha,\beta)} = a_{kp}^{(\alpha)} b_{lq}^{(\beta)}, \qquad c_{ij;kl}^{(\alpha,\beta)} = a_{ik}^{(\alpha)} b_{jl}^{(\beta)}.$$

If we assume that G_α is the unit element of the group $\mathcal{G}$, then $A^{(\alpha)}$ is the unit matrix, i.e., $a_{kp}^{(\alpha)} = 0$ for $k \neq p$ and $a_{pp}^{(\alpha)} = 1$, so that (69) gives

$$\sum_{l=1}^{m} x_{ij;pl} b_{lq}^{(\beta)} = \sum_{l=1}^{m} b_{jl}^{(\beta)} x_{il;pq}. \tag{70}$$

Similarly, if we assume that $B^{(\beta)}$ corresponds to the unit element of the group $\mathfrak{H}$, we find that

$$\sum_{k=1}^{n} x_{ij;kq} a_{kp}^{(\alpha)} = \sum_{k=1}^{n} a_{ik}^{(\alpha)} x_{kj;pq}. \tag{71}$$

If we take the $(nm)^2$ elements $x_{ij;kl}$ and fix the indices i and k, we obtain m^2 elements

$$x_{ij;kl} \qquad (j, l = 1, 2, \ldots, m),$$

which form a matrix of order m, which we denote by $X_1^{(i,k)}$. In just the same way, holding the indices j and l fixed in $x_{ij;kl}$, we obtain a matrix $X_2^{(j,l)}$ of order n. By (70), all the matrices $X_1^{(i,k)}$ commute with all the matrices $B^{(\beta)}$ forming an irreducible representation of the group $\mathfrak{H}$, and therefore, by Theorem 3 of Sec. 72, all the matrices $X_1^{(i,k)}$ are multiples of the unit matrix, i.e., for fixed i and k, $x_{ij;kl} = 0$ if $j \neq l$, and all the elements $x_{ij;kl}$ have the same value if $j = l$. This fact can be written as

$$x_{ij;kl} = x_{i1;k1} \delta_{jl}, \tag{72a}$$

where, as always, $\delta_{jl} = 0$ if $j \neq l$ and $\delta_{jj} = 1$. In just the same way, by (71), all the matrices $X_2^{(j,l)}$ commute with all the matrices $A^{(\alpha)}$ forming an irreducible representation of the group $\mathcal{G}$, and we find that

$$x_{ij;kl} = x_{1j;1l} \delta_{ik}. \tag{72b}$$

A comparison of (72a) and (72b) shows that $x_{ij;kl}$ is different from zero only if $i = k$ and $j = l$, in which case all the elements $x_{ij;ij}$ are the same, i.e., the matrix X which commutes with all the matrices $C^{(\alpha,\beta)}$ must be a multiple of the unit matrix. From this it follows immediately that the representation of the group $\mathfrak{F}$ defined by the direct product $A^{(\alpha)} \times B^{(\beta)}$ is irreducible. Moreover, it can be shown that *all representations of the group $\mathfrak{F}$ are obtained in this way.*

Now let $\mathcal{G}$ and $\mathcal{H}$ be two groups of linear transformations in the same number of variables, and suppose that any two matrices G_α and H_β commute:

$$G_\alpha H_\beta = H_\beta G_\alpha. \tag{73}$$

Whereas previously we defined an element $F_{\alpha\beta}$ of the group $\mathfrak{F}$ as a pair of elements (G_α, H_β) with the multiplication law given by (68), we can now regard the element $F_{\alpha\beta}$ of $\mathfrak{F}$ as simply the matrix product (73), which is independent of the order of the factors; this new group $\mathfrak{F}$ is isomorphic to the old group $\mathfrak{F}$.† If G_{α_0} and H_{β_0} are both the unit matrix, then the product $G_{\alpha_0}H_{\beta_0} = H_{\beta_0}G_{\alpha_0}$ is also the unit matrix. Moreover, the matrix $G_\alpha^{-1}H_\beta^{-1} = H_\beta^{-1}G_\alpha^{-1}$ is obviously the inverse of the matrix $G_\alpha H_\beta$, and by (73), we have the multiplication law

$$G_{\alpha_2}H_{\beta_2} \cdot G_{\alpha_1}H_{\beta_1} = (G_{\alpha_2}G_{\alpha_1})(H_{\beta_2}H_{\beta_1}).$$

Thus, all the properties which obtain when $F_{\alpha\beta}$ is defined as (G_α, H_β) are also satisfied when $F_{\alpha\beta}$ is defined as $G_\alpha H_\beta$, i.e., the products (73) actually form a group $\mathfrak{F}$, as asserted.

As an example, consider the case where $\mathcal{G}$ is the three-dimensional rotation group and $\mathcal{H}$ is the group of order 2, consisting of the identity transformation I and the operation S of reflection in the origin (Sec. 63). Then the condition (73) is satisfied, since clearly $G_\alpha S = SG_\alpha$ if G_α is any three-dimensional rotation. In this case, $\mathfrak{F}$ is the group of all real orthogonal transformations in three dimensions. As we saw in Sec. 73, the group $\mathcal{H}$ has two one-dimensional representations; one is the identity representation, in which the number $+1$ is associated with both I and S, and the other is the antisymmetric representation, in which $+1$ is associated with I and -1 is associated with S. Given any representation $D_j\{\alpha, \beta, \gamma\}$ of the rotation group, we can form the direct product of the matrices $D_j\{\alpha, \beta, \gamma\}$ with both representations of $\mathcal{H}$. In one case, we obtain a representation of the group $\mathfrak{F}$ of real orthogonal transformations in which the same matrix $D_j\{\alpha, \beta, \gamma\}$ corresponds to the pure rotation $\{\alpha, \beta, \gamma\}$, with Eulerian angles α, β, γ, and to the rotation $\{\alpha, \beta, \gamma\}$ plus reflection in the origin; we denote this representation of the group $\mathfrak{F}$ by $D_j^+\{\alpha, \beta, \gamma\}$. In the other case, the matrix $D_j\{\alpha, \beta, \gamma\}$ corresponds to the pure rotation $\{\alpha, \beta, \gamma\}$, while the matrix $-D_j\{\alpha, \beta, \gamma\}$ corresponds

† It is assumed here that $\mathcal{G}$ and $\mathcal{H}$ have only the identity transformation in common.

to the rotation $\{\alpha, \beta, \gamma\}$ plus reflection in the origin; we denote this representation of $\mathfrak{F}$ by $D_j^-\{\alpha, \beta, \gamma\}$.

As another example of the direct product of two groups, let (x_1, y_1, z_1) and (x_2, y_2, z_2) be two points, and again suppose that $\mathcal{G}$ is the three-dimensional rotation group. Then our variables undergo the linear transformations

$$\begin{aligned} x_k' &= g_{11}x_k + g_{12}y_k + g_{13}z_k, \\ y_k' &= g_{21}x_k + g_{22}y_k + g_{23}z_k, \\ z_k' &= g_{31}x_k + g_{32}y_k + g_{33}z_k, \qquad (k = 1, 2) \end{aligned}$$

where the matrix $\|g_{ik}\|$ corresponds to some rotation. Let $\mathcal{H}$ be the group consisting of the identity transformation and the transformation S which interchanges the indices 1 and 2 of the two points, i.e., let

$$S = \begin{pmatrix} 1 & 2 \\ 2 & 1 \end{pmatrix}.$$

Obviously, $S^2 = I$, so that $\mathcal{H}$ consists only of the two transformations I and S. Clearly, $G_\alpha S = SG_\alpha$ if G_α is any rotation, since it does not matter whether we renumber the points before or after the rotation. In this example, we obtain the same representations of the group $\mathfrak{F}$ of all real orthogonal transformations in three dimensions as in the preceding example. We can also consider the case of n points instead of just two points, choosing $\mathcal{H}$ to be the group of permutations of the indices of these points. Then $\mathcal{H}$ is a group of linear transformations in n variables, which is isomorphic to the group of permutations of n symbols. In this case, the operations of rotation and permutation of indices again commute, and we obtain a representation of the group $\mathfrak{F}$ by forming the direct product of a representation of the rotation group and a representation of the permutation group.

81. Reduction of the Direct Product $D_j \times D_{j'}$ of Two Representations of the Rotation Group. We return now to the example mentioned at the end of Sec. 79, i.e., the Schrödinger equation for two electrons, where the interaction between electrons is neglected. As already noted, in this case, the eigenfunctions of the Schrödinger equation generate a representation of the rotation group which is the direct product of two representations of the rotation group. In this section, we shall be concerned with the important problem of decomposing such a representation into irreducible components (cf. remarks at the end of Sec. 79).

Thus, given two irreducible representations $D_j\{\alpha, \beta, \gamma\}$ and $D_{j'}\{\alpha, \beta, \gamma\}$ of the rotation group, we form the direct product $D_j \times D_{j'}$, which is also a representation of the rotation group (Sec. 79). Our problem is to find the irreducible components of which $D_j \times D_{j'}$ consists. As we have seen (Sec. 74), the quantities

$$U_m = \frac{u_1^{j+m} u_2^{j-m}}{\sqrt{(j+m)!(j-m)!}} \qquad (m = -j, -j+1, \ldots, j-1, j) \quad (74)$$

are the components of a vector in the carrier space of the $(2j+1)$-dimensional representation D_j, and the corresponding quantities for the representation $D_{j'}$ are

$$V_{m'} = \frac{v_1^{j'+m'} v_2^{j'-m'}}{\sqrt{(j'+m')!(j'-m')!}} \qquad (m' = -j', -j'+1, \ldots, j'-1, j'), \quad (75)$$

where u_1, u_2 and v_1, v_2 undergo the same unitary transformation with determinant 1 (Sec. 74). Moreover, the $(2j+1)(2j'+1)$ quantities

$$W_{mm'} = U_m V_{m'} = \frac{u_1^{j+m} u_2^{j-m} v_1^{j'+m'} v_2^{j'-m'}}{\sqrt{(j+m)!(j-m)!(j'+m')!(j'-m')!}}$$
$$(m = -j, -j+1, \ldots, j-1, j;\ m' = -j', -j'+1, \ldots, j'-1, j') \quad (76)$$

are the components of a vector in the carrier space of the representation $D_j \times D_{j'}$ of the rotation group (cf. Sec. 79). Here, the indices j and j' can take either integral or half-integral values, i.e., strictly speaking, we regard D_j and $D_{j'}$ as irreducible representations of the two-dimensional unitary group with determinant 1.

Now let k be an integer (or half integer), obeying the inequality

$$|j - j'| \leq k \leq j + j'. \quad (77)$$

Then, *we can form* $2k + 1$ *linear combinations of the quantities* (76) *which generate the representation* D_k *of the rotation group.* To prove this, we first form the expression

$$L = (u_1 v_2 - u_2 v_1)^l (u_1 x_1 + u_2 x_2)^{2j-l} (v_1 x_1 + v_2 x_2)^{2j'-l}, \quad (78)$$

where l is a fixed integer obeying the inequalities

$$l \geq 0, \qquad l \leq 2j, \qquad l \leq 2j'. \quad (79)$$

If the variables u_1, u_2 and v_1, v_2 undergo the same linear transformation with determinant 1

$$u_1' = a_{11} u_1 + a_{12} u_2, \qquad v_1' = a_{11} v_1 + a_{12} v_2,$$
$$u_2' = a_{21} u_1 + a_{22} u_2, \qquad v_2' = a_{21} v_1 + a_{22} v_2,$$

(where $a_{11}a_{22} - a_{12}a_{21} = 1$), then the first factor in (78) remains unchanged, since

$$u_1'v_2' - u_2'v_1' = (a_{11}a_{22} - a_{12}a_{21})(u_1v_2 - u_2v_1).$$

The expression (78) is obviously a homogeneous polynomial in x_1 and x_2 of degree $2(j + j' - l)$ and therefore consists of terms of the form

$$a_s x_1^s x_2^{2(j+j'-l)-s} \qquad (s = 0, 1, \ldots, 2(j + j' - l)).$$

Introducing the notation

$$k = j + j' - l \tag{80}$$

and

$$y_{m''} = \frac{x_1^{k+m''}x_2^{k-m''}}{\sqrt{(k + m'')!(k - m'')!}} \qquad (m'' = -k, -k + 1, \ldots, k - 1, k), \tag{81}$$

we can write (78) in the form

$$L = \sum_{m''=-k}^{k} c_{m''}y_{m''}, \tag{82}$$

where the coefficients $c_{m''}$ depend on the variables u_1, u_2, v_1, v_2. It is an immediate consequence of (78) that $c_{m''}$ is a homogeneous polynomial of degree $2j$ in u_1, u_2 and a homogeneous polynomial of degree $2j'$ in v_1, v_2, i.e., $c_{m''}$ consists of terms of the form

$$b_{pq}u_1^p u_2^{2j-p} v_1^q v_2^{2j'-q}.$$

Then, in view of (74) and (75), $c_{m''}$ is a linear combination of the products $U_mV_{m'}$, of the form

$$c_{m''} = \sum_m \sum_{m'} d_{mm'}^{(m'')} U_m V_{m'} \qquad (m'' = -k, -k + 1, \ldots, k - 1, k), \tag{83}$$

where the coefficients $d_{mm'}^{(m'')}$ now no longer contain u_1, u_2, v_1, v_2. It should be noted that the variables u_1 and v_1 appear in the expression (78) either in conjunction with the variable x_1 or in the first factor of (78); in the latter case, the exponents of u_1 and v_1 always have the sum l. Thus, since $y_{m''}$ contains $x_1^{k+m''}$, we see that the sum of the exponents of u_1 and v_1 in (82) is $k + m'' + l$, which by (79) equals $j + j' + m''$. Since U_m contains u_1^{j+m} and $V_{m'}$ contains $v_1^{j'+m'}$, it follows at once that each of the expressions (83) contains only products $U_mV_{m'}$, for which $m + m' = m''$.

We now assert that the expressions (83) are the required $2k + 1$ linear combinations of the quantities (76), i.e., the linear combinations which generate the representation D_k of the rotation group. To prove this, we first recall the following fact (cf. Secs. 21 and 40): Given two linear transformations $\mathbf{x}' = T_1\mathbf{x}$ and $\mathbf{y}' = T_2\mathbf{y}$, between the vectors

$$\mathbf{x} = (x_1, x_2, \ldots, x_n), \qquad \mathbf{x}' = (x_1', x_2', \ldots, x_n'),$$
$$\mathbf{y} = (y_1, y_2, \ldots, y_n), \qquad \mathbf{y}' = (y_1', y_2', \ldots, y_n'),$$

a necessary and sufficient condition for the relation

$$x_1'y_1' + x_2'y_2' + \cdots + x_n'y_n' = x_1y_1 + x_2y_2 + \cdots + x_ny_n$$

to hold is that T_2 be *contragredient* to T_1, i.e., that $T_2 = (\tilde{T}_1)^{-1}$. Next, suppose that the variables u_1, u_2 and v_1, v_2 undergo the same unitary transformation A with determinant 1, and suppose that the variables x_1 and x_2 simultaneously undergo the transformation $(\tilde{A})^{-1}$ contragredient to A. Then, it follows that the sums

$$u_1x_1 + u_2x_2 \qquad \text{and} \qquad v_1x_1 + v_2x_2$$

remain unchanged. Moreover, as shown above, the first factor in the expression (78) for L, and hence the entire sum L, is invariant under these transformations. Thus, by (82), when the variables $y_{m''}$ undergo the transformation C induced by the transformation $(\tilde{A})^{-1}$ on the variables x_1, x_2, the variables $c_{m''}$ undergo the transformation B contragredient to C. Introducing the new variables

$$z_{m''} = \frac{u_1^{k+m''}u_2^{k-m''}}{\sqrt{(k+m'')!(k-m'')!}} \qquad (m'' = -k, -k+1, \ldots, k-1, k),$$

and applying the binomial theorem, we find that

$$(u_1x_1 + u_2x_2)^{2k} = (2k)! \sum_{m''=-k}^{k} z_{m''}y_{m''}. \tag{84}$$

The left-hand side of (84) is invariant under our transformations, so that the same is true of the right-hand side, i.e., the variables $z_{m''}$ undergo the same transformation B, contragredient to C, as the variables $c_{m''}$. But, as we know, the variables $z_{m''}$ generate the representation D_k of the rotation group, when the variables u_1, u_2 undergo unitary transformations with determinant 1. Thus, our assertion has finally been proved.

It will be recalled that the $(2j+1)(2j'+1)$ variables $U_mV_{m'}$ defined by (76), can be interpreted as the components of a vector in the carrier space of the representation $D_j \times D_{j'}$ of the rotation group. We have just shown that we can construct $2k+1$ linear combinations of the $U_mV_{m'}$ which generate the representation D_k of the rotation group. In view of the formula (80) and the inequalities (79), we see that the number k can take the following values:

$$k = j + j', j + j' - 1, \ldots, |j - j'| \tag{85}$$

(cf. (77)). Thus, the total number of linear combinations of the quantities $U_m V_{m'}$ given by the formula (83), equals the sum

$$(2j + 2j' + 1) + (2j + 2j' - 1) + \cdots + (2j - 2j' + 1), \tag{86}$$

where, for definiteness, we have assumed that $j \geq j'$; (86) is the sum of an arithmetic series with

$$\frac{(2j + 2j' + 1) - (2j - 2j' + 1)}{2} + 1 = 2j' + 1$$

terms and equals $(2j + 1)(2j' + 1)$, which is just the total number of quantities $U_m V_{m'}$ given by (76). (The same result is obtained if $j < j'$.) Writing $r = (2j + 1)(2j' + 1)$, we denote by

$$w_1, w_2, \ldots, w_r \tag{87}$$

the linear combinations (83), arranged in the order corresponding to the representations

$$D_{j+j'}, D_{j+j'-1}, \ldots, D_{|j-j'|}.$$

When the variables u_1, u_2 and v_1, v_2 undergo a unitary transformation with determinant 1, the variables $U_m V_{m'}$ take new values $U'_m V'_{m'}$, and correspondingly, we obtain new values w'_s of the variables w_s ($s = 1, 2, \ldots, r$). The w'_s are obtained from the w_s by using the linear transformation which has the quasi-diagonal matrix

$$[D_{j+j'}, D_{j+j'-1}, \ldots, D_{|j-j'|}], \tag{88}$$

where each D_k corresponds to the same unitary transformation of the variables u_1, u_2 and v_1, v_2. (Below, we shall show that the linear combinations (87) of the quantities $U_m V_{m'}$ are linearly independent.)

The result just obtained can be expressed as follows: Form the direct product $D_j \times D_{j'}$, which is the matrix of a linear transformation in the variables $U_m V_{m'}$, and let T be the matrix of the linear transformation expressing the variables w_s in terms of the variables $U_m V_{m'}$. Then, we have the formula

$$[D_{j+j'}, D_{j+j'-1}, \ldots, D_{|j-j'|}] = T(D_j \times D_{j'})T^{-1}, \tag{89}$$

which gives the decomposition of the direct product into its irreducible components. The formula (89) is often written symbolically as†

$$D_j \times D_{j'} = D_{j+j'} + D_{j+j'-1} + \cdots + D_{|j-j'|}.$$

† This result is known as the *Clebsch-Gordan formula.*

We recall that each D_k depends on a unitary transformation, i.e., more precisely, D_k stands for

$$D_k\begin{Bmatrix} a & b \\ -\bar{b} & \bar{a} \end{Bmatrix}.$$

This result can be generalized to the case of several factors. Thus, for example, we can write

$$\begin{aligned} D_1 \times D_1 \times D_1 &= (D_2 + D_1 + D_0) \times D_1 \\ &= D_3 + D_2 + D_1 + D_2 + D_1 + D_0 + D_1 \\ &= D_3 + 2D_2 + 3D_1 + D_0. \end{aligned} \tag{90}$$

Here, since D_1 is a matrix of order 3 (Sec. 74), $D_1 \times D_1$ is a matrix of order 9 and $D_1 \times D_1 \times D_1$ is a matrix of order 27. The formula (90) shows that the matrix $D_1 \times D_1 \times D_1$ is equivalent to the quasi-diagonal matrix

$$[D_3, D_2, D_2, D_1, D_1, D_1, D_0],$$

for any unitary transformation. The order of this matrix is

$$(2 \cdot 3 + 1) + 2(2 \cdot 2 + 1) + 3(2 \cdot 1 + 1) + (2 \cdot 0 + 1) = 27,$$

as required.

Finally, we prove the linear independence of the quantities w_s, which are linear forms in the variables $U_m V_{m'}$. The w_s are just the quantities denoted by $c_{m''}$ in the early part of this section; since in constructing the $c_{m''}$, we can choose various values of k, or equivalently, various values of l, it is more accurate to write $c_{m''}^{(l)}$. As already noted, the expression for each $c_{m''}^{(l)}$ contains only products $U_m V_{m'}$ for which $m + m' = m''$. This immediately implies that there is no possibility of linear dependence between the coefficients $c_{m''}^{(l)}$ unless the values of l are different and the values of m'' are the same. If we expand the last two terms in parentheses in the expression (78) and collect coefficients of $x_1^{k+m''}x_2^{k-m''}$, then, except for a constant factor, we obtain an expression for $c_{m''}^{(l)}$ in terms of u_1, u_2, v_1, v_2; this expression is the product of $(u_1v_2 - u_2v_1)^l$ with a certain polynomial in u_1, u_2, v_1, v_2, with positive integral coefficients. It is easy to see that expressions of this kind cannot be linearly dependent for different l. For example, suppose that we have the linear relation

$$\alpha_1 c_{m''}^{(l_1)} + \alpha_2 c_{m''}^{(l_2)} + \alpha_3 c_{m''}^{(l_3)} = 0,$$

where $l_1 < l_2 < l_3$, and the α_k are nonzero constants. This relation must hold identically for all u_1, u_2, v_1, v_2. In view of what has just been said about the $c_{m''}^{(l)}$, it is clear that if we set $u_2 = v_1 = v_2 = 1$ (for example), we obtain a relation of the form

$$\alpha_1(u_1 - 1)^{l_1}p_1(u_1) + \alpha_2(u_1 - 1)^{l_2}p_2(u_1) + \alpha_3(u_1 - 1)^{l_3}p_3(u_1) = 0,$$

where the $p_k(u_1)$ are polynomials in u_1 with positive integral coefficients. Dividing this relation by $(u_1 - 1)^{l_1}$ and then setting $u_1 = 1$, we obtain $\alpha_1 = 0$, which contradicts our assumption. In this way, we prove that linear dependence between the coefficients $c_{m''}^{(l)}$, and hence between the quantities w_s, is impossible. (Of course, by expanding (78), we can obtain explicit expressions for the w_s in terms of the variables (76).)

82. The Orthogonality Property. The matrices forming inequivalent, irreducible, unitary representations have a certain *orthogonality property*, which is often used in the applications of group theory to physics. This property is stated in the following theorem:

THEOREM. *Given a group $\mathcal{G}$ of order m with elements $G_1, G_2, \ldots, G_m$, let $A^{(1)}, A^{(2)}, \ldots, A^{(m)}$ and $B^{(1)}, B^{(2)}, \ldots, B^{(m)}$ be two inequivalent, irreducible, unitary representations of $\mathcal{G}$, with matrix elements denoted by the corresponding small letters with two subscripts. Then the relation*

$$\sum_{s=1}^{m} a_{ij}^{(s)} \overline{b_{kl}^{(s)}} = 0 \tag{91}$$

holds for any choice of the subscripts. A similar relation holds for a single irreducible representation; if $A^{(s)}$ is a p-dimensional, irreducible, unitary representation of $\mathcal{G}$, the relation

$$\sum_{s=1}^{m} a_{ij}^{(s)} \overline{a_{kl}^{(s)}} = \frac{m}{p}\, \delta_{ik}\delta_{jl} \tag{92}$$

holds, i.e., the sum on the left equals m/p if the pairs (i, j) and (k, l) are the same and vanishes if they are different.

To prove this theorem, we need a preliminary lemma. First, we remind the reader of the definition of multiplication for rectangular matrices (Sec. 7). Suppose that E is a matrix with n_1 rows and n_2 columns, and F is a matrix with n_2 rows and n_3 columns, i.e., that the number of columns of E equals the number of rows of F. Then the elements of the product EF are given by the formula

$$\{EF\}_{ik} = \sum_{s=1}^{n_2} \{E\}_{is}\{F\}_{sk};$$

obviously, the matrix EF has n_1 rows and n_3 columns.

We can now state the required lemma:

LEMMA. *If the matrices $A^{(s)}$ of order p and $B^{(s)}$ of order q form two inequivalent, irreducible, unitary representations of the group $\mathcal{G}$, and if the* *rectangular matrix C with p rows and q columns satisfies the relations*

$$A^{(s)}C = CB^{(s)} \qquad (s = 1, 2, \ldots, m) \tag{93}$$

for every s, then C is a zero matrix, i.e., all the elements of C vanish.

PROOF. We first consider the case $p = q$, where C is a square matrix. If C has a nonvanishing determinant, then C^{-1} exists, and (93) implies that

$$A^{(s)} = CB^{(s)}C^{-1},$$

i.e., our two representations are equivalent, contrary to hypothesis. Thus, the determinant of C must vanish. Suppose that some of the elements c_{ik} of C are nonvanishing. Then, as we know from Sec. 14, for arbitrary x_s, the linear forms

$$c_{i1}x_1 + c_{i2}x_2 + \cdots + c_{ip}x_p \qquad (i = 1, 2, \ldots, p)$$

define a subspace R of dimension equal to the rank of C, which in this case must be a subspace of dimension ≥ 1 and $<p$. If now we write (93) in the form of a linear transformation

$$A^{(s)}C\mathbf{x} = CB^{(s)}\mathbf{x}$$

on the vector $\mathbf{x} = (x_1, x_2, \ldots, x_p)$, we see that the vector $C\mathbf{x}$ on the left is an arbitrary vector in R, while the right-hand side is also a vector in R, being the result of applying the linear transformation C to the vector $B^{(s)}\mathbf{x}$. In other words, when applied to an arbitrary vector in R, the unitary matrix $A^{(s)}$ again gives a vector in R. But then, by Theorem 1 of Sec. 71, the matrices $A^{(s)}$ form a reducible representation of $\mathcal{G}$, contrary to hypothesis.

This proof is still valid in $p > q$. In this case, the rank of the matrix C is automatically less than p, and the linear forms

$$c_{i1}x_1 + c_{i2}x_2 + \cdots + c_{iq}x_q \qquad (i = 1, 2, \ldots, p)$$

define a subspace R (of the p-dimensional carrier space of the $A^{(s)}$) whose dimension is less than p; thus, the proof just given remains valid. Finally, if $p < q$, we take the transpose of (93), obtaining

$$B^{(s)*}C^* = C^*B^{(s)*}.$$

Here, the order q of the matrices $B^{(s)*}$ is greater than the order p of the matrices $A^{(s)*}$. It follows by the argument just given that the unitary matrices $B^{(s)*}$ transform a certain subspace into itself, and hence are reducible, i.e., the $B^{(s)*}$ can be simultaneously reduced to quasi-diagonal form by a suitable choice of basis vectors. Therefore, the same is true of the unitary matrices $B^{(s)}$, i.e., the $B^{(s)}$ form a reducible representation of $\mathcal{G}$, contrary to hypothesis. This completes the proof of the lemma.

It should be noted that the lemma is true even if the matrices $A^{(s)}$ and $B^{(s)}$ are not unitary. As we know from Theorem 1 of Sec. 72, we can always make $A^{(s)}$ and $B^{(s)}$ unitary by transforming to equivalent representations $T_1A^{(s)}T_1^{-1}$ and $T_2B^{(s)}T_2^{-1}$. As a result, the matrix C appearing in (93) is replaced by a new matrix C_1, related to C by the formula

$$C = T_1^{-1}C_1T_2.$$

Then, if C_1 is a zero matrix, so is C.

We can now prove the orthogonality theorem stated at the beginning of this section. The proof goes as follows: Instead of $A^{(s)}$ and $B^{(s)}$, we now write $A(G_s)$ and $B(G_s)$, where G_s is the element of the group $\mathcal{G}$, to which the matrices $A^{(s)}$ and $B^{(s)}$ are assigned. We introduce the matrix

$$C = \sum_{s=1}^{m} A(G_s)XB(G_s)^{-1} \tag{94}$$

and show that it satisfies the relation (93).

Let G_t be any fixed element of $\mathcal{G}$; then we have

$$A(G_t)C = \sum_{s=1}^{m} A(G_t)A(G_s)XB(G_s)^{-1}.$$

By the definition of a representation,

$$A(G_t)A(G_s) = A(G_tG_s), \qquad B(G_t)B(G_s) = B(G_tG_s),$$

so that

$$A(G_t)C = \sum_{s=1}^{m} A(G_tG_s)XB(G_tG_s)^{-1}B(G_t). \tag{95}$$

As the element G_s runs through the whole group, so does the product G_tG_s (Sec. 62); hence, (95) can be written in the form

$$A(G_t)C = CB(G_t),$$

i.e., the matrix C defined by (94) actually satisfies the relation (93). Therefore, by the lemma, C is a zero matrix. It follows that

$$\sum_{s=1}^{m} A(G_s)XB(G_s)^{-1} = 0 \tag{96}$$

for any choice of X.

Now suppose that a fixed element $\{X\}_{jl}$ of X equals 1, while all the other elements vanish. Then, (96) reduces to

$$\sum_{s=1}^{m} \{A(G_s)\}_{ij}\{B(G_s)^{-1}\}_{lk} = 0. \tag{97}$$

The matrix $B(G_s)$, being unitary, is obtained from $B(G_s)^{-1}$ by interchanging rows and columns and then taking the complex conjugate. Thus, finally, (97) becomes

$$\sum_{s=1}^{m} a_{ij}^{(s)}\overline{b_{kl}^{(s)}} = 0,$$

for any indices i, j, k, l; this is just the required formula (91).

In just the same way, by constructing the matrix

$$D = \sum_{s=1}^{m} A(G_s)XA(G_s)^{-1},$$

where X is any square matrix of order p, we can show that

$$A(G_s)D = DA(G_s) \qquad (s = 1, 2, \ldots, m).$$

Then, it follows from Theorem 3 of Sec. 74, that D is a multiple of the unit matrix I, i.e.,

$$\sum_{s=1}^{m} A(G_s)XA(G_s)^{-1} = cI,$$

where the number c depends on the choice of X. Again, suppose that $\{X\}_{jl} = 1$, while all the other elements of X vanish, and let c_j be the value of c corresponding to this matrix. Then we can write

$$\sum_{s=1}^{m} \{A(G_s)\}_{ij}\{A(G_s)^{-1}\}_{lk} = c_{jl}\delta_{ik}. \tag{98}$$

To determine c_{jl}, we set $i = k$ and sum over i from 1 to p, obtaining

$$pc_{jl} = \sum_{s=1}^{m}\sum_{i=1}^{p} \{A(G_s)^{-1}\}_{li}\{A(G_s)\}_{ij} = \sum_{s=1}^{m} \{I\}_{lj}. \tag{99}$$

The right-hand side of (99) equals m if $l = j$ and vanishes if $l \neq j$. Therefore

$$c_{jl} = \frac{m}{p}\,\delta_{jl},$$

and (98) can be written in the form

$$\sum_{s=1}^{m} \{A(G_s)\}_{ij}\{A(G_s)^{-1}\}_{lk} = \frac{m}{p}\,\delta_{ik}\delta_{jl}, \tag{100}$$

which agrees with (92), since the matrices $A(G_s)$ are unitary. This completes the proof of the orthogonality theorem.

It is not hard to see that the relation (91) holds for two inequivalent irreducible representations of $\mathcal{G}$, even if they are not unitary. To see this, let $A'(G_s)$, $B'(G_s)$ be two inequivalent, irreducible, *nonunitary* representations of $\mathcal{G}$ (of order p and q, respectively), and let $A(G_s)$, $B(G_s)$ be the equivalent unitary *representations*, i.e.,

$$A(G_s) = C_1 A'(G_s) C_1^{-1}, \qquad B(G_s) = C_2 B'(G_s) C_2^{-1},$$

where C_1 and C_2 are certain matrices which do not depend on s. Since $B(G_s)$ is unitary, we have

$$B(G_s)^{-1} = \overline{B(G_s)^*} = \overline{(C_2^{-1})^*}\,\overline{B'(G_s)^*}\,\overline{C_2^*},$$

and (96) can be written in the form

$$\sum_{s=1}^{m} C_1 A'(G_s) C_1^{-1} X \overline{(C_2^{-1})^*}\,\overline{B'(G_s)^*}\,\overline{C_2^*} = 0.$$

Multiplying from the left by C_1^{-1} and from the right by $\overline{(C_2^*)^{-1}}$, and introducing the matrix $Y = C_1^{-1} X \overline{(C_2^{-1})^*}$ (with p rows and q columns), we obtain

$$\sum_{s=1}^{m} A'(G_s) Y \overline{B'(G_s)^*} = 0.$$

Since X is arbitrary, so is Y, and we have as before

$$\sum_{s=1}^{m} a_{ij}^{(s)} \overline{b_{kl}^{(s)}} = 0,$$

for any i, j, k, l, where this time $a_{ij}^{(s)}$, $b_{kl}^{(s)}$ denote the elements of $A'(G_s)$, $B'(G_s)$. We also observe that the formula (100) holds for any irreducible representation $A(G_s)$ of $\mathcal{G}$, whether or not it is unitary. This follows from the proof of (100), in particular, from the fact that the use of Theorem 3 of Sec. 72 does not require that $A(G_s)$ be a unitary representation.

83. Characters. As in the preceding section, let $A(G_s)$ and $B(G_s)$ be two inequivalent irreducible representations (of order p and q, respectively) of a group $\mathcal{G}$ with elements $G_1, G_2, \ldots, G_m$. We denote by $\chi(G_s)$ and $\chi'(G_s)$ the traces of the matrices $A(G_s)$ and $B(G_s)$, i.e., the sums of their diagonal elements:

$$\chi(G_s) = \sum_{i=1}^{p} \{A(G_s)\}_{ii}, \qquad \chi'(G_s) = \sum_{k=1}^{q} \{B(G_s)\}_{kk}.$$

The numbers $\chi(G_s)$, $\chi'(G_s)$ are called the *characters* of the representations $A(G_s)$, $B(G_s)$. Since equivalent representations obviously have identical characters (Sec. 27), the orthogonality relation (91) gives

$$\sum_{s=1}^{m} \{A(G_s)\}_{ii} \overline{\{B(G_s)\}_{kk}} = 0.$$

Summing this formula over i and k, we obtain the following orthogonality relation for characters:

$$\sum_{s=1}^{m} \chi(G_s)\overline{\chi'(G_s)} = 0. \tag{101}$$

In just the same way, (98) gives

$$\sum_{s=1}^{m} \{A(G_s)\}_{ii}\overline{\{A(G_s)\}_{kk}} = \frac{m}{p}\,\delta_{ik},$$

and summing this formula over i and k, we obtain

$$\sum_{s=1}^{m} \chi(G_s)\overline{\chi(G_s)} = m. \tag{102}$$

Using (101) and (102), we now prove some theorems about characters.

THEOREM. 1. *A necessary and sufficient condition for two irreducible representations to be equivalent is that their characters be identical.*

PROOF. We have already remarked that equivalent representations, whether reducible or irreducible, have the same characters; this proves the necessity of the condition. To prove the sufficiency, we assume that two irreducible representations have the same characters, i.e., $\chi(G_s) = \chi'(G_s)$ for $s = 1, 2, \ldots, m$. Then the two representations are equivalent, since by (102) we have

$$\sum_{s=1}^{m} \chi(G_s)\overline{\chi'(G_s)} = m,$$

whereas if the representations were inequivalent, we would have (101), QED. (Obviously, the two equivalent representations consist of matrices of the same order.)

Next, corresponding to each irreducible representation of $\mathcal{G}$, we introduce the vector with components

$$\frac{1}{\sqrt{m}}\,\chi(G_1), \frac{1}{\sqrt{m}}\,\chi(G_2), \ldots, \frac{1}{\sqrt{m}}\,\chi(G_m) \tag{103}$$

in the complex m-dimensional space R_m. By (102), the vector (103) is normalized, and by (101), the vectors of the form (103) corresponding to inequivalent representations are orthogonal. It follows that *there cannot exist more than m inequivalent irreducible representations of a group* $\mathcal{G}$ *of order m*. Later (Theorem 7 of Sec. 84), we shall find the *exact* number of inequivalent irreducible representations of a group, a number which, in the meantime, we denote by l. Thus, let $\omega^{(i)}$ $(i = 1, 2, \ldots, l)$ denote the inequivalent irreducible representations of a group $\mathcal{G}$, and let

$$\chi^{(i)}(G_1), \chi^{(i)}(G_2), \ldots, \chi^{(i)}(G_m) \qquad (i = 1, 2, \ldots, l)$$

denote the corresponding characters. Let ω be any other (reducible) representation of $\mathcal{G}$, with characters

$$\chi(G_1), \chi(G_2), \ldots, \chi(G_m). \tag{104}$$

Then, after being reduced, ω consists of quasi-diagonal matrices made up of the matrices in the representations $\omega^{(l)}$. Therefore, the characters (104) satisfy the relation

$$\chi(G_s) = \sum_{i=1}^{l} a_i\chi^{(i)}(G_s), \tag{105}$$

where the nonnegative integer a_i $(i = 1, 2, \ldots, l)$ gives the number of times the representation $\omega^{(i)}$ appears in the representation ω after the reduction. To derive a formula for the coefficient a_i in terms of the characters of the representation ω, we multiply both sides of (105) by $\chi^{(i)}(G_s)$ and sum over s. Then, using (101) and (102), we obtain

$$\sum_{s=1}^{m} \chi(G_s)\overline{\chi^{(i)}(G_s)} = a_i m,$$

so that

$$a_i = \frac{1}{m}\sum_{s=1}^{m} \chi(G_s)\overline{\chi^{(i)}(G_s)} \qquad (i = 1, 2, \ldots, l). \tag{106}$$

This formula gives a unique value for every a_i. Thus, we have proved the following theorem:

THEOREM 2. *Every reducible representation decomposes into a unique set of irreducible representations.*

Using (106), we can easily generalize Theorem 1 to the case of arbitrary representations, i.e., representations which are not necessarily irreducible:

THEOREM 3. *A necessary and sufficient condition for two representations to be equivalent is that their characters be identical.*

PROOF. The necessity of the condition has already been noted in the proof of Theorem 1. To prove the sufficiency, suppose that two representations have identical characters $\chi(G_s)$. Then, according to (106), we obtain the same values for the numbers a_k, and hence, both representations reduce to sets of quasi-diagonal matrices made up of the same irreducible representations. Going over to equivalent representations if necessary, we can assume that these irreducible representations appear in the same order in the two sets of quasi-diagonal matrices. (By Theorem 2 of Sec. 71, applying the same permutation to both the rows and columns of a matrix amounts to transforming to an equivalent representation.) Thus, representations with identical characters can be reduced to identical quasi-diagonal matrices, i.e., the representations are equivalent, QED.

We now turn to the problem of determining l, the total number of inequivalent irreducible representations of the group $\mathcal{G}$. Suppose that $\mathcal{G}$ decomposes into r different classes $C_1, C_2, \ldots, C_r$† (Sec. 64), where each class consists of the elements obtained from any element G_t of the class by using the formula

$$G_s G_t G_s^{-1} \qquad (s = 1, 2, \ldots, m).$$

Clearly, every representation of $\mathcal{G}$ assigns similar matrices, with identical traces, to the elements of a given class. It follows that every representation has no more than r different characters, where each character corresponds not just to one element but to all the elements of a given class. Thus, the terms in the sum (101) are the same for all the elements of a given class; let $\chi(C_k)$, $\chi'(C_k)$ denote the characters corresponding to the elements of the class C_k in two inequivalent irreducible representations. Then, if the classes $C_1, C_2, \ldots, C_r$ contain $g_1, g_2, \ldots, g_r$ elements, respectively, (101) becomes

$$\sum_{k=1}^{r} \chi(C_k)\overline{\chi'(C_k)}g_k = 0,$$

and (102) becomes

$$\sum_{k=1}^{r} \chi(C_k)\overline{\chi(C_k)}g_k = m.$$

Thus, the characters $\chi^{(i)}(C_k)$ of the inequivalent irreducible representa-

† Note that in the case of classes, we depart from our previous notation and use roman letters $C_1, C_2, \ldots$ to denote *sets* of group elements.

tions $\omega^{(i)}$, $(i = 1, 2, \ldots, l)$, obey the formulas

$$\begin{aligned} \sum_{k=1}^{r} \chi^{(i)}(C_k)\overline{\chi^{(j)}(C_k)}g_k &= 0 \qquad \text{for } i \neq j, \\ \sum_{k=1}^{r} \chi^{(i)}(C_k)\overline{\chi^{(i)}(C_k)}g_k &= m. \end{aligned} \tag{107}$$

If we introduce l vectors, with components

$$\sqrt{\frac{g_1}{m}}\,\chi^{(i)}(C_1),\ \sqrt{\frac{g_2}{m}}\,\chi^{(i)}(C_2),\ \ldots,\ \sqrt{\frac{g_r}{m}}\,\chi^{(i)}(C_r) \qquad (i = 1, 2, \ldots, l)$$

in the r-dimensional space R_r, then (107) shows that these l vectors are orthonormal, and hence linearly independent. It follows that the number of these vectors cannot exceed the dimension of the space R_r, i.e., $l \leq r$, and we have proved the following theorem:

THEOREM 4. *The number of inequivalent irreducible representations of a group $\mathcal{G}$ does not exceed the number of classes of $\mathcal{G}$.*

In the next section, we shall show that l actually equals r. To prove this, it is sufficient to prove that $l \geq r$, since we have just shown that $l \leq r$. In proving that $l \geq r$, we shall introduce some new concepts and some new relations between characters, which are of independent interest.

We conclude this section by deriving another relation between the characters of any irreducible representation. If the class C_k consists of the elements

$$G_1^{(k)},\ G_2^{(k)},\ \ldots,\ G_{g_k}^{(k)},$$

and if G_s is any element of the group $\mathcal{G}$, then the elements

$$G_s G_i^{(k)} G_s^{-1} \qquad (i = 1, 2, \ldots, g_k)$$

are again just the elements of C_k, but generally in a different order. It follows that the set denoted by C_pC_q, consisting of all products

$$G_u^{(p)}G_v^{(q)} \qquad (u = 1, 2, \ldots, g_p;\ v = 1, 2, \ldots, g_q)$$

of an element from C_p with an element from C_q, is the same as the set of all elements of the form

$$G_sG_u^{(p)}G_v^{(q)}G_s^{-1} = (G_sG_u^{(p)}G_s^{-1})(G_sG_v^{(q)}G_s^{-1}).$$

Thus, the set C_pC_q has the following property: If an element G belongs to C_pC_q, then the whole class containing G also belongs to C_pC_q, and every element of this class appears in C_pC_q the same number of times.

Now let the nonnegative integer a_{pqk} be the number of times that the elements of the class C_k appear in the set C_pC_q. Then, in a purely formal way, we can write†

$$C_pC_q = \sum_{k=1}^{r} a_{pqk}C_k, \tag{108}$$

or more explicitly,

$$(G_1^{(p)} + G_2^{(p)} + \cdots + G_{g_p}^{(p)})(G_1^{(q)} + G_2^{(q)} + \cdots + G_{g_q}^{(q)}) = \sum_{k=1}^{r} a_{pqk}(G_1^{(k)} + G_2^{(k)} + \cdots + G_{g_k}^{(k)}). \tag{109}$$

Let $A(G_s)$ be the matrices of some n-dimensional irreducible representation of the group $\mathcal{G}$ and form the sum $A(C_k)$ of the matrices corresponding to the class C_k:

$$A(C_k) = \sum_{j=1}^{g_k} A(G_j^{(k)}).$$

Since for any $i = 1, 2, \ldots, g_k$ and any G_s in $\mathcal{G}$, the elements $G_sG_i^{(k)}G_s^{-1}$ are again just the elements in the class C_k, we see that the matrix $A(C_k)$ commutes with all the matrices $A(G_s)$. It follows by Theorem 3 of Sec. 74 that $A(C_k)$ is a multiple of the unit matrix; i.e.,

$$A(C_k) = b_kI \qquad (k = 1, 2, \ldots, r), \tag{110}$$

where the b_k are certain constants. In view of the symbolic formula (108) or (109), defining the numbers a_{pqk}, we obtain the following relation between the numbers b_k:

$$b_pb_q = \sum_{k=1}^{r} a_{pqk}b_k. \tag{111}$$

The trace of the matrix $A(C_k)$ equals the sum of the traces of the matrices $A(G_i^{(k)})$, $(i = 1, 2, \ldots, g_k)$, i.e., equals $g_k\chi(C_k)$. On the other hand, it follows from (110) that the trace of $A(C_k)$ equals nb_k. Therefore, we have

$$nb_k = g_k\chi(C_k),$$

or

$$b_k = \frac{g_k}{n}\chi(C_k)$$

and (111) leads to the following theorem:

THEOREM 5. *The characters of any irreducible representation consisting of matrices of order n, satisfy the relation*

$$g_p\chi(C_p)g_q\chi(C_q) = n\sum_{k=1}^{r} a_{pqk}g_k\chi(C_k). \tag{112}$$

† The numbers a_{pqk} are called the *class multiplication coefficients* of the group $\mathcal{G}$.

We note that one of the classes C_k is the class consisting only of the unit element E of the group $\mathcal{G}$; in any representation, the image of E is the unit matrix, with trace equal to its order n. We shall always denote this class by C_1, so that $\chi(C_1) = n$; thus, (112) can be rewritten as

$$g_p\chi(C_p)g_q\chi(C_q) = \chi(C_1) \sum_{k=1}^{r} a_{pqk}g_k\chi(C_k). \tag{113}$$

We now find the values of the constants a_{pq1}. Given a class C_p, the set $C_{p'}$ consisting of the inverses of the elements of C_p is also a class; this follows at once from the definition of a class and the fact that $G_sG_tG_s^{-1} = G_u$ implies $G_sG_t^{-1}G_s^{-1} = G_u^{-1}$. The class $C_{p'}$ may coincide with C_p, i.e., it may happen that $p' = p$. In any event, C_p and $C_{p'}$ contain the same number of elements, so that $g_p = g_{p'}$. Thus, if we set $q = p'$ in (108), the class C_1 appears g_p times in the right-hand side, while if $q \neq p'$, the right-hand side does not contain C_1, i.e.,

$$a_{pq1} = \begin{cases} 0 & \text{if } q \neq p', \\ g_p & \text{if } q = p'. \end{cases} \tag{114}$$

84. The Regular Representation of a Group. Any permutation group can be regarded as a group of linear transformations. For example, the permutation

$$\begin{pmatrix} 1 & 2 & 3 & 4 \\ 2 & 4 & 3 & 1 \end{pmatrix}$$

can be written in the form of a linear transformation taking x_1 into y_2, x_2 into y_4, x_3 into y_3, and x_4 into y_1:

$$\begin{aligned} y_1 &= 0x_1 + 0x_2 + 0x_3 + 1x_4, \\ y_2 &= 1x_1 + 0x_2 + 0x_3 + 0x_4, \\ y_3 &= 0x_1 + 0x_2 + 1x_3 + 0x_4, \\ y_4 &= 0x_1 + 1x_2 + 0x_3 + 0x_4. \end{aligned}$$

We now consider the following way of representing a finite group $\mathcal{G}$ by a permutation group (cf. Example 3 of Sec. 67). Multiply the elements $G_1, G_2, \ldots, G_m$ of $\mathcal{G}$ from the right by an element G_s. This causes a permutation of the elements and thereby leads, in the way just indicated, to a matrix P_s, which we regard as the image of G_s. This representation is called the *regular representation* of $\mathcal{G}$. One of the elements G_s is the unit element of $\mathcal{G}$, which we denote by E. The matrix corresponding to E in the regular representation is the unit matrix of order m; since the trace of this matrix is m, we have $\chi(E) = m$. However, if we multiply the elements $G_1, G_2, \ldots, G_m$ by any element $G_s \neq E$, none of the elements remains in place, i.e., the corresponding matrix has all its diagonal elements equal to zero. Thus, in the regular representation, we have $\chi(G_s) = 0$ if $G_s \neq E$.

Suppose now that when the regular representation is in reduced form, it contains each of the irreducible representations $\omega^{(k)}$ (discussed in Sec. 83) h_k times. Then, according to what has just been said, we have

$$\sum_{t=1}^{l} h_t \chi^{(t)}(G_s) = \begin{cases} 0 & \text{for } G_s \neq E, \\ m & \text{for } G_s = E. \end{cases} \tag{115}$$

Multiplying both sides of (115) by $\overline{\chi^{(k)}(G_s)}$, summing over s, and using (101) and (102), we obtain

$$h_k m = m\overline{\chi^{(k)}(E)}. \tag{116}$$

Let n_k be the dimension of the representation $\omega^{(k)}$. Then

$$\overline{\chi^{(k)}(E)} = \chi^{(k)}(E) = n_k,$$

and (116) gives $h_k = n_k$. This allows us to write (115) in the form

$$\sum_{t=1}^{l} \chi^{(t)}(E)\chi^{(t)}(G_s) = \sum_{t=1}^{l} n_t \chi^{(t)}(G_s) = \begin{cases} 0 & \text{for } G_s \neq E, \\ m & \text{for } G_s = E. \end{cases} \tag{117}$$

Thus, we arrive at the following theorem:

THEOREM 6. *The regular representation contains each irreducible representation $\omega^{(k)}$ a number of times equal to the order of the matrices in the representation $\omega^{(k)}$. Moreover, the characters of the representations $\omega^{(k)}$ satisfy the relation* (117).

Next, we write the formula (113) for the case of the representation $\omega^{(t)}$, i.e.,

$$g_p \chi^{(t)}(C_p) g_q \chi^{(t)}(C_q) = \chi^{(t)}(C_1) \sum_{k=1}^{r} a_{pqk} g_k \chi^{(t)}(C_k),$$

and then sum over t from 1 to l:

$$g_p g_q \sum_{t=1}^{l} \chi^{(t)}(C_p)\chi^{(t)}(C_q) = \sum_{k=1}^{r} a_{pqk} g_k \sum_{t=1}^{l} \chi^{(t)}(C_1)\chi^{(t)}(C_k).$$

Using (117), we obtain

$$g_p g_q \sum_{t=1}^{l} \chi^{(t)}(C_p)\chi^{(t)}(C_q) = a_{pq1} m,$$

so that, by (114),

$$\sum_{t=1}^{l} \chi^{(t)}(C_p)\chi^{(t)}(C_q) = \begin{cases} 0 & \text{for } q \neq p', \\ \dfrac{m}{g_{p'}} & \text{for } q = p'. \end{cases}$$

Now, the system of l homogeneous linear equations in the variables $x_1, x_2, \ldots, x_r$

$$\sum_{q=1}^{r} x_q \chi^{(k)}(C_q) = 0 \qquad (k = 1, 2, \ldots, l), \tag{118}$$

has only the trivial solution $x_1 = x_2 = \cdots = x_r = 0$, since by multiplying both sides of (118) by $\chi^{(k)}(C_p)$ and summing over k from 1 to l, we obtain $x_{p'} = 0$, where p' is any of the numbers $1, 2, \ldots, r$. Since the system (118) has only the trivial solution, the number of equations in the system cannot be less than the number of unknowns (Sec. 10), i.e., $l \geq r$. Therefore, since $l \leq r$ (as proved in the preceding section), we have $l = r$, i.e., the following theorem holds:

THEOREM 7. *The total number of inequivalent irreducible representations of a finite group $\mathcal{G}$ equals the number of classes of $\mathcal{G}$.*

Finally, we note an interesting consequence of Theorem 6. The regular representation of $\mathcal{G}$ consists of matrices of order m. On the other hand, according to Theorem 6, each representation $\omega^{(k)}$, consisting of matrices of order n_k, appears n_k times in the regular representation. This implies that

$$\sum_{k=1}^{r} n_k^2 = m, \tag{119}$$

i.e., we have the following theorem:

THEOREM 8. *The sum of the squares of the dimensions of the inequivalent irreducible representations $\omega^{(k)}$ equals the order of the group $\mathcal{G}$.*

85. Examples of Representations of Finite Groups

1. Consider an Abelian group $\mathcal{G}$ consisting of the elements

$$A_2^k A_1^i \qquad (k = 0, 1, 2, \ldots, n-1; i = 0, 1, 2, \ldots, m-1),$$

where (*a*) A_1 and A_2 commute, (*b*) $A_1^m = E$, $A_2^n = E$, and (*c*) $A_2^0 A_1^0 = E$ by definition. Every element of $\mathcal{G}$ forms a class by itself, and all the irreducible representations of $\mathcal{G}$ are one-dimensional. Let α be any mth root of unity, and let β be any nth root of unity. Then, if we assign the number $\beta^k \alpha^i$ to the element $A_2^k A_1^i$, it is easy to see that we obtain a representation of $\mathcal{G}$. There are in all mn different one-dimensional representations of $\mathcal{G}$, obtained by giving various values to α and β (the number of different nth roots of unity is n). Since the total number of classes (i.e., elements) is also mn, this gives all the inequivalent irreducible representations of $\mathcal{G}$. In the case where the number of *generators* (i.e., elements A_s) is greater than 2, the representations are constructed in a similar way.

2. Consider the dihedral group corresponding to a regular n-gon; this group consists of the $2n$ elements

$$E,\ A^i,\ T,\ TA^i \qquad (i = 1, 2, \ldots, n-1),$$

where $$A^n = E, \qquad T^2 = E, \qquad T^{-1} = T, \qquad TAT^{-1} = A^{-1} \tag{120}$$

(cf. Sec. 59). The relations (120) are immediate consequences of the geometric meaning of the rotations A and T. It follows from the last of (120) that $TA^iT^{-1} = A^{-i}$.

First suppose that n is an odd number, i.e., $n = 2m + 1$. Then the group consists of $m + 2$ classes. One of these classes consists of E only, each of m classes consists of a pair of elements A^s and A^{-s} $(s = 1, 2, \ldots, m)$, where $A^s = A^{2m+1-s}$, and one class consists of the element T and all the elements TA^i. All this is easily verified by using the relations (120). In this case, there are two one-dimensional representations, one assigning the number 1 to both A and T and the other assigning 1 to A and -1 to T. Now let

$$\epsilon = \cos\frac{2\pi}{n} + i\sin\frac{2\pi}{n}.$$

Then, we can construct m two-dimensional representations in which the following matrices correspond to the elements A and T:

$$A \to \begin{Vmatrix} \epsilon^s & 0 \\ 0 & \epsilon^{-s} \end{Vmatrix}, \qquad T \to \begin{Vmatrix} 0 & 1 \\ 1 & 0 \end{Vmatrix} \qquad (s = 1, 2, \ldots, m). \tag{121}$$

These matrices satisfy the relations (120) and therefore lead to a representation of the group, since every relation involving the elements A and T is a consequence of (120). Moreover, each of these m representations is irreducible, for otherwise each representation would reduce to two one-dimensional representations; then, the matrices A and T would have to commute, but it is easily verified that A and T do not commute for any s. Finally, the inequivalence of the representations (121) follows from the fact that the matrix corresponding to the element A has the eigenvalues ϵ^s and ϵ^{-s}, and these eigenvalues are different for different s. Thus, we have constructed all $m + 2$ inequivalent irreducible representations. In this case, the formula (119) reduces to

$$2 \cdot 1^2 + m \cdot 2^2 = 4m + 2 = 2n.$$

Next suppose that n is an even number, i.e., $n = 2m$. Then the representation (121) corresponding to the value $s = m$ is

$$A \to \begin{Vmatrix} -1 & 0 \\ 0 & -1 \end{Vmatrix}, \qquad T \to \begin{Vmatrix} 0 & 1 \\ 1 & 0 \end{Vmatrix},$$

which can be reduced to two one-dimensional representations

$$A \to -1, \qquad T \to +1,$$
$$A \to -1, \qquad T \to -1.$$

This reduction can be accomplished by using any matrix S for which STS^{-1} has diagonal form, since the eigenvalues of T are obviously ± 1. Thus, for $n = 2m$, there are four one-dimensional representations and $m - 1$ two-dimensional representations; in this case, the formula (119) becomes

$$4 \cdot 1^2 + (m - 1)2^2 = 4m = 2n.$$

3. We now consider the representations of the tetrahedral group, or equivalently, of the alternating group of degree 4, which is isomorphic to the tetrahedral group, as shown in Example 2 of Sec. 65. This group is of order 12 and consists of four classes. Therefore, it has four inequivalent irreducible representations, whose dimensions must satisfy the equation

$$n_1^2 + n_2^2 + n_3^2 + n_4^2 = 12.$$

Except for the order of the terms, this equation has the unique solution

$$n_1 = n_2 = n_3 = 1, \; n_4 = 3$$

in positive integers, i.e., the group has three one-dimensional representations and one three-dimensional representation. The one-dimensional representations assign the same number to every element in a given class. It is not hard to show that the three one-dimensional representations assign the following numbers to the various classes:

$$\begin{array}{llll} \mathrm{I} \to 1, & \mathrm{II} \to 1, & \mathrm{III} \to 1, & \mathrm{IV} \to 1, \\ \mathrm{I} \to 1, & \mathrm{II} \to 1, & \mathrm{III} \to \epsilon, & \mathrm{IV} \to \epsilon^2, \\ \mathrm{I} \to 1, & \mathrm{II} \to 1, & \mathrm{III} \to \epsilon^2, & \mathrm{IV} \to \epsilon, \end{array}$$

where

$$\epsilon = \cos \frac{2\pi}{3} + i \sin \frac{2\pi}{3},$$

and the classes I, II, III, IV are the same as in Example 2 of Sec. 65. The irreducible three-dimensional representation is the tetrahedral group itself, i.e., the group of spatial rotations (third order matrices) carrying the tetrahedron into itself. If this representation were reducible, it would have to reduce to three one-dimensional representations; this is impossible, since the tetrahedral group is not an Abelian group.

The theory presented in the last few sections pertains to finite groups. To extend the theory to the rotation group, for example, we have to make a more detailed study of infinite groups which depend on parameters. Before studying groups of this kind, we shall discuss the representations of the Lorentz group. In fact, the rotation group and the Lorentz group, together with their representations, will serve as our basic examples of infinite groups depending on parameters.

86. Representations of the Two-Dimensional Unimodular Group. In Sec. 74 we constructed representations of the two-dimensional unitary group; this led us to representations of the rotation group. Similarly, we can construct representations of the group of linear transformations

$$\begin{aligned} x_1' &= ax_1 + bx_2, \\ x_2' &= cx_1 + dx_2, \qquad (ad - bc = 1), \end{aligned} \tag{122}$$

with determinant 1, i.e., the two-dimensional unimodular group. According to Sec. 70, this will lead us to single-valued and two-valued representations of the group of proper orthochronous Lorentz transformations.

However, the situation here is quite different from that in Sec. 74. In fact, in the case of the group of unitary unimodular transformations,

$$\begin{aligned} x_1' &= ax_1 + bx_2, \\ x_2' &= -\bar{b}x + \bar{a}x, \qquad (a\bar{a} + b\bar{b} = 1), \end{aligned} \tag{123}$$

one possible representation of the group consists in assigning the transformation (123) itself to each transformation (123). Moreover, as is easily seen, another possible representation is to assign the transformation

$$\begin{aligned} y_1' &= \bar{a}y_1 + \bar{b}y_2, \\ y_2' &= -by_1 + ay_2 \end{aligned} \tag{124}$$

to each transformation (123), where the coefficients of (124) are the complex conjugates of those of (123). These two representations are equivalent, as follows at once from the formula

$$\begin{Vmatrix} 0 & 1 \\ -1 & 0 \end{Vmatrix} \begin{Vmatrix} a & b \\ -\bar{b} & \bar{a} \end{Vmatrix} = \begin{Vmatrix} \bar{a} & \bar{b} \\ -b & a \end{Vmatrix} \begin{Vmatrix} 0 & 1 \\ -1 & 0 \end{Vmatrix}.$$

On the other hand, for the group (122), the conjugate representation

$$\begin{aligned} y_1' &= \bar{a}y_1 + \bar{b}y_2, \\ y_2' &= \bar{c}y_1 + \bar{d}y_2 \end{aligned} \tag{125}$$

is not equivalent to the group itself. To see this, it is sufficient to consider the case $b = c = 0$. Then, the matrix of the transformation (122) has the eigenvalues a and d, while the matrix of the transformation (125) has the eigenvalues $\bar{a}$ and $\bar{d}$. Obviously, we can choose complex numbers a and d, satisfying the condition $ad = 1$, such that the pair of numbers a, d differs from the pair of numbers $\bar{a}$, $\bar{d}$. Therefore, the corresponding matrices cannot be similar. Thus, in the case of the unimodular group, we already have two inequivalent two-dimensional representations, i.e., the group (122) itself and the group (125). Below, we shall discuss the question of whether or not these and other representations of the unimodular group are reducible.

We can also construct representations of the group (122) by using the same method as in Sec. 74. It is necessary only to substitute d for $\bar{a}$ and $-c$ for $\bar{b}$ in the formula (99). This leads to the $(2j + 1)$-dimensional representation

$$D_j\begin{Bmatrix} a & b \\ c & d \end{Bmatrix}_{ls} = \sum_k \frac{\sqrt{(j+l)!(j-l)!(j+s)!(j-s)!}}{k!(j-k-s)!(j+l-k)!(k+s-l)!} \times a^{j+l-k}b^k c^{k+s-l} d^{j-k-s} \qquad (j = 0, \tfrac{1}{2}, 1, \tfrac{3}{2}, \ldots), \tag{126}$$

where j takes nonnegative integral and half-integral values, and we set $0! = 1$ and $0^0 = 1$. As before, l and s can take the values $-j, -j+1, \ldots, j-1, j$, and the summation over k is subject to the conditions

$$k \geq 0, \qquad k \geq l - s, \qquad k \leq j - s, \qquad k \leq j + l.$$

For $j = 0$, we get the representation which is identically 1.

In view of the remarks made above, we can immediately write down another representation of (122) by replacing the numbers a, b, c, d in the right-hand side of (126) by their complex conjugates; we denote this representation by

$$\bar{D}_{j'}\begin{Bmatrix} a & b \\ c & d \end{Bmatrix} \qquad (j' = 0, \tfrac{1}{2}, 1, \tfrac{3}{2}, \ldots). \tag{127}$$

We can then form the direct product of the representations (126) and (127), which gives a new representation of dimension $(2j + 1)(2j' + 1)$; we denote this representation by

$$E_{j,j'}\begin{Bmatrix} a & b \\ c & d \end{Bmatrix}. \tag{128}$$

Using (126), we can easily write down the matrix elements corresponding to (128).

We now show that representations of the type (128) *are inequivalent for different choices of j and j'.* Clearly, the only relevant case is when the two representations have the same dimension. Thus, let

$$E_{p,q}\begin{Bmatrix} a & b \\ c & d \end{Bmatrix} \qquad \text{and} \qquad E_{p_1,q_1}\begin{Bmatrix} a & b \\ c & d \end{Bmatrix}$$

be two representations of the type (128), of the same dimension

$$(2p + 1)(2q + 1) = (2p_1 + 1)(2q_1 + 1).$$

If we set $b = c = 0$, the matrix (126) becomes a diagonal matrix with elements

$$D_j \begin{Bmatrix} a & 0 \\ 0 & d \end{Bmatrix}_{ll} = a^{j+l}d^{j-l} \qquad (l = -j, -j+1, \ldots, j-1, j).$$

Since the direct product of two diagonal matrices is a diagonal matrix (Sec. 78), the matrices $E_{p,q}$ and E_{p_1,q_1} have the following eigenvalues (when $b = c = 0$):

$$E_{p,q}\colon a^{p+l}d^{p-l}\bar{a}^{q+m}\bar{d}^{q-m} \qquad (l = -p, -p+1, \ldots, p-1, p; \; m = -q, -q+1, \ldots, q-1, q),$$

$$E_{p_1,q_1}\colon a^{p_1+l_1}d^{p_1-l_1}\bar{a}^{q_1+m_1}\bar{d}^{q_1-m_1} \qquad (l_1 = -p_1, -p_1+1, \ldots, p_1-1, p_1; \; m_1 = -q_1, -q_1+1, \ldots, q_1-1, q_1). \tag{129}$$

Since $ad = 1$, (129) becomes

$$E_{p,q}\colon a^{2l}\bar{a}^{2m},$$
$$E_{p_1,q_1}\colon a^{2l_1}\bar{a}^{2m_1}.$$

The number a, which can be any nonzero complex number, can obviously be chosen in such a way as to make the eigenvalues of the matrix $E_{p,q}$ different from the eigenvalues of E_{p_1,q_1}. Thus, the representations (128) are inequivalent for different choices of j and j', as asserted. We note that for $j' = 0$, (128) is just the representation (126), while for $j = 0$, (128) is the representation obtained from (126) by setting $j = j'$ and replacing a, b, c, d by their complex conjugates.

It should be pointed out that the representation (128) is not equivalent to a unitary representation. For if it were equivalent to a unitary representation, all the eigenvalues of any matrix of the representation would have absolute values equal to 1 (Sec. 42). However, as we have just seen, when $b = c = 0$, the eigenvalues of the representation $E_{p,q}$ are $a^{2l}\bar{a}^{2m}$ and can obviously have absolute values different from 1. The only exception is the representation $E_{0,0}$, which is the trivial representation assigning the number 1 to every element of the group (122).

Now suppose we have a representation (which is not necessarily equivalent to a unitary representation). Then, according to Theorem 2 of Sec. 72, if the representation is reducible, i.e., equivalent to a representation consisting of quasi-diagonal matrices of identical structure, then there must exist a matrix which is not a multiple of the unit matrix and which commutes with all the matrices of the representation. Thus, to prove that the representation (128) is irreducible, for any j and j', it is sufficient to show that a matrix commuting with all the matrices of (128) must be a multiple of the unit matrix. This can be shown by a method just like that used in Sec. 74. Thus, no two of the representations (128) are equivalent and they are all irreducible.

A definition of reducibility different from the one given in Sec. 71 is often used, i.e., a representation is said to be *reducible* if all its matrices (of order n, say) leave a certain subspace L_k invariant $(0 < k < n)$.† As we saw in Sec. 71, if a representation which is reducible in this new sense consists of unitary matrices, then it is also reducible in the usual sense, i.e., it is equivalent to a quasi-diagonal representation. However, if a representation is not unitary, then the invariance of a subspace does not imply reducibility in the usual sense. It can be shown that every representation (128) of the group (122) is not only irreducible in the usual sense, but is also irreducible in the new sense, i.e., it leaves no subspace L_k invariant $(0 < k < n)$. Moreover, it can be shown that every representation of the group (122) either is equivalent to one of the representations (128) or can be reduced to a form consisting of several of the representations (128).

In Sec. 79, we saw that if the variables ξ_i and η_k are the components of vectors in the carrier spaces of two representations of a group, then the variables $\xi_i\eta_k$ are the components of a vector in the carrier space of the direct product of the representations. Thus, we can assert that the quantities

$$\frac{x_1^{j+k}x_2^{j-k}}{\sqrt{(j+k)!(j-k)!}}\,\frac{y_1^{j'+k'}y_2^{j'-k'}}{\sqrt{(j'+k')!(j'-k')!}}$$
$$(k = -j, -j+1, \ldots, j-1, j;\ k' = -j', -j'+1, \ldots, j'-1, j')$$

are the components of a vector in the carrier space of the representation (128), where x_1, x_2 undergo the transformation (122) and y_1, y_2 undergo the transformation (125).

So far, we have discussed representations of the group consisting of proper orthochronous Lorentz transformations (Sec. 70), which are only a subset of Lorentz transformations with determinant 1. Moreover, there are also Lorentz transformations with determinant -1. The study of the structure of these more general sets of Lorentz transformations, and the extension of representations of the group of proper orthochronous Lorentz transformations to the full Lorentz group, exhibit certain features not encountered in the case of the group of real orthogonal transformations in three dimensions. In defining the full Lorentz group, we can still impose the restriction that the time direction be preserved; we then have only to add the reflection

$$x_1' = -x_1,\ x_2' = -x_2,\ x_3' = -x_3,\ x_4' = x_4$$

to the group of proper orthochronous Lorentz transformations. A dis-

† See Probs. 7–9.

cussion of all these matters can be found, for example, in the books of Cartan and van der Waerden.†

87. Proof That the Lorentz Group Is Simple. We now prove that the Lorentz group is simple, using a method similar to that used in Sec. 76 to prove that the rotation group is simple. It is sufficient to show that the group $\mathcal{G}$, consisting of the transformations (122), has no nontrivial normal subgroups other than the normal subgroup $\mathcal{H}$, consisting of the matrices E and $-E$. In proving this, we shall exploit our basic results on the reduction of matrices to canonical form and the fact that the determinant of the matrix V reducing any matrix in $\mathcal{G}$ to canonical form can always be chosen to be 1 (Sec. 27), i.e., V can be chosen to be in $\mathcal{G}$.

Thus, suppose that $\mathcal{G}$ has a normal subgroup $\mathcal{H}_1$ containing a matrix

$$A = \begin{Vmatrix} a & b \\ c & d \end{Vmatrix} \qquad (ad - bc = 1)$$

different from E and $-E$. Then we have to show that $\mathcal{H}_1$ coincides with $\mathcal{G}$. If $\mathcal{H}_1$ contains a matrix B, then $\mathcal{H}_1$ also contains all the matrices VBV^{-1}, where V is any matrix in $\mathcal{G}$. Since the product of the eigenvalues of any matrix in $\mathcal{G}$ is 1, to prove that $\mathcal{H}_1$ coincides with $\mathcal{G}$, it suffices to show that $\mathcal{H}_1$ contains the matrix with eigenvalues t and t^{-1}, where t is any complex number different from 0 and ± 1, that $\mathcal{H}_1$ contains E and $-E$, and finally, that $\mathcal{H}_1$ contains the matrices

$$\begin{Vmatrix} 1 & 0 \\ 1 & 1 \end{Vmatrix} \quad \text{and} \quad \begin{Vmatrix} -1 & 0 \\ 1 & -1 \end{Vmatrix}. \tag{130}$$

This third requirement takes care of the case where the eigenvalues are equal and the elementary divisors are of multiplicity 2 (cf. Sec. 27).

Now let X be a variable matrix in $\mathcal{G}$, of the form

$$X = \begin{Vmatrix} x & y \\ z & x \end{Vmatrix} \qquad (x^2 - yz = 1).$$

Then, since $\mathcal{H}_1$ is a normal subgroup and contains A, it must also contain the matrix

$$Y = A(XA^{-1}X^{-1}).$$

For the trace s of the matrix Y, we find the expression

$$s = 2 + b^2z^2 + c^2y^2 - [(a - d)^2 + 2bc]yz.$$

† E. Cartan, "Leçons sur la théorie des spineurs," Hermann et Cie., Paris, 1938 (in two volumes). B. L. van der Waerden, "Die Gruppentheoretische Methode in der Quantenmechanik," Springer-Verlag, Berlin, 1932.

Since A is different from E and $-E$, we cannot simultaneously have $b = c = 0$ and $a = d$. Therefore, s is not a constant, and by varying y and z, we can give arbitrary complex values to s. Thus, since the eigenvalues of Y are the roots of the quadratic equation

$$\lambda^2 - s\lambda + 1 = 0,$$

we can obtain any values t and t^{-1} for these roots, i.e., $\mathcal{H}_1$ contains all matrices with different eigenvalues and unit determinants. Being a subgroup, $\mathcal{H}_1$ obviously contains E, and moreover $\mathcal{H}_1$ contains $-E$, since $-E$ can be written as the product

$$-E = [t, t^{-1}][-t^{-1}, -t]$$

of two factors, each of which is an element of $\mathcal{H}_1$. Finally, $\mathcal{H}_1$ contains the matrices (130), since we can easily write each of these matrices as a product of two matrices with unit determinants and different eigenvalues:

$$\begin{Vmatrix} 1 & 0 \\ 1 & 1 \end{Vmatrix} = \begin{Vmatrix} \beta^{-1} & 0 \\ 0 & \beta \end{Vmatrix} \begin{Vmatrix} \beta & 0 \\ \beta^{-1} & \beta^{-1} \end{Vmatrix}$$
$$\begin{Vmatrix} -1 & 0 \\ 1 & -1 \end{Vmatrix} = \begin{Vmatrix} \beta^{-1} & 0 \\ 0 & \beta \end{Vmatrix} \begin{Vmatrix} -\beta & 0 \\ \beta^{-1} & -\beta^{-1} \end{Vmatrix} \qquad (\beta \neq 0, \pm 1).$$

Thus, we have shown that $\mathcal{H}_1$ must coincide with $\mathcal{G}$, i.e., $\mathcal{G}$ has no nontrivial normal subgroups other than the normal subgroup $\mathcal{H}$ consisting of E and $-E$. This proves that the group of proper orthochronous Lorentz transformations is simple, and therefore, by the same argument as in Sec. 76, this group cannot have homomorphic (unfaithful) representations.

PROBLEMS†

1. Let $\mathcal{G}$ be a cyclic group of order m. Find an isomorphism between $\mathcal{G}$ and a group of *singular* matrices of order n.
2. The multiplicative group $\mathfrak{H}_n$ of all nonsingular (complex) matrices of order n is called the *full homogeneous linear group*. Show that the set of all n-dimensional linear transformations of the form

$$\mathbf{x}' = \{D, \mathbf{a}\}\mathbf{x} \equiv D\mathbf{x} + \mathbf{a},$$

where $\mathbf{a}$ is an arbitrary n-dimensional vector, and $D \in \mathfrak{H}_n$ is a group $\mathfrak{J}_n$ (called the *full inhomogeneous linear group*). Show that the mapping

$$\{D, \mathbf{a}\} \rightarrow \begin{Vmatrix} D & \mathbf{a} \\ 0 & 1 \end{Vmatrix}$$

is a faithful representation of $\mathfrak{J}_n$.

† All equation numbers (unless otherwise specified) refer to the corresponding equations of Chap. 8. For hints and answers, see page 447. These problems are mostly by J. S. Lomont, with some supplementation and modification by the translator and J. T. Schwartz.

3. Let $\mathcal{G}$ be a finite group with elements G_α, and let $\mathcal{A}$ be a representation of $\mathcal{G}$ assigning the matrix A_α to G_α. Prove that

$$\left(\sum_\alpha A_\alpha^2\right) A_\beta = A_\beta \left(\sum_\alpha A_\alpha^2\right)$$

for any A_β in $\mathcal{A}$.

4. Let a group have two equivalent representations $\mathcal{A}$ and $\mathcal{B}$, consisting of the real matrices A_α and B_α, respectively. Show that there is a real nonsingular matrix Y such that $YA_\alpha Y^{-1} = B_\alpha$.
5. Prove that the group of real 4×4 Lorentz matrices is not equivalent to a group of unitary matrices.
6. Let a group have a unitary representation $\mathcal{U}$ which is equivalent to a representation $\mathcal{R}$ consisting of real matrices. Prove that $\mathcal{U}$ is equivalent to a representation consisting of real orthogonal matrices.

In the next two problems, an n-dimensional representation is said to be *irreducible* if it has no *proper* invariant subspaces (cf. Sec. 86), i.e., no subspaces of dimension k, where $0 < k < n$.

7. Let a group have two irreducible representations $\mathcal{A}$ and $\mathcal{B}$, consisting of $m \times m$ matrices A_α and $n \times n$ matrices B_α, respectively, and suppose that

$$A_\alpha S = SB_\alpha$$

for all α, where S is an $m \times n$ matrix. Show that either $S = 0$ or S is square and nonsingular. ($\mathcal{A}$ and $\mathcal{B}$ can be the same.)

Comment: This result is known as *Schur's lemma,* and is essentially another version of the theorem of Sec. 82.†

8. Let a group have an irreducible representation $\mathcal{A}$, consisting of the matrices A_α. Using Schur's lemma, prove that the only matrices commuting with all the matrices A_α are multiples of the unit matrix (cf. Sec. 72).
9. Find a group which is irreducible in the sense of Sec. 71, but not in the sense of Prob. 7.
10. Prove that the number of different one-dimensional representations of a finite group $\mathcal{G}$ is equal to the index of its commutator subgroup (see Prob. 44 of Chap. 7).
11. Prove that the symmetric group of degree n ($n > 1$) has exactly two one-dimensional representations (cf. Example 1 of Sec. 73).
12. Let $\mathcal{G}$ be the additive group of all real numbers x. Prove that every finite-dimensional, continuous, unitary, irreducible‡ representation of $\mathcal{G}$ has the form

$$x \to e^{i\lambda x},$$

where λ is a real number which depends on the representation.

13. Let A be an $n \times n$ matrix and B an $m \times m$ matrix. Show that

(a) $(A \times B)^{-1} = A^{-1} \times B^{-1}$; (b) $\operatorname{tr}(A \times B) = \operatorname{tr} A \operatorname{tr} B$;
(c) $\det(A \times B) = (\det A)^m (\det B)^n$.

(By tr A is meant the trace of A; cf. Sec. 27.)

14. If we set $c_{ik;pq}^{(\alpha)} = a_{ip}^{(\alpha)} a_{kq}^{(\alpha)}$ and $m = n$ in (65), we see that the *Kronecker square* of A (i.e., $A \times A$) is the transformation matrix of a second rank contravariant tensor $b^{ik}(i, k = 1, 2, \ldots, n)$, where A is the matrix of a transformation from one

† In this problem (and some others), the requirement that $\mathcal{A}$ and $\mathcal{B}$ be representations is superfluous.

‡ In the sense of Prob. 7.

(affine) coordinate system to another. (Cf. formula (50) of Sec. 24.) Suppose that b^{ik} is symmetric ($b^{ik} = b^{ki}$); then the $\mu \times \mu$ transformation matrix of the $\mu = \frac{1}{2}n(n+1)$ independent components b^{ik} for which $i \leq k$ is called the *symmetrized Kronecker square* of A, written $A \circledS A$. Similarly, suppose that b^{ik} is skew-symmetric ($b^{ik} = -b^{ki}$); then the $\nu \times \nu$ transformation matrix of the $\nu = \frac{1}{2}n(n-1)$ independent components b^{ik} for which $i < k$ is called the *antisymmetrized Kronecker square* of A, written $A \circleda A$. Show that $A \times A$ is the direct sum of $A \circledS A$ and $A \circleda A$. (If the matrix B is similar to the quasi-diagonal matrix $[B_1, B_2]$, then B is said to be the *direct sum* of B_1 and B_2, written $B = B_1 \oplus B_2$.)

15. Find $A \times A$, $A \circledS A$, and $A \circleda A$ for the matrix

$$A = \begin{Vmatrix} 1 & 0 \\ a & 1 \end{Vmatrix},$$

and verify that $A \times A = (A \circledS A) \oplus (A \circleda A)$ by explicitly reducing $A \times A$.

16. Show that if A and B are $n \times n$ matrices, then

(a) $(A \circledS A)(B \circledS B) = AB \circledS AB$; (b) $(A \circleda A)(B \circleda B) = AB \circleda AB$;
(c) $\operatorname{tr}(A \times A) = \operatorname{tr}(A \circledS A) + \operatorname{tr}(A \circleda A)$;
(d) $\operatorname{tr}(A \circledS A) = \frac{1}{2}[(\operatorname{tr} A)^2 + \operatorname{tr}(A^2)]$;
(e) $\operatorname{tr}(A \circleda A) = \frac{1}{2}[(\operatorname{tr} A)^2 - \operatorname{tr}(A^2)]$.

17. Show that if every element of a finite group $\mathcal{G}$ other than the unit element I is of order 2, then $\mathcal{G}$ is an Abelian group of order 2^m, which is isomorphic to the direct product $\mathcal{C} \times \mathcal{C} \times \cdots \times \mathcal{C}$ (m times), where $\mathcal{C}$ is the cyclic group of order 2. Show that the groups in Probs. 4 and 5 of Chap. 7 are isomorphic to $\mathcal{C} \times \mathcal{C}$ and $\mathcal{C} \times \mathcal{C} \times \mathcal{C}$, respectively.

18. Let $\mathcal{G}$ be a finite group of order g, let $\mathcal{H}$ be a subgroup of $\mathcal{G}$ of order h and index $\nu = g/h$, and let $G_1 = I, G_2, \ldots, G_\nu$ be ν elements of $\mathcal{G}$ such that the left cosets $G_1\mathcal{H}, G_2\mathcal{H}, \ldots, G_\nu\mathcal{H}$ are distinct (cf. Sec. 63). Let

$$\sigma_{ij}(G, H) = \begin{cases} 1 & \text{if } G_iHG_j^{-1} = G \\ 0 & \text{otherwise} \end{cases} \qquad (i, j = 1, 2, \ldots, \nu),$$

where $G \in \mathcal{G}$, $H \in \mathcal{H}$, and let $\sigma(G, H)$ be the $\nu \times \nu$ matrix $\|\sigma_{ij}(G, H)\|$. Finally, let $\mathcal{B}$ be any n-dimensional representation of $\mathcal{H}$ assigning the matrix $B(H)$ to H, and let $\mathcal{A}$ be the matrix function on $\mathcal{G}$ assigning the matrix

$$A(G) = \sum_H \sigma(G, H) \times B(H)$$

to A, where the sum is over all $H \in \mathcal{H}$. Prove that $\mathcal{A}$ is a representation of $\mathcal{G}$ of dimension νn.

Comment: The representation $\mathcal{A}$ of $\mathcal{G}$ is called the representation of $\mathcal{G}$ *induced* by the representation $\mathcal{B}$ of $\mathcal{H}$.

19. Let $\mathcal{G}$ be a group with elements G_α, and let $\chi(G_\alpha)$ be a character belonging to an n-dimensional unitary representation of $\mathcal{G}$. Prove that $|\chi(G_\alpha)| \leq n$ for all α.

20. Let $\mathcal{A}_1$ and $\mathcal{A}_2$ be two representations of a group $\mathcal{G}$, with characters χ_1 and χ_2, respectively. Prove that $\chi_1 + \chi_2$ and $\chi_1\chi_2$ are also characters of representations of $\mathcal{G}$.

21. Let $\mathcal{G}$ be a group of order g, whose center $\mathcal{Z}$ is of order k (see Prob. 36 of Chap. 7). Let d be the dimension of an irreducible representation $\mathcal{A}$ of $\mathcal{G}$. Prove that $d \leq \sqrt{g/h}$.

22. Let $\mathcal{G}$ be a group of order m with r classes $C_1, C_2, \ldots, C_r$, where C_i consists of g_i elements ($i = 1, \ldots, r$).† Let $\chi^{(k)}$ be the character of the kth irreducible representation $\mathcal{A}_k$ of $\mathcal{G}$ ($k = 1, \ldots, r$). Prove that

$$\sum_{k=1}^{r} \chi^{(k)}(C_i)\overline{\chi^{(k)}(C_j)}g_i = m\delta_{ij}, \tag{I}$$

where δ_{ij} is the Kronecker delta (cf. (107)). Let a_{pqk} be the class multiplication coefficients of $\mathcal{G}$ (cf. (108)), and let $\mathcal{A}_k$ have dimension d_k. Prove that

$$a_{pqk} = \frac{1}{m}\sum_{i=1}^{r} \frac{g_p g_q \chi^{(i)}(C_p)\chi^{(i)}(C_q)\overline{\chi^{(i)}(C_k)}}{d_i}. \tag{II}$$

23. A unitary representation of the symmetric group $\mathcal{S}$ of degree 3 is given in Example 2 of Sec. 67; call this representation $\mathcal{A}$. Prove that $\mathcal{A}$ is irreducible, and verify that it satisfies (92). What are the other irreducible representations of $\mathcal{S}$? Verify that the characters of each of these representations satisfy (107) and formula (I) of the preceding problem. Find the class multiplication coefficients a_{pqk} of $\mathcal{S}$, by using both (108) and formula (II) of the preceding problem.

24. Let the notation be the same as in Prob. 22. Prove that

$$\sum_{p=1}^{r}\sum_{q=1}^{r} a_{pqk}\overline{\chi^{(i)}(C_p)\chi^{(j)}(C_q)} = \frac{m}{d_i}\delta_{ij}\overline{\chi^{(i)}(C_k)}.$$

25. Let $\mathcal{G}$ be a finite group of order g with elements G_α, let χ be the character of a representation of $\mathcal{G}$, and let $P(z)$ be any polynomial in z with integral coefficients. Prove that

$$\Phi = \sum_{\alpha=1}^{g} P[\chi(G_\alpha)]$$

is divisible by g.

26. Let $\mathcal{G}$ be a finite group with elements G_α, and let χ be the character of a representation of $\mathcal{G}$. Prove that $\chi(G_\alpha) = \overline{\chi(G_\alpha^{-1})}$.

27. A representation of a group is said to be of the *first kind* if it is equivalent to a representation by real matrices, of the *second kind* if it is equivalent to its own complex conjugate but is not of the first kind, and of the *third kind* if it is not of the first or second kind. Prove that a representation of a finite group is of the third kind if and only if its character is not real.

28. Let $\mathcal{A}$ be any irreducible representation of a finite group $\mathcal{G}$, and let m be the number of times the trivial one-dimensional representation (i.e., the representation assigning the number 1 to every element of $\mathcal{G}$) appears as a component of $\mathcal{A} \times \mathcal{A}$. Prove that m is either 0 or 1, and that $m = 0$ if and only if $\mathcal{A}$ is of the third kind.

29. Let $\mathcal{U}$ be the group of 2×2 unitary matrices with determinant 1 (cf. Secs. 69 and 74). Prove that $\mathcal{U}$ is of the second kind (regarded as a representation of itself).

† Cf. footnote on p. 362.

30. Prove that the number of irreducible representations of the first and second kinds of a finite group $\mathcal{G}$ equals the number of ambivalent classes of $\mathcal{G}$ (see Prob. 34 of Chap. 7).

31. Let $p_n(x)$ be a sequence of real polynomials converging uniformly on every finite subinterval of the positive real axis to the function $f(x) = x^{1/2}$. Prove that if H is a positive definite Hermitian matrix (i.e., $(H\mathbf{x}, \mathbf{x}) > 0$ for all $\mathbf{x}$), then the matrices $p_n(H)$ converge to a positive definite Hermitian square root $H^{1/2}$ of H. Show that $H^{1/2}$ commutes with every matrix commuting with H, and show that $H^{1/2}$ is real if H is real.

32. Let a group have two equivalent representations $\mathcal{A}$ and $\mathcal{B}$ consisting of unitary matrices A_α and B_α, respectively. Show that there is a unitary matrix U such that $UA_\alpha U^{-1} = B_\alpha$.

33. Let a group have two equivalent representations $\mathcal{A}$ and $\mathcal{B}$ consisting of real orthogonal matrices A_α and B_α, respectively. Show that there is a real orthogonal matrix R such that $RA_\alpha R^{-1} = B_\alpha$.

34.† Let $\mathcal{A}$ be an irreducible representation which is *unitarily equivalent* to its complex conjugate $\bar{\mathcal{A}}$, i.e., such that $\bar{\mathcal{A}} = U^{-1}\mathcal{A}U$, where U is unitary. Prove that $U^* = \pm U$.

35. Let $\mathcal{A}$ be an irreducible unitary representation which is unitarily equivalent to its complex conjugate $\bar{\mathcal{A}}$, i.e., such that $\bar{\mathcal{A}} = U^{-1}\mathcal{A}U$, where U is unitary. Prove that $\mathcal{A}$ is of the first kind if and only if U is symmetric and of the second kind if and only if U is skew-symmetric.

36. Let $\mathcal{A}$ be an irreducible representation of dimension d of a finite group of order g with elements G_α, and let χ be the character of $\mathcal{A}$. Let

$$\zeta = \frac{1}{g}\sum_\alpha \chi(G_\alpha^2),$$

where the sum is over all G_α in $\mathcal{G}$. Prove that

(**a**) $\zeta = 1$ if and only if $\mathcal{A}$ is of the first kind;
(**b**) $\zeta = -1$ if and only if $\mathcal{A}$ is of the second kind;
(**c**) $\zeta = 0$ if and only if $\mathcal{A}$ is of the third kind.

37. Let $\mathcal{G}$ be a finite group of order g with elements G_α, and let $\rho(G_\alpha)$ be the number of square roots of G_α (cf. Probs. 34 and 35 of Chap. 7). Let $\mathcal{A}^{(p)}$ be the pth irreducible representation of $\mathcal{G}$, with character $\chi^{(p)}$, and let $c_p = 1, -1$, or 0, depending on whether $\mathcal{A}^{(p)}$ is of the first, second, or third kind, respectively. Prove that

$$\rho(G_\alpha) = \sum_p c_p\chi^{(p)}(G_\alpha).$$

38. Let a group have an irreducible unitary representation $\mathcal{A}$ consisting of matrices A_α, and let F be a matrix such that $A_\alpha^* F A_\alpha = F$ for all α. Prove that

(**a**) If $\mathcal{A}$ is of the first kind, then $F^* = F$;
(**b**) If $\mathcal{A}$ is of the second kind, then $F^* = -F$;
(**c**) If $\mathcal{A}$ is of the third kind, then $F = 0$;

† In the problems that follow, we are again using the definition of irreducibility preceding Prob. 7 (cf. Sec. 86).

(d) If $F \neq 0$ and $F^* = F$, then $\mathcal{A}$ is of the first kind;
(e) If $F \neq 0$ and $F^* = -F$, then $\mathcal{A}$ is of the second kind;
(f) If only the zero matrix satisfies $A_\alpha^* F A_\alpha = F$, then $\mathcal{A}$ is of the third kind.

39. A real representation of a finite group $\mathcal{G}$ is called *real-irreducible* if the real vector space on which it acts has no proper invariant subspace. Two real representations $\mathcal{A}$ and $\mathcal{B}$, consisting of the matrices A_α and B_α, respectively, are said to be *real-equivalent* if there is a real matrix S such that $SA_\alpha S^{-1} = B_\alpha$ for all α. Prove that

(a) Every real representation of $\mathcal{G}$ is real-equivalent to a representation by orthogonal matrices;
(b) Every real representation of $\mathcal{G}$ can be decomposed into a direct sum of a finite number of real-irreducible representations of $\mathcal{G}$;
(c) Two real-irreducible representations $\mathcal{A}$ and $\mathcal{B}$ of $\mathcal{G}$ which are not real-equivalent satisfy the orthogonality relation

$$\sum_\alpha \{A_\alpha\}_{ij}\{B_\alpha\}_{kl} = 0.$$

40. Let $\mathcal{G}$ be a finite group, and let $\mathcal{A}_1$, $\mathcal{A}_2$, and $\mathcal{A}_3$ denote any irreducible representations of $\mathcal{G}$ of the first, second, and third kinds, respectively. Show that

(a) The representations of the form $\mathcal{A}_1$, $\mathcal{A}_2 \oplus \mathcal{A}_2$, and $\mathcal{A}_3 \oplus \bar{\mathcal{A}}_3$ are all equivalent to real-irreducible representations of $\mathcal{G}$;
(b) Every real-irreducible representation of $\mathcal{G}$ is real-equivalent to a representation of the form $\mathcal{A}_1$, $\mathcal{A}_2 \oplus \mathcal{A}_2$, or $\mathcal{A}_3 \oplus \bar{\mathcal{A}}_3$.

Chapter 9

Continuous Groups

88. Continuous Groups. Structure Constants. The three-dimensional rotation group and the group of proper orthochronous Lorentz transformations are examples of groups whose elements depend on continuously varying parameters. (In the case of the rotation group, the Eulerian angles are a possible choice of these parameters.) Both of these groups consist of linear transformations, and the dependence of the groups on parameters results from the fact that the elements of the matrices corresponding to the linear transformations depend on parameters. The continuous groups considered in this chapter are all of this type.

Suppose that the matrix elements t_{ik} of the linear transformations forming a group $\mathcal{G}$ are functions of r real parameters $\alpha_1, \alpha_2, \ldots, \alpha_r$ which satisfy the following conditions: (*a*) The unit element of $\mathcal{G}$, for which $t_{ik} = 0$ if $i \neq k$, $t_{ii} = 1$, corresponds to the parameter values $\alpha_1 = \alpha_2 = \cdots = \alpha_r = 0$; (*b*) the t_{ik} are single-valued functions of the parameters α_s for all values of the α_s in a sufficiently small neighborhood of the origin of an r-dimensional real space W_r; (*c*) every element of $\mathcal{G}$ in a sufficiently small neighborhood of the unit element corresponds to certain parameter values α_s lying in a neighborhood of the origin of W_r. (A group element is said to lie *in a neighborhood of the unit element* if the elements t_{ik} of the corresponding matrix lie in a neighborhood of 0 for $i \neq k$ and in a neighborhood of 1 if $i = k$.) If these conditions are met, then there exists a one-to-one correspondence between a certain neighborhood of the origin of W_r and the elements of $\mathcal{G}$ lying in a certain neighborhood of the unit element. Subsequently, we shall deal not only with a "local" one-to-one correspondence of this type, but also with a "global" one-to-one correspondence, in which every point of the space W_r lying in a certain region V containing the origin corresponds to a definite element of $\mathcal{G}$, and conversely, every element of $\mathcal{G}$ corresponds to a definite point of V. However, for the time being, we shall be concerned only with a "local" correspondence of the type just described; in all considerations based on this "local" point of view, it will be assumed that the parameter values are sufficiently close to zero and that the group elements are sufficiently close to the unit element. We denote the elements of the group corresponding to the parameter values α_s, β_s, γ_s, etc. ($s = 1, 2, \ldots, r$), by the symbols G_α, G_β, G_γ, etc.

Consider now the product $G_\gamma = G_\beta G_\alpha$ of any two elements of $\mathcal{G}$. The parameters γ_s of the element $G_\gamma = G_\beta G_\alpha$ are single-valued functions

$$\gamma_s = \varphi_s(\beta_1, \beta_2, \ldots, \beta_r; \alpha_1, \alpha_2, \ldots, \alpha_r) \qquad (s = 1, 2, \ldots, r) \quad (1)$$

of the parameters α_s and β_s, and when α_s and β_s are near zero, so is γ_s. We assume that these functions and their partial derivatives up to order 4 are continuous for all α_s and β_s sufficiently close to zero. Since the unit element of $\mathcal{G}$ has parameter values equal to zero, we have

$$\begin{aligned} &\varphi_s(\beta_1, \beta_2, \ldots, \beta_r; 0, 0, \ldots, 0) = \beta_s, \\ &\varphi_s(0, 0, \ldots, 0; \alpha_1, \alpha_2, \ldots, \alpha_r) = \alpha_s, \qquad (s = 1, 2, \ldots, r), \end{aligned} \quad (2)$$

so that

$$\begin{aligned} \frac{\partial \varphi_i}{\partial \beta_k} &= \delta_{ik} \qquad \text{for } \alpha_s = 0, \\ \frac{\partial \varphi_i}{\partial \alpha_k} &= \delta_{ik} \qquad \text{for } \beta_s = 0, \qquad (s = 1, 2, \ldots, r), \end{aligned} \quad (3)$$

where, as usual, $\delta_{ik} = 0$ if $i \neq k$ and $\delta_{ii} = 1$. The parameters $\tilde{\alpha}_s$ corresponding to the inverse element G_α^{-1} are defined by the relations

$$\varphi_s(\tilde{\alpha}_1, \tilde{\alpha}_2, \ldots, \tilde{\alpha}_r; \alpha_1, \alpha_2, \ldots, \alpha_r) = 0 \qquad (s = 1, 2, \ldots, r); \quad (4)$$

obviously (4) holds if we set all the α_s and $\tilde{\alpha}_s$ equal to zero. By (3), the Jacobian of the left-hand side of (4) with respect to the $\tilde{\alpha}_s$ equals 1 when $\alpha_s = \tilde{\alpha}_s = 0$ ($s = 1, 2, \ldots, r$). Thus, by the implicit function theorem (Sec. 19), when the α_s are sufficiently close to zero, the equations (4) define the $\tilde{\alpha}_s$ as continuous functions of $\alpha_1, \alpha_2, \ldots, \alpha_r$, which vanish if $\alpha_1 = \alpha_2 = \cdots = \alpha_r = 0$.

Next, using Maclaurin's theorem, we expand (1) in powers of α_s and β_s up to terms of order 3. In view of (1) and (2), we obtain

$$\gamma_s = \alpha_s + \beta_s + \sum_{i,k} a_{ik}^{(s)} \alpha_i \beta_k + \sum_{i,k,l} a_{ikl}^{(s)} \alpha_i \alpha_k \beta_l + \sum_{i,k,l} b_{ikl}^{(s)} \alpha_i \beta_k \beta_l + \epsilon^{(s)}$$
$$(s = 1, 2, \ldots, r), \quad (5)$$

where $a_{ik}^{(s)}$, $a_{ikl}^{(s)}$, $b_{ikl}^{(s)}$ are numerical coefficients, and $\epsilon^{(s)}$ is an infinitesimal of order no less than 4 with respect to α_s and β_s. The summation in (5) extends over values of i, k, l from 1 to r. The numbers

$$C_{ik}^{(s)} = a_{ik}^{(s)} - a_{ki}^{(s)} \qquad (s, i, k = 1, 2, \ldots, r) \quad (6)$$

are usually called the *structure constants* of the group $\mathcal{G}$ for the parameters α_s. If we introduce new parameters α'_s related to the old parameters α_s by the formulas

$$\begin{aligned} \alpha_s &= \omega_s(\alpha'_1, \alpha'_2, \ldots, \alpha'_r), \\ \omega_s(0, 0, \ldots, 0) &= 0, \qquad (s = 1, 2, \ldots, r), \end{aligned} \tag{7}$$

where the functions ω_s have a sufficiently large number of derivatives and the formulas (7) can be solved uniquely in terms of the α'_s for all α_s in a sufficiently small neighborhood of zero, then, in general, the new parameters α'_s lead to new structure constants.

It follows from the definition (6) that

$$C_{ki}^{(s)} = -C_{ik}^{(s)}. \tag{8}$$

Moreover, using (4) and the fact that multiplication of elements of $\mathcal{G}$ is associative, we can also prove the following relation between the structure constants:

$$\sum_{s=1}^{r} (C_{is}^{(t)}C_{jk}^{(s)} + C_{js}^{(t)}C_{ki}^{(s)} + C_{ks}^{(t)}C_{ij}^{(s)}) = 0 \qquad (i, j, k, t = 1, 2, \ldots, r). \tag{9}$$

We shall not need this relation, and therefore we omit its proof.†

Returning to the formula (1), we observe that it follows from (3) and the theorem on implicit functions that (1) can be solved in a sufficiently small neighborhood of the origin for the parameters

$$\beta_s = \psi_s(\gamma_1, \gamma_2, \ldots, \gamma_r; \alpha_1, \alpha_2, \ldots, \alpha_r) \qquad (s = 1, 2, \ldots, r). \tag{10}$$

We observe that here the condition

$$\beta_s = 0 \qquad (s = 1, 2, \ldots, r)$$

is equivalent to the condition

$$\gamma_s = \alpha_s \qquad (s = 1, 2, \ldots, r).$$

We now use (5) and (10) to form two square matrices $S(\alpha_s)$ and $T(\alpha_s)$ of order r, whose elements $S_{ik}(\alpha_s)$ and $T_{ik}(\alpha_s)$ depend on the parameters α_s and are defined by

$$S_{ik}(\alpha_s) = \left(\frac{\partial \gamma_i}{\partial \beta_k}\right)_{\beta_s=0}, \qquad T_{ik}(\alpha_s) = \left(\frac{\partial \beta_i}{\partial \gamma_k}\right)_{\gamma_s=\alpha_s} \qquad (s, i, k = 1, 2, \ldots, r). \tag{11}$$

† See Prob. 3.

Using the rule for differentiation of composite functions, and calculating the derivative of γ_i with respect to γ_k and the derivative of β_i with respect to β_k, we obtain

$$S(\alpha_s)T(\alpha_s) = E \quad \text{and} \quad T(\alpha_s)S(\alpha_s) = E, \tag{12}$$

where E is the unit matrix of order r. It follows from (3) that $S(\alpha_s)$ reduces to E when $\alpha_1 = \alpha_2 = \cdots = \alpha_r = 0$; moreover, it follows from (12) that $T(\alpha_s)$ has the same property.

It is not hard to prove that the structure constants $C_{ik}^{(p)}$ can be expressed in terms of the elements of $S(\alpha_s)$ and $T(\alpha_s)$ by the formulas

$$C_{ik}^{(p)} = \left(\frac{\partial S_{pk}(\alpha_s)}{\partial \alpha_i} - \frac{\partial S_{pi}(\alpha_s)}{\partial \alpha_k}\right)_{\alpha_s=0} \tag{13}$$

and

$$C_{ik}^{(p)} = \left(\frac{\partial T_{pi}(\alpha_s)}{\partial \alpha_k} - \frac{\partial T_{pk}(\alpha_s)}{\partial \alpha_i}\right)_{\alpha_s=0}. \tag{14}$$

In fact, from (5) and (11), we obtain

$$a_{ik}^{(p)} = \left(\frac{\partial^2 \gamma_p}{\partial \alpha_i \partial \beta_k}\right)_{\alpha_s=\beta_s=0} = \left(\frac{\partial S_{pk}(\alpha_s)}{\partial \alpha_i}\right)_{\alpha_s=0}. \tag{15}$$

Interchanging the indices i and k gives

$$a_{ki}^{(p)} = \left(\frac{\partial S_{pi}(\alpha_s)}{\partial \alpha_k}\right)_{\alpha_s=0}, \tag{16}$$

which, together with (15), implies (13). Moreover, in view of (12), we have

$$\sum_{j=1}^{r} S_{pj}(\alpha_s)T_{jk}(\alpha_s) = \delta_{pk}. \tag{17}$$

We differentiate (17) with respect to α_i and then equate all the α_s to zero. Since the matrices $S(\alpha_s)$ and $T(\alpha_s)$ reduce to the unit matrix when $\alpha_s = 0$ ($s = 1, 2, \ldots, r$), we obtain

$$\left(\frac{\partial S_{pk}(\alpha_s)}{\partial \alpha_i}\right)_{\alpha_s=0} + \left(\frac{\partial T_{pk}(\alpha_s)}{\partial \alpha_i}\right)_{\alpha_s=0} = 0.$$

Thus, by (15), we have

$$a_{ik}^{(p)} = -\left(\frac{\partial T_{pk}(\alpha_s)}{\partial \alpha_i}\right)_{\alpha_s=0},$$

which, together with the relation obtained by interchanging i and k, implies (14).

The formula (5) defines the basic group operation, i.e., (5) gives the parameters γ_s of the product $G_\gamma = G_\beta G_\alpha$ of two group elements in terms of the parameters α_s and β_s of the factors G_α and G_β. It is clear from (5) that if α_s and β_s are sufficiently small, then, to a first approximation, the group operation reduces to

$$\gamma_s = \alpha_s + \beta_s \qquad (s = 1, 2, \ldots, r),$$

so that, to a first approximation, the group $\mathcal{G}$ is Abelian. If $\mathcal{G}$ is actually Abelian, then

$$\varphi_s(\beta_1, \beta_2, \ldots, \beta_r; \alpha_1, \alpha_2, \ldots, \alpha_r) = \varphi_s(\alpha_1, \alpha_2, \ldots, \alpha_r; \beta_1, \beta_2, \ldots, \beta_r) \qquad (s = 1, 2, \ldots, r),$$

and $a_{ki}^{(s)} = a_{ik}^{(s)}$ in the expansion (5), i.e., all the structure constants of an Abelian group vanish. For more general groups, the structure constants are nonzero, and the second order terms in (5) already exhibit a departure from commutativity. Using (5), we can easily obtain an expression for the parameters $\tilde{\alpha}_s$ corresponding to the element G_α^{-1}, by setting $\gamma_s = 0$ and replacing β_s by $\tilde{\alpha}_s$. Then, applying the usual rule for differentiation of an implicit function, we obtain

$$\tilde{\alpha}_s = -\alpha_s + \sum_{i,k} a_{ik}^{(s)} \alpha_i \alpha_k + \epsilon_1^{(s)} \qquad (s = 1, 2, \ldots, r),$$

where $\epsilon_1^{(s)}$ is an infinitesimal of at least the third order with respect to $\alpha_1, \alpha_2, \ldots, \alpha_r$.

89. Infinitesimal Transformations. Let $\mathcal{G}$ be a continuous group of n-dimensional linear transformations, parameterized by $\alpha_1, \alpha_2, \ldots, \alpha_r$, and write a typical linear transformation

$$\mathbf{x} = G_\alpha \mathbf{u}, \tag{18}$$

where G_α is the matrix corresponding to the parameter values α_s, and $\mathbf{u}$, $\mathbf{x}$ are vectors in an n-dimensional complex vector space R_n. We introduce the following definition: If the elements of a matrix A are differentiable functions of a parameter t, then by *the derivative of the matrix* A with respect to t, we mean the matrix whose elements are obtained by differentiating the elements of A, i.e.,

$$\left\{\frac{dA}{dt}\right\}_{ik} = \frac{d}{dt}\{A\}_{ik}.$$

If the elements of A depend on several variables, we define the partial derivatives of A similarly. Moreover, if the components of a vector $\mathbf{z} = (z_1, z_2, \ldots, z_n)$ in R_n are differentiable functions of t, then $d\mathbf{z}/dt$ is defined as the vector with components dz_i/dt, i.e., differentiation of a vector means differentiation of its components.

We now introduce the *infinitesimal transformations of the group* $\mathcal{G}$, defined by

$$I_k = \left(\frac{\partial G_\alpha}{\partial \alpha_k}\right)_{\alpha_s=0} \qquad (k = 1, 2, \ldots, r). \tag{19}$$

Clearly, I_k is a matrix of order n, with numerical elements. Returning to the formula (18), we assume that $\mathbf{u}$ is a fixed vector, whose components do not depend on the parameters α_s. Then, in general, the transformed vector $\mathbf{x}$ still depends on the α_s and obeys certain basic differential equations, which we now derive. We begin by applying the linear transformation G_β to both sides of (18), obtaining

$$G_\beta \mathbf{x} = G_\gamma \mathbf{u}, \tag{20}$$

where $G_\gamma = G_\beta G_\alpha$ and the parameters γ_s are expressed in terms of α_s and β_s by the basic formula (5), which specifies the group operation. Differentiating (20) with respect to β_p, setting $\beta_s = 0$, i.e., $\gamma_s = \alpha_s$ ($s = 1, 2, \ldots, r$), and then using the definition (19), we obtain

$$I_p \mathbf{x} = \sum_{j=1}^{r} \left(\frac{\partial(G_\gamma \mathbf{u})}{\partial \gamma_j}\right)_{\gamma_s=\alpha_s} \left(\frac{\partial \gamma_j}{\partial \beta_p}\right)_{\beta_s=0}. \tag{21}$$

The first factor in the sum is clearly the derivative of the right-hand side of (18) with respect to α_j. Thus, using (11), we can write (21) in the form

$$I_p \mathbf{x} = \sum_{j=1}^{r} S_{jp}(\alpha_s) \frac{\partial \mathbf{x}}{\partial \alpha_j} \qquad (p = 1, 2, \ldots, r). \tag{22}$$

Introducing the vectors

$$\mathbf{X} = \left(\frac{\partial \mathbf{x}}{\partial \alpha_1}, \frac{\partial \mathbf{x}}{\partial \alpha_2}, \ldots, \frac{\partial \mathbf{x}}{\partial \alpha_r}\right),$$

and

$$\mathbf{Y} = (I_1 \mathbf{x}, I_2 \mathbf{x}, \ldots, I_r \mathbf{x}),$$

we can write (22) as a linear transformation

$$\mathbf{Y} = S^*(\alpha_s)\mathbf{X}, \tag{23}$$

where $S^*(\alpha_s)$ denotes the transpose of the matrix $S(\alpha_s)$. Multiplying (23) by $T^*(\alpha_s)$ from the left and using (12), we obtain

$$\mathbf{X} = T^*(\alpha_s)\mathbf{Y},$$

or in expanded form,

$$\frac{\partial \mathbf{x}}{\partial \alpha_p} = \sum_{j=1}^{r} T_{jp}(\alpha_s) I_j \mathbf{x} \qquad (p = 1, 2, \ldots, r). \tag{24}$$

Thus, the components x_k of the vector $\mathbf{x}$ defined by (18) satisfy the relations

$$\frac{\partial x_k}{\partial \alpha_p} = \sum_{j=1}^{r} T_{jp}(\alpha_s) \sum_{t=1}^{n} \{I_j\}_{kt} x_t \qquad (k = 1, 2, \ldots, n;\ p = 1, 2, \ldots, r), \tag{25}$$

where the elements of the matrix I_j are denoted by $\{I_j\}_{kt}$. The equation (24) for $\mathbf{x}$ must be supplemented by the initial condition

$$\mathbf{x}\Big|_{\alpha_s=0} = \mathbf{u}, \tag{26}$$

where $\mathbf{u}$ is an arbitrary given vector; this initial condition is an immediate consequence of (18). We note that the quantities $T_{jp}(\alpha_s)$ appearing as coefficients in (24) are defined directly in terms of the group operation (5).

By using the fact that the second derivative of $\mathbf{x}$ with respect to α_p and α_q does not depend on the order of differentiation, we can derive relations between the quantities I_p. Thus, differentiating (24) with respect to α_q, we obtain

$$\frac{\partial^2 \mathbf{x}}{\partial \alpha_p \partial \alpha_q} = \sum_{j=1}^{r} \left(\frac{\partial T_{jp}(\alpha_s)}{\partial \alpha_q} I_j \mathbf{x} + T_{jp}(\alpha_s) I_j \frac{\partial \mathbf{x}}{\partial \alpha_q} \right),$$

or

$$\frac{\partial^2 \mathbf{x}}{\partial \alpha_p \partial \alpha_q} = \sum_{j=1}^{r} \frac{\partial T_{jp}(\alpha_s)}{\partial \alpha_q} I_j \mathbf{x} + \sum_{j=1}^{r} \sum_{k=1}^{r} T_{jp}(\alpha_s) T_{kq}(\alpha_s) I_j I_k \mathbf{x}, \tag{27}$$

where we have substituted for $\partial \mathbf{x}/\partial \alpha_q$ the expression obtained from (24) by replacing p by q. Equating (27) to the expression obtained from (27) by interchanging p and q, we obtain

$$\left[\sum_{j=1}^{r} \left(\frac{\partial T_{jp}(\alpha_s)}{\partial \alpha_q} - \frac{\partial T_{jq}(\alpha_s)}{\partial \alpha_p} \right) I_j + \sum_{j=1}^{r} \sum_{k=1}^{r} (T_{jp}(\alpha_s) T_{kq}(\alpha_s) - T_{jq}(\alpha_s) T_{kp}(\alpha_s)) I_j I_k \right] \mathbf{x} = 0, \tag{28}$$

which is a direct consequence of (24). We now set all the α_s to zero in (28) and use (14), (26), and the fact that $T(\alpha_s) = E$ if all the $\alpha_s = 0$; the result is

$$\left[\sum_{j=1}^{r} C_{pq}^{(j)} I_j + (I_p I_q - I_q I_p) \right] \mathbf{u} = 0. \tag{29}$$

Since the vector $\mathbf{u}$ is arbitrary, (29) implies the following *commutation relations* between the infinitesimal transformations:

$$I_q I_p - I_p I_q = \sum_{j=1}^{r} C_{pq}^{(j)} I_j \qquad (p, q = 1, 2, \ldots, r). \tag{30}$$

Starting with a given continuous group $\mathcal{G}$ and using (24), we have defined the I_j and proved the relation (30). We now show that the equation (24), or equivalently, the system (25), has a unique solution satisfying the initial condition (26). Suppose there were two such solutions. Then, by the linearity of (24), the difference between these two solutions must also satisfy (24), and moreover, by (26), it must reduce to the zero vector for $\alpha_s = 0$. Thus, we have to show that a solution $\mathbf{x}$ of (24), which vanishes for $\alpha_s = 0$, is identically zero. For simplicity, we assume that $r = 3$, and we let $\mathbf{x}(\alpha_1, \alpha_2, \alpha_3)$ be the solution of (24), which vanishes for $\alpha_1 = \alpha_2 = \alpha_3 = 0$. Then, we write (24) for $p = 1$ and set $\alpha_2 = \alpha_3 = 0$. The result is an ordinary differential equation (with independent variable α_1) for the vector $\mathbf{x}(\alpha_1, 0, 0)$, with the initial condition $\mathbf{x}(0, 0, 0) = 0$. It follows by a familiar uniqueness theorem that the solution of this differential equation vanishes identically, so that we have $\mathbf{x}(\alpha_1, 0, 0) \equiv 0$. Next, we write (24) for $p = 2$ and set $\alpha_3 = 0$. As a result, for every fixed α_1, we obtain an ordinary differential equation (with independent variable α_2) for the vector $\mathbf{x}(\alpha_1, \alpha_2, 0)$, where now the initial condition is $\mathbf{x}(\alpha_1, 0, 0) = 0$ for every α_1. By the same uniqueness theorem, it follows that $\mathbf{x}(\alpha_1, \alpha_2, 0) \equiv 0$. Finally, we write (24) for $p = 3$. Then, for any fixed α_1, α_2, we again obtain an ordinary differential equation (with independent variable α_3), this time for the vector $\mathbf{x}(\alpha_1, \alpha_2, \alpha_3)$, where the initial condition is $\mathbf{x}(\alpha_1, \alpha_2, 0) = 0$ for every α_1, α_2. By using the uniqueness theorem once more, we find that $x(\alpha_1, \alpha_2, \alpha_3) \equiv 0$, as was to be shown.

Thus, given the infinitesimal transformations I_j and the coefficients $T_{jp}(\alpha_s)$ specified by the group operation (5), the equation (24) and the initial condition (26) lead to a unique finite transformation (18). In other words, the infinitesimal transformations determine the group, a fact of importance in our subsequent considerations. The proof that a solution of (24) actually exists is based on a general theorem on partial differential equations, which, as applied to equation (24), goes as follows: A necessary and sufficient condition for the equation (24) to have a solution for any initial condition (26) is that the term in brackets in (28) should vanish identically in the variables α_s, for any choice of p and q. This matter will be discussed further in Sec. 94.

90. The Rotation Group. To illustrate the preceding considerations, we return once again to the group of three-dimensional rotations about the origin of coordinates. The corresponding third order matrices depend on three parameters, and the Eulerian angles are a possible choice of these parameters. However, for all further calculations, we shall use other parameters α_1, α_2, α_3, which are defined as follows: We regard every rotation as a counterclockwise rotation through an angle no greater than π about a certain directed axis drawn from the origin (cf. Sec. 42). (Two rotations through the angle π about oppositely directed axes lead to the same final position.) Then, we can represent every rotation by a vector drawn from the origin and directed along the axis of rotation, with the length of the vector being equal to the angle of rotation. Our new parameters $\alpha_1, \alpha_2, \alpha_3$ are just the projections of this vector on the coordinate axes. This establishes a one-to-one correspondence between the elements of the rotation group and the points $(\alpha_1, \alpha_2, \alpha_3)$ of a sphere V of radius π with center at the origin, where we identify opposite ends of every diameter of V. In the present case, this one-to-one correspondence holds for the whole rotation group and the whole sphere V, and not just in a neighborhood of the origin and in a corresponding neighborhood of the unit element of the group. Moreover, by writing down all the matrices of the rotation group in terms of the parameters α_1, α_2, α_3, we can prove that the requirements of Sec. 88, concerning the continuity and differentiability of the functions φ_s, are satisfied.

Instead of deriving the formula (5) for the basic group operation, we shall determine the structure constants by direct calculation of the matrices of the infinitesimal transformations. To calculate I_1, we set $\alpha_2 = \alpha_3 = 0$, obtaining the transformation

$$\begin{aligned} x_1' &= x_1, \\ x_2' &= x_2 \cos \alpha_1 - x_3 \sin \alpha_1, \\ x_3' &= x_2 \sin \alpha_1 + x_3 \cos \alpha_1, \end{aligned}$$

which corresponds to a rotation through the angle α_1 about the x-axis. Then, we differentiate the matrix of this transformation with respect to α_1 and set $\alpha_1 = 0$; the result is

$$I_1 = \begin{Vmatrix} 0 & 0 & 0 \\ 0 & 0 & -1 \\ 0 & 1 & 0 \end{Vmatrix}.$$

Similarly, we find

$$I_2 = \begin{Vmatrix} 0 & 0 & 1 \\ 0 & 0 & 0 \\ -1 & 0 & 0 \end{Vmatrix}, \qquad I_3 = \begin{Vmatrix} 0 & -1 & 0 \\ 1 & 0 & 0 \\ 0 & 0 & 0 \end{Vmatrix}.$$

We can now calculate directly the quantities $I_qI_p - I_pI_q$ appearing in

(30) and thereby determine the structure constants. An elementary calculation gives the commutation relations

$$I_1I_2 - I_2I_1 = I_3, \qquad I_2I_3 - I_3I_2 = I_1, \qquad I_3I_1 - I_1I_3 = I_2. \quad (31)$$

Expanding the right-hand side of (18) in powers of α_s and retaining only first order terms, we find

$$\mathbf{x} \doteq \mathbf{u} + (\alpha_1 I_1 + \alpha_2 I_2 + \alpha_3 I_3)\mathbf{u},$$

where the symbol $\doteq$ denotes approximate equality. Thus, as a result of an infinitesimal transformation, the vector $\mathbf{u}$ changes by an amount

$$\delta u \doteq \alpha_1 I_1 \mathbf{u} + \alpha_2 I_2 \mathbf{u} + \alpha_3 I_3 \mathbf{u},$$

where each term on the right gives the change in $\mathbf{u}$ due to a small rotation about one of the coordinate axes. For example, as a result of a rotation about the x-axis through the small angle α_1, we find that the changes in the components u_1, u_2, u_3 of $\mathbf{u}$ are

$$\delta u_1 \doteq 0, \qquad \delta u_2 \doteq -u_3\alpha_1, \qquad \delta u_3 \doteq u_2\alpha_1,$$

where, as before, we consider only first order terms in α_1.

91. Infinitesimal Transformations and Representations of the Rotation Group. We now examine the relation between the infinitesimal transformations discussed in the preceding section and representations of the rotation group. Suppose that the matrices $F(\alpha_1, \alpha_2, \alpha_3)$ of order n are in one-to-one correspondence with the elements of the rotation group in the neighborhood of the unit element, and let the elements of the matrices $F(\alpha_1, \alpha_2, \alpha_3)$ be continuous and differentiable functions of the parameters $\alpha_1, \alpha_2, \alpha_3$. Every rotation D can be written as the product of a finite number of rotations lying in the neighborhood of the unit element, and the product of the corresponding matrices $F(\alpha_1, \alpha_2, \alpha_3)$ gives a matrix representing D. However, this procedure may lead in the large to a multiple-valued representation of the rotation group, since by continuously varying the parameters of the rotation, we can leave D and then return to it and find that it is represented by a different matrix. For example, this situation has already been encountered in the case of the two-valued representation of the rotation group (Sec. 75).

For the matrices $F(\alpha_1, \alpha_2, \alpha_3)$, we have the same group operation and hence the same structure constants as for the rotations themselves. Thus, if we form infinitesimal transformations I_k of the group $\mathcal{G}$ of matrices $F(\alpha_1, \alpha_2, \alpha_3)$, these transformations will be matrices of order n obeying the relations (31). After finding such matrices, we can write the differential equation (24) for a vector $\mathbf{x}$ in R_n, since the $T_{jp}(\alpha_s)$ are determined only by the group operation. This equation will have a unique solution for the given initial condition (26), and this solution is just the transformation

$$\mathbf{x} = F(\alpha_1, \alpha_2, \alpha_3)\mathbf{u},$$

which represents the rotation group in the neighborhood of the unit element. In the present case, there are three parameters, and (24) is equivalent to $3n$ equations in the n components of the vector

$$\mathbf{x} = (x_1, x_2, \ldots, x_n).$$

The essential point for what follows is that (24) has only one solution obeying the initial condition (26), i.e., in our previous language, *a representation of the rotation group is completely determined by its infinitesimal transformations* I_1, I_2, I_3.

Thus, the whole problem reduces to determining the infinitesimal transformations, a matter that we now consider. Instead of the required matrices I_1, I_2, I_3, we introduce the new matrices

$$A_1 = -I_2 + iI_1, \qquad A_2 = I_2 + iI_1, \qquad A_3 = iI_3. \tag{32}$$

It can easily be verified that these matrices satisfy the commutation relations

$$\begin{aligned} A_3A_1 - A_1A_3 &= A_1, \\ A_3A_2 - A_2A_3 &= -A_2, \\ A_1A_2 - A_2A_1 &= 2A_3, \end{aligned} \tag{33}$$

instead of (31). The representation $\mathcal{G}$ has to include a representation of the Abelian subgroup of rotations about the z-axis, whose elements correspond to the matrices $F(0, 0, \alpha_3)$. By appropriately choosing the basis vectors, all these matrices can be reduced simultaneously to diagonal form, since the irreducible representations of an Abelian group are all one-dimensional. Then, for any of these basis vectors, the transformation $F(0, 0, \alpha_3)$ takes the form

$$F(0, 0, \alpha_3)\mathbf{v} = e^{l\alpha_3}\mathbf{v}. \tag{34}$$

If we set $l = -im$ and denote $\mathbf{v}$ by $\mathbf{v}_m$, (34) becomes

$$F(0, 0, \alpha_3)\mathbf{v}_m = e^{-im\alpha_3}\mathbf{v}_m. \tag{35}$$

Since we require only that the representation be single-valued in the neighborhood of $\alpha_1 = \alpha_2 = \alpha_3 = 0$, we do not have to assume that m is an integer. It follows from (35) and the definition of I_3 that

$$A_3\mathbf{v}_m = iI_3\mathbf{v}_m = i\left(\frac{\partial}{\partial\alpha_3} F(0, 0, \alpha_3)\mathbf{v}_m\right) = i\left(\frac{\partial}{\partial\alpha_3} e^{-im\alpha_3}\mathbf{v}_m\right) = m\mathbf{v}_m,$$

so that

$$A_3\mathbf{v}_m = m\mathbf{v}_m, \tag{36}$$

i.e., $\mathbf{v}_m$ is an eigenvector of the operator A_3 corresponding to the eigenvalue m. (If there are several such eigenfactors, then $\mathbf{v}_m$ denotes any one of them.) Next, we prove the following lemma:

LEMMA. *If the vector* $\mathbf{v}$ *is an eigenvector of the operator* A_3, *corresponding to the eigenvalue* M, *then the vector* $A_1\mathbf{v}$ *(if it is nonzero) is an eigenvector of* A_3, *corresponding to the eigenvalue* $M + 1$, *and similarly,* $A_2\mathbf{v}$ *is an eigenvector of* A_3, *corresponding to the eigenvalue* $M - 1$.

PROOF. By hypothesis, $A_3\mathbf{v} = M\mathbf{v}$. It follows from (33) that

$$A_3(A_1\mathbf{v}) = (A_1A_3 + A_1)\mathbf{v} = A_1(A_3\mathbf{v}) + A_1\mathbf{v} = A_1(M\mathbf{v}) + A_1\mathbf{v} = (M + 1)A_1\mathbf{v}.$$

Similarly, we have

$$A_3(A_2\mathbf{v}) = (A_2A_3 - A_2)\mathbf{v} = A_2(A_3\mathbf{v}) - A_2\mathbf{v} = A_2(M\mathbf{v}) - A_2\mathbf{v} = (M - 1)A_2\mathbf{v},$$

which proves the lemma.

Now, the matrix A_3 has at most n different eigenvalues. Let j be an eigenvalue (there may be several) such that no other eigenvalue has a larger real part (at this point, we still allow for the possibility that j may be complex), and let $\mathbf{v}_j$ be an eigenvector corresponding to j (there may be several). By the lemma, the eigenvalue corresponding to $A_1\mathbf{v}_j$ must be $j + 1$, but by definition, A_3 can have no such eigenvalue. Therefore, we must have

$$A_1\mathbf{v}_j = 0. \tag{37}$$

Using the lemma again, we see that if the vectors

$$\mathbf{v}_{j-1} = A_2\mathbf{v}_j,\ \mathbf{v}_{j-2} = A_2\mathbf{v}_{j-1},\ \ldots \tag{38}$$

are nonzero, then they belong to the eigenvalues $j - 1, j - 2, \ldots$ of the operator A_3. The sequence of vectors (38) must vanish after a certain vector, since the number of different eigenvalues of A_3 cannot exceed n.

We now prove the formula

$$A_1\mathbf{v}_k = \rho_k\mathbf{v}_{k+1} \qquad (k = j, j - 1, j - 2, \ldots), \tag{39}$$

where ρ_k is an integer to be determined below. By (37), the formula (39) holds for $k = j$, with $\rho_j = 0$, and we can take v_{j+1} to be the zero vector, say. Moreover, if (39) holds for some k, it also holds for $k - 1$, since

$$A_1\mathbf{v}_{k-1} = A_1(A_2\mathbf{v}_k) = (A_2A_1 + 2A_3)\mathbf{v}_k = A_2(A_1\mathbf{v}_k) + 2A_3\mathbf{v}_k = A_2(\rho_k\mathbf{v}_{k+1}) + 2k\mathbf{v}_k = (\rho_k + 2k)\mathbf{v}_k,$$

by (33), (36), and (38). Note that since $A_1\mathbf{v}_j = 0$, we have $A_1\mathbf{v}_{j-1} = 2j\mathbf{v}_j$. Thus, ρ_j is defined by the recursion formula

$$\rho_{k-1} = \rho_k + 2k, \qquad \rho_j = 0 \qquad (k = j, j - 1, \ldots),$$

whose solution is

$$\rho_k = j(j+1) - k(k+1),$$

i.e., $$A_1\mathbf{v}_k = [j(j+1) - k(k+1)]\mathbf{v}_{k+1} \qquad (k = j, j-1, \ldots). \tag{40}$$

Next, we determine the subscript s of the first of the vectors (38) which vanishes, i.e., we determine the value of s for which $\mathbf{v}_s = 0$ and $\mathbf{v}_{s+1} \neq 0$. It follows from (38) that $\rho_s = 0$, so that

$$j(j+1) - s(s+1) = 0.$$

This quadratic equation in s has the roots $s = j$ and $s = -(j+1)$. The value $s = j$ is excluded, since the vector $\mathbf{v}_j \neq 0$ and does not appear in the sequence (38). Thus, $s = -(j+1)$, and the vectors

$$\mathbf{v}_j, \mathbf{v}_{j-1}, \ldots, \mathbf{v}_{-j+1}, \mathbf{v}_{-j} \tag{41}$$

of the sequence (38) are nonzero, while $A_2\mathbf{v}_{-j} = \mathbf{v}_{-j-1} = 0$. The number of vectors (41) is $2j+1$, from which it is clear that $j \geq 0$ and that j takes either integral or half-integral values. If $2j+1 = n$, we can choose the vectors (41) as basis vectors in the space R_n, while if $2j+1 < n$, these vectors span a subspace L_{2j+1} of R_n. Moreover, if $2j+1 < n$, every vector in the sequence (41) satisfies (40) and the equation

$$A_3\mathbf{v}_k = k\mathbf{v}_k \qquad (k = j, j-1, \ldots, -j+1, -j),$$

and in addition, $A_2\mathbf{v}_k = \mathbf{v}_{k-1}$, with $\mathbf{v}_{-j-1} = 0$. Therefore, the operators A_1, A_2, and A_k map the subspace L_{2j+1} into itself, and the formulas just cited completely define these operators in L_{2j+1}. Moreover, it follows at once from (38) and (40) that no subspace L_k of L_{2j+1} $(0 < k < 2j+1)$ is invariant under the operators A_1, A_2, and A_3.

Having defined the operators A_1, A_2, A_3, we can solve (32) for the operators I_1, I_2, I_3, and then construct the equations (24) for vectors in the subspace L_{2j+1}. The solution of (24) is a $(2j+1)$-dimensional vector

$$\mathbf{x} = F_j(\alpha_1, \alpha_2, \alpha_3)\mathbf{u},$$

where $F_j(\alpha_1, \alpha_2, \alpha_3)$ is a $(2j+1)$-dimensional representation of the rotation group. Moreover, no subspace L_k of L_{2j+1} $(0 < k < 2j+1)$ is invariant under the matrices of this representation, since L_k is not invariant under the operators A_1, A_2, A_3; therefore, the representation is irreducible. If $2j+1 = n$, these considerations apply to the whole space R_n, but if $2j+1 < n$, we have separated out from the representation of the whole space R_n an irreducible component of dimension $2j+1$, i.e., a representation which leaves no subspace L_k of its carrier space invariant $(0 < k < 2j + 1)$. It is an immediate consequence of our argument that to within an equivalence, there exists only one irreducible representation of a given dimension $2j+1$. However, an irreducible (unitary) representation of every dimension $2j + 1$ has already been constructed in Sec. 75. It follows that the representations given there exhaust all possibilities, i.e., the representations given in this section, based on constructing the operators A_1, A_2, A_3 in L_{2j+1}, must be equivalent to the representations of Sec. 75.

The vectors (41) can be multiplied by arbitrary nonzero constants; if this is done, numerical factors appear in the formulas (38) and (40). The constants can be chosen in such a way that we have

$$\begin{aligned} A_1\mathbf{v}_k &= \sqrt{j(j+1) - k(k+1)}\,\mathbf{v}_{k+1}, \\ A_2\mathbf{v}_k &= \sqrt{j(j+1) - k(k-1)}\,\mathbf{v}_{k-1}, \\ A_3\mathbf{v}_k &= k\mathbf{v}_k, \end{aligned} \tag{42}$$

where $\mathbf{v}_{j+1} = 0$ and $\mathbf{v}_{-j-1} = 0$. With this choice of constants, we obtain the representations constructed in Secs. 74 and 75 by using the quantities

$$\eta_l = \frac{x_1^{j+l}x_2^{j-l}}{\sqrt{(j+l)!(j-l)!}}.$$

The considerations just given allow us to decompose any representation of the rotation group into its irreducible components by finding the eigenvectors of the operators A_3 corresponding to the largest eigenvalue and then using the construction (38).

92. Representations of the Lorentz Group. We now consider the two-dimensional unimodular group (Sec. 70):

$$\begin{aligned} x_1' &= ax_1 + bx_2, \\ x_2' &= cx_1 + dx_2, \qquad (ad - bc = 1). \end{aligned} \tag{43}$$

The matrix of this transformation contains four complex parameters, which satisfy one constraint. Thus, there are only three arbitrary complex parameters, or equivalently, six arbitrary real parameters. Introducing these real parameters, we write the matrix of the transformation (43) as

$$A = \begin{Vmatrix} 1 + \alpha_1 + i\alpha_2 & \alpha_3 + i\alpha_4 \\ \alpha_5 + i\alpha_6 & d(\alpha_s) \end{Vmatrix}, \tag{44}$$

where $$d(\alpha_s) = \frac{1 + (\alpha_3 + i\alpha_4)(\alpha_5 + i\alpha_6)}{1 + \alpha_1 + i\alpha_2}.$$

This group has six infinitesimal transformations I_k', which are easily constructed. For example, I_1' is obtained by setting all the α_s except α_1 equal to zero in (44), differentiating with respect to α_1 and then setting

α_1 equal to zero. In this way, we obtain

$$I_1' = \begin{Vmatrix} 1 & 0 \\ 0 & -1 \end{Vmatrix}, \quad I_2' = \begin{Vmatrix} i & 0 \\ 0 & -i \end{Vmatrix}, \quad I_3' = \begin{Vmatrix} 0 & 1 \\ 0 & 0 \end{Vmatrix},$$
$$I_4' = \begin{Vmatrix} 0 & i \\ 0 & 0 \end{Vmatrix}, \quad I_5' = \begin{Vmatrix} 0 & 0 \\ 1 & 0 \end{Vmatrix}, \quad I_6' = \begin{Vmatrix} 0 & 0 \\ i & 0 \end{Vmatrix}. \tag{45}$$

The I_k' satisfy commutation relations of the form (cf. (30))

$$I_q'I_p' - I_p'I_q' = \sum_{j=1}^{6} C_{pq}^{(j)} I_j' \qquad (p < q;\ p, q = 1, 2, \ldots, 6),$$

where, by their very definition, the structure constants $C_{pq}^{(j)}$ are real. It should be noted that there is no linear dependence (with real coefficients) between the matrices I_k', which satisfy the following 15 commutation relations:

$$\begin{array}{lll} I_1'I_3' - I_3'I_1' = 2I_3', & I_1'I_4' - I_4'I_1' = 2I_4', & I_2'I_3' - I_3'I_2' = 2I_4', \\ I_1'I_5' - I_5'I_1' = -2I_5', & I_1'I_6' - I_6'I_1' = -2I_6', & I_2'I_5' - I_5'I_2' = -2I_6', \\ I_3'I_5' - I_5'I_3' = I_1', & I_3'I_6' - I_6'I_3' = I_2', & I_4'I_5' - I_5'I_4' = I_2', \end{array}$$
$$\begin{array}{ll} I_2'I_4' - I_4'I_2' = -2I_3', & I_1'I_2' - I_2'I_1' = 0, \\ I_2'I_6' - I_6'I_2' = 2I_5', & I_3'I_4' - I_4'I_3' = 0, \\ I_4'I_6' - I_6'I_4' = -I_1', & I_5'I_6' - I_6'I_5' = 0. \end{array}$$

Now, dropping the primes, we denote by I_k $(k = 1, 2, \ldots, 6)$ the infinitesimal transformations of a representation of the unimodular group. Clearly, the I_k satisfy the same 15 relations of the form

$$I_qI_p - I_pI_q = \sum_{j=1}^{6} C_{pq}^{(j)} I_j$$

as the I_k', with the same coefficients $C_{pq}^{(j)}$. If we introduce the new quantities

$$\begin{array}{lll} I_3 + iI_4 = 2A_1, & I_5 + iI_6 = 2A_2, & I_1 + iI_2 = 4A_3, \\ I_3 - iI_4 = 2B_1, & I_5 - iI_6 = 2B_2, & I_1 - iI_2 = 4B_3, \end{array} \tag{46}$$

then the 15 relations given above can be written as

$$A_pB_q - B_qA_p = 0 \qquad (p, q = 1, 2, 3), \tag{47}$$

$$\begin{aligned} A_3A_1 - A_1A_3 &= A_1, \\ A_3A_2 - A_2A_3 &= -A_2, \\ A_1A_2 - A_2A_1 &= 2A_3, \end{aligned} \tag{48}$$

and

$$\begin{aligned} B_3B_1 - B_1B_3 &= B_1, \\ B_3B_2 - B_2B_3 &= -B_2, \\ B_1B_2 - B_2B_1 &= 2B_3. \end{aligned} \tag{49}$$

These relations are satisfied by the infinitesimal transformations of any representation of the group (43). It should be noted that (47) and (48) hold trivially if we take the I_k to be the matrices I'_k given by (45), for then the matrices A_1, A_2, and A_3 vanish.

Since the relations (48) and (49) are the same as the relations (33), the considerations of the preceding section can be applied here. Thus, if $\mathbf{v}_j$ is an eigenvector (there may be several) of the operator A_3 belonging to the largest eigenvalue of A_3, there are $2j + 1$ eigenvectors

$$\mathbf{v}_k \qquad (k = j, j - 1, \ldots, -j + 1, -j)$$

of A_3, which are transformed by the operators A_1, A_2, A_3 in the way described by the formulas (42), where $\mathbf{v}_{j+1} = \mathbf{v}_{-j-1} = 0$. Now let $L^{(j)}$ be the subspace spanned by *all* the eigenvectors of A_3 belonging to the eigenvalue j. Then, if a vector $\mathbf{v}$ belongs to $L^{(j)}$, the vector $B_q\mathbf{v}$ ($q = 1, 2, 3$) also belongs to $L^{(j)}$. In fact, by (47), we have

$$A_3(B_q\mathbf{v}) = B_q(A_3\mathbf{v}) = B_q(j\mathbf{v}) = jB_q\mathbf{v},$$

so that $B_q\mathbf{v}$ is an eigenvector of A_3 corresponding to the eigenvalue j (or possibly the zero vector), i.e., $B_q\mathbf{v}$ is an element of $L^{(j)}$.

We now repeat for $L^{(j)}$ the argument given in the preceding section, replacing the operators A_k by the operators B_k. Thus, we can construct in $L^{(j)}$ a sequence of vectors

$$\mathbf{v}_{jk'} \qquad (k' = j', j' - 1, \ldots, -j' + 1, -j')$$

which transform by the formulas obtained from (42) by changing j to j' and A_k to B_k. By applying A_2 repeatedly to any vector $\mathbf{v}_{jk'}$, we obtain $2j + 1$ vectors

$$\mathbf{v}_{kk'} \qquad (k = j, j - 1, \ldots, -j + 1, -j).$$

Thus, we finally obtain $(2j + 1)(2j' + 1)$ vectors $\mathbf{v}_{kk'}$, which satisfy the relations

$$\begin{aligned}
A_1\mathbf{v}_{kk'} &= \sqrt{j(j+1) - k(k+1)}\;\mathbf{v}_{k+1,k'},\\
A_2\mathbf{v}_{kk'} &= \sqrt{j(j+1) - k(k-1)}\;\mathbf{v}_{k-1,k'},\\
A_3\mathbf{v}_{kk'} &= k\mathbf{v}_{kk'},\\
B_1\mathbf{v}_{kk'} &= \sqrt{j(j+1) - k'(k'+1)}\;\mathbf{v}_{k,k'+1},\\
B_2\mathbf{v}_{kk'} &= \sqrt{j(j+1) - k'(k'-1)}\;\mathbf{v}_{k,k'-1},\\
B_3\mathbf{v}_{kk'} &= k'\mathbf{v}_{kk'}.
\end{aligned}$$

These relations define the operators A_p and B_q in a space of dimension $(2j + 1)(2j' + 1)$. We can then solve (46) for the infinitesimal transformations which appear in (24). Then, (24) leads to a unique representation of the unimodular group (and hence of the Lorentz group), which is just the representation constructed in Sec. 86.

The last few sections are based on the treatment given by van der Waerden in the book cited in Sec. 86.

93. Auxiliary Formulas. Returning to the formulas of Sec. 88, we have

$$G_\gamma = G_\beta G_\alpha, \tag{50}$$

where the parameters γ_s are given in terms of the α_s and β_s by the formulas (1) and (5), which define the basic group operation. We form the matrix $S(G_\beta, G_\alpha)$, whose elements

$$S_{ik}(G_\beta, G_\alpha) = \frac{\partial \gamma_i}{\partial \beta_k} \qquad (i, k = 1, 2, \ldots, r) \tag{51}$$

depend on the variables α_k and β_k, i.e., on the group elements G_α and G_β. This matrix has already been considered in Sec. 88 for the case

$$\beta_1 = \beta_2 = \cdots = \beta_r = 0,$$

i.e., $G_\beta = E$, where E is the unit element of the group. We now study the properties of the matrix $S(G_\beta, G_\alpha)$ in some detail.

First, we note that by definition

$$S(G_\beta, E) = I,$$

where I is the unit matrix of order r. Next, we prove the formula

$$S(G_\beta, G_\alpha)S(E, G_\beta) = S(E, G_\beta, G_\alpha). \tag{52}$$

To do this, we get $G_\alpha = G_{\alpha'}G_{\alpha''}$, so that

$$G_\gamma = G_\beta G_\alpha = (G_\beta G_{\alpha''})G_{\alpha'} = G_\delta G_{\alpha'},$$

where $G_\delta = G_\beta G_{\alpha''}$. We then apply the rule for differentiating composite functions, obtaining

$$\frac{\partial \gamma_i}{\partial \beta_k} = \sum_{s=1}^{r} \frac{\partial \gamma_i}{\partial \delta_s}\frac{\partial \delta_s}{\partial \beta_k} = \sum_{s=1}^{r} S_{is}(G_\delta, G_{\alpha'})S_{sk}(G_\beta, G_{\alpha''}),$$

which implies

$$S(G_\beta, G_{\alpha''}G_{\alpha'}) = S(G_\delta, G_{\alpha'})S(G_\beta, G_{\alpha''}). \tag{53}$$

Setting $G_\beta = E$, $G_{\alpha''} = G_\beta$, and $G_{\alpha'} = G_\alpha$ in (53), we obtain (52).

For $G_\alpha = G_\beta^{-1}$, we obtain the following expression for the matrix which is the inverse of the matrix $S(E, G_\beta)$:

$$S^{-1}(E, G_\beta) = S(G_\beta, G_\beta^{-1}). \tag{54}$$

In the notation of Sec. 88, we would denote the matrix $S(E, G_\beta)$ by $S(\beta_s)$ and the inverse matrix (54) by $T(\beta_s)$, but now we denote these matrices by $S(G_\beta)$ and $T(G_\beta)$:

$$S(E, G_\beta) = S(G_\beta), \qquad S^{-1}(E, G_\beta) = T(G_\beta).$$

Then we have

$$S(G_\beta)T(G_\beta) = T(G_\beta)S(G_\beta) = E. \tag{55}$$

Since by (52) we have

$$S(G_\beta, G_\alpha) = S(E, G_\gamma)S^{-1}(E, G_\beta) = S(G_\gamma)S^{-1}(G_\beta),$$

the relation (51) can be written in the form

$$\frac{\partial \gamma_i}{\partial \beta_k} = \sum_{s=1}^{r} S_{is}(G_\gamma)T_{sk}(G_\beta). \tag{56}$$

Multiplying both sides of (56) by $T_{mi}(G_\gamma)$, summing over i, and using (55), we obtain

$$\sum_{i=1}^{r} T_{mi}(G_\gamma)\frac{\partial \gamma_i}{\partial \beta_k} = T_{mk}(G_\beta).$$

Differentiating (56) with respect to β_l, we find that

$$\frac{\partial^2 \gamma_i}{\partial \beta_k \partial \beta_l} = \sum_{s,p=1}^{r} \frac{\partial S_{is}(G_\gamma)}{\partial \gamma_p}\frac{\partial \gamma_p}{\partial \beta_l} T_{sk}(G_\beta) + \sum_{s=1}^{r} S_{is}(G_\gamma)\frac{\partial T_{sk}(G_\beta)}{\partial \beta_l}$$

or

$$\frac{\partial^2 \gamma_i}{\partial \beta_k \partial \beta_l} = \sum_{s,p,q=1}^{r} \frac{\partial S_{is}(G_\gamma)}{\partial \gamma_p} S_{pq}(G_\gamma)T_{ql}(G_\beta)T_{sk}(G_\beta) + \sum_{s=1}^{r} S_{is}(G_\gamma)\frac{\partial T_{sk}(G_\beta)}{\partial \beta_l}, \tag{57}$$

where we have substituted for $\partial \gamma_p / \partial \beta_l$ from (56). Interchanging the subscripts k, l and the summation indices s, q in (57), and using the fact that the second partial derivative in (57) does not depend on the order of differentiation, we obtain

$$\sum_{s,p,q=1}^{r} \left(\frac{\partial S_{is}(G_\gamma)}{\partial \gamma_p} S_{pq}(G_\gamma) - \frac{\partial S_{iq}(G_\gamma)}{\partial \gamma_p} S_{ps}(G_\gamma)\right) T_{ql}(G_\beta)T_{sk}(G_\beta)$$
$$= -\sum_{s=1}^{r} S_{is}(G_\gamma)\left(\frac{\partial T_{sk}(G_\beta)}{\partial \beta_l} - \frac{\partial T_{sl}(G_\beta)}{\partial \beta_k}\right) \qquad (i, k, l = 1, 2, \ldots, r). \tag{58}$$

We multiply both sides of (58) by the product $S_{lf}(G_\beta)S_{kg}(G_\beta)T_{hi}(G_\gamma)$ and sum over i, k, and l from 1 to r. Then, using (55), we obtain the following set of equations, which are equivalent to (58):

$$\sum_{i,p=1}^{r}\left(\frac{\partial S_{ig}(G_\gamma)}{\partial\gamma_p}S_{pf}(G_\gamma)-\frac{\partial S_{if}(G_\gamma)}{\partial\gamma_p}S_{pg}(G_\gamma)\right)T_{hi}(G_\gamma)$$
$$=-\sum_{k,l=1}^{r}S_{lf}(G_\beta)S_{kg}(G_\beta)\left(\frac{\partial T_{hk}(G_\beta)}{\partial\beta_l}-\frac{\partial T_{hl}(G_\beta)}{\partial\beta_k}\right)$$
$$(f,\ g,\ h=1,\ 2,\ \ldots,\ r).\qquad(59)$$

It is easy to go back from (59) to (58); this is accomplished by multiplying both sides of (59) by the product $T_{fl_1}(G_\beta)T_{gk_1}(G_\beta)S_{i_1h}(G_\gamma)$ and then summing over f, g, and h.

The left-hand side of (59) depends only on γ_s, while the right-hand side depends on β_s. Thus, since G_α is arbitrary in the formula (50), β_s and γ_s are effectively independent and both sides of (59) must equal the same constant. In particular, we can write

$$-\sum_{k,l=1}^{r}S_{lf}(G_\beta)S_{kg}(G_\beta)\left(\frac{\partial T_{hk}(G_\beta)}{\partial\beta_l}-\frac{\partial T_{hl}(G_\beta)}{\partial\beta_k}\right)=C_{fg}^{(h)},$$

which becomes

$$-\sum_{s,t=1}^{r}S_{ti}(G_\alpha)S_{sk}(G_\alpha)\left(\frac{\partial T_{ps}(G_\alpha)}{\partial\alpha_t}-\frac{\partial T_{pt}(G_\alpha)}{\partial\alpha_s}\right)=C_{ik}^{(p)},\qquad(60)$$

if we make suitable changes of indices. If we set $G_\alpha = E$ in (60), i.e., $\alpha_1 = \alpha_2 = \cdots = \alpha_r = 0$, and bear in mind that $S(E) = I$, we obtain

$$C_{ik}^{(p)}=\left(\frac{\partial T_{pi}(G_\alpha)}{\partial\alpha_k}-\frac{\partial T_{pk}(G_\alpha)}{\partial\alpha_i}\right)_{\alpha_s=0}.\qquad(61)$$

Comparing (61) with formula (14) of Sec. 88, we see that the $C_{ik}^{(p)}$ are the structure constants defined previously. Multiplying both sides of (60) by $T_{il}(G_\alpha)T_{km}(G_\alpha)$, summing over i and k, and using (55), we obtain

$$\frac{\partial T_{pm}(G_\alpha)}{\partial\alpha_l}-\frac{\partial T_{pl}(G_\alpha)}{\partial\alpha_m}=-\sum_{i,k=1}^{r}C_{ik}^{(p)}T_{il}(G_\alpha)T_{km}(G_\alpha).\qquad(62)$$

We now return to formulas (28) and (30) of Sec. 88. It will be recalled that (30) was obtained by equating to zero the term in brackets in (28), for $\alpha_s = 0$ $(s = 1, 2, \ldots, r)$. Using (62), we can easily show that *it follows from* (30) *that the term in brackets in* (28) *vanishes for any* α_s. To see this, we write the second term in brackets in (28) in the form

$$\sum_{j,k=1}^{r} T_{jp}T_{kq}I_jI_k - \sum_{j,k=1}^{r} T_{jq}T_{kp}I_jI_k, \tag{63}$$

where we omit the argument G_α. Interchanging j and k in the second term of (63) and using (30), we obtain

$$\sum_{j,k=1}^{r} T_{jp}T_{kq}(I_jI_k - I_kI_j) = \sum_{j,k,s}^{r} T_{jp}T_{kq}C_{kj}^{(s)}I_s. \tag{64}$$

On the other hand, using (62), we find that the first term in brackets in (28), i.e.,

$$\sum_{j=1}^{r} \left(\frac{\partial T_{jp}}{\partial \alpha_q} - \frac{\partial T_{jq}}{\partial \alpha_p}\right) I_j,$$

is just the negative of the right-hand side of (64). Thus, the term in brackets in (28) vanishes for any α_s, QED.

Finally, along with the matrix $S(G_\beta, G_\alpha)$, we consider the matrix $S'(G_\beta, G_\alpha)$, whose elements are defined by the formula

$$S'_{ik}(G_\beta, G_\alpha) = \frac{\partial \gamma_i}{\partial \alpha_k}. \tag{65}$$

In just the same way as before, we can prove the formulas

$$\begin{aligned} S'(E, G_\alpha) &= I, \\ S'(G_\beta G_\alpha, E) &= S'(G_\beta, G_\alpha)S'(G_\alpha, E), \\ S'^{-1}(G_\alpha, E) &= S'(G_\alpha^{-1}, G_\alpha), \end{aligned} \tag{66}$$

which will be needed later.

94. Construction of a Group from Its Structure Constants. In this section, we discuss in general terms the problem of constructing a group operation and a corresponding group of linear transformations from given structure constants $C_{ik}^{(p)}$, which satisfy the relations (8) and (9) of Sec. 88. This construction is based on a theorem (referred to at the end of Sec. 89) from the theory of partial differential equations. We now formulate this theorem.

Suppose we have the following systems of partial differential equations:

$$\frac{\partial z_i}{\partial x_k} = X_{ik}(x_1, \ldots, x_n; z_1, \ldots, z_m)$$

$$(i = 1, 2, \ldots, m;\ k = 1, 2, \ldots, n). \tag{67}$$

Then we write the condition

$$\frac{\partial^2 z_i}{\partial x_k \partial x_l} = \frac{\partial^2 z_i}{\partial x_l \partial x_k} \qquad (k \neq l),$$

using the system (67). The result is

$$\frac{\partial X_{ik}}{\partial x_l} + \sum_{s=1}^{m} \frac{\partial X_{ik}}{\partial z_s}\frac{\partial z_s}{\partial x_l} = \frac{\partial X_{il}}{\partial x_k} + \sum_{s=1}^{m} \frac{\partial X_{il}}{\partial z_s}\frac{\partial z_s}{\partial x_k}$$

or

$$\frac{\partial X_{ik}}{\partial x_l} + \sum_{s=1}^{m} \frac{\partial X_{ik}}{\partial z_s} X_{sl} = \frac{\partial X_{il}}{\partial x_k} + \sum_{s=1}^{m} \frac{\partial X_{il}}{\partial z_s} X_{sk} \qquad (k \neq l), \qquad (68)$$

where we have substituted for $\partial z_s/\partial x_l$ and $\partial z_s/\partial x_k$ from (67). This equation is a relation between the variables x_k and z_i.

THEOREM. *Let the functions X_{ik} and all their partial derivatives appearing in the relations* (68) *be continuous in a neighborhood of the point $x_k = x_k^{(0)}$, $z_i = z_i^{(0)}$ as well as at the point itself, and let the relations* (68) *hold identically in all the x_k and z_i; these conditions are called the (complete) integrability conditions for the system* (67). *Then, the system* (67) *has a unique solution corresponding to the initial conditions*

$$z_i \Big|_{x_k = x_k^{(0)}} = z_i^{(0)}.$$

We now describe the construction of a group operation and a corresponding group of linear transformations from given structure constants $C_{ik}^{(p)}$ $(i, k, p = 1, 2, \ldots, r)$, satisfying the relations (8) and (9) of Sec. 88. It can be verified that solving the system (62) for the partial derivatives gives a system of partial differential equations for which (8) and (9) are the integrability conditions. Thus, by the theorem just cited, there exists a unique matrix $T(G_\alpha)$ with elements $T_{pq}(G_\alpha)$, where $p, q = 1, 2, \ldots, r$, which satisfies the system (62) and reduces to the unit matrix when $G_\alpha = E$, i.e., when $\alpha_1 = \alpha_2 = \cdots = \alpha_r = 0$. Once we have $T(G_\alpha)$, we can construct its inverse matrix

$$S(G_\alpha) = T^{-1}(G_\alpha).$$

To construct the group operation, we turn to the system (56). The right-hand side of (56) now consists of known functions of β_s and γ_s $(s = 1, 2, \ldots, r)$. Moreover, it can be verified that the system (62) gives the integrability conditions for the system (56). Therefore, there exists a unique solution of the system (56), which satisfies the initial conditions

$$\gamma_i \Big|_{\beta_k = 0} = \alpha_i,$$

and it is just this solution which gives the group operation. The initial conditions express the fact that the element G_γ defined by (50) reduces to G_α when $\beta_1 = \beta_2 = \cdot \cdot \cdot = \beta_r = 0$.

Next, we consider the construction of a group of linear transformations (i.e., a group of matrices of a given order) from given structure constants. As just shown, we already have the matrix $T(G_\alpha)$. Moreover, as remarked at the end of Sec. 89, the vanishing of the term in square brackets in (28), for any choice of indices and any α_s, gives the integrability conditions for the system (24) or (25), and as shown in Sec. 94, this term vanishes if the matrices I_s satisfy the relations (30). Thus, the first step in the solution of our problem is to construct matrices I_s of a given order which satisfy (30). This is a complicated algebraic problem. Then, once we have the matrices I_s, the corresponding system (25) has a unique solution satisfying the initial condition (26), and this solution gives a continuous group of matrices with the specified structure constants $C_{ik}^{(p)}$.

It can be shown that integration of the system (62) with the initial condition $T(E) = I$ reduces to integration of a system of ordinary linear differential equations with constant coefficients, i.e., we have the following result (cited without proof): *Construct the system*

$$\frac{dw_{ik}(t)}{dt} = \delta_{ik} + \sum_{p,q=1}^{r} C_{pq}^{(i)} \alpha_p w_{qk}(t)$$

of ordinary linear differential equations with constant coefficients, where $\delta_{ik} = 0$ *if* $i \neq k$, $\delta_{ii} = 1$, *and* $\alpha_1, \alpha_2, \ldots, \alpha_r$ *are given constants. Then, the functions* $T_{ik}(\alpha_s) = w_{ik}(1)$ *satisfy the system* (62) *and the initial condition* $T(E) = I$.

A detailed treatment of the problem of constructing continuous groups from given structure constants and an investigation of various other topics in the theory of continuous groups can be found in Pontryagin's book.†

95. Integration on a Group. The Orthogonality Property. In Secs. 82 and 83, we proved a variety of relations involving sums of quantities depending on group elements, where the summation extends over all the elements of a discrete group. In the case of a continuous group, summation is replaced by integration with respect to the parameters specifying the group elements. Suppose that we have a continuous group

† L. S. Pontryagin, "Continuous Groups," 2d Russian ed., Gostekhizdat, Moscow, 1954. Translated into German by V. Ziegler as "Topologische Gruppen," Teubner Verlagsgesellschaft, Stuttgart, 1957 (vol. 1), 1958 (vol. 2). The first Russian edition translated into English by E. Lehmer as "Topological Groups," Princeton University Press, Princeton, N.J., 1946.

$\mathcal{G}$, which for some choice of parameters $\alpha_1, \alpha_2, \ldots, \alpha_r$ is in one-to-one correspondence with a bounded closed region V of the real r-dimensional space W_r consisting of the points $(\alpha_1, \alpha_2, \ldots, \alpha_r)$, i.e., a definite point of V corresponds to each element of $\mathcal{G}$, and vice versa. Moreover, we assume that the functions $\varphi_j(\beta_1, \ldots, \beta_r; \alpha_1, \ldots, \alpha_r)$ defining the group operation are continuous and differentiable the required number of times, both within V and on the boundary of V. We also assume that the parameters $\tilde{\alpha}_s$ corresponding to the inverse element G_α^{-1} are continuous functions of the parameters α_s. A group with these properties is said to be *compact*.

To define integration on the group $\mathcal{G}$, we consider the determinant of the matrix $S'(G_\beta, G_\alpha)$ defined by (65) and denote this determinant by

$$\Delta'(G_\beta, G_\alpha) = \left|\frac{\partial \gamma_i}{\partial \alpha_k}\right|. \tag{69}$$

It is an immediate consequence of (66) that

$$\begin{aligned} \Delta'(E, G_\alpha) &= 1, \\ \Delta'(G_\beta G_\alpha, E) &= \Delta'(G_\beta, G_\alpha)\Delta'(G_\alpha, E). \end{aligned}$$

If we write $\delta'(G_\beta) = \Delta'(G_\beta, E)$, we find that

$$\Delta'(G_\beta, G_\alpha) = \frac{\delta'(G_\beta G_\alpha)}{\delta'(G_\alpha)}. \tag{70}$$

Then, since $\delta'(E) = \Delta'(E, E) = 1$, we obtain

$$\Delta'(G_\alpha^{-1}, G_\alpha) = \frac{1}{\delta'(G_\alpha)}. \tag{71}$$

For brevity, we write

$$u'(G_\alpha) = \Delta'(G_\alpha^{-1}, G_\alpha). \tag{72}$$

By the assumptions made above, $u'(G_\alpha)$ is a continuous function in the closed region V, and moreover, it does not vanish in V, since

$$\frac{1}{u'(G_\alpha)} = \delta'(G_\alpha) = \Delta'(G_\alpha, E)$$

is also a continuous function in V. Thus, since $u'(E) = 1$, we see that $u'(G_\alpha)$ and $\delta'(G_\alpha)$ are both positive functions. By (70), the same is true of $\Delta'(G_\beta, G_\alpha)$.

Now let $f(G_\alpha) = f(\alpha_1, \ldots, \alpha_r)$ be any function which is continuous in the closed region V. We define the integral of $f(G_\alpha)$ over the group $\mathcal{G}$ by the formula

$$\int_{\mathcal{G}} f(G_\alpha)\, dG_\alpha = \int_V f(\alpha_1, \ldots, \alpha_r) u'(G_\alpha)\, d\alpha_1 \cdots d\alpha_r, \tag{73}$$

where the integral on the right is an ordinary integral over the region V. Next, we prove that the integral (73) has the following property of *left invariance:*

$$\int_{\mathcal{G}} f(G_\alpha)\, dG_\alpha = \int_{\mathcal{G}} f(G_\beta G_\alpha)\, dG_\alpha, \tag{74}$$

or, in coordinate form,

$$\int_V f(\alpha_1, \ldots, \alpha_r) u'(G_\alpha)\, d\alpha_1 \cdots d\alpha_r = \int_V f(\gamma_1, \ldots, \gamma_r) u'(G_\alpha)\, d\alpha_1 \cdots d\alpha_r, \tag{75}$$

where G_β is any fixed element of $\mathcal{G}$. To see this, we replace the variable element G_α appearing in the left-hand side of (75) by the variable element G_δ, where $G_\alpha = G_\beta G_\delta$; the parameters $\delta_1, \delta_2, \ldots, \delta_r$ still range over the domain V. In the present notation, the Jacobian of this transformation is

$$\left|\frac{\partial \alpha_i}{\partial \delta_k}\right| = \Delta'(G_\beta, G_\delta) = \frac{\delta'(G_\beta G_\delta)}{\delta'(G_\delta)} = \frac{u'(G_\delta)}{u'(G_\beta G_\delta)} = \frac{u'(G_\delta)}{u'(G_\alpha)}$$

so that

$$\begin{aligned}\int_V f(\alpha_1, \ldots, \alpha_r) u'(G_\alpha)\, d\alpha_1 \cdots d\alpha_r &= \int_V f(\alpha_1, \ldots, \alpha_r) u'(G_\alpha) \frac{u'(G_\delta)}{u'(G_\alpha)}\, d\delta_1 \cdots d\delta_r \\ &= \int_{\mathcal{G}} f(G_\beta G_\delta)\, dG_\delta,\end{aligned}$$

which coincides with (74) or (75) after replacing G_δ by G_α.

Similarly, we can construct a *right-invariant* integral. To do this, we introduce the determinant

$$\Delta(G_\beta, G_\alpha) = \left|\frac{\partial \gamma_i}{\partial \beta_k}\right| \tag{76}$$

of the matrix $S(G_\beta, G_\alpha)$ defined by (51). This time we have the relations

$$\begin{aligned}\Delta(G_\alpha, E) &= 1, \\ \Delta(E, G_\beta G_\alpha) &= \Delta(G_\beta, G_\alpha)\Delta(E, G_\beta), \\ \Delta(G_\beta, G_\alpha) &= \frac{\delta(G_\beta, G_\alpha)}{\delta(G_\beta)},\end{aligned}$$

where $\delta(G_\alpha) = \Delta(E, G_\alpha)$. Introducing the positive function

$$u(G_\alpha) = \frac{1}{\delta(G_\alpha)} = \Delta(G_\alpha, G_\alpha^{-1}), \tag{77}$$

we define an integral over $\mathcal{G}$ by the formula

$$\int_V f(\alpha_1, \ldots, \alpha_r) u(G_\alpha)\, d\alpha_1 \cdots d\alpha_r = \int_{\mathcal{G}} f(G_\alpha)\, \widetilde{dG_\alpha}, \tag{78}$$

where the tilde over the differential distinguishes this integral from the integral (73). It is easy to verify that the integral (77) is right invariant, i.e.,

$$\int_{\mathcal{G}} f(G_\alpha)\, \widetilde{dG_\alpha} = \int_{\mathcal{G}} f(G_\alpha G_\beta)\, \widetilde{dG_\alpha}.$$

We now assert that changing G_α to G_α^{-1} in the function being integrated changes a left-invariant integral into a right-invariant integral, and vice versa. To prove this, we differentiate the relation $G_\lambda = G_\alpha G_\beta$ (written in terms of parameters) with respect to α_k and then set $G_\beta = G_\alpha^{-1}$ in all subsequent formulas. The result is

$$\frac{\partial \lambda_i}{\partial \alpha_k} + \sum_{s=1}^{r} \frac{\partial \lambda_i}{\partial \beta_s} \frac{\partial \beta_s}{\partial \alpha_k} = 0,$$

so that

$$\left| \frac{\partial \lambda_i}{\partial \alpha_k} \right| = (-1)^r \left| \frac{\partial \lambda_i}{\partial \beta_k} \right| \left| \frac{\partial \beta_i}{\partial \alpha_k} \right|;$$

then, using (69), (72), (76), and (77), we obtain

$$\left| \frac{\partial \beta_i}{\partial \alpha_k} \right| = (-1)^r \frac{\Delta(G_\alpha, G_\alpha^{-1})}{\Delta'(G_\alpha, G)_\alpha^{-1}} = (-1)^r \frac{u(G_\alpha)}{u'(G_\alpha^{-1})}. \tag{79}$$

We also note that the identity

$$\left| \frac{\partial \beta_i}{\partial \alpha_k} \right| \left| \frac{\partial \alpha_i}{\partial \beta_k} \right| = 1$$

implies that

$$\left| \frac{\partial \alpha_i}{\partial \beta_k} \right| = (-1)^r \frac{u'(G_\alpha^{-1})}{u(G_\alpha)}. \tag{80}$$

Now consider the integral (75) with the variable element G_α replaced by the variable element G_β. Using (79) to replace $d\beta_1 \cdots d\beta_r$ by $d\alpha_1 \cdots d\alpha_r$ in the usual way, we obtain

$$\begin{aligned} \int_V f(\beta_1, \ldots, \beta_r) u'(G_\beta)\, d\beta_1 \cdots d\beta_r \\ &= \int_V f(\tilde{\alpha}_1, \ldots, \tilde{\alpha}_r) u'(G_\beta) \left\| \frac{\partial \beta_i}{\partial \alpha_k} \right\| d\alpha_1 \cdots d\alpha_r \\ &= \int_V f(\tilde{\alpha}_1, \ldots, \tilde{\alpha}_r) u'(G_\beta) \frac{u(G_\alpha)}{u'(G_\beta)}\, d\alpha_1 \cdots d\alpha_r, \end{aligned}$$

where

$$\left\| \frac{\partial \beta_i}{\partial \alpha_k} \right\|$$

denotes the absolute value of the determinant (79). Canceling $u'(G_\beta)$ and replacing β by α in the left-hand side, we obtain

$$\int_V f(\alpha_1, \ldots, \alpha_r)u'(G_\alpha)\, d\alpha_1 \cdots d\alpha_r = \int_V f(\tilde{\alpha}_1, \ldots, \tilde{\alpha}_r)u(G_\alpha)\, d\alpha_1 \cdots d\alpha_r, \tag{81}$$

as asserted. Similarly, if we start with the integral (78) and use (80) to replace $d\alpha_1 \cdots d\alpha_r$ by $d\beta_1 \cdots d\beta_r$, we obtain

$$\begin{aligned}\int_V f(\alpha_1, \ldots, \alpha_r)u(G_\alpha)\, d\alpha_1 \cdots d\alpha_r &= \int_V f(\tilde{\beta}_1, \ldots, \tilde{\beta}_r)u(G_\alpha) \left\| \frac{\partial \alpha_i}{\partial \beta_k} \right\| d\beta_1 \cdots d\beta_r \\ &= \int_V f(\tilde{\beta}_1, \ldots, \tilde{\beta}_r)u(G_\alpha) \frac{u'(G_\beta)}{u(G_\alpha)}\, d\beta_1 \cdots d\beta_r.\end{aligned}$$

Canceling $u(G_\alpha)$ and replacing β by α in the right-hand side, we obtain

$$\int_V f(\alpha_1, \ldots, \alpha_r)u(G_\alpha)\, d\alpha_1 \cdots d\alpha_r = \int_V f(\tilde{\alpha}_1, \ldots, \tilde{\alpha}_r)u'(G_\alpha)\, d\alpha_1 \cdots d\alpha_r, \tag{82}$$

as asserted.

So far, we have not used the compactness of the group $\mathcal{G}$, and the region V can even be infinite, provided that the function $f(\alpha_1, \ldots, \alpha_r)$ is such that all the integrals just written are actually meaningful. We now use the compactness of $\mathcal{G}$ to prove that $u(G_\alpha) = u'(G_\alpha)$. The proof goes as follows: Consider the determinant

$$D(G_\beta, G_\alpha) = \left| \frac{\partial \mu_i}{\partial \beta_k} \right|$$

where

$$G_\mu = G_\alpha^{-1} G_\beta G_\alpha. \tag{83}$$

Setting $G_\alpha = G_{\alpha''}G_{\alpha'}$, we can write (83) as

$$G_\mu = (G_{\alpha''}G_{\alpha'})^{-1}G_\beta(G_{\alpha''}G_{\alpha'}) = G_{\alpha'}^{-1}G_\nu G_{\alpha'},$$

where $G_\nu = G_{\alpha''}^{-1}G_\beta G_{\alpha''}$. Therefore

$$\left| \frac{\partial \mu_i}{\partial \beta_k} \right| = \left| \frac{\partial \mu_i}{\partial \nu_k} \right| \left| \frac{\partial \nu_i}{\partial \beta_k} \right| = D(G_\nu, G_{\alpha'})D(G_\beta, G_{\alpha''}),$$

so that

$$D(G_\beta, G_{\alpha''}G_{\alpha'}) = D(G_{\alpha''}^{-1}G_\beta G_{\alpha''}, G_{\alpha'})D(G_\beta, G_{\alpha''}). \tag{84}$$

Setting $G_\beta = E$ in (84), we obtain

$$D(E, G_{\alpha''}G_{\alpha'}) = D(E, G_{\alpha'})D(E, G_{\alpha''}). \tag{85}$$

If we introduce the numerical function

$$\eta(G_\alpha) = D(E, G_\alpha),$$

then (85) becomes

$$\eta(G_{\alpha''}G_{\alpha'}) = \eta(G_{\alpha''})\eta(G_{\alpha'}), \tag{86}$$

i.e., multiplying group elements corresponds to multiplying corresponding values of the function $\eta(G_\alpha)$. Obviously, we have

$$\eta(E) = 1 \quad \text{and} \quad \eta(G_\alpha)\eta(G_\alpha^{-1}) = 1, \tag{87}$$

and the function $\eta(G_\alpha)$ is continuous and positive in the closed region V.

Next, we use the compactness of $\mathcal{G}$ to show that $\eta(G_\alpha) = 1$ for any element G_α. Suppose that $\eta(G_\alpha) \neq 1$ for some element G_α. If $\eta(G_\alpha) < 1$, then by (87), $\eta(G_\alpha^{-1}) > 1$; thus, we can always assume that $\eta(G_\alpha) > 1$. Then we have

$$\eta(G_\alpha^n) = [\eta(G_\alpha)]^n \to \infty \qquad \text{as } n \to \infty.$$

This contradicts the fact that the function $\eta(G_\alpha)$, being continuous in the closed domain V, must be bounded in V. We can now establish the relation between $u(G_\alpha)$ and $u'(G_\alpha)$. Let

$$G_\gamma = G_\beta G_\alpha = G_\alpha^{-1}(G_\alpha G_\beta)G_\alpha = G_\alpha^{-1}G_\rho G_\alpha,$$

where $G_\rho = G_\alpha G_\beta$. Then we have

$$\left|\frac{\partial \gamma_i}{\partial \beta_k}\right| = \Delta(G_\beta, G_\alpha),$$

while, on the other hand,

$$\left|\frac{\partial \gamma_i}{\partial \beta_k}\right| = \left|\frac{\partial \gamma_i}{\partial \rho_k}\right|\left|\frac{\partial \rho_i}{\partial \beta_k}\right| = D(G_\rho, G_\alpha)\Delta'(G_\alpha, G_\beta),$$

i.e.,

$$\Delta(G_\beta, G_\alpha) = D(G_\alpha G_\beta, G_\alpha)\Delta'(G_\alpha, G_\beta).$$

Setting $G_\beta = G_\alpha^{-1}$, we obtain

$$\Delta(G_\alpha^{-1}, G_\alpha) = D(E, G_\alpha)\Delta'(G_\alpha, G_\alpha^{-1}),$$

i.e.,

$$u(G_\alpha^{-1}) = \eta(G_\alpha)u'(G_\alpha^{-1}),$$

or

$$u(G_\alpha^{-1}) = u'(G_\alpha^{-1})$$

for any G_α, since $\eta(G_\alpha) = 1$. Thus, for compact groups, the left-invariant integral (73) coincides with the right-invariant integral (78). Moreover, it follows from (81) and (82) that this integral also coincides with

$$\int_V f(\bar{\alpha}_1, \ldots, \bar{\alpha}_r)u(G_\alpha)\,d\alpha_1 \cdots d\alpha_r = \int_V f(\bar{\alpha}_1, \ldots, \bar{\alpha}_r)u'(G_\alpha)\,d\alpha_1 \cdots d\alpha_r.$$

For noncompact groups, the left-invariant integral can be different from the right-invariant integral. As an example, consider the group of linear transformations of the form

$$z' = e^{\alpha_1}z + \alpha_2,$$

where α_1 and α_2 vary from $-\infty$ to $+\infty$. In this case, $r = 2$ and V is the whole plane. The product of two successive transformations

$$z' = e^{\alpha_1}z + \alpha_2, \qquad z'' = e^{\beta_1}z' + \beta_2$$

is

$$z'' = e^{\beta_1+\alpha_1}z + (\alpha_2 e^{\beta_1} + \beta_2),$$

i.e.,

$$\gamma_1 = \varphi_1(\beta_1, \beta_2; \alpha_1, \alpha_2) = \beta_1 + \alpha_1,$$
$$\gamma_2 = \varphi_2(\beta_1, \beta_2; \alpha_1, \alpha_2) = \alpha_2 e^{\beta_1} + \beta_2.$$

The unit element has parameters $\alpha_1 = \alpha_2 = 0$, and the element G_α^{-1} has parameters $\tilde{\alpha}_1 = -\alpha_1$, $\tilde{\alpha}_2 = -\alpha_2 e^{-\alpha_1}$. Moreover, elementary calculations show that

$$\Delta'(G_\beta, G_\alpha) = \begin{vmatrix} 1 & 0 \\ 0 & e^{\beta_1} \end{vmatrix} = e^{\beta_1}, \qquad \delta'(G_\alpha) = e^{\alpha_1}, \qquad u'(G_\alpha) = e^{-\alpha_1},$$

$$\Delta(G_\beta, G_\alpha) = \begin{vmatrix} 1 & 0 \\ \alpha_2 e^{\beta_1} & 1 \end{vmatrix} = 1, \qquad \delta(G_\alpha) = u(G_\alpha) = 1.$$

Thus, the left-invariant integral has the form

$$\int_{-\infty}^{\infty}\int_{-\infty}^{\infty} f(\alpha_1, \alpha_2)e^{-\alpha_1}\,d\alpha_1\,d\alpha_2,$$

and the right-invariant integral has the form

$$\int_{-\infty}^{\infty}\int_{-\infty}^{\infty} f(\alpha_1, \alpha_2)\,d\alpha_1\,d\alpha_2.$$

We now show that in proving the equality of the right-invariant and left-invariant integrals, i.e., the relation $u(G_\alpha) = u'(G_\alpha)$, we can replace the requirement that the group $\mathcal{G}$ be compact by other requirements. Let $\mathcal{G}'$ be the subgroup of $\mathcal{G}$, consisting either of elements of the form

$$G_\alpha G_\beta G_\alpha^{-1} G_\beta^{-1} \tag{88}$$

or of products of any number of such elements, where G_α and G_β are elements of $\mathcal{G}$; $\mathcal{G}'$ is called the *commutator subgroup* of $\mathcal{G}$.† It is easy to see that if G_γ is an element of the form (88), then so is G_γ^{-1}. Moreover, if G_γ is an element of the form (88), then so is $G_\delta G_\gamma G_\delta^{-1}$, where G_δ is any element of $\mathcal{G}$. Thus, the subgroup $\mathcal{G}'$ generated by the elements (88) is a normal subgroup of $\mathcal{G}$. $\mathcal{G}'$ reduces to the unit element if and only if all the elements (88) equal E, i.e., if and only if $\mathcal{G}$ is an Abelian group. $\mathcal{G}'$ may coincide with $\mathcal{G}$, and then $\mathcal{G}$ is said to be *perfect;* in particular, this is the case if $\mathcal{G}$ is a simple non-Abelian group.

† Cf. Prob. 44 of Chap. 7.

It follows from (86) and (87) that $\eta(G_\alpha G_\beta G_\alpha^{-1} G_\beta^{-1}) = 1$, that $\eta(G_\gamma) = 1$ for all G_γ in $\mathcal{G}'$, and that $\eta(G_\alpha)$ has the same value for all the elements belonging to a coset of $\mathcal{G}'$. Thus, $\eta(G_\alpha)$ has a well-defined value for every element of the factor group of $\mathcal{G}'$ in $\mathcal{G}$. If $\mathcal{G}'$ coincides with $\mathcal{G}$, then $\eta(G_\alpha) = 1$ for any $\mathcal{G}_\alpha$ in $\mathcal{G}$. Moreover, if the factor group of $\mathcal{G}'$ in $\mathcal{G}$ is compact, we also have $\eta(G_\alpha) = 1$ for all G_α in $\mathcal{G}$, and consequently $u(G_\alpha) = u'(G_\alpha)$.

The left-invariance and right-invariance of the integral over a continuous group are analogous to the following property of a finite group: If G_t is a fixed element of the group and G_s is a variable element, then the products G_sG_t and G_tG_s run through all the elements of the group just once (Sec. 62). We used this property to prove that every representation of a finite group is equivalent to a unitary representation (Sec. 72) and to prove various orthogonality relations for finite groups (Secs. 82, 83). By using the invariant integral, we can prove analogous results for compact continuous groups. Thus, if $A(G_\alpha)$ and $B(G_\alpha)$ are two inequivalent, irreducible, unitary representations of a compact group, we have the following *orthogonality relation:*

$$\int_V \{A(G_\alpha)\}_{ij}\{B(G_\alpha)\}_{kl}\, u(G_\alpha)\, d\alpha_1 \cdots d\alpha_r = 0,$$

involving the matrix elements $\{A(G_\alpha)\}_{ij}$ and $\{B(G_\alpha)\}_{kl}$. For a single irreducible representation, we have

$$\int_V \{A(G_\alpha)\}_{ij}\{\overline{A(G_\alpha)}\}_{kl}\, u(G_\alpha)\, d\alpha_1 \cdots d\alpha_r = \frac{\delta_{ik}\delta_{jl}}{p}\int_V u(G_\alpha)\, d\alpha_1 \cdots d\alpha_r,$$

where p is the order of the matrices. Moreover, the *characters* of the representations $A(G_\alpha)$ and $B(G_\alpha)$, defined by

$$\chi(G_\alpha) = \sum_{i=1}^{p} \{A(G_\alpha)\}_{ii} \qquad \text{and} \qquad \chi'(G_\alpha) = \sum_{i=1}^{q} \{B(G_\alpha)\}_{ii},$$

where p, q are the orders of the matrices $A(G_\alpha)$, $B(G_\alpha)$, respectively, obey the relations

$$\int_V \chi(G_\alpha)\overline{\chi'(G_\alpha)}u(G_\alpha)\, d\alpha_1 \cdots d\alpha_r = 0 \tag{89}$$

and
$$\int_V \chi(G_\alpha)\overline{\chi(G_\alpha)}u(G_\alpha)\, d\alpha_1 \cdots d\alpha_r = \int_V u(G_\alpha)\, d\alpha_1 \cdots d\alpha_r. \tag{90}$$

96. Examples

1. Consider the Abelian group consisting of the rotations of a plane about the origin. Then $r = 1$, and the single parameter α gives the angle of rotation. We assume that α lies in the interval $[0, 2\pi]$, where the end points 0 and 2π are identified. Consecutive rotations through

the angles α and β lead to a rotation through the angle $\beta + \alpha$, where, if necessary, we subtract 2π to make $\beta + \alpha$ lie in the interval $[0, 2\pi]$, i.e., we reduce $\beta + \alpha$ modulo 2π. Thus, we have

$$\Delta(G_\beta, G_\alpha) = \frac{\partial}{\partial\beta}(\beta + \alpha) = 1, \qquad \Delta'(G_\beta, G_\alpha) = \frac{\partial}{\partial\alpha}(\beta + \alpha) = 1,$$

so that $u(G_\alpha) = u'(G_\alpha) = 1$. We know from Sec. 73 that this group has the one-dimensional, irreducible, unitary representations $e^{im\alpha}$ ($m = 0, \pm1, \pm2, \ldots$). In this case, (89) and (90) give the familiar formulas

$$\int_0^{2\pi} e^{im_1\alpha}\overline{e^{im_2\alpha}}\,d\alpha = \int_0^{2\pi} e^{i(m_1-m_2)\alpha}\,d\alpha = \begin{cases} 0 & \text{for } m_1 \neq m_2, \\ 2\pi & \text{for } m_1 = m_2. \end{cases}$$

We note that as a result of being reduced modulo 2π, the sum $\beta + \alpha$ fails to be continuous or to have derivatives when $\beta + \alpha = 2\pi$.

2. Next consider the three-dimensional rotation group, which we parameterize somewhat differently than in Sec. 90, i.e., a rotation through the angle ω about an axis making angles α, β, and γ with the x, y, and z-axes is described by the four parameters

$$\begin{aligned} a_0 &= \cos \tfrac{1}{2}\omega, \\ a_1 &= \cos\alpha \sin \tfrac{1}{2}\omega, \\ a_2 &= \cos\beta \sin \tfrac{1}{2}\omega, \\ a_3 &= \cos\gamma \sin \tfrac{1}{2}\omega, \end{aligned} \tag{91}$$

which obey the relation

$$a_0^2 + a_1^2 + a_2^2 + a_3^2 = 1. \tag{92}$$

Thus, we can choose a_1, a_2, a_3 as our parameters and regard a_0 as a function of a_1, a_2, and a_3. The identity transformation has the parameter values $a_0 = 1$, $a_1 = a_2 = a_3 = 0$.

If we first perform the rotation with parameters a_0, a_1, a_2, a_3 and then perform the rotation with parameters b_0, b_1, b_2, b_3, then it is not hard to verify that the parameters c_0, c_1, c_2, c_3 of the resulting rotation are given by

$$\begin{aligned} c_0 &= a_0b_0 - a_1b_1 - a_2b_2 - a_3b_3, \\ c_1 &= a_0b_1 + a_1b_0 + a_2b_3 - a_3b_2, \\ c_2 &= a_0b_2 - a_1b_3 + a_2b_0 + a_3b_1, \\ c_3 &= a_0b_3 + a_1b_2 - a_2b_1 + a_3b_0. \end{aligned} \tag{93}$$

Regarding a_0 as a function of a_1, a_2, a_3 and using (92), we obtain

$$a_0\frac{\partial a_0}{\partial a_j} + a_j = 0 \qquad (j = 1, 2, 3),$$

so that $\partial a_0/\partial a_j = 0$ when the group element is E. Using this fact, we can easily calculate the Jacobian (76) for $b_0 = 1$, $b_1 = b_2 = b_3 = 0$. The result is

$$\frac{\partial(c_1, c_2, c_3)}{\partial(b_1, b_2, b_3)} = \begin{vmatrix} a_0 & -a_3 & a_2 \\ a_3 & a_0 & -a_1 \\ -a_2 & a_1 & a_0 \end{vmatrix} = a_0(a_0^2 + a_1^2 + a_2^2 + a_3^2)$$
$$= a_0 \sqrt{1 - a_1^2 - a_2^2 - a_3^2}.$$

Thus, the (right-) invariant integral has the form

$$\int_V f(a_1, a_2, a_3) \frac{1}{\sqrt{1 - a_1^2 - a_2^2 - a_3^2}} \, da_1 \, da_2 \, da_3, \tag{94}$$

where the region V is a sphere of radius 1 with its center at the origin of coordinates. We note that the formulas (93) are an immediate consequence of the rule for multiplying *quaternions*, i.e.,

$$c_0 + c_1 i + c_2 j + c_3 k = (a_0 + a_1 i + a_2 j + a_3 k)(b_0 + b_1 i + b_2 j + b_3 k),$$

where the quantities i, j, and k obey the multiplication rules

$$i^2 = j^2 = k^2 = -1, \quad ij = -ji = k, \quad jk = -kj = i, \quad ki = -ik = j.$$

It is not hard to show that the parameters a_0, a_1, a_2, a_3 are related to the Eulerian angles α, β, γ by the formulas

$$a_0 = \cos \tfrac{1}{2}\beta \cos \tfrac{1}{2}(\alpha + \gamma), \qquad a_1 = \sin \tfrac{1}{2}\beta \cos \tfrac{1}{2}(\gamma - \alpha),$$
$$a_2 = \sin \tfrac{1}{2}\beta \sin \tfrac{1}{2}(\gamma - \alpha), \qquad a_3 = \cos \tfrac{1}{2}\beta \sin \tfrac{1}{2}(\alpha + \gamma).$$

In terms of the parameters α, β, γ, the invariant integral (94) becomes

$$\int_V f(\alpha, \beta, \gamma) \sin \beta \sin^2 \tfrac{1}{2}(\alpha - \gamma) \, d\alpha \, d\beta \, d\gamma,$$

and the region V is suitably modified, i.e., $0 \le \alpha < 2\pi$, $0 \le \beta < \pi$, $0 \le \gamma < 2\pi$. We observe that in the integral (94), the function

$$\frac{1}{a_0} = \frac{1}{\sqrt{1 - a_1^2 - a_2^2 - a_3^2}}$$

becomes infinite if $\omega = \pi$. This is related to the fact that the formulas (91) for a_1, a_2, and a_3 contain $\sin \tfrac{1}{2}\omega$ instead of ω. However, in this regard, it should be noted that the properties discussed in Sec. 95 in connection with compactness need hold only for a suitable choice of parameters and may be lost for another choice of parameters. Moreover, the three-dimensional rotation group exhibits the same singularities in continuity and differentiability as pointed out at the end of Example 1, for the case of the two-dimensional rotation group. Finally, we note that the left-invariant and right-invariant integrals of the three-dimensional rotation group coincide; this is an immediate consequence of the fact that the rotation group is a simple non-Abelian group.

3. Now consider the Lorentz group, which, as we saw in Sec. 70, is homomorphic to the group of linear transformations

$$\begin{aligned} x_1' &= a_0x_1 + a_1x_2, \\ x_2' &= a_2x_1 + a_3x_2, \qquad (a_0a_3 - a_1a_2 = 1), \end{aligned} \tag{95}$$

with determinant 1. The unit element has the parameters $a_0 = a_3 = 1$, $a_1 = a_2 = 0$. We can regard a_0 as a function of a_1, a_2, a_3 and choose as our parameters the real and imaginary parts of a_1, a_2, a_3. The group operation reduces to multiplication of second order matrices, and we have

$$\begin{aligned} c_0 &= b_0a_0 + b_1a_2, \qquad c_1 = b_0a_1 + b_1a_3, \\ c_2 &= b_2a_0 + b_3a_2, \qquad c_3 = b_2a_1 + b_3a_3. \end{aligned} \tag{96}$$

If we write $a_k = \alpha_k' + i\alpha_k''$ $(k = 1, 2, 3)$, the parameters of the group are $\alpha_1', \alpha_1'', \alpha_2', \alpha_2'', \alpha_3', \alpha_3''$; we also write $b_k = \beta_k' + i\beta_k''$ and $c_k = \gamma_k' + i\gamma_k''$. To define the invariant integrals, we have to calculate the Jacobians

$$\frac{\partial(\gamma_1', \gamma_1'', \gamma_2', \gamma_2'', \gamma_3', \gamma_3'')}{\partial(\beta_1', \beta_1'', \beta_2', \beta_2'', \beta_3', \beta_3'')}$$

for $\beta_1' = \beta_1'' = \beta_2' = \beta_2'' = \beta_3'' = 0$, $\beta_3' = 1$, and

$$\frac{\partial(\gamma_1', \gamma_1'', \gamma_2', \gamma_2'', \gamma_3', \gamma_3'')}{\partial(\alpha_1', \alpha_1'', \alpha_2', \alpha_2'', \alpha_3', \alpha_3'')}$$

for $\alpha_1' = \alpha_1'' = \alpha_2' = \alpha_2'' = \alpha_3'' = 0$, $\alpha_3' = 1$. (It does not matter that we have $\alpha_3' = 1$ instead of $\alpha_3' = 0$ when the group element is E.) In both cases, we obtain the same invariant integral,

$$\int_V f(\alpha_1', \alpha_1'', \alpha_2', \alpha_2'', \alpha_3', \alpha_3'') \frac{1}{\alpha_3'^2 + \alpha_3''^2}\, d\alpha_1'\, d\alpha_1''\, d\alpha_2'\, d\alpha_2''\, d\alpha_3'\, d\alpha_3'', \tag{97}$$

where the region V is the whole six-dimensional space. The fact that the left-invariant and right-invariant integrals coincide follows from the fact that the commutator $\mathcal{G}'$ of the group $\mathcal{G}$ of transformations (95), i.e., the subgroup generated by the elements of the form $G_\alpha G_\beta G_\alpha^{-1} G_\beta^{-1}$ (cf. Sec. 95), coincides with the whole group $\mathcal{G}$, so that $\mathcal{G}$ is perfect (Sec. 95). In fact, it is not hard to show that $\mathcal{G}'$ does not reduce to either the unit element E or the normal subgroup consisting of E and $-E$.

A simple method for calculating the density $u(G_\alpha) = u'(G_\alpha)$ appearing in the invariant integral (97) is based on the following lemma, which uses the idea of an analytic function of several complex variables:

LEMMA. *Let $w_s = u_s + iv_s$ $(s = 1, 2, \ldots, k)$ be analytic functions of the complex variables $z_s = x_s + iy_s$ $(s = 1, 2, \ldots, k)$. Then the Jacobian*

$$\frac{\partial(u_1, v_1, \ldots, u_k, v_k)}{\partial(x_1, y_1, \ldots, x_k, y_k)}$$

equals the square of the absolute value of the Jacobian

$$\frac{\partial(w_1, \ldots, w_k)}{\partial(z_1, \ldots, z_k)}.$$

PROOF. Using the *Cauchy-Riemann Equations*

$$\frac{\partial u_i}{\partial x_k} = \frac{\partial v_i}{\partial y_k}, \qquad \frac{\partial v_i}{\partial x_k} = -\frac{\partial u_i}{\partial y_k},$$

we can write

$$\frac{\partial(u_1, v_1, \ldots, u_k, v_k)}{\partial(x_1, y_1, \ldots, x_k, y_k)} = \begin{vmatrix} a_{11} & -b_{11} & a_{12} & -b_{12} & \cdots & a_{1k} & -b_{1k} \\ b_{11} & a_{11} & b_{12} & a_{12} & \cdots & b_{1k} & a_{1k} \\ \cdots & \cdots & \cdots & \cdots & \cdots & \cdots & \cdots \\ a_{k1} & -b_{k1} & a_{k2} & -b_{k2} & \cdots & a_{kk} & -b_{kk} \\ b_{k1} & a_{k1} & b_{k2} & a_{k2} & \cdots & b_{kk} & a_{kk} \end{vmatrix},$$

where

$$a_{ik} = \frac{\partial u_i}{\partial x_k}, \qquad b_{ik} = \frac{\partial v_i}{\partial x_k}.$$

If to each odd column we add the next column multiplied by $-i$, we obtain the determinant

$$\begin{vmatrix} c_{11} & -b_{11} & c_{12} & -b_{12} & \cdots & c_{1k} & -b_{1k} \\ -ic_{11} & a_{11} & -ic_{12} & a_{12} & \cdots & -ic_{1k} & a_{1k} \\ \cdots & \cdots & \cdots & \cdots & \cdots & \cdots & \cdots \\ c_{k1} & -b_{k1} & c_{k2} & -b_{k2} & \cdots & c_{kk} & -b_{kk} \\ -ic_{k1} & a_{k1} & -ic_{k2} & a_{k2} & \cdots & -ic_{kk} & a_{kk} \end{vmatrix},$$

where

$$c_{ik} = a_{ik} + ib_{ik}.$$

Next, to each even row we add the preceding row multiplied by i; the result is

$$\begin{vmatrix} c_{11} & -b_{11} & c_{12} & -b_{12} & \cdots & c_{1k} & -b_{1k} \\ 0 & \overline{c_{11}} & 0 & \overline{c_{12}} & \cdots & 0 & \overline{c_{1k}} \\ \cdots & \cdots & \cdots & \cdots & \cdots & \cdots & \cdots \\ c_{k1} & -b_{k1} & c_{k2} & -b_{k2} & \cdots & c_{kk} & -b_{kk} \\ 0 & \overline{c_{k1}} & 0 & \overline{c_{k2}} & \cdots & 0 & \overline{c_{kk}} \end{vmatrix}.$$

Finally, we move all the odd columns to the left and all the odd rows upward. The result is

$$\begin{vmatrix} c_{11} & c_{12} & \cdots & c_{1k} & -b_{11} & -b_{12} & \cdots & -b_{1k} \\ c_{21} & c_{22} & \cdots & c_{2k} & -b_{21} & -b_{22} & \cdots & -b_{2k} \\ \cdots & \cdots & \cdots & \cdots & \cdots & \cdots & \cdots & \cdots \\ c_{k1} & c_{k2} & \cdots & c_{kk} & -b_{k1} & -b_{k2} & \cdots & -b_{kk} \\ 0 & 0 & \cdots & 0 & \overline{c_{11}} & \overline{c_{12}} & \cdots & \overline{c_{1k}} \\ 0 & 0 & \cdots & 0 & \overline{c_{21}} & \overline{c_{22}} & \cdots & \overline{c_{2k}} \\ \cdots & \cdots & \cdots & \cdots & \cdots & \cdots & \cdots & \cdots \\ 0 & 0 & \cdots & 0 & \overline{c_{k1}} & \overline{c_{k2}} & \cdots & \overline{c_{kk}} \end{vmatrix},$$

from which it follows that

$$\frac{\partial(u_1, v_1, \ldots, u_k, v_k)}{\partial(x_1, y_1, \ldots, x_k, y_k)} = \begin{vmatrix} c_{11} & \cdots & c_{1k} \\ \cdots & \cdots & \cdots \\ c_{k1} & \cdots & c_{kk} \end{vmatrix} \begin{vmatrix} \overline{c_{11}} & \cdots & \overline{c_{1k}} \\ \cdots & \cdots & \cdots \\ \overline{c_{k1}} & \cdots & \overline{c_{kk}} \end{vmatrix} = \left| \frac{\partial(w_1, \ldots, w_k)}{\partial(z_1, \ldots, z_k)} \right|^2,$$

QED.

We can now find the function $u(G_\alpha) = u'(G_\alpha)$ appearing in the invariant integral (97). According to the lemma, we have to calculate the Jacobians

$$\frac{\partial(c_1, c_2, c_3)}{\partial(b_1, b_2, b_3)} \quad \text{for } b_0 = b_3 = 1,\ b_1 = b_2 = 0, \tag{98}$$

and

$$\frac{\partial(c_1, c_2, c_3)}{\partial(a_1, a_2, a_3)} \quad \text{for } a_0 = a_3 = 1,\ a_1 = a_2 = 0. \tag{99}$$

It follows from $a_0a_3 - a_1a_2 = 1$ that

$$-a_2 + a_3 \frac{\partial a_0}{\partial a_1} = 0, \qquad -a_1 + a_3 \frac{\partial a_0}{\partial a_2} = 0, \qquad a_0 + a_3 \frac{\partial a_0}{\partial a_3} = 0. \tag{100}$$

Moreover, it follows from (96) that

$$\begin{aligned} &\frac{\partial c_1}{\partial a_1} = b_0, && \frac{\partial c_1}{\partial a_2} = 0, && \frac{\partial c_1}{\partial a_3} = b_1, \\ &\frac{\partial c_2}{\partial a_1} = b_2 \frac{\partial a_0}{\partial a_1}, && \frac{\partial c_2}{\partial a_2} = b_2 \frac{\partial a_0}{\partial a_2} + b_3, && \frac{\partial c_2}{\partial a_3} = b_2 \frac{\partial a_0}{\partial a_3}, \\ &\frac{\partial c_3}{\partial a_1} = b_2, && \frac{\partial c_3}{\partial a_2} = 0, && \frac{\partial c_3}{\partial a_3} = b_3. \end{aligned}$$

Using (100), we find that (99) equals

$$(b_0b_3 - b_1b_2)b_3 = b_3.$$

It follows from the lemma that

$$u'(G_\alpha) = \frac{1}{|a_3|^2} = \frac{1}{\alpha_3'^2 + \alpha_3''^2},$$

which implies (97). We obtain the same result for $u(G_\alpha)$ by using (98).

PROBLEMS†

1. How many structure constants does an r-parameter group have? (Do not count the structure constants $C_{ii}^{(s)}$, which vanish, and if $i \neq j$, count $C_{ij}^{(s)}$ and $C_{ji}^{(s)}$ as one, since $C_{ij}^{(s)} = -C_{ji}^{(s)}$.)
2. Let the coefficients $a_{ik}^{(s)}$, $a_{ikl}^{(s)}$, and $b_{ikl}^{(s)}$ be defined as in (5), and set

$$A_{ikl}^{(s)} = a_{ikl}^{(s)} + a_{kil}^{(s)} = A_{kil}^{(s)},$$
$$B_{ikl}^{(s)} = b_{ikl}^{(s)} + b_{ilk}^{(s)} = B_{ilk}^{(s)}.$$

Show that

$$\sum_{s=1}^{r} (a_{is}^{(t)}a_{kl}^{(s)} - a_{ik}^{(s)}a_{sl}^{(t)}) = A_{ikl}^{(t)} - B_{ikl}^{(t)},$$

where $i, k, l, t = 1, 2, \ldots, r$.
3. Prove formula (9) of Sec. 88.

Henceforth, if A and B are any two $n \times n$ matrices (or the corresponding transformations), we denote the quantity $AB - BA$ by $[A, B]$.
4. Show that if A, B, and C are any $n \times n$ matrices, then

$$[[A, B], C] + [[B, C], A] + [[C, A], B] = 0.$$

This result is known as *Jacobi's identity.*
5. Let A and B be any two $n \times n$ matrices. Show that

$$e^{-A}Be^{A} = B + \frac{1}{1!}[B, A] + \frac{1}{2!}[[B, A], A] + \cdots.$$

(For the theory of e^A, see Prob. 4 of the Appendix.)
6. Let $\mathcal{G}_1$ be the *Euclidean group,* i.e., the group of all real inhomogeneous linear transformations of the form

$$x' = x\cos\theta - y\sin\theta + a,$$
$$y' = x\sin\theta + y\cos\theta + b \qquad (0 \le \theta < 2\pi,\ -\infty < a, b < \infty).$$

Let $\mathcal{G}_2$ be the real, proper, orthochronous, inhomogeneous Lorentz group in two variables, i.e., the group of all real inhomogeneous linear transformations of the form

$$x' = x\cosh\theta + y\sinh\theta + a,$$
$$y' = x\sinh\theta + y\cosh\theta + b \qquad (-\infty < \theta < \infty,\ -\infty < a, b < \infty).$$

† All equation numbers refer to the corresponding equations of Chap. 9. For hints and answers, see p. 453. These problems are by J. S. Lomont.

Note that both $\mathcal{G}_1$ and $\mathcal{G}_2$ are subgroups of the group $\mathfrak{J}_2$ of Prob. 2 of Chap. 8. Using the faithful three-dimensional representation of $\mathfrak{J}_2$ given there, find the corresponding infinitesimal transformations of $\mathcal{G}_1$ and $\mathcal{G}_2$, together with the appropriate commutation relations (30).

7. Let I_1, I_2, and I_3 be the infinitesimal transformations of the three-dimensional rotation group $\mathcal{R}$, which satisfy the commutation relations (31), and let $J = I_1^2 + I_2^2 + I_3^2$. Show that J commutes with I_1, I_2, I_3, and evaluate J in every irreducible representation of $\mathcal{R}$.

8. Let $I_1, I_2, \ldots, I_r$ be a set of infinitesimal operators (transformations) of a group, which satisfy the commutation relations (30):

$$[I_q, I_p] = \sum_{j=1}^{r} C_{pq}^{(j)} I_j \qquad (p, q = 1, 2, \ldots, r).$$

An operator function J of the I_s which commutes with all the I_s purely by virtue of the above commutation relations is called a *Casimir operator* of the group. Let $\mathcal{G}_1$ and $\mathcal{G}_2$ be the groups defined in Prob. 6. Show that

$$J_1 = I_2^2 + I_3^2, \qquad J_2 = e^{2\pi I_1}$$

are Casimir operators of $\mathcal{G}_1$, while

$$J = I_2^2 - I_3^2$$

is a Casimir operator of $\mathcal{G}_2$ (using the infinitesimal operators of Prob. 6).

9. Show that every 3×3 rotation matrix R can be written in the form

$$R = e^{\boldsymbol{\lambda}\cdot\mathbf{I}},$$

where $\boldsymbol{\lambda} = (\lambda_1, \lambda_2, \lambda_3)$ is a real three-dimensional vector and

$$\mathbf{I} = (I_1, I_2, I_3)$$

is the "vector" whose components are the infinitesimal transformations

$$I_1 = \begin{Vmatrix} 0 & 0 & 0 \\ 0 & 0 & -1 \\ 0 & 1 & 0 \end{Vmatrix}, \qquad I_2 = \begin{Vmatrix} 0 & 0 & 1 \\ 0 & 0 & 0 \\ -1 & 0 & 0 \end{Vmatrix}, \qquad I_3 = \begin{Vmatrix} 0 & -1 & 0 \\ 1 & 0 & 0 \\ 0 & 0 & 0 \end{Vmatrix}$$

of the rotation group. (Here $\boldsymbol{\lambda} \cdot \mathbf{I} \equiv (\boldsymbol{\lambda}, \mathbf{I})$ denotes the scalar product of $\boldsymbol{\lambda}$ and $\mathbf{I}$, i.e., $\boldsymbol{\lambda} \cdot \mathbf{I} = \lambda_1 I_1 + \lambda_2 I_2 + \lambda_3 I_3$.) Show that the angle of rotation φ of R is related to $\boldsymbol{\lambda}$ by the formula

$$\varphi = \|\boldsymbol{\lambda}\| + 2n\pi$$

where n is an integer, and show that the axis of rotation is parallel to $\boldsymbol{\lambda}$ (unless $\boldsymbol{\lambda} = 0$). (Cf. Example in Sec. 42.)

10. Show that the 4×4 Lorentz matrix corresponding to the 2×2 unimodular matrix

$$\begin{Vmatrix} a & b \\ c & d \end{Vmatrix}$$

is

$$\begin{Vmatrix} \text{Re}\,(\bar{a}d + \bar{b}c) & -\text{Im}\,(\bar{a}d - \bar{b}c) & \text{Re}\,(\bar{a}c - \bar{b}d) & \text{Re}\,(\bar{a}c + \bar{b}d) \\ \text{Im}\,(\bar{a}d + \bar{b}c) & \text{Re}\,(\bar{a}d - \bar{b}c) & \text{Im}\,(\bar{a}c - \bar{b}d) & \text{Im}\,(\bar{a}c + \bar{b}d) \\ \text{Re}\,(\bar{a}b - \bar{c}d) & -\text{Im}\,(\bar{a}b - \bar{c}d) & \tfrac{1}{2}(\bar{a}a - \bar{b}b - \bar{c}c + \bar{d}d) & \tfrac{1}{2}(\bar{a}a + \bar{b}b - \bar{c}c - \bar{d}d) \\ \text{Re}\,(\bar{a}b + \bar{c}d) & -\text{Im}\,(\bar{a}b + \bar{c}d) & \tfrac{1}{2}(\bar{a}a - \bar{b}b + \bar{c}c - \bar{d}d) & \tfrac{1}{2}(\bar{a}a + \bar{b}b + \bar{c}c + \bar{d}d) \end{Vmatrix} \quad \text{(I)}.$$

(Re z and Im z denote the real and imaginary parts of z, respectively.)

11. Prove that the infinitesimal transformations of the proper orthochronous Lorentz group (i.e., the transformations given by formula (I) of the preceding problem) are given by

$$I_1 = 2\begin{Vmatrix} 0 & 0 & 0 & 0 \\ 0 & 0 & 0 & 0 \\ 0 & 0 & 0 & 1 \\ 0 & 0 & 1 & 0 \end{Vmatrix}, \quad I_2 = 2\begin{Vmatrix} 0 & 1 & 0 & 0 \\ -1 & 0 & 0 & 0 \\ 0 & 0 & 0 & 0 \\ 0 & 0 & 0 & 0 \end{Vmatrix}, \quad I_3 = \begin{Vmatrix} 0 & 0 & -1 & 1 \\ 0 & 0 & 0 & 0 \\ 1 & 0 & 0 & 0 \\ 1 & 0 & 0 & 0 \end{Vmatrix},$$

$$I_4 = \begin{Vmatrix} 0 & 0 & 0 & 0 \\ 0 & 0 & 1 & -1 \\ 0 & -1 & 0 & 0 \\ 0 & -1 & 0 & 0 \end{Vmatrix}, \quad I_5 = \begin{Vmatrix} 0 & 0 & 1 & 1 \\ 0 & 0 & 0 & 0 \\ -1 & 0 & 0 & 0 \\ 1 & 0 & 0 & 0 \end{Vmatrix}, \quad I_6 = \begin{Vmatrix} 0 & 0 & 0 & 0 \\ 0 & 0 & 1 & 1 \\ 0 & -1 & 0 & 0 \\ 0 & 1 & 0 & 0 \end{Vmatrix}.$$

Verify that these matrices satisfy the 15 commutation relations of Sec. 92.

12. Let A be any real 4×4 skew-symmetric matrix, and let

$$G = \begin{Vmatrix} 1 & 0 & 0 & 0 \\ 0 & 1 & 0 & 0 \\ 0 & 0 & 1 & 0 \\ 0 & 0 & 0 & -1 \end{Vmatrix},$$

as in Prob. 11 of Chap. 7. Prove that AG can be expressed in the form

$$AG = \lambda_1 I_1 + \lambda_2 I_2 + \cdots + \lambda_6 I_6,$$

where the λ_i are real numbers, and the I_i are the infinitesimal transformations of the preceding problem.

13. Prove that every real, proper, orthochronous, 4×4 Lorentz matrix L can be expressed in the form

$$L = e^{AG},$$

where A is a real 4×4 skew-symmetric matrix, and G is defined as in the preceding problem.

14. Prove that any real, proper, orthochronous, 4×4 Lorentz matrix L can be written in the form

$$L = \exp\,(\lambda_1 I_1 + \lambda_2 I_2 + \cdots + \lambda_6 I_6),$$

where the λ_s are real numbers, and the I_s are the infinitesimal transformations of the Lorentz group (see Prob. 11).

15. Prove that there exist real, proper, orthochronous, 4×4 Lorentz matrices which do not have 1 as an eigenvalue.

16. Prove that the full homogeneous linear group $\mathfrak{H}_2$ in two variables (cf. Prob. 2 of Chap. 8) is not perfect.
17. Consider the real, proper, orthochronous, *inhomogeneous* Lorentz group $\mathfrak{L}_i$, i.e., the group consisting of the four-dimensional linear transformations of the form

$$\mathbf{x}' = \{L, \mathbf{a}\}\mathbf{x} \equiv L\mathbf{x} + \mathbf{a},$$

where $\mathbf{a}$ is an arbitrary four-dimensional vector, and L is a transformation of the *homogeneous* Lorentz group $\mathfrak{L}_h$ in four variables, i.e., a real, proper, orthochronous, 4×4 Lorentz matrix (cf. Prob. 2 of Chap. 8). Prove that $\mathfrak{L}_i$ is perfect.
18. Let $\mathfrak{L}_i$ be the group defined in the previous problem. Prove that the left-invariant and right-invariant integrals of $\mathfrak{L}_i$ are the same.
19. Find the left-invariant and right-invariant integrals of the following groups:

(a) The additive group of real numbers;
(b) The group $\mathcal{G}_1$ of Prob. 6;
(c) The group $\mathcal{G}_2$ of Prob. 6.

20. Find the left-invariant and right-invariant integrals for the group $\mathfrak{H}_2$ of Prob. 16.
21. Give an example of a continuous group whose left-invariant and right-invariant integrals are the same, but which contains a subgroup whose left-invariant and right-invariant integrals are different.

Appendix

SYSTEMS OF FIRST ORDER LINEAR DIFFERENTIAL EQUATIONS

In this appendix we apply the theory of residues† to the problem of integrating systems of first order linear differential equations with constant coefficients (see Sec. 17). Consider the system

$$\begin{aligned} \dot{x}_1 &= a_{11}x_1 + a_{12}x_2 + \cdots + a_{1n}x_n, \\ \dot{x}_2 &= a_{21}x_1 + a_{22}x_2 + \cdots + a_{2n}x_n, \\ &\cdots\cdots\cdots\cdots\cdots\cdots \\ \dot{x}_n &= a_{n1}x_1 + a_{n2}x_2 + \cdots + a_{nn}x_n. \end{aligned} \tag{1}$$

Here the x_j are functions of t, which we are trying to determine, the $\dot{x}_j$ are their derivatives, and the a_{ij} are given constants. We look for a solution of the system (1) in the form

$$x_j = \sum_R \varphi_j(z)e^{tz} \qquad (j = 1, 2, \ldots, n), \tag{2}$$

where the $\varphi_j(z)$ are certain rational functions (to be determined) of the complex variable z, and the symbol

$$\sum_R f(z)$$

denotes the sum of the residues of the function $f(z)$ at all its singularities which lie at a finite distance from the origin (of the complex plane). The function $\varphi_j(z)e^{tz}$ appearing in (2) depends not only on the complex variable z with respect to which the residues are calculated, but also on the real parameter t. Thus, in general, the sum of the residues will be a function of t. Since z and t vary independently, we can carry out the differentiation of the functions (2) behind the summation sign. In other words, the result of first differentiating the function

$$\varphi_j(z)e^{tz} \tag{3}$$

† Here, we presuppose a knowledge of the rudiments of complex variable theory.

with respect to t and then taking the sum of its residues is the same as first taking the sum of its residues and then differentiating the sum with respect to t. Thus, together with (2), we have the formulas

$$\dot{x}_j = \sum_R z\varphi_j(z)e^{tz} \qquad (j = 1, 2, \ldots, n). \tag{4}$$

Substituting (2) and (4) into the system (1) and transposing all the terms to the left-hand side, we obtain

$$\begin{aligned}
&\sum_R [(a_{11} - z)\varphi_1(z) + a_{12}\varphi_2(z) + \cdots + a_{1n}\varphi_n(z)]e^{tz} = 0,\\
&\sum_R [a_{21}\varphi_1(z) + (a_{22} - z)\varphi_2(z) + \cdots + a_{2n}\varphi_n(z)]e^{tz} = 0,\\
&\cdots\cdots\cdots\cdots\cdots\cdots\cdots\cdots\cdots\cdots\\
&\sum_R [a_{n1}\varphi_1(z) + a_{n2}\varphi_2(z) + \cdots + (a_{nn} - z)\varphi_n(z)]e^{tz} = 0.
\end{aligned}$$

These equations will certainly be satisfied if we equate the expressions in brackets to arbitrary constants, since then we shall have functions of the form Ce^{tz} behind the sign $\sum\limits_R$, and these functions have no singularities at a finite distance from the origin. Denoting the arbitrary constants by $-C_1, -C_2, \ldots, -C_n$, we obtain the following system of linear *algebraic* equations for determining the functions $\varphi_j(z)$:

$$\begin{aligned}
(a_{11} - z)\varphi_1(z) + a_{12}\varphi_2(z) + \cdots + a_{1n}\varphi_n(z) &= -C_1,\\
a_{21}\varphi_1(z) + (a_{22} - z)\varphi_2(z) + \cdots + a_{2n}\varphi_n(z) &= -C_2,\\
\cdots\cdots\cdots\cdots\cdots\cdots\cdots\cdots\cdots&\\
a_{n1}\varphi_1(z) + a_{n2}\varphi_2(z) + \cdots + (a_{nn} - z)\varphi_n(z) &= -C_n.
\end{aligned}$$

Next, we solve this system by Cramer's rule (Sec. 8). The result is

$$\varphi_j(z) = \frac{\Delta_j(z)}{\Delta(z)} \qquad (j = 1, 2, \ldots, n), \tag{5}$$

where

$$\Delta(z) = \begin{vmatrix} a_{11} - z & a_{21} & \cdots & a_{1n} \\ a_{21} & a_{22} - z & \cdots & a_{2n} \\ \cdots & \cdots & \cdots & \cdots \\ a_{n1} & a_{n2} & \cdots & a_{nn} - z \end{vmatrix}, \tag{6}$$

and $\Delta_j(z)$ is obtained by replacing the elements of the jth column of $\Delta(z)$ by the constant terms $-C_j$. We note that the equation $\Delta(z) = 0$ is the familiar *secular equation* of Sec. 17. Substituting (5) into (2), we find that the solution of the system (2) is given by

$$x_j = \sum_R \frac{\Delta_j(z)}{\Delta(z)} e^{tz} \qquad (j = 1, 2, \ldots, n). \tag{7}$$

We now show that this solution satisfies the initial conditions

$$x_1(0) = C_1,\ x_2(0) = C_2,\ \ldots,\ x_n(0) = C_n. \tag{8}$$

It is sufficient to verify this for x_1. We have

$$x_1(0) = \sum_R \frac{\Delta_1(z)}{\Delta(z)}, \tag{9}$$

where the denominator is given by (6) and obviously represents a polynomial of degree n with leading coefficient $(-1)^n z^n$. The numerator of the rational function appearing in (9) has the form

$$\Delta_1(z) = \begin{vmatrix} -C_1 & a_{12} & \cdots & a_{1n} \\ -C_2 & a_{22} - z & \cdots & a_{2n} \\ \cdots & \cdots & \cdots & \cdots \\ -C_n & a_{n2} & \cdots & a_{nn} - z \end{vmatrix}.$$

By expanding this determinant with respect to elements of the first column, it is easily seen that $\Delta_1(z)$ is a polynomial of degree $n - 1$ with leading term $(-1)^n C_1 z^{n-1}$, and hence we can rewrite (9) in the form

$$x_1(0) = \sum_R \frac{(-1)^n C_1 z^{n-1} + \cdots}{(-1)^n z^n + \cdots}, \tag{10}$$

where the dots indicate terms of the polynomials of lower degree, which play no role in the present calculation.

To show that $x_1(0) = C_1$, we first prove the following general result concerning the sum of the residues of a rational function:

LEMMA. *The sum of the residues of a rational function at its poles lying at a finite distance from the origin equals the coefficient of z^{-1} in the expansion of the rational function in a neighborhood of the point at infinity.*

PROOF. Suppose the rational function has an expansion of the form

$$f(z) = \sum_k b_k z^k \tag{11}$$

in a neighborhood of the point at infinity, and consider the integral

$$\frac{1}{2\pi i} \int_{C_R} f(z)\, dz,$$

where C_R is a circle of radius R with center at the origin. Then for sufficiently large R, all the poles of $f(z)$ lie inside C_R and the integral gives the sum of the residues of $f(z)$ at these poles. On the other hand, for sufficiently large R, the circle C_R lies in a neighborhood of the point at infinity. Then, we can use the expansion (11) to calculate the integral, from which it immediately follows that the integral equals b_{-1}, QED.

We now apply the lemma to the rational function (10), noting that it has the following expansion in a neighborhood of the point at infinity:

$$\frac{(-1)C_1z^{n-1} + \cdots}{(-1)^n z^n + \cdots} = \frac{C_1}{z} + \frac{\beta_2}{z^2} + \cdots,$$

where β_2 is some constant. Thus, the lemma shows at once that

$$x_1(0) = C_1,$$

as asserted, and the fact that $x_j(0) = C_j$ $(j = 2, \ldots, n)$ is proved in just the same way. In other words, the solution given by (7) satisfies the initial conditions (8), i.e., the constants C_j specify the initial conditions. Consequently, the formulas (7) give the general integral of the system (1).

EXAMPLE. Consider the system

$$\begin{aligned} \dot{x}_1 &= \phantom{x_1 + {}} x_2 + x_3, \\ \dot{x}_2 &= x_1 \phantom{{}+ x_2} + x_3, \\ \dot{x}_3 &= x_1 + x_2 \end{aligned}$$

In this case

$$\Delta(z) = \begin{vmatrix} -z & 1 & 1 \\ 1 & -z & 1 \\ 1 & 1 & -z \end{vmatrix}, \tag{12}$$

or $\quad \Delta(z) = -z(z^2 - 1) + 2(z + 1) = (z + 1)(-z^2 + z + 2).$

Thus, x_1 is given by the formula

$$x_1 = \sum_R \frac{\Delta_1(z)}{(z + 1)(-z^2 + z + 2)} e^{tz},$$

where

$$\Delta_1(z) = \begin{vmatrix} -C_1 & 1 & 1 \\ -C_2 & -z & 1 \\ -C_3 & 1 & -z \end{vmatrix}.$$

Expanding $\Delta_1(z)$ and dividing by $1 + z$, we obtain

$$x_1 = \sum_R \frac{C_1(1 - z) - C_2 - C_3}{-z^2 + z + 2} e^{tz}.$$

The denominator has the roots $z = -1$ and $z = 2$. Calculating the residues at these points by the usual rule, i.e., evaluating the ratio of the numerator to the derivative of the denominator, we obtain

$$x_1 = (\tfrac{2}{3}C_1 - \tfrac{1}{3}C_2 - \tfrac{1}{3}C_3)e^{-t} + (\tfrac{1}{3}C_1 + \tfrac{1}{3}C_2 + \tfrac{1}{3}C_3)e^{2t}.$$

We note that in this example the polynomial $\Delta(z)$ has a double root $z = -1$, but that nevertheless, the quantity multiplying e^{-t} in the expression for x_1 is just a constant, and not a polynomial of the first degree in t (cf. remark in Sec. 17).

Next, we consider the case of a system of *inhomogeneous* equations (corresponding to *forced oscillations*):

$$\dot{x}_j = a_{j1}x_1 + \cdots + a_{jn}x_n + f_j(t) \qquad (j = 1, 2, \ldots, n), \tag{13}$$

where the $f_j(t)$ are given functions of t. Suppose we look for a solution of the form

$$x_j(t) = -\sum_R \frac{C_1(t)A_{1j}(z) + \cdots + C_n(t)A_{nj}(z)}{\Delta(z)} e^{tz}, \tag{14}$$

where the $A_{ij}(z)$ are the cofactors of the determinant $\Delta(z)$ and the $C_j(t)$ are functions of t to be determined (this is the so-called *method of variation of parameters*). Substituting (14) into (13) and bearing in mind that (14) gives a solution of the *homogeneous* system (1) when the $C_j(t)$ are arbitrary constants, we obtain the following equations involving the derivatives $\dot{C}_j(t)$:

$$-\sum_R \frac{\dot{C}_1(t)A_{1j}(z) + \cdots + \dot{C}_n(t)A_{nj}(z)}{\Delta(z)} e^{tz} = f_j(t)$$

$$(s = 1, 2, \ldots, n). \tag{15}$$

We now show that this system of equations can be satisfied by setting

$$\dot{C}_1(t) = e^{-tz}f_1(t), \ \ldots, \ \dot{C}_n(t) = e^{-tz}f_n(t). \tag{16}$$

After making this substitution, the left-hand side of (15) becomes

$$\sum_R \frac{f_1(t)A_{1j}(z) + \cdots + f_n(t)A_{nj}(z)}{\Delta(z)}. \tag{17}$$

If $i \neq j$, then in forming the cofactor $A_{ij}(z)$ we have to omit the elements $a_{ii} - z$ and $a_{jj} - z$ appearing on the principal diagonal of the determinant $\Delta(z)$, and hence $A_{ij}(z)$ is a polynomial of degree $n - 2$ in z. Then, by the lemma proved above,

$$\sum_R \frac{A_{ij}(z)}{\Delta(z)} = 0 \qquad (i \neq j),$$

since in a neighborhood of the point at infinity the rational function $A_{ij}(z)/\Delta(z)$ begins with a term of the form az^{-2}, i.e., contains no term in z^{-1}. On the other hand, the cofactor $A_{ii}(z)$ is a polynomial of degree $n - 1$ with leading coefficient $(-1)^{n-1}z^{n-1}$, and therefore

$$-\sum_R \frac{A_{ii}(z)}{\Delta(z)} = 1.$$

It follows at once that the expression (17) is equal to $f_j(t)$. Integrating

(16), we obtain

$$C_j(t) = \int_0^t e^{-\tau z} f_j(\tau)\, d\tau \qquad (j = 1, 2, \ldots, n), \tag{18}$$

where we have chosen the constants of integration so as to make $C_j(0) = 0$, i.e., $x_j(0) = 0$ $(j = 1, 2, \ldots, n)$. Substituting (18) into (14), we finally obtain

$$x_j = -\sum_R \int_0^t \frac{f_1(\tau)A_{1j}(z) + \cdots + f_n(\tau)A_{nj}(z)}{\Delta(z)} e^{(t-\tau)z}\, d\tau$$

$$(j = 1, 2, \ldots, n).$$

We now indicate another approach to the solution of the *homogeneous* system (1). Let the functions x_j, which we are trying to determine, be the components of a vector

$$\mathbf{x} = (x_1, x_2, \ldots, x_n),$$

i.e., $\mathbf{x}$ is a vector whose components are functions of t. We define the derivative of $\mathbf{x}$ with respect to t as the vector with components $\dot{x}_j$:

$$\frac{d\mathbf{x}}{dt} = (\dot{x}_1, \dot{x}_2, \ldots, \dot{x}_n),$$

and we introduce the matrix A whose elements are the coefficients of the system (1). Then (1) can be written in the form

$$\frac{d\mathbf{x}}{dt} = A\mathbf{x}. \tag{19}$$

Suppose that we want the solution of (19) that satisfies the initial conditions

$$x_j(0) = x_j^{(0)} \qquad (j = 1, 2, \ldots, n). \tag{20}$$

These initial conditions form a vector which we denote by

$$\mathbf{x}^{(0)} = (x_1^{(0)}, x_2^{(0)}, \ldots, x_n^{(0)}).$$

It is not hard to see that the solution of the system (19) satisfying the initial conditions (20) has the form

$$\mathbf{x} = \left(I + \frac{At}{1!} + \frac{A^2t^2}{2!} + \cdots\right)\mathbf{x}^{(0)}, \tag{21}$$

where I is the unit matrix of order n; (21) can be written as

$$\mathbf{x} = e^{At}\mathbf{x}^{(0)}, \tag{22}$$

if we introduce the matrix

$$e^{At} = I + \frac{At}{1!} + \frac{A^2t^2}{2!} + \cdots.$$

(The convergence of e^{At}, and related matters, are discussed in the problems at the end of this Appendix.) In fact, according to (21), we have

$$\mathbf{x} = \mathbf{x}^{(0)} + \frac{t}{1!} A\mathbf{x}^{(0)} + \frac{t^2}{2!} A^2\mathbf{x}^{(0)} + \cdots . \tag{23}$$

Differentiating (23) with respect to t, we obtain

$$\frac{d\mathbf{x}}{dt} = A\mathbf{x}^{(0)} + \frac{t}{1!} A^2\mathbf{x}^{(0)} + \frac{t^2}{2!} A^3\mathbf{x}^{(0)} + \cdots ,$$

or

$$\frac{d\mathbf{x}}{dt} = A\left(I + \frac{t}{1!} A + \frac{t^2}{2!} A^2 + \cdots\right)\mathbf{x}^{(0)},$$

so that

$$\frac{d\mathbf{x}}{dt} = A\mathbf{x}$$

by (21). Moreover, (21) and (22) obviously satisfy the initial conditions, i.e., $\mathbf{x}(0) = \mathbf{x}^{(0)}$.

There is still another way in which the system (1) can be written by using matrices. First we develop the basic rules for differentiating matrices. Suppose that the elements of a matrix X are functions of the variable t. Then, as in Sec. 89, we define the derivative dX/dt as the matrix whose elements are obtained by differentiating the elements of X:

$$\left\{\frac{dX}{dt}\right\}_{ik} = \frac{d}{dt}\{X\}_{ik}.$$

It is an immediate consequence of this definition that the derivative of the sum or product of two matrices is given by the usual rules, i.e., if X and Y are two matrices whose elements are functions of t, then

$$\frac{d}{dt}(X + Y) = \frac{dX}{dt} + \frac{dY}{dt}$$

and

$$\frac{d}{dt}(XY) = \frac{dX}{dt} Y + X\frac{dY}{dt}, \tag{24}$$

where it should be kept in mind that in general we cannot change the order of the factors in (24). The proof of (24) follows at once from the fact that

$$\{XY\}_{ik} = \sum_{s=1}^{n} \{X\}_{is}\{Y\}_{sk}$$

implies

$$\frac{d}{dt}\{XY\}_{ik} = \sum_{s=1}^{n} \frac{d\{X\}_{is}}{dt}\{Y\}_{sk} + \sum_{s=1}^{n} \{X\}_{is}\frac{d\{Y\}_{sk}}{dt}.$$

Moreover, (24) can be easily generalized to the case of any number of factors; for example, we have

$$\frac{d}{dt}(XYZ) = \frac{dX}{dt}YZ + X\frac{dY}{dt}Z + XY\frac{dZ}{dt}.$$

Next, we derive the formula for differentiating the inverse of a matrix X. Suppose that $\det X \neq 0$, so that there exists a matrix X^{-1} such that

$$XX^{-1} = I. \tag{25}$$

Differentiating (25) with respect to t gives

$$\frac{dX}{dt}X^{-1} + X\frac{dX^{-1}}{dt} = 0,$$

whence

$$\frac{dX^{-1}}{dt} = -X^{-1}\frac{dX}{dt}X^{-1},$$

which gives the required derivative of X^{-1}.

Returning to the system (1), we now think of the n solutions of (1) as forming a square matrix

$$X(t) = \left\| \begin{matrix} x_{11}(t) & x_{12}(t) & \cdots & x_{1n}(t) \\ x_{21}(t) & x_{22}(t) & \cdots & x_{2n}(t) \\ \cdots & \cdots & \cdots & \cdots \\ x_{n1}(t) & x_{n2}(t) & \cdots & x_{nn}(t) \end{matrix} \right\|, \tag{26}$$

consisting of n^2 functions, where $x_{ij}(t)$ denotes the ith function belonging to the jth solution. In other words, the system (1) now takes the form

$$\dot{x}_{ik}(t) = a_{i1}x_{1k}(t) + a_{i2}x_{2k}(t) + \cdots + a_{in}x_{nk}(t) \quad (i, k = 1, 2, \ldots, n),$$

or in matrix form,

$$\frac{dX}{dt} = AX, \tag{27}$$

where X is given by (26), and every column of X gives a solution of (19). The initial conditions are specified by giving the matrix X at the time $t = 0$, i.e., by writing

$$X(0) = X^{(0)}, \tag{28}$$

where $X^{(0)}$ is an arbitrary matrix with constant elements. Then, just as before, it can be shown that the solution of (27) satisfying the initial conditions (28) has the form

$$X(t) = e^{At}X^{(0)}. \tag{29}$$

Next, we show that $\det X(t) \neq 0$ for all t, provided that $\det X^{(0)} \neq 0$. First we observe that if Y is any matrix, then the determinant of

$$e^Y = I + \frac{Y}{1!} + \frac{Y^2}{2!} + \cdots + \frac{Y^n}{n!} + \cdots \tag{30}$$

is always different from zero. To see this, form the matrix

$$e^{-Y} = I - \frac{Y}{1!} + \frac{Y^2}{2!} - \cdots + (-1)^n \frac{Y^n}{n!} + \cdots. \tag{31}$$

The terms in the two infinite series (30) and (31) commute, since they involve only ordinary numbers and powers of the same matrix Y. Thus, the result of formally multiplying (30) and (31) is the same as if Y were a number z, and hence the identity $e^z e^{-z} = 1$ implies the corresponding identity

$$e^Y e^{-Y} = I,$$

which is valid for any matrix Y, i.e., e^Y has the inverse e^{-Y}, so that $\det e^Y \neq 0$, as asserted.† Hence it follows from (29) that if $\det X^{(0)} \neq 0$, then

$$\det X(t) = \det e^{At} \det X^{(0)} \neq 0$$

for all t. Thus, *the matrix $X(t)$ gives n linearly independent solutions of the system* (102).

Finally, we show that if Y is a matrix which gives any n solutions of (27), then Y can be written as

$$Y = XB, \tag{32}$$

where X is the matrix given above, and B is a matrix with constant elements. It is clear that (32) expresses the fact that any solution of the system (1) is a linear combination of n linearly independent solutions of the system. To prove (32), we first note that by hypothesis, Y must satisfy (27), so that

$$\frac{dY}{dt} = AY. \tag{33}$$

Moreover, by hypothesis, X also satisfies (27) and $\det X \neq 0$, i.e., the inverse matrix X^{-1} exists. By the rule proved above for differentiation of an inverse matrix, we have

$$\frac{dX^{-1}}{dt} = -X^{-1}\frac{dX}{dt}X^{-1},$$

so that by (27)

$$\frac{dX^{-1}}{dt} = -X^{-1}AXX^{-1} = -X^{-1}A. \tag{34}$$

Forming the derivative of the product $X^{-1}Y$, and using (33) and (34), we find

$$\frac{d}{dt}(X^{-1}Y) = \frac{dX^{-1}}{dt}Y + X^{-1}\frac{dY}{dt} = -X^{-1}AY + X^{-1}AY = 0.$$

Therefore, $X^{-1}Y$ is a constant matrix, whose elements do not depend on t. This proves (32).

† Note that if Y and Z are two matrices which do not commute, then in general $e^Y e^Z \neq e^{Y+Z}$.

PROBLEMS†

1. Solve the following systems of differential equations:

(a) $\dot{x} + 3x + y = 0, \quad \dot{y} - x + y = 0,$ where $x(0) = y(0) = 1$;
(b) $\dot{x} + 7x - y = 0, \quad \dot{y} + 2x + 5y = 0,$ where $x(0) = y(0) = 1$;
(c) $\dot{x} = y + z, \quad \dot{y} = z + x, \quad \dot{z} = x + y,$ where $x(0) = -1, y(0) = 1, z(0) = 0$;
(d) $\dot{x} = y - z, \quad \dot{y} = z - 2x, \quad \dot{z} = 2x - y.$

2. Solve Prob. 50 of Chap. 2 by the method given in the text.

3. Solve the following systems of differential equations:

(a) $\dot{x} = 3y - x, \quad \dot{y} = x + y + e^{at}$;
(b) $\dot{x} = y, \quad \dot{y} = x + e^{t} + e^{-t}$;
(c) $\dot{x} + n^2 y = \cos nt, \quad \dot{y} + n^2 x = \sin nt.$

In each case $x(0) = y(0) = 0$.

4. Let X be any matrix of order n. *Define* e^X as the infinite series

$$e^X = I + \frac{1}{1!}X + \frac{1}{2!}X^2 + \cdots + \frac{1}{m!}X^m + \cdots,$$

where I is the unit matrix of order n. This formula is equivalent to the n^2 ordinary infinite series

$$\{e^X\}_{ik} = \delta_{ik} + \frac{1}{1!}\{X\}_{ik} + \frac{1}{2!}\{X^2\}_{ik} + \cdots + \frac{1}{m!}\{X^m\}_{ik} + \cdots.$$

Show that each of these series converges for any X, so that the matrix

$$e^X = \|\{e^X\}_{ik}\|$$

is meaningful. Give a rigorous proof of the formula $e^X e^{-X} = I$ used in the text. Show that

$$Se^X S^{-1} = e^{SXS^{-1}},$$

and in particular, show that if X is diagonalizable, so that

$$TXT^{-1} = [\lambda_1, \lambda_2, \ldots, \lambda_n],$$

then

$$e^X = T^{-1}[e^{\lambda_1}, e^{\lambda_2}, \ldots, e^{\lambda_n}]T.$$

Comment: We can use this last formula to *define* e^X for any diagonalizable matrix X (cf. Prob. 6 of Chap. 6). However, the infinite series definition of e^X applies to any X, diagonalizable or not.

5. Find e^A, where

(a) $A = \begin{Vmatrix} 3 & -1 \\ 1 & 1 \end{Vmatrix}$; (b) $A = \begin{Vmatrix} 4 & -2 \\ 6 & -3 \end{Vmatrix}$; (c) $A = \begin{Vmatrix} 4 & 2 & -5 \\ 6 & 4 & -9 \\ 5 & 3 & -7 \end{Vmatrix}$.

6. Show that

$$\det e^A = e^{\operatorname{tr} A}.$$

7. Solve the system of differential equations

$$\dot{x}_1 = x_2 + x_3, \qquad \dot{x}_2 = x_1 + x_3, \qquad \dot{x}_3 = x_1 + x_2$$

given in the text, using the method of (22) or (29).

† These problems are by the translator, making some use of problem collections by Gyunter and Kuzmin and by Proskuryakov. The overdot denotes differentiation with respect to t. For hints and answers, see page 457.

Bibliography

Aitken, A. C., "Determinants and Matrices," Oliver & Boyd, Ltd., London, 1959.

Bellman, R., "Introduction to Matrix Algebra," McGraw-Hill Book Co., Inc., New York, 1960.

Birkhoff, G. and MacLane, S., "A Survey of Modern Algebra," The Macmillan Co., New York, 1941.

Cooke, R. G., "Infinite Matrices and Sequence Spaces," The Macmillan Co., London, 1950.

Gantmacher, F. R., "The Theory of Matrices," in 2 vols., translated by K. A. Hirsch, Chelsea Publishing Co., New York, 1959.

Gelfand, I. M., "Lectures on Linear Algebra," translated by A. Shenitzer, Interscience Publishers, Inc., New York, 1961.

Hall, M., Jr., "The Theory of Groups," The Macmillan Co., New York, 1959.

Halmos, P. R., "Finite-dimensional Vector Spaces," D. Van Nostrand Co., Inc., Princeton, 1958.

Halmos, P. R., "Introduction to Hilbert Space," Chelsea Publishing Co., New York, 1951.

Hamburger, H. L. and Grimshaw, M. E., "Linear Transformations in *n*-Dimensional Vector Space," Cambridge University Press, Cambridge, 1956.

Hoffman, K. and Kunze, R., "Linear Algebra," Prentice-Hall, Inc., Englewood Cliffs, N.J., 1961.

Jacobson, N., "Lectures in Abstract Algebra," vol. 1, "Basic Concepts," vol. 2, "Linear Algebra," D. Van Nostrand Co., Inc., Princeton, 1951, 1953.

Kurosh, A. G., "The Theory of Groups," in 2 vols., translated by K. A. Hirsch, Chelsea Publishing Co., New York, 1956.

Ledermann, W., "Introduction to the Theory of Finite Groups," Oliver & Boyd, Ltd., London, 1957.

Littlewood, D. E., "The Theory of Group Characters and Matrix Representation of Groups," 2d ed., Oxford University Press, Oxford, 1958.

Lomont, J. S., "Applications of Finite Groups," Academic Press, Inc., New York, 1959.

Mirsky, L., "An Introduction to Linear Algebra," Oxford University Press, Oxford, 1955.

Murnaghan, F. D., "The Theory of Group Representations," Johns Hopkins Press, Baltimore, 1938.

Perlis, S., "The Theory of Matrices," Addison-Wesley Publishing Co., Reading, 1952.

Schreier, O. and Sperner, E., "Introduction to Modern Algebra and Matrix Theory," translated by M. Davis and M. Hauser, Chelsea Publishing Co., New York, 1959.

Shilov, G. E., "An Introduction to the Theory of Linear Spaces," translated by R. A. Silverman, Prentice-Hall Inc., Englewood Cliffs, N.J., 1961.

Sokolnikoff, I. S., "Tensor Analysis, Theory and Applications," John Wiley & Sons, Inc., New York, 1951.

Spain, Barry, "Tensor Calculus," Oliver & Boyd, Ltd., London, 1960.

Stoll, R. R., "Linear Algebra and Matrix Theory," McGraw-Hill Book Co., Inc., New York, 1952.

Synge, J. L. and Schild, A., "Tensor Calculus," University of Toronto Press, Toronto, 1956.

Thrall, R. M. and Tornheim, L., "Vector Spaces and Matrices," John Wiley & Sons, Inc., New York, 1957.

Wigner, E. P., "Group Theory and Its Applications to the Quantum Mechanics of Atomic Spectra," translated by J. J. Griffith, Academic Press, Inc., New York, 1959.

Hints and Answers

CHAPTER 1

1. (a) 1; (c) -1; (e) 0; (g) $4ab$; (i) 1; (k) $\cos(\alpha + \beta)$; (m) 1; (o) $ab - c^2 - d^2$; (q) 0.
2. (a) $x = 3, y = -1$; (c) $x = \frac{2}{5}, y = \frac{1}{5}$; (e) $x = \cos(\beta - \alpha), y = \sin(\beta - \alpha)$.
5. (a) 40; (c) 100; (e) 0; (g) 1; (i) 4; (k) 6; (m) 0; (o) $a^3 + b^3 + c^3 - 3abc$; (q) $2x^3 - (a + b + c)x^2 + abc$; (s) $1 + \alpha^2 + \beta^2 + \gamma^2$;
(u) $\sin(\beta - \gamma) + \sin(\gamma - \alpha) + \sin(\alpha - \beta)$;
(w) $xyz + 2(ace - bcf + adf + bde) - x(e^2 + f^2) - y(c^2 + d^2) - z(a^2 + b^2)$;
(y) $2abc(a + b + c)^3$.
6. (a) $x = 3, y = -2, z = 2$; (c) $x = 1, y = 2, z = -1$.
7. $x = \frac{1}{3}(a + b + c)$, $y = \frac{1}{3}(a + b\epsilon^2 + c\epsilon)$, $z = \frac{1}{3}(a + b\epsilon + c\epsilon^2)$.
8. Add the equations. **9.** $x = -abc$, $y = ab + ac + bc$, $z = -(a + b + c)$.
10. 4. **11.** (a) 5; (c) 13; (e) $\frac{1}{2}n(n - 1)$. **12.** $a_{11}a_{22}a_{33} \cdots a_{nn}$.
13. (a) Minus. **14.** (a) $j = 3, k = 4$.
15. The permutation $n, n - 1, n - 2, \ldots, 3, 2, 1$, which contains $\frac{1}{2}n(n - 1)$ inversions.
16. In the permutation $a_1, a_2, \ldots, a_n$, move the element b_1 to the first position, then move b_2 to the second position, etc. The permutation $n, 1, 2, \ldots, n - 1$ cannot be brought into the form $1, 2, \ldots, n$ by using fewer than $n - 1$ transpositions.
17. Cf. solution to Prob. 16. **18.** $\binom{n}{2} - k$.† **19.** $\frac{1}{2}n!\binom{n}{2}$.
20. Move 1 to the first position by transposing neighboring elements; move 2 to the second position in the same way; etc.
21. It does not change. **22.** It does not change. **23.** Transpose the determinant.
24. (b) It does not change. **25.** $x = 0, 1, 2, \ldots, n - 1$.
26. 0. Begin by subtracting the first column from the other columns.
28. (a) $abcd$; (c) $abcd + ab + ad + 1$.
30. Set $a_1 = \lambda(1 + \sigma_1)$, $a_2 = \lambda(1 + \sigma_2)$, $\ldots$, $a_n = \lambda(1 + \sigma_n)$.
32. (a) 10; (c) 60. **33.** Use Laplace's theorem.
34. Expand the determinant with respect to its rows.
35. The determinant does not change if x is replaced by $-x$, and it vanishes for $x = 0$. The same is true for z.
36. To show that $x + y + z$ is a factor, add the other columns to the first column; to show that $x - y - z$ is a factor, add the second column to the first and subtract the third and fourth, etc.

† $\binom{n}{k} = \dfrac{n!}{k!(n - k)!}$.

37. $(-1)^{n-1}$. **40.** Use the law of cosines.
43. Show that D_n satisfies the recursion formula $D_n = a_nD_{n-1} + a_1a_2 \cdots a_{n-1}$.
44. Expand the determinant with respect to its columns.

45. (a) 2; (c) $6x - 4y + 3z - 12 = 0$; (e) $\frac{1}{6}\begin{vmatrix} 2 & 0 & 0 \\ 0 & 2 & 0 \\ 0 & 0 & 2 \end{vmatrix} = \frac{4}{3}$.

46. $(x_1 + x_2 + \cdots + x_n) \prod\limits_{n \geq i > k \geq 1} (x_i - x_k)$. Calculate the determinant

$$\begin{vmatrix} 1 & x_1 & x_1^2 & \cdots & x_1^n \\ 1 & x_2 & x_2^2 & \cdots & x_2^n \\ \cdot & \cdot & \cdot & \cdot & \cdot \\ 1 & x_n & x_n^2 & \cdots & x_n^n \\ 1 & z & z^2 & \cdots & z^n \end{vmatrix}$$

both as a Vandermonde determinant (in reverse order) and by expanding it with respect to its last row. Then compare the coefficients of z^{n-1} in the two expressions.
47. Cf. (8) and (10) of Sec. 1.
48. Note that $d^{(k)}(0) = 0$ $(k = 0, 1, \ldots, n-2)$, and add all the other columns to the first.
49. -50. **50.** $(a^2 + b^2 + c^2 + d^2)^2$.
51. (a) 0 for $n > 2$, $(x_2 - x_1)(y_2 - y_1)$ for $n = 2$. Represent the determinant as the product

$$\begin{vmatrix} 1 & x_1 & 0 & \cdots & 0 \\ 1 & x_2 & 0 & \cdots & 0 \\ \cdot & \cdot & \cdot & \cdot & \cdot \\ 1 & x_n & 0 & \cdots & 0 \end{vmatrix} \begin{vmatrix} 1 & y_1 & 0 & \cdots & 0 \\ 1 & y_2 & 0 & \cdots & 0 \\ \cdot & \cdot & \cdot & \cdot & \cdot \\ 1 & y_n & 0 & \cdots & 0 \end{vmatrix};$$

(c) 0 if $n > 2$, $-\sin^2(\alpha_1 - \alpha_2)$ if $n = 2$.
52. Study the product

$$\begin{vmatrix} a_0 & a_1 & a_2 & \cdots & a_{n-1} \\ a_{n-1} & a_0 & a_1 & \cdots & a_{n-2} \\ \cdot & \cdot & \cdot & \cdot & \cdot \\ a_1 & a_2 & a_3 & \cdots & a_0 \end{vmatrix} \begin{vmatrix} 1 & 1 & \cdots & 1 \\ 1 & \epsilon_1 & \cdots & \epsilon_{n-1} \\ \cdot & \cdot & \cdot & \cdot \\ 1 & \epsilon_1^{n-1} & \cdots & \epsilon_{n-1}^{n-1} \end{vmatrix}.$$

53. $[x + (n-1)a](x - a)^{n-1}$. **54.** $A_{ij} = \sum\limits_{k=1}^{n} a_{ik}a_{kj}$ $\quad (i, j = 1, 2, \ldots, n)$.

55. Let $\varphi_1 = \alpha_2 - \alpha_3$, $\varphi_2 = \alpha_3 - \alpha_1$, $\varphi_3 = \alpha_1 - \alpha_2$. **56.** (b) 4.
57. First consider the case where M appears in the upper left-hand corner of D. Using the row-by-row rule to multiply D by the minor M', written in the form

$$\begin{vmatrix} A_{11} & \cdots & A_{1m} & A_{1,m+1} & \cdots & A_{1n} \\ \cdot & \cdot & \cdot & \cdot & \cdot & \cdot \\ A_{m1} & \cdots & A_{mm} & A_{m,m+1} & \cdots & A_{mn} \\ 0 & \cdots & 0 & 1 & \cdots & 0 \\ \cdot & \cdot & \cdot & \cdot & \cdot & \cdot \\ 0 & \cdots & 0 & 0 & \cdots & 1 \end{vmatrix},$$

prove that $DM' = D^mC$, i.e., $M' = D^{m-1}C$ if $D \neq 0$. (Do not neglect to study the case $D = 0$ (cf. Sec. 6).) If M does not appear in the upper left-hand corner of D, move it there by making appropriate row and column interchanges.

58. Use Prob. 57. **61. (b)** $\begin{Vmatrix} 4 & 2 & 1 \\ 8 & 4 & 2 \\ 12 & 6 & 3 \end{Vmatrix}$, rank $= 1$. **62. (a)** 1; **(c)** 2.

63. (a) Rank $= 2$ for $\lambda = 0$; rank $= 3$ for $\lambda \neq 0$.

CHAPTER 2

1. (a) $x = 1, y = 0, z = 1$; **(c)** $x = 1, y = -1, z = -1, t = 1$.

2. (a) and **(c)**. **3.** The determinant

$$\begin{vmatrix} 1 & x_0 & x_0^2 & \cdots & x_0^n \\ 1 & x_1 & x_1^2 & \cdots & x_1^n \\ \cdot & \cdot & \cdot & \cdots & \cdot \\ 1 & x_n & x_n^2 & \cdots & x_n^n \end{vmatrix}$$

is nonvanishing.

4. $f(x) = 3 - 5x + x^2$.

5. (b) Add the equations. $x = \frac{1}{3}(a + b + c - 2d)$, $y = \frac{1}{3}(a + b + d - 2c)$, $z = \frac{1}{3}(a + c + d - 2b)$, $t = \frac{1}{3}(b + c + d - 2a)$;

(d) Set $x_1 + x_2 + \cdots + x_n = s$.
Then $s - x_2 = 2, s - x_3 = 3, \ldots, s - x_n = n$.
Therefore $x_2 = -1, x_3 = -2, \ldots, x_n = -(n - 1), x_1 = 1 + \frac{1}{2}n(n - 1)$.

6. (b) $x_1 = -\frac{5}{17}, x_2 = \frac{23}{17}$; **(d)** $x_2 = -\frac{1}{5} - \frac{2}{5}x_1, x_3 = -\frac{8}{5} + \frac{9}{5}x_1, x_4 = 0$.

7. (b) General solution: $x_3 = -x_1\frac{5}{2} + 5x_2, x_4 = \frac{7}{2}x_1 - 7x_2$.
Complete system of solutions:

$$x_1 = 1,\ x_2 = 0,\ x_3 = -\tfrac{5}{2},\ x_4 = \tfrac{7}{2},$$
$$x_1 = 0,\ x_2 = 1,\ x_3 = 5,\quad x_4 = -7;$$

(d) No nontrivial solution; **(f)** No nontrivial solution;

(h) General solution: $x_1 = 0, x_2 = \frac{1}{3}(x_3 - 2x_5), x_4 = 0$.
Complete system of solutions:

$$x_1 = 0,\ x_2 = \tfrac{1}{3},\quad x_3 = 1,\ x_4 = 0,\ x_5 = 0,$$
$$x_1 = 0,\ x_2 = -\tfrac{2}{3},\ x_3 = 0,\ x_4 = 0,\ x_5 = 1.$$

9. Use Prob. 48 of Chap. 1, with $a_1 = a_2 = a_3 = a_4 = 1, x = \lambda - 1$.

10. The rows of A form a complete system of solutions; those of B do not.

11. (a) The forms are linearly independent;
(c) $y_1 + 3y_2 - y_3 = 0, 2y_1 - y_2 - y_4 = 0$.

12. If $\lambda = 10$, then $3y_1 + 2y_2 - 5y_3 - y_4 = 0$.

13. (a) No; **(b)** No; **(c)** Yes; **(d)** No; **(e)** No; **(f)** Yes; **(g)** No.

14. (b), **(d)**, and **(e)** are linearly dependent.

16. Form any linear combination $\sum_{i=1}^{s} \lambda_i \mathbf{a}_i$, where not all the λ_i are equal to zero. Choose among the λ_i a coefficient λ_j exceeded by no other coefficient in absolute value. Then, the jth component of the linear combination cannot vanish.

17. (b) Dimension equal to the largest integer not exceeding $\frac{1}{2}(n + 1)$. Let $\mathbf{e}_k$ be the vector with 1 in the $(2k - 1)$th position and zeros elsewhere. Then, the vectors $\mathbf{e}_1, \mathbf{e}_2, \ldots$ form a basis;

(d) Dimension 2. The vectors $(1, 0, 1, 0, 1, 0, \ldots)$ and $(0, 1, 0, 1, 0, 1, \ldots)$ form a basis.

18. (c) Let l_1 and l_2 be the dimensions of L_1 and L_2, respectively. Let k be the dimension of $L_1 \cap L_2$, and choose a basis $\mathbf{e}_1, \ldots, \mathbf{e}_k$ in $L_1 \cap L_2$. Complete this basis by the vectors $\mathbf{f}_{k+1}, \ldots, \mathbf{f}_{l_1}$ to form a basis for L_1 and by the vectors

$\mathbf{g}_{k+1}, \ldots, \mathbf{g}_{l_2}$ to form a basis for L_2. Any vector in $L_1 + L_2$ is a linear combination of the vectors $\mathbf{e}_1, \ldots, \mathbf{e}_k, \mathbf{f}_{k+1}, \ldots, \mathbf{f}_{l_1}, \mathbf{g}_{k+1}, \ldots, \mathbf{g}_{l_2}$. Show that these $l_1 + l_2 - k$ vectors are linearly independent, and hence form a basis for $L_1 + L_2$.

20. (a) The lines $\mathbf{x}_0 + t\mathbf{x}_1$ and $\mathbf{y}_0 + t\mathbf{y}_1$ lie in the hyperplane

$$\mathbf{x}_0 + t(\mathbf{y}_0 - \mathbf{x}_0) + t_1\mathbf{x}_1 + t_2\mathbf{y}_1;$$

(b) The planes $\mathbf{x}_0 + t_1\mathbf{x}_1 + t_2\mathbf{x}_2$ and $\mathbf{y}_0 + t_1\mathbf{y}_1 + t_2\mathbf{y}_2$ lie in the hyperplane

$$\mathbf{x}_0 + t(\mathbf{y}_0 - \mathbf{x}_0) + t_1\mathbf{x}_1 + t_2\mathbf{x}_2 + t_3\mathbf{y}_1 + t_4\mathbf{y}_2.$$

21. (b) For example, $(\frac{1}{2}, -\frac{1}{2}, \frac{1}{2}, -\frac{1}{2}), (\frac{1}{2}, -\frac{1}{2}, -\frac{1}{2}, \frac{1}{2})$.

22. (b) 45°. **23.** All the angles equal φ_n, where $\cos \varphi_n = 1/\sqrt{n}$.

25. For odd n, there are none. For $n = 2m$, the number of diagonals is $\binom{2m-1}{m-1}$.

26. The coordinates of the points are given by the rows of the following matrix:

$$\left\|\begin{matrix} 1 & 0 & 0 & \cdots & 0 & 0 \\ \frac{1}{2} & \sqrt{\frac{3}{4}} & 0 & \cdots & 0 & 0 \\ \frac{1}{2} & \frac{1}{\sqrt{12}} & \sqrt{\frac{4}{6}} & \cdots & 0 & 0 \\ \cdots & \cdots & \cdots & \cdots & \cdots & \cdots \\ \frac{1}{2} & \frac{1}{\sqrt{12}} & \frac{1}{\sqrt{24}} & \cdots & \frac{1}{\sqrt{2n(n-1)}} & \sqrt{\frac{n+1}{2n}} \end{matrix}\right\|.$$

27. (a) $\mathbf{y} = 3\mathbf{a}_1 - 2\mathbf{a}_2 = (1, -1, -1, 5)$, $\mathbf{z} = (3, 0, -2, -1)$;
(c) $\mathbf{y} = (5, -5, -2, -1)$, $\mathbf{z} = (2, 1, 1, 3)$.

32. Let the matrix be

$$A = \left\|\begin{matrix} a_{11} & \cdots & a_{1n} \\ \cdots & \cdots & \cdots \\ a_{s1} & \cdots & a_{sn} \end{matrix}\right\|.$$

If all the elements of A are zero, then it is already in diagonal form. If A has nonzero elements, then by interchanging rows and columns we can make the element a_{11} nonzero. Multiplying the first row by a_{11}^{-1} replaces a_{11} by 1. Subtracting the first column multiplied by a_{1j} from the jth column replaces a_{1j} by 0 ($j = 2, \ldots, n$). Subtracting the first row multiplied by a_{i1} from the ith row replaces a_{i1} by 0 ($i = 2, \ldots, s$). As a result, A goes into a new matrix

$$A' = \left\|\begin{matrix} 1 & 0 & \cdots & 0 \\ 0 & a'_{22} & \cdots & a'_{2n} \\ \cdots & \cdots & \cdots & \cdots \\ 0 & a'_{s2} & \cdots & a'_{sn} \end{matrix}\right\|,$$

which has the same rank as A. Now apply the process just described to the matrix appearing in the lower right-hand corner of A', etc.

$$\text{Rank} \left\|\begin{matrix} 0 & 2 & -4 \\ -1 & -4 & 5 \\ 3 & 1 & 7 \\ 0 & 5 & -10 \\ 2 & 3 & 0 \end{matrix}\right\| = \text{rank} \left\|\begin{matrix} 1 & 0 & 0 \\ 0 & 1 & 0 \\ 0 & 0 & 0 \\ 0 & 0 & 0 \\ 0 & 0 & 0 \end{matrix}\right\| = 2.$$

33. In the example given, expand the determinants

$$\begin{vmatrix} 5 & 3 & 4 \\ 5 & 3 & 4 \\ 6 & 5 & 6 \end{vmatrix} \quad \text{and} \quad \begin{vmatrix} 6 & 5 & 6 \\ 5 & 3 & 4 \\ 6 & 5 & 6 \end{vmatrix}$$

with respect to the first row. Generalize this result.

34. (b) and (d) are compatible.

36. For $\lambda = -3$, the system is incompatible. For $\lambda = 1$, the general solution has the form $x = 1 - y - z - w$, where y, z, and w are arbitrary. For $\lambda \neq 1$, $\lambda \neq -3$, the system has the unique solution $x = y = z = 1/(\lambda + 3)$.

37. The rank of the matrix

$$\begin{Vmatrix} a_1 & b_1 & c_1 \\ a_2 & b_2 & c_2 \\ a_3 & b_3 & c_3 \end{Vmatrix}$$

must not change if the last column is deleted.

38. The rank of the matrix

$$\begin{Vmatrix} a_1 & b_1 & c_1 & d_1 \\ a_2 & b_2 & c_2 & d_2 \\ \cdot & \cdot & \cdot & \cdot \\ a_n & b_n & c_n & d_n \end{Vmatrix}$$

must be 2 and must not change if the last column is deleted.

39. The rank of the augmented matrix must decrease by 1 when the kth column is deleted.

40. Let r be the rank of the coefficient matrix, and let r_1 be the rank of the augmented matrix. If $r = 2$, $r_1 = 3$, the system has no solution, and the lines do not pass through a common point, but at least two lines are different and intersect. If $r = r_1 = 2$, the system has a unique solution and the lines have a common point, but at least two lines are different. If $r = 1$, $r_1 = 2$, the system has no solutions, and the lines are parallel or coincide, but at least two lines are different. If $r = r_1 = 1$, the solution depends on one parameter and all the lines coincide.

41. (a) 4.

43. For example,

$$\Delta = \begin{vmatrix} (\mathbf{a}_1, \mathbf{a}_1) & (\mathbf{a}_1, \mathbf{a}_2) \\ (\mathbf{a}_2, \mathbf{a}_1) & (\mathbf{a}_2, \mathbf{a}_2) \end{vmatrix},$$

where $\mathbf{a}_1 = (1, 0)$, $\mathbf{a}_2 = (\delta, \sqrt{\Delta})$.

44. Let $M = 1$. Then, equality holds in (44) if and only if the rows are orthogonal and all $a_{ik} = \pm 1$. This is impossible for odd $n > 1$. For $n = 2^m$, the equality holds for the determinants

$$A_1 = |1|, \; A_2 = \begin{vmatrix} 1 & 1 \\ -1 & 1 \end{vmatrix}, \; A_4 = \begin{vmatrix} A_2 & A_2 \\ -A_2 & A_2 \end{vmatrix}, \; \ldots,$$

$$A_{2^{m+1}} = \begin{vmatrix} A_{2^m} & A_{2^m} \\ -A_{2^m} & A_{2^m} \end{vmatrix}, \; \ldots.$$

45. If $\mathbf{w}$ is in L, then

$$\|\mathbf{x} - \mathbf{w}\|^2 = \|\mathbf{y} + \mathbf{z} - \mathbf{w}\|^2 = \|\mathbf{y} - \mathbf{w}\|^2 + \|\mathbf{z}\|^2 \geq \|\mathbf{z}\|^2.$$

46. Cf. Prob. 45 and the condition for equality in formula (40) of Sec. 16.

47. Cf. Prob. 46. To verify $\sqrt{G} = |D|$, square D using the row-by-row rule.

49. Apply Prob. 46.

50. (a) $x = \frac{1}{3}e^{-t} + \frac{1}{6}e^{2t} + \frac{1}{2}e^{-2t}$, $y = \frac{1}{3}e^{-t} + \frac{1}{6}e^{2t} - \frac{1}{2}e^{-2t}$, $z = -\frac{1}{3}e^{-t} + \frac{1}{3}e^{2t}$;
(b) $x = e^{-6t}(C_1 \cos t + C_2 \sin t)$, $y = e^{-6t}[(C_1 + C_2) \cos t + (C_2 - C_1) \sin t]$.

52. $x = e^{mt}(Ae^{-imt} + Be^{imt}) + e^{-mt}(Ce^{imt} + De^{-imt})$,
$y = ie^{mt}(Ae^{-imt} - Be^{imt}) + ie^{-mt}(Ce^{imt} - De^{-imt})$.
The simplest approach is to write $x = e^{\lambda t}$, $y = be^{\lambda t}$, instead of $x = a \cos \lambda t$, $y = b \sin \lambda t$, as suggested in Sec. 17. Then, one obtains the secular equation (56) with λ^2 replaced by $-\lambda^2$, which has solutions for $\lambda = m(\pm 1 \pm i)$.

53. (a) $x_1x_2^2$. **56.** $\varphi_2^2 - \varphi_1 - 2\varphi_3 = 0$. **58.** Yes.

CHAPTER 3

3. (a) $\begin{Vmatrix} -1 & 0 & 0 \\ 0 & 1 & 0 \\ 0 & 0 & 1 \end{Vmatrix}$; (c) $\begin{Vmatrix} 0 & 1 & 0 \\ -1 & 0 & 0 \\ 0 & 0 & 1 \end{Vmatrix}$ or $\begin{Vmatrix} 0 & -1 & 0 \\ 1 & 0 & 0 \\ 0 & 0 & 1 \end{Vmatrix}$, depending on the direction of the rotation.

4. Rotate about $\mathbf{x} \times \mathbf{x}'$ to bring $\mathbf{x}$ into coincidence with $\mathbf{x}'$. Then rotate about $\mathbf{x}'$ to bring $\mathbf{y}$ into coincidence with $\mathbf{y}'$.

5. (a) $\begin{Vmatrix} 3 & -1 \\ 5 & -1 \end{Vmatrix}$; (c) $\begin{Vmatrix} 6 & 2 & -1 \\ 6 & 1 & 1 \\ 8 & -1 & 4 \end{Vmatrix}$; (e) $\begin{Vmatrix} 1 & 7 & 0 \\ 1 & 5 & -2 \\ 6 & 12 & 2 \end{Vmatrix}$.

6. (b) $\begin{Vmatrix} 15 & 20 \\ 20 & 35 \end{Vmatrix}$; (d) $\begin{Vmatrix} 1 & n \\ 0 & 1 \end{Vmatrix}$.

7. Since

$$A = \begin{Vmatrix} 1 & \frac{\alpha}{n} \\ -\frac{\alpha}{n} & 1 \end{Vmatrix} = \sqrt{1 + \frac{\alpha^2}{n^2}} \begin{Vmatrix} \cos\varphi & \sin\varphi \\ -\sin\varphi & \cos\varphi \end{Vmatrix}, \qquad \text{where } \tan\varphi = \frac{\alpha}{n},$$

we have

$$A^n = \left(1 + \frac{\alpha^2}{n^2}\right)^{n/2} \begin{Vmatrix} \cos n\varphi & \sin n\varphi \\ -\sin n\varphi & \cos n\varphi \end{Vmatrix}.$$

Then, since $\lim\limits_{n\to\infty} n\varphi = \alpha \lim\limits_{\varphi\to 0} \frac{\varphi}{\tan\varphi} = \alpha$, $\quad \lim\limits_{n\to\infty} A^n = \begin{Vmatrix} \cos\alpha & \sin\alpha \\ -\sin\alpha & \cos\alpha \end{Vmatrix}$.

8. (b) $\begin{Vmatrix} 0 & 0 & 0 \\ 0 & 0 & 0 \\ 0 & 0 & 0 \end{Vmatrix}$. **9.** (b) $\begin{Vmatrix} x & y \\ 0 & x \end{Vmatrix} = (x - y)I + yA$, where x and y are arbitrary.

10. (b) $\begin{Vmatrix} 0 & 0 \\ 0 & 0 \end{Vmatrix}$.

12. (a) $\begin{Vmatrix} a & b \\ c & -a \end{Vmatrix}$, where $bc = -a^2$; (b) Use Prob. 11 to show that $A^3 = 0$ implies that $A^2 = 0$; (c) $\pm \begin{Vmatrix} 1 & 0 \\ 0 & 1 \end{Vmatrix}$ and $\begin{Vmatrix} a & b \\ c & -a \end{Vmatrix}$, where $a^2 = 1 - bc$.

13. (a) If $A = 0$, X is arbitrary. If $|A| \neq 0$, $X = 0$. If $|A| = 0$, but $A \neq 0$, then the rows of A are proportional. Let α/β be the ratio of the elements in the first and second rows of A. Then

$$X = \begin{Vmatrix} -\beta x & \alpha x \\ -\beta y & \alpha y \end{Vmatrix},$$

where x and y are arbitrary.

(b) $X = \pm\begin{Vmatrix} 1 & 0 \\ 0 & 1 \end{Vmatrix}$ or $X = \begin{Vmatrix} x & y \\ z & -x \end{Vmatrix}$, where x, y, z satisfy $x^2 + yz = 1$ but are otherwise arbitrary; (c) $X = \pm\begin{Vmatrix} 1 & \frac{1}{2} \\ 0 & 1 \end{Vmatrix}$; (d) $X = \pm\frac{1}{\sqrt{2}}\begin{Vmatrix} 1 & 1 \\ 1 & 1 \end{Vmatrix}$;

(e) $X = \pm\begin{Vmatrix} 1 & 1 \\ 1 & 0 \end{Vmatrix}$ or $X = \pm\frac{1}{\sqrt{5}}\begin{Vmatrix} 3 & 1 \\ 1 & 2 \end{Vmatrix}$.

14. (b) $\begin{Vmatrix} 1 & -2 & 7 \\ 0 & 1 & -2 \\ 0 & 0 & 1 \end{Vmatrix}$.

15. (d) $X = \begin{Vmatrix} 1+a & b \\ -2a & 1-2b \end{Vmatrix}$; X does not exist. **18.** Use Prob. 17.

19. Let X be the matrices with 1 in one position and 0 elsewhere. **21.** (a) -1.

22. $A_{ij}a_{ik} = A_{ji}a_{ki} = \Delta\delta_{jk}$, where $\delta_{jk} = 1$ if $j = k$, $\delta_{jk} = 0$ if $j \neq k$.

24. For example, if T_j^i is a mixed tensor of rank 2,

$$T'^i_j = T^\alpha_\beta \frac{\partial x'^i}{\partial x^\alpha}\frac{\partial x^\beta}{\partial x'^j}.$$

27. $T'^m_{nij} = T^r_{stu}\frac{\partial x'^m}{\partial x^r}\frac{\partial x^s}{\partial x'^n}\frac{\partial x^t}{\partial x'^i}\frac{\partial x^u}{\partial x'^j}$. Therefore we have

$$T'^m_{mij} = T^r_{stu}\frac{\partial x'^m}{\partial x^r}\frac{\partial x^s}{\partial x'^m}\frac{\partial x^t}{\partial x'^i}\frac{\partial x^u}{\partial x'^j} = T^r_{stu}\,\delta^s_r\frac{\partial x^t}{\partial x'^i}\frac{\partial x^u}{\partial x'^j} = T^r_{rtu}\frac{\partial x^t}{\partial x'^i}\frac{\partial x^u}{\partial x'^j}.$$

28. 12.

31. The formula $T'^{ijk}X'^p_{ij} = U'^{kp}$ gives

$$T'^{ijk}X^r_{st}\frac{\partial x'^p}{\partial x^r}\frac{\partial x^s}{\partial x'^i}\frac{\partial x^t}{\partial x'^j} = U^{mn}\frac{\partial x'^k}{\partial x^m}\frac{\partial x'^p}{\partial x^n} = T^{abm}X^n_{ab}\frac{\partial x'^k}{\partial x^m}\frac{\partial x'^p}{\partial x^n}$$

or

$$\frac{\partial x'^p}{\partial x^r}\left(T'^{ijk}\frac{\partial x^s}{\partial x'^i}\frac{\partial x^t}{\partial x'^j} - T^{stm}\frac{\partial x'^k}{\partial x^m}\right)X^r_{st} = 0.$$

Taking the inner product with $\partial x^\gamma/\partial x'^p$, we obtain

$$\left(T'^{ijk}\frac{\partial x^s}{\partial x'^i}\frac{\partial x^t}{\partial x'^j} - T^{stm}\frac{\partial x'^k}{\partial x^m}\right)X^\gamma_{st} = 0.$$

Since X^γ_{st} is arbitrary, the expression in parentheses vanishes identically, i.e.,

$$T'^{ijk}\frac{\partial x^s}{\partial x'^i}\frac{\partial x^t}{\partial x'^j} = T^{stm}\frac{\partial x'^k}{\partial x^m}.$$

Taking the inner product with $\frac{\partial x'^\alpha}{\partial x^s}\frac{\partial x'^\beta}{\partial x^t}$, we find that

$$T'^{\alpha\beta k} = T^{stm}\frac{\partial x'^\alpha}{\partial x^s}\frac{\partial x'^\beta}{\partial x^t}\frac{\partial x'^k}{\partial x^m}, \text{ QED.}$$

33. For example, if $T'^k = T^i\frac{\partial x'^k}{\partial x^i}$, then

$$\frac{\partial T'^k}{\partial x'^s} = \frac{\partial T^i}{\partial x'^s}\frac{\partial x'^k}{\partial x^i} + T^i\frac{\partial}{\partial x'^s}\left(\frac{\partial x'^k}{\partial x^i}\right) = \frac{\partial T^i}{\partial x^j}\frac{\partial x^j}{\partial x'^s}\frac{\partial x'^k}{\partial x^i} + T^i\frac{\partial^2 x'^k}{\partial x^l\partial x^i}\frac{\partial x^l}{\partial x'^s}.$$

The presence of the extra term prevents $(\partial T^k/\partial x^s)$ from being a mixed tensor of rank 2. However, for the special case of orthogonal coordinate transformations, this term vanishes.

34. If $A = \|a_{ik}\|$, det $A = e_{rst}a_{1r}a_{2s}a_{3t} = e_{rst}a_{r1}a_{s2}a_{t3}$. Use this fact to prove that e_{ijk} is an oriented Cartesian tensor.

35. (a) $(\mathbf{A} \times \mathbf{B})_i = e_{ijk}A^iB^k$, where $\mathbf{A} = (A^1, A^2, A^3)$, $\mathbf{B} = (B^1, B^2, B^3)$.

37. Consider various special rotations, e.g.,

$$\left\|\begin{matrix} -1 & 0 & 0 \\ 0 & -1 & 0 \\ 0 & 0 & 1 \end{matrix}\right\|, \quad \left\|\begin{matrix} 0 & 1 & 0 \\ -1 & 0 & 0 \\ 0 & 0 & 1 \end{matrix}\right\|, \quad \text{etc.}$$

38. (b) $\left\|\begin{matrix} 8 & 6 & 4 & 2 \\ 5 & 0 & -5 & -10 \\ 7 & 7 & 7 & 7 \\ 10 & 9 & 8 & 7 \end{matrix}\right\|$. **39.** $\left\|\begin{matrix} a & b & c & d \\ 0 & a & b & c \\ 0 & 0 & a & b \\ 0 & 0 & 0 & a \end{matrix}\right\|$.

43. $\left\|\begin{matrix} 1 & 1 \\ 0 & 0 \end{matrix}\right\| \left\|\begin{matrix} 0 & 1 \\ 0 & 1 \end{matrix}\right\|$ is not idempotent.

44. (a) $\left\|\begin{matrix} 1 & -1 & 0 & \cdots & 0 \\ 0 & 1 & -1 & \cdots & 0 \\ 0 & 0 & 1 & \cdots & 0 \\ \cdots & \cdots & \cdots & \cdots & \cdots \\ 0 & 0 & 0 & \cdots & 1 \end{matrix}\right\|$; (c) $\left\|\begin{matrix} 2-n & 1 & 1 & \cdots & 1 \\ 1 & -1 & 0 & \cdots & 0 \\ 1 & 0 & -1 & \cdots & 0 \\ \cdots & \cdots & \cdots & \cdots & \cdots \\ 1 & 0 & 0 & \cdots & -1 \end{matrix}\right\|$.

45. $c = -1/3$.

47. Use formula (49) of Chap. 1, or else show that $A\tilde{A}$ is a Gram determinant and use Prob. 42 of Chap. 2.

48. Use formula (49) of Chap. 1.

51. Perform the given elementary transformation on the unit matrix whose order is the same as the number of rows of A, in the case of a row transformation, or on the unit matrix whose order is the same as the number of columns of A, in the case of a column transformation. Verify that the matrices so obtained are the required matrices.

52. Use the preceding problem and Prob. 32 of Chap. 2.

53. Interpret the rows of A as vectors (cf. Prob. 28 of Chap. 2). Then at most r rows of A are linearly independent, and at most $m - s$ of these linearly independent rows lie outside a given group of s rows of A. Thus, at least $r + s - m$ of the s rows must be linearly independent. The result also holds for columns, since interchanging rows and columns of a matrix does not change its rank.

54. Use the result of Prob. 53.

55. We have already proved in Sec. 7 that $r_{AB} \le r_A, r_B$. To prove that $r_A + r_B - n \le r_{AB}$, use Probs. 52 and 54 and the fact that multiplying a matrix by a nonsingular matrix does not change its rank (Sec. 26).

56. $AB = \left\|\begin{matrix} C_{11} & C_{12} \\ C_{21} & C_{22} \end{matrix}\right\|$, where $C_{11} = \left\|\begin{matrix} 6 \\ 9 \end{matrix}\right\|$, $C_{12} = \left\|\begin{matrix} 2 & 4 \\ 9 & 6 \end{matrix}\right\|$, $C_{21} = \|8\|$, $C_{22} = \|9 \quad 1\|$.

57. (b) $\lambda_1 = 7, \quad \lambda_2 = -2, \quad V = \left\|\begin{matrix} 1 & 4 \\ 1 & -5 \end{matrix}\right\|$;

(d) $\lambda_1 = 0, \quad \lambda_{2,3} = \pm\sqrt{-14}, \quad V = \left\|\begin{matrix} 3 & \dfrac{3+2\sqrt{-14}}{13} & \dfrac{3-2\sqrt{-14}}{13} \\ -1 & & \\ 2 & 2-3\sqrt{-14} & 2+3\sqrt{-14} \end{matrix}\right\|$.

58. The eigenvalues of A^{-1} are the reciprocals of those of A; the eigenvalues of A^2 are the squares of those of A^2.

59. $\lambda_k = a_0 + a_1\epsilon_k + a_2\epsilon_k^2 + \cdots + a_{n-1}\epsilon_k^{n-1}$ $(k = 0, 1, \ldots, n-1)$, where

$$\epsilon_k = \cos\frac{2k\pi}{n} + i\sin\frac{2k\pi}{n}.$$

61. Apply formula (45) of Sec. 7. 62. Apply Prob. 57 of Chap. 1.

63. Apply the Schwarz inequality to

$$\text{(a)}\ \sum_{k=1}^{n} 1\cdot a_k;\quad \text{(b)}\ \sum_{k=1}^{n}\sqrt{k}\,\frac{a_k}{\sqrt{k}}.$$

64. $s_n < 1 + \frac{1}{2} + \frac{1}{3} + \cdots + 1/(n+1)$.

70. (a) $(1/\sqrt{2}, 1/\sqrt{2}, 0)$, $(1/\sqrt{2}, -1/\sqrt{2}, 0)$, $(0, 0, 1)$.

71. We have $\mathbf{y}^{(1)} = \mathbf{x}^{(1)}$ and $\mathbf{y}^{(2)} = \alpha\mathbf{x}^{(1)} + \mathbf{x}^{(2)}$, where $\alpha = -(\mathbf{x}^{(1)}, \mathbf{x}^{(2)})/(\mathbf{x}^{(1)}, \mathbf{x}^{(1)})$. In $G(\mathbf{x}^{(1)}, \ldots, \mathbf{x}^{(n)})$, replace $\mathbf{x}^{(1)}$ by $\mathbf{y}^{(1)}$. Then add the first row multiplied by α to the second row, and add the first column multiplied by $\bar{\alpha}$ to the second column. The result is to replace $\mathbf{x}^{(2)}$ by $\mathbf{y}^{(2)}$ everywhere in $G(\mathbf{x}^{(1)}, \ldots, \mathbf{x}^{(n)})$. Continue this process.

72. Suppose the orthogonalization process takes the vectors $\mathbf{a}_1, \ldots, \mathbf{a}_k, \mathbf{b}_1, \ldots, \mathbf{b}_l$ into $\mathbf{c}_1, \ldots, \mathbf{c}_k, \mathbf{d}_1, \ldots, \mathbf{d}_l$ and the vectors $\mathbf{b}_1, \ldots, \mathbf{b}_l$ *by themselves* into $\mathbf{e}_1, \ldots, \mathbf{e}_l$. Apply the preceding problem, and bear in mind Prob. 49 of Chap. 2.

CHAPTER 4

1. Given the equation $(A(\mathbf{x}_0 + \mathbf{x}), \mathbf{x}_0 + \mathbf{x}) + 2(\mathbf{b}, \mathbf{x}_0 + \mathbf{x}) + c = 0$, the equation $(A(\mathbf{x}_0 - \mathbf{x}), \mathbf{x}_0 - \mathbf{x}) + 2(\mathbf{b}, \mathbf{x}_0 - \mathbf{x}) + c = 0$ implies $A\mathbf{x}_0 + \mathbf{b} = 0$, and conversely. Then use Prob. 29 of Chap. 2.

2. (b) $9y_1^2 + 18y_2^2 - 9y_3^2$, $x_1 = \frac{2}{3}y_1 + \frac{2}{3}y_2 - \frac{1}{3}y_3$, $x_2 = -\frac{1}{3}y_1 + \frac{2}{3}y_2 + \frac{2}{3}y_3$, $x_3 = \frac{2}{3}y_1 - \frac{1}{3}y_2 + \frac{2}{3}y_3$;

(d) $5y_1^2 - y_2^2 - y_3^2$, $x_1 = \frac{1}{3}\sqrt{3}\,y_1 + \frac{1}{6}\sqrt{6}\,y_2 + \frac{1}{2}\sqrt{2}\,y_3$, $x_2 = \frac{1}{3}\sqrt{3}\,y_1 + \frac{1}{6}\sqrt{6}\,y_2 - \frac{1}{2}\sqrt{2}\,y_3$, $x_3 = \frac{1}{3}\sqrt{3}\,y_1 - \frac{1}{3}\sqrt{6}\,y_2$;

(f) $3y_1^2 - 6y_2^2$, $x_1 = \frac{2}{3}y_1 + \frac{1}{6}\sqrt{2}\,y_2 + \frac{1}{2}\sqrt{2}\,y_3$, $x_2 = \frac{1}{3}y_1 - \frac{2}{3}\sqrt{2}\,y_2$, $x_3 = \frac{2}{3}y_1 + \frac{1}{6}\sqrt{2}\,y_2 - \frac{1}{2}\sqrt{2}\,y_3$;

(h) $2y_1^2 + 4y_2^2 - 2y_3^2 - 4y_4^2$, $x_1 = \frac{1}{2}(y_1 + y_2 + y_3 + y_4)$, $x_2 = \frac{1}{2}(-y_1 + y_2 + y_3 - y_4)$, $x_3 = \frac{1}{2}(-y_1 - y_2 + y_3 + y_4)$, $x_4 = \frac{1}{2}(y_1 - y_2 + y_3 - y_4)$;

(j) $5y_1^2 - 5y_2^2 + 5y_3^2$, $x_1 = \frac{1}{5}\sqrt{5}\,(2y_1 + y_2)$, $x_2 = \frac{1}{5}\sqrt{5}\,(y_1 - 2y_2)$, $x_3 = \frac{1}{5}\sqrt{5}\,(2y_3 + y_4)$, $x_4 = \frac{1}{5}\sqrt{5}\,(-y_3 + 2y_4)$;

(l) $9y_1^2 + 9y_2^2 + 9y_3^2$, $x_1 = y_1$, $x_2 = \frac{1}{3}(y_2 - 2y_3 + 2y_4)$, $x_3 = \frac{1}{3}(2y_2 + y_3 - 2y_4)$, $x_4 = \frac{1}{3}(2y_2 - 2y_3 + y_4)$;

(n) $4y_1^2 + 4y_2^2 + 4y_3^2 - 6y_4^2 - 6y_5^2$, $x_1 = y_1$, $x_2 = \frac{1}{5}\sqrt{5}\,(y_2 + 2y_4)$, $x_3 = \frac{1}{5}\sqrt{5}\,(-2y_2 + y_4)$, $x_4 = \frac{1}{10}\sqrt{10}\,(y_3 + 3y_5)$, $x_5 = \frac{1}{10}\sqrt{10}\,(3y_3 - y_5)$;

(p) $$\frac{n+1}{2}y_1^2 + \frac{1}{2}y_2^2 + \cdots + \frac{1}{2}y_n^2,\qquad y_1 = \frac{1}{\sqrt{n}}(x_1 + x_2 + \cdots + x_n),$$

$$y_i = \frac{1}{\sqrt{i(i-1)}}\,[x_1 + x_2 + \cdots + x_{i-1} - (i-1)x_i]\quad (i = 2, 3, \ldots n).$$

4. (a) The forms f and h are orthogonally equivalent, but neither is orthogonally equivalent to g.

5. (b) $B = \begin{Vmatrix} \frac{2}{3} & \frac{2}{3} & \frac{1}{3} \\ -\frac{2}{3} & \frac{1}{3} & \frac{2}{3} \\ -\frac{1}{3} & \frac{2}{3} & -\frac{2}{3} \end{Vmatrix}$, $\quad B^{-1}AB = \begin{Vmatrix} 1 & 0 & 0 \\ 0 & 4 & 0 \\ 0 & 0 & 7 \end{Vmatrix}$.

6. See Prob. 51 of Chap. 3.

7. (b) $y_1^2 + y_2^2 - y_3^2$, $\quad x_1 = \frac{1}{2}y_1 + y_2$, $\quad x_2 = y_2 + y_3$, $\quad x_3 = -y_2 + y_3$;
   (d) $y_1^2 + y_2^2 - y_3^2$, $\quad x_1 = \frac{1}{2}\sqrt{2}\,y_1 - \frac{5}{3}\sqrt{3}\,y_2 + \frac{1}{3}\sqrt{3}\,y_3$,
   $x_2 = -\frac{1}{3}\sqrt{3}\,y_2 + \frac{1}{3}\sqrt{3}\,y_3$, $\quad x_3 = \frac{1}{3}\sqrt{3}\,y_2 + \frac{1}{3}\sqrt{3}\,y_3$;
   (f) $y_1^2 - y_2^2$, $\quad x_1 = y_1 - y_2 - y_3$, $\quad x_2 = y_1 + y_2 - y_3$, $\quad x_3 = y_3$, $\quad x_4 = y_4$.

8. (b) $x_1 = \frac{3}{4}\sqrt{2}\,y_1 - \frac{1}{4}\sqrt{2}\,y_2 + \frac{5}{4}y_3$, $\quad x_2 = \frac{1}{4}\sqrt{2}\,y_1 - \frac{1}{4}\sqrt{2}\,y_2 + \frac{1}{4}y_3$,
   $x_3 = -\frac{1}{4}\sqrt{2}\,y_1 - \frac{1}{4}\sqrt{2}\,y_2 + \frac{1}{4}y_3$.

9. (a) If $A = C^*C$, where $C = \|c_{ik}\|$, then $(Ax, x) = \sum_{k=1}^{n}\sum_{i=1}^{n} (c_{ki}x_i)^2 \geq 0$. If $(Ax, x) \geq 0$, then there exists an orthogonal matrix B such that $B^{-1}AB$ is a diagonal matrix $[\lambda_1, \ldots, \lambda_n]$, where $\lambda_i \geq 0$ $(i = 1, 2, \ldots, n)$. Then $A = B[\sqrt{\lambda_1}, \ldots, \sqrt{\lambda_n}][\sqrt{\lambda_1}, \ldots, \sqrt{\lambda_n}]B^{-1}$. Now set $C = [\sqrt{\lambda_1}, \ldots, \sqrt{\lambda_n}]B^{-1}$.

10. (b) $|\lambda| < \frac{5}{3}$; (d) No such values of λ exist.

11. By Prob. 9, $B = C^*C$, where $C = \|c_{ik}\|$. Therefore, $\sum_{i,j=1}^{n} a_{ij}b_{ij}x_ix_j$

$$= \sum_{k=1}^{n}\sum_{i,j=1}^{n} a_{ij}(c_{ki}x_i)(c_{kj}x_j) \geq 0.$$ Investigate the positive definite case.

12. By Prob. 9, $A = \|a_{ik}\| = C^*C$, where all the row vectors of C are linearly independent, i.e., A is a Gram determinant formed from linearly independent vectors. Therefore, all the principal minors of A, in particular, the determinants $\Delta_1, \Delta_2, \ldots, \Delta_n$ of Sec. 36, are positive, being themselves Gram determinants formed from linearly independent vectors. (Cf. Prob. 42 of Chap. 2.)

13. Consider the quadratic form $-x_2^2 + 2x_1x_3$.

14. Let $g = f + l^2$, where $l = c_1x_1 + \cdots + c_nx_n$. By renumbering the variables if necessary, make $c_n \neq 0$. Then make the nonsingular transformation $y_i = x_i$ $(i = 1, 2, \ldots, n - 1)$, $y_n = l/c_n$. In the new variables, the forms become f_1 and g_1, say, with discriminants D_{f_1} and D_{g_1}, and we have $D_{g_1} = D_{f_1} + c_n^2D_{n-1}$, where D_{n-1} is the minor obtained by deleting the last row and the last column of the coefficient matrix of f_1. By Prob. 12, D_{n-1} is positive. Now use the fact that under a linear transformation, the discriminant of a quadratic form is multiplied by the square of the determinant of the matrix which transforms from the new to the old variables (Sec. 37).

15. Write f as

$$f = a_{11}\left(x_1 + \frac{a_{12}}{a_{11}}x_2 + \cdots + \frac{a_{1n}}{a_{11}}x_n\right)^2 + f_1(x_2, \ldots, x_n),$$

and then use the result of Prob. 14 to show that

$$D_f = a_{11}D_{f_1} \leq a_{11}D_g.$$

16. Use the result of Prob. 15.

17. (b) $f_1 = y_1^2 + y_2^2, \quad g_1 = 4y_1^2 - 2y_2^2, \quad x_1 = -2\sqrt{2}\,y_1 + 3\sqrt{2}\,y_2,$
$x_2 = \frac{1}{2}\sqrt{2}\,y_1 - \frac{1}{2}\sqrt{2}\,y_2;$
(d) $f_1 = y_1^2 + 2y_2^2 - y_3^2, \quad g_1 = y_1^2 + y_2^2 + y_3^2, \quad x_1 = \frac{1}{3}y_2 + \frac{1}{3}y_3,$
$x_2 = \frac{1}{3}y_2 - \frac{2}{3}y_3, \quad x_3 = \frac{1}{2}y_1 - \frac{1}{6}y_2 + \frac{1}{3}y_3;$
(f) $f_1 = y_1^2 + 2y_2^2 - 3y_3^2, \quad g_1 = y_1^2 + y_2^2 + y_3^2, \quad x_1 = y_1 - y_2,$
$x_2 = -y_2 + y_3, \quad x_3 = -3y_2 + 2y_3.$

18. No. In each case, the discriminant of the quadratic form $f - \lambda g$ has complex roots.

19. Let $\mu = m_A/m_B$. Then

$$\lambda_1 = 0, \qquad p_1 = \frac{\sqrt{m_A}}{2} \frac{1}{\sqrt{1 + 1/(2\mu)}} \left(q_1 + \frac{1}{\mu} q_2 + q_3\right),$$

corresponding to motion of the center of mass of the molecule with uniform velocity (there are no external forces),

$$\lambda_2 = \sqrt{\frac{k}{m_A}}, \qquad p_2 = \frac{\sqrt{m_A}}{2} (q_1 - q_3),$$

$$\lambda_3 = \sqrt{\frac{k}{m_A}(1 + 2\mu)}, \qquad p_3 = \frac{\sqrt{m_A}}{2} \frac{1}{\sqrt{1 + 2\mu}} (q_1 - 2q_2 + q_3).$$

21. As in Sec. 37, find a nonsingular transformation B such that

$$B^*AB = [\lambda_1, \lambda_2, \ldots, \lambda_n], \qquad B^*CB = [1, 1, \ldots, 1].$$

22. As in Sec. 33, reduce A to diagonal form by an orthogonal transformation.

23. Obviously, the right-hand side does not exceed the left-hand side. Reduce A to diagonal form by an orthogonal transformation, and let λ_j be the eigenvalue of largest absolute value (not necessarily unique). By the Schwarz inequality, the left-hand side is bounded by $|\lambda_j|$. Now choose $\mathbf{x}$ so that the right-hand side equals $|\lambda_j|$.

24. (d) Use Probs. 22 and 23 and the fact that

$$\max_{\|\mathbf{y}\|=1} |(\mathbf{z}, \mathbf{y})| = \|\mathbf{z}\|;$$

(e) $\|A\| = \max\limits_{\|\mathbf{x}\|=1} \|A\mathbf{x}\| = \max\limits_{\|\mathbf{x}\|=1, \|\mathbf{y}\|=1} |(A\mathbf{x}, \mathbf{y})| = \max\limits_{\|\mathbf{x}\|=1, \|\mathbf{y}\|=1} |(\mathbf{x}, A^*\mathbf{y})|$
$\le \max\limits_{\|\mathbf{y}\|=1} \|A^*\mathbf{y}\| = \|A^*\|$. Similarly, $\|A^*\| \le \|A\|$.

25. Use Prob. 58 of Chap. 3 and Prob. 22.

26. (a) Use Prob. 21 and the fact that $(A\mathbf{z}, \mathbf{z}) \le (A\mathbf{u}, \mathbf{u})$;
(b) Let $A\mathbf{w} = \lambda_k \mathbf{w}$ $(k \ne 1, \|\mathbf{w}\| = 1)$. The components w_i of $\mathbf{w}$ are of mixed sign, since $(\mathbf{w}, \mathbf{u}) = 0$ (Sec. 32). Let $\mathbf{w}'$ be the vector with components $|w_i|$. Then $|\lambda_k| = |(A\mathbf{w}, \mathbf{w})| < (A\mathbf{w}', \mathbf{w}') \le \lambda_1$.
(c) Otherwise, $\lambda_1 = (A\mathbf{z}, \mathbf{z}) = (A\mathbf{u}, \mathbf{u})$ is impossible.
(d) By Part (c), there cannot be two orthogonal eigenvectors associated with λ_1.

27. Use Probs. 22 and 25. (a) The eigenvalues are $a \pm c$; (b) The eigenvalues are $\alpha - 1$ and $n + \alpha - 1$ (see Prob. 48 of Chap. 1).

28. (a) Use preceding problem; (b) Replace y by $-y$.

29. Use the fact that a symmetric matrix is uniquely determined by its quadratic form.

30. Apply preceding problem to $A - \lambda$. **31.** Apply preceding problem.

32. $B = \frac{1}{2}(A + \tilde{A})$, $C = (1/2i)(A - \tilde{A})$.

34. First show that

$$B = U^{-1}AU = \left\| \begin{matrix} \lambda & C \\ 0 & A_1 \end{matrix} \right\|,$$

where $C = (c_1, \ldots, c_{n-1})$. It follows from $\tilde{B}B = B\tilde{B}$ that $C = 0$ and $\tilde{A}_1 A_1 = A_1 \tilde{A}_1$.

35. Use preceding problem and mathematical induction.

36. (b) $U = \left\| \begin{matrix} \frac{1}{2}\sqrt{2} & \frac{1}{2}\sqrt{2} \\ \frac{i}{2}\sqrt{2} & -\frac{i}{2}\sqrt{2} \end{matrix} \right\|, \qquad U^{-1} \left\| \begin{matrix} a & b \\ -b & a \end{matrix} \right\| U = \left\| \begin{matrix} a+ib & 0 \\ 0 & a-ib \end{matrix} \right\|.$

37. An $n \times n$ matrix can be diagonalized by a unitary matrix if and only if it has n pairwise orthogonal eigenvectors. Now use Prob. 35.

38. Normalize the k eigenvectors of A, and let them be the first k columns of U.

40. $\displaystyle\sum_{i,j=1}^{n} |a_{ij}|^2 = \operatorname{tr} \tilde{A}A.$

41. If $A\mathbf{x} = \lambda\mathbf{x}$, then $(\lambda - a_{rr})x_r = \displaystyle\sum_{s \neq r} a_{rs}x_s \; (r = 1, \ldots, n)$. Choose k such that $|x_k| \geq |x_j|$ for all j. Then

$$|\lambda - a_{kk}|\,|x_k| = \left| \sum_{s \neq k} a_{ks}x_s \right| \leq |x_k| \sum_{s \neq k} |a_{ks}| = \rho_k |x_k|.$$

43. Parallel the proof given in Sec. 41 for commuting Hermitian matrices.

44. Use Prob. 30.

45. (b) If $C = \dfrac{1}{\sqrt{6}} \left\| \begin{matrix} 2-i & 1 \\ -1 & 2+i \end{matrix} \right\|$, then $C^{-1}HC = [2, 8]$.

46. (b) Let H_1 be the first matrix and H_2 the second. If

$$U = \left\| \begin{matrix} \frac{1}{2}\sqrt{3} & -\frac{1}{2} \\ \frac{1}{2} & \frac{1}{2}\sqrt{3} \end{matrix} \right\|,$$

then $U^{-1}H_1U = [2, -2]$, $U^{-1}H_2U = [3, -1]$.

47. (b) If $W = \left\| \begin{matrix} \frac{1}{3}\sqrt{3} & \frac{1}{3}\sqrt{6} & 0 \\ \frac{1}{3}\sqrt{3} & -\frac{1}{6}\sqrt{6} & -\frac{1}{2}\sqrt{2} \\ \frac{1}{3}\sqrt{3} & \frac{1}{6} - \sqrt{6} & \frac{1}{2}\sqrt{2} \end{matrix} \right\|,$

$$\text{then } W^{-1}VW = \left\| \begin{matrix} 1 & 0 & 0 \\ 0 & \frac{1}{2} & \frac{1}{2}\sqrt{3} \\ 0 & -\frac{1}{2}\sqrt{3} & \frac{1}{2} \end{matrix} \right\|.$$

48. Use the fact that U has one real column and two complex conjugate columns (Sec. 42).

49. (b) If $V = \left\| \begin{matrix} \frac{2}{3} & \frac{2}{3} & -\frac{i}{3} \\ -\frac{2i}{3} & \frac{i}{3} & \frac{2}{3} \\ \frac{i}{3} & -\frac{2i}{3} & \frac{2}{3} \end{matrix} \right\|$, then $V^{-1}UV = [1, i, -i]$.

50. First generalize Prob. 9 to the case of positive *Hermitian* matrices. Then write $B = \sqrt{\bar{A}A}$, $U = AB^{-1}$.

52. Let $U = J + iK$, where J and K are real. It follows from the symmetry of U that J and K are symmetric and from Prob. 32 that J and K commute. Hence by Sec. 41, J and K can be simultaneously diagonalized by a real orthogonal transformation A, which also diagonalizes U, i.e., $A^*UA = [\lambda_1, \ldots, \lambda_n]$. Now let $V = A[\sqrt{\lambda_1}, \ldots, \sqrt{\lambda_n}]A^*$. Then V is symmetric and unitary, and $V^2 = U$.

CHAPTER 5

1. Let $0 = N_0 < N_1 < N_2 < \cdots < N_i < \cdots$ be such that

$$d_i^2 = \sum_{n=N_{i-1}+1}^{N_i} |x_n|^2 \geq 4^i,$$

and define $$a_n = \frac{\bar{x}_n}{2^i d_i} \qquad (N_{i-1} + 1 \leq n \leq N_i),$$

where $i = 1, 2, \ldots$.

2. Cf. finite-dimensional case in Secs. 29 and 30 and Prob. 65 of Chap. 3.

3. Use the Schwarz inequality, and consider the special case $\mathbf{y} = \mathbf{x}/\|\mathbf{x}\|$.

5. Use the Schwarz inequality and Prob. 2. **6.** Use the Schwarz inequality.

8. Consider $(\mathbf{x}_n - \mathbf{x}, \mathbf{x}_n - \mathbf{x})$.

10. Examine in detail the system $(\mathbf{x}, \mathbf{x}^{(n)}) = 0$, $n = 1, 2, \ldots$.

11. $M = (1 + a)^2$. Use Prob. 5, and then consider $\mathbf{x} = (1, c, c^2, \ldots)$, $c < 1$, as $c \to 1$.

14. $\|A\mathbf{x}\|^2 \leq \sum_{n=1}^{\infty} \left(\sum_{m=1}^{\infty} \sqrt{|a_{nm}|} \sqrt{|a_{nm}|}\, |x_m| \right)^2$. Then use the Schwarz inequality, and change the order of summation.

16. Use Prob. 14. **20.** Consider $f_n(x) = x^n$, $0 \leq x \leq 1$. **23.** (a) and (b).

24. $F(x) = 1 - x \log [(1 + x)/x]$. Yes.

26. The coefficients in formula (88) of Sec. 51 are $a_{kj} = \delta_{-k,j}$.

CHAPTER 6

5. Let L_1 be the set of all $\mathbf{x}$ in R_n such that $T\mathbf{x} = \mathbf{x}$, and let L_2 be the orthogonal complement of L_1. Then $T\mathbf{x} = -\mathbf{x}$ for all $\mathbf{x}$ in L_2.

7. Use Probs. 35 and 44 of Chap. 4.

8. Let $f(\lambda)$ be a polynomial such that $f(\lambda_i) = \bar{\lambda}_i$, where the λ_i are the eigenvalues of A. Then $f(A) = \bar{A}$.

9. The two coordinate axes.

11. $\lambda = 0$ is an eigenvalue of multiplicity n. The only eigenvectors are constants.

13. $3x_1 - 3x_2 + x_3 = 0$. **16.** Use Prob. 3. **17.** Use Probs. 10 and 12.

18. Use the preceding problem and induction. Then choose a basis such that the first k basis vectors come from M_k $(1 \leq k \leq n)$.

19. Use the preceding problem.

20. Let λ be an eigenvalue of A. Show that the set of all eigenvectors of A corresponding to the eigenvalue λ is an invariant subspace of B. Then use Prob. 10.

21. Follow the proof of Prob. 17, using induction and Probs. 12 and 19.

23. Reduce A to canonical form, and apply Prob. 46 of Chap. 3.

24. (a) $\lambda - 1$; (c) $(\lambda - \alpha)^p$.

25. For scalar matrices $A = \lambda I$ and only for such matrices. **27.** (a) $\lambda^2 - 4\lambda + 4$.

29. For the matrices

$$\begin{Vmatrix} 1 & 0 & 0 & 0 \\ 0 & 1 & 0 & 0 \\ 0 & 0 & 1 & 0 \\ 0 & 0 & 1 & 1 \end{Vmatrix} \quad \text{and} \quad \begin{Vmatrix} 1 & 0 & 0 & 0 \\ 1 & 1 & 0 & 0 \\ 0 & 0 & 1 & 1 \\ 0 & 0 & 1 & 1 \end{Vmatrix},$$

$\varphi(\lambda) = (\lambda - 1)^4$, $\mu(\lambda) = (\lambda - 1)^2$, but these matrices are not similar.

30. Let $A = TBT^{-1}$, where A is the given matrix and where $B = [B_1, B_2, \ldots, B_m]$ is its canonical form involving the submatrices

$$B_i = \begin{Vmatrix} \lambda_i & 0 & \cdots & 0 & 0 \\ 1 & \lambda_i & \cdots & 0 & 0 \\ \cdot & \cdot & \cdots & \cdot & \cdot \\ 0 & 0 & \cdots & \lambda_i & 0 \\ 0 & 0 & \cdots & 1 & \lambda_i \end{Vmatrix}.$$

Then $A^* = (T^*)^{-1}B^*T^*$. Let

$$H_i = \begin{Vmatrix} 0 & 0 & \cdots & 0 & 1 \\ 0 & 0 & \cdots & 1 & 0 \\ \cdot & \cdot & \cdots & \cdot & \cdot \\ 0 & 1 & \cdots & 0 & 0 \\ 1 & 0 & \cdots & 0 & 0 \end{Vmatrix}$$

have the same order as B_i, and let $H = [H_1, H_2, \ldots, H_m]$. Then $B_i^* = H_i^{-1}B_iH_i$, i.e., $B^* = H^{-1}BH$. Therefore,
$A^* = (T^*)^{-1}H^{-1}BHT^* = (T^*)^{-1}H^{-1}T^{-1}ATHT^* = C^{-1}AC$, where $C = THT^*$.

31. The matrix $C = THT^*$ in the solution of the preceding problem is symmetric and nonsingular. Let $D = C^{-1}A$. Then $D^* = A^*(C^*)^{-1} = C^{-1}ACC^{-1} = D$. Thus, the matrix D is also symmetric and $A = CD$.

33. (b) A and B are similar; (d) B and C are similar to each other, but not to A.

36. (b) $(\lambda + 1), (\lambda + 1)^2, \begin{Vmatrix} -1 & 0 & 0 \\ 1 & -1 & 0 \\ 0 & 0 & -1 \end{Vmatrix}$; (d) $(\lambda - 1), (\lambda - 1)^2, \begin{Vmatrix} 1 & 0 & 0 \\ 1 & 1 & 0 \\ 0 & 0 & 1 \end{Vmatrix}$;

(f) $(\lambda - \alpha), (\lambda - \alpha)^2, \begin{Vmatrix} \alpha & 0 & 0 \\ 0 & \alpha & 0 \\ 0 & 1 & \alpha \end{Vmatrix}$; (h) $(\lambda - 1), (\lambda - 1)^3, \begin{Vmatrix} 1 & 0 & 0 & 0 \\ 0 & 1 & 0 & 0 \\ 0 & 1 & 1 & 0 \\ 0 & 0 & 1 & 1 \end{Vmatrix}$;

(j) $(\lambda - n)^n, \begin{Vmatrix} n & 0 & \cdots & 0 & 0 \\ 1 & n & \cdots & 0 & 0 \\ \cdot & \cdot & \cdots & \cdot & \cdot \\ 0 & 0 & \cdots & n & 0 \\ 0 & 0 & \cdots & 1 & n \end{Vmatrix}.$

37. (b) If $S = \begin{Vmatrix} 1 & -3 & -2 \\ 1 & 0 & 0 \\ 1 & 0 & 1 \end{Vmatrix}$, then $SAS^{-1} = \begin{Vmatrix} 0 & 0 & 0 \\ 1 & 0 & 0 \\ 0 & 0 & 0 \end{Vmatrix}$;

(d) If $S = \begin{Vmatrix} 3 & 1 & 1 \\ 1 & 0 & 0 \\ 5 & 0 & 1 \end{Vmatrix}$, then $SAS^{-1} = \begin{Vmatrix} 3 & 0 & 0 \\ 1 & 3 & 0 \\ 0 & 0 & 3 \end{Vmatrix}$.

38. Use the fact that if $SAS^{-1} = N$, then $SA^mS^{-1} = N^m$, and choose S so that N is the canonical form of A.

39. (c) Suppose that an eigenvalue λ of absolute value 1 is associated with a $k \times k$ submatrix in the canonical representation of T, i.e., with the submatrix

$$\begin{Vmatrix} \lambda & 0 & \cdots & 0 & 0 \\ 1 & \lambda & \cdots & 0 & 0 \\ \cdot & \cdot & \cdots & \cdot & \cdot \\ 0 & 0 & \cdots & \lambda & 0 \\ 0 & 0 & \cdots & 1 & \lambda \end{Vmatrix} = \lambda I + N,$$

where I is the unit matrix of order k. Then if $m \geq k - 1$, we have

$$(\lambda I + N)^m = \lambda^m I + m\lambda^{m-1}N + \frac{m(m-1)}{1 \cdot 2}\lambda^{m-2}N^2 + \cdots$$
$$+ \frac{m(m-1)\cdots(m-k+2)}{1 \cdot 2 \cdots (k-1)}\lambda^{m-k+1}N^{k-1}.$$

Unless $k = 1$, the norm of $(\lambda I + N)^m$ becomes infinite as $m \to \infty$.

40. Use the preceding problem to analyze the case where T has an eigenvalue of absolute value 1. In the case of an eigenvalue λ of absolute value < 1, study the expression $(\lambda I + N)^m$ further.

CHAPTER 7

1. (b) and (e) are groups.

2. Let $P_{12}(\theta)$ be a rotation of the 1, 2-plane (i.e., the xy-plane) about the z-axis, and show that θ can be chosen so that the 1, 2-matrix element of the product $RP_{12}(\theta)$ is zero. Continue with P_{13} and P_{23}.

3. Use Prob. 48 of Chap. 4 and Prob. 6 of Chap. 6.

10. In units such that $c = 1$, formula (17) of Sec. 60 can be written as

$$x' = \alpha(x - vt), \qquad y' = y, \qquad z' = z, \qquad t' = \alpha(t - vx),$$

where $\alpha = (1 - v^2)^{-1/2}$. Write $\mathbf{x} = \mathbf{x}_p + \mathbf{x}_o$, where $\mathbf{x}_p$ is parallel to $\mathbf{v}$ and $\mathbf{x}_o$ is orthogonal to $\mathbf{v}$. Then

$$\mathbf{x}_p' = \alpha(\mathbf{x}_p - \mathbf{v}t), \qquad \mathbf{x}_o' = \mathbf{x}_o, \qquad t' = \alpha(t - (\mathbf{v}, \mathbf{x}_p)).$$

Now use $\mathbf{x}' = \mathbf{x}_p' + \mathbf{x}_o'$, $\quad \mathbf{x}_p = v^{-2}(\mathbf{x}, \mathbf{v})\mathbf{v}$, $\quad \mathbf{x}_o = \mathbf{x} - \mathbf{x}_p$.

11. Cf. comment to Prob. 8. L_0 has only two linearly independent eigenvectors.

12. $L_1 = \begin{Vmatrix} 4/3 & -1/3 & 1/3 & 1 \\ -1/3 & 4/3 & -1/3 & -1 \\ 1/3 & -1/3 & 4/3 & 1 \\ 1 & -1 & 1 & 2 \end{Vmatrix}, \; R = \begin{Vmatrix} 2/3 & -2/3 & -1/3 & 0 \\ 2/3 & 1/3 & 2/3 & 0 \\ -1/3 & -2/3 & 2/3 & 0 \\ 0 & 0 & 0 & 1 \end{Vmatrix}.$

16. $X = \begin{pmatrix} 1 & 2 & 3 & 4 & 5 & 6 & 7 \\ 4 & 2 & 6 & 7 & 1 & 3 & 5 \end{pmatrix}.$

17. Use the fact that $(i_1, i_2, \ldots, i_m) = (i_1, i_m)(i_1, i_{m-1}) \cdots (i_1, i_2)$, so that even (odd) cycles are odd (even) permutations.

18. $\begin{pmatrix} 1 & 2 & 3 & 4 \\ 1 & 2 & 3 & 4 \end{pmatrix}, \begin{pmatrix} 1 & 2 & 3 & 4 \\ 1 & 2 & 4 & 3 \end{pmatrix}, \begin{pmatrix} 1 & 2 & 3 & 4 \\ 2 & 1 & 3 & 4 \end{pmatrix}, \begin{pmatrix} 1 & 2 & 3 & 4 \\ 2 & 1 & 4 & 3 \end{pmatrix}, \begin{pmatrix} 1 & 2 & 3 & 4 \\ 3 & 4 & 1 & 2 \end{pmatrix},$
$\begin{pmatrix} 1 & 2 & 3 & 4 \\ 3 & 4 & 2 & 1 \end{pmatrix}, \begin{pmatrix} 1 & 2 & 3 & 4 \\ 4 & 3 & 1 & 2 \end{pmatrix}, \begin{pmatrix} 1 & 2 & 3 & 4 \\ 4 & 3 & 2 & 1 \end{pmatrix}.$

19. No. $A(BC) \neq (AB)C$.

22. Use the fact that if m and n are relatively prime, then there exist integers r and s such that $rm + sn = 1$.

23. If $rm + sn = 1$, let $A = C^{sn}$, $B = C^{rm}$.

25. Use the fact that if $i, j > 1$, $i \neq j$, then $(1, i)(1, j)(1, i) = (i, j)$.

26. Let the indicated cycles generate a subgroup $\mathfrak{A}$ of the symmetric group of degree n, and let i, j, and k be different numbers ≥ 3 (if such are compatible with the given value of n). Then, if $(1, 2, i)$ belongs to $\mathfrak{A}$, so does its inverse $(i, 2, 1) = (2, 1, i)$. Thus, $\mathfrak{A}$ contains $(j, 2, 1)(1, 2, i)(1, 2, j) = (1, i, j)$, $(1, 2, j)(2, 1, i)(2, 1, j) = (2, i, j)$, $(k, 2, 1)(1, i, j)(1, 2, k) = (i, j, k)$, i.e., $\mathfrak{A}$ contains *all* cycles of length 3, and hence is identical with the alternating group of degree n (see Sec. 61).

28. $n = 6$. The group has the multiplication table of Sec. 62.

29. A subgroup of index 2 must contain the square of every element in the group. But any cycle of length 3 is the square of a cycle of length 3.

30. Use the result at the end of Sec. 61.

31. $$\frac{n!}{1^{\alpha_1}\alpha_1!\,2^{\alpha_2}\alpha_2! \cdots n^{\alpha_n}\alpha_n!}$$

34. $\rho(A) = \sum_B \delta_{A,B^2}$, where $\delta_{A,B^2} = 1$ if $A = B^2$ and 0 otherwise. Thus

$$\sum_A \rho^2(A) = \sum_{A,B} \rho(A)\delta_{A,B^2} = \sum_B \rho(B^2) = \sum_{B,C} \delta_{B^2,C^2} = \sum_{A,C} \delta_{CACA,C^2} = \sum_{A,C} \delta_{A,CA^{-1}C^{-1}}.$$

But if $A = CA^{-1}C^{-1}$, then A belongs to an ambivalent class, and the number of elements C satisfying $A = CA^{-1}C^{-1}$ is equal to the *index* of the ambivalent class containing A (i.e., the order of the group divided by the number of elements in the class). Hence $\sum_A \rho^2(A) = \sum_A' \mathfrak{s}(A)$, where $\mathfrak{s}(A)$ is the index of the ambivalent class containing A, and the sum extends only over elements in ambivalent classes. Finally, one easily sees that $\sum_A' \mathfrak{s}(A) = \alpha g$.

35. Suppose $\sum_A n(A) = \sum_A \rho^2(A)$. Then by Prob. 34, $\sum_A n(A) = \alpha g$. Moreover, by Prob. 32, $n(A) = g/m(A)$, where $m(A)$ is the number of elements in the class containing A. Therefore, $\sum_A n(A) = g \sum_A \frac{1}{m(A)} = gm$, where m is the total number of classes, so that $\alpha = m$, i.e., every class is ambivalent. Conversely, suppose that $\alpha = m$. By Prob. 32, the number of elements transforming A into A^{-1} is $n(A)$, i.e.,

$$n(A) = \sum_C \delta_{A,CA^{-1}C^{-1}} = \sum_A \rho^2(A),$$

where the last equality follows from the solution of Prob. 34.

36. (d) Suppose $\mathcal{G}$ consists of the classes C_i $(i = 1, 2, \ldots, k)$, where C_i contains α_i elements. Then $p^m = \alpha_1 + \alpha_2 + \cdots + \alpha_k$, since the classes do not overlap. Now use Part (c) of Prob. 32 and the fact that every element of $\mathcal{Z}$ forms a class by itself. ($\mathcal{Z}$ is non-empty, since it contains at least the unit element.)

43. Suppose that $\mathcal{G}/\mathcal{Z}$ is cyclic, and let A be an element of a coset of $\mathcal{Z}$ which generates $\mathcal{G}/\mathcal{Z}$. Then $\mathcal{G}$ is generated by A and the elements of $\mathcal{Z}$, so that $\mathcal{G}$ is Abelian, contrary to hypothesis.

45. According to Prob. 26, the alternating group of degree n is generated by the cycles of the form $(1, 2, 3), (1, 2, 4), \ldots, (1, 2, n)$. But $(1, 2, k)$ is the commutator of $(1, 2)$ and $(k, 1)$.

46. If H is any element of $\mathcal{H}$ and K is any element of $\mathcal{K}$, then $HKH^{-1}K^{-1}$ belongs to both $\mathcal{H}$ and $\mathcal{K}$.

CHAPTER 8

1. Let $\mathcal{G}$ consist of the elements $A, A^2, \ldots, A^m = I$, and let ϵ be an mth root of unity. Then write

$$A^p \to \begin{Vmatrix} \epsilon^p & 0 & \cdots & 0 \\ \epsilon^p & 0 & \cdots & 0 \\ \cdots & \cdots & \cdots & \cdots \\ \epsilon^p & 0 & \cdots & 0 \end{Vmatrix}.$$

2. $\{D, \mathbf{a}\}\{D', \mathbf{a}'\} = \{DD', D\mathbf{a}' + \mathbf{a}\}, \qquad \{D, \mathbf{a}\}^{-1} = \{D^{-1}, -D^{-1}\mathbf{a}\}.$

3. $\left(\sum_\alpha A_\alpha^2\right) A_\beta = \sum_\alpha A_\alpha^2 A_\beta = A_\beta \sum_\alpha (A_\beta^{-1}A_\alpha A_\beta)(A_\beta^{-1}A_\alpha A_\beta) = A_\beta\left(\sum_\alpha A_\alpha^2\right).$

4. Since $\mathcal{A}$ and $\mathcal{B}$ are equivalent, $XA_\alpha X^{-1} = B_\alpha$ for some nonsingular X. Let $X = X' + iX''$, where X' and X'' are real. Then $X'A_\alpha = B_\alpha X'$, $X''A_\alpha = B_\alpha X''$. Now let $Y = X' + \mu X''$, where μ is any real number such that $\det(X' + \mu X'') \neq 0$.

5. As shown in Prob. 11 of Chap. 7, not every 4×4 Lorentz matrix is diagonalizable.

6. Let $\mathcal{U}$ consist of the unitary matrices U_α, and let $\mathcal{R}$ consist of the real matrices R_α; then $R_\alpha = S^{-1}U_\alpha S$. Let $H = \tilde{\bar{S}}S$, so that H is positive definite and Hermitian. It follows that $\tilde{\bar{R}}_\alpha HR_\alpha = R_\alpha^* HR_\alpha = H$. Then if $H = J + iK$, we have $R_\alpha^* JR_\alpha = J$, where J is positive definite and symmetric. Now let $B_\alpha = J^{1/2}R_\alpha J^{-1/2}$. Then $B_\alpha^* B_\alpha = I$, and the B_α are equivalent to the U_α.

7. Write the n columns of S as vectors $\mathbf{d}_1, \ldots, \mathbf{d}_n$. Then $AS = \|A\mathbf{d}_1 \cdots A\mathbf{d}_n\|$, while

$$SB = \left\| \sum_{j=1}^n b_{j1}\mathbf{d}_j \cdots \sum_{j=1}^n b_{jn}\mathbf{d}_j \right\|,$$

where $B = \|b_{ik}\|$, so that $A\mathbf{d}_k = \sum_{j=1}^n b_{jk}\mathbf{d}_k$. (We drop the subscripts on A_α and B_α.) Thus, the space spanned by $\mathbf{d}_1, \ldots, \mathbf{d}_n$ is invariant under A and hence is either the space consisting only of the zero vector or the whole m-dimensional carrier space of $\mathcal{A}$. In the first case $S = 0$, and in the second case $n \geq m$. Applying the same argument to the transposed equation $S^*A^* = B^*S^*$, we find in the second case that $m \geq n$, so that $m = n$. Since then $\mathbf{d}_1, \ldots, \mathbf{d}_n$ are linearly independent, it follows that S is nonsingular. (This proof is closely related to that of the theorem of Sec. 82.)

8. Let $A_\alpha S = SA_\alpha$ for all α, and let λ be an eigenvalue of S. Then $S - \lambda I$ commutes with every A_α in $\mathcal{A}$ and moreover is singular. Hence, by Schur's lemma, $S = \lambda I$.

9. The group of all nonsingular triangular matrices of order n (see Prob. 41 of Chap. 3).

10. Let $\mathcal{H}$ be the commutator subgroup (of order h) of the group $\mathcal{G}$ (of order g). Then $\mathcal{G}$ is homomorphic to the Abelian quotient group $\mathcal{G}/\mathcal{H}$ (cf. Prob. 44 of Chap. 7). Every representation of $\mathcal{G}/\mathcal{H}$ is one-dimensional, and there are g/h of them (cf. Sec. 73). Since $\mathcal{G}$ is homomorphic to $\mathcal{G}/\mathcal{H}$, every representation of $\mathcal{G}/\mathcal{H}$ is a representation of $\mathcal{G}$. Hence $\mathcal{G}/\mathcal{H}$ has at least g/h one-dimensional representations. Moreover, any other one-dimensional representation $\mathcal{A}$ of $\mathcal{G}$ would be isomorphic to an Abelian quotient group of some other normal subgroup $\mathcal{K}$ of $\mathcal{G}$.

But then $\mathcal{K}$ would have to contain $\mathcal{H}$ (by the problem cited), and hence $\mathfrak{a}$ is a representation of the quotient group $\mathcal{G}/\mathcal{H}$ (cf. Prob. 38 of Chap. 7), i.e., $\mathfrak{a}$ is one of the g/h one-dimensional representations of $\mathcal{G}$ already counted.

11. By Prob. 45 of Chap. 7, the commutator subgroup of the symmetric group is the alternating group, which has index 2.

12. Since $\mathcal{G}$ is Abelian, every irreducible representation must be one-dimensional, by Schur's lemma. An irreducible unitary representation of $\mathcal{G}$ must therefore have the form $x \to e^{i\varphi(x)}$, where φ is a real function. The value $e^{i\varphi(x)}$ defines the value $\varphi(x)$ only up to a multiple of 2π; however, if we demand that φ be continuous, the function φ is defined up to an additive constant, which is a multiple of 2π. Let φ be a continuous function determined in this way. In order for $x \to e^{i\varphi(x)}$ to be a representation, we must have

$$x + y \to e^{i\varphi(x+y)} = e^{i\varphi(x)}e^{i\varphi(y)}.$$

Therefore, $\varphi(x + y) = \varphi(x) + \varphi(y) + 2\pi n(x, y)$, where $n(x, y)$ is an integer-valued function. Since φ is continuous, $n(x, y)$ must be continuous and hence independent of x and y, i.e., $\varphi(x + y) = \varphi(x) + \varphi(y) + 2\pi n$. Let $\psi(x) = \varphi(x) + 2\pi n$; then $\psi(x + y) = \psi(x) + \psi(y)$. Thus $\psi(nx) = n\psi(x)$, and setting $x = y/m$, we obtain $\psi(ny/m) = (n/m)\psi(y)$. Thus $\psi(r) = r\psi(1)$ for every rational r, so that by continuity, $\psi(x) = \lambda x$, where $\lambda = \psi(1)$. Hence $e^{i\varphi(x)} = e^{i\lambda x}$.

13. **(c)** The determinant of the matrix $A \times B$ with elements $a_{ik}b_{jl}$ does not depend on the arrangement of the index pairs (i,j) and (k,l). Moreover, by formula (56) of Sec. 78, $A \times B = (A \times I_m)(I_n \times B)$, where I_m and I_n are the unit matrices of orders m and n, respectively. By making a suitable choice of indices, the matrix $A \times I_m$ has the form $[A, A, \ldots, A]$, where A is repeated m times. Therefore, $\det(A \times I_m) = (\det A)^m$. Similarly, with another choice of indices, $\det(I_n \times B) = (\det B)^n$, and the desired result follows at once.

14. Use the fact that

$$b^{ik} = \frac{b^{ik} + b^{ki}}{2} + \frac{b^{ik} - b^{ki}}{2} \qquad (i, k = 1, 2, \ldots, n).$$

15. $$A \times A = \begin{Vmatrix} 1 & 0 & 0 & 0 \\ a & 1 & 0 & 0 \\ a & 0 & 1 & 0 \\ a^2 & a & a & 1 \end{Vmatrix}, \quad A \circledS A = \begin{Vmatrix} 1 & 0 & 0 \\ a & 1 & 0 \\ a^2 & 2a & 1 \end{Vmatrix}, \quad A \circledast A = \|1\|.$$

$$S(A \times A)S^{-1} = \begin{Vmatrix} 1 & 0 & 0 & 0 \\ a & 1 & 0 & 0 \\ a^2 & 2a & 1 & 0 \\ 0 & 0 & 0 & 1 \end{Vmatrix}, \qquad \text{if } S = \begin{Vmatrix} 1 & 0 & 0 & 0 \\ 0 & \frac{1}{2} & \frac{1}{2} & 0 \\ 0 & 0 & 0 & 1 \\ 0 & \frac{1}{2} & -\frac{1}{2} & 0 \end{Vmatrix}.$$

16. Formulas **(a)** and **(b)** are the restrictions of formula (56) of Sec. 78 to the collections of symmetric and skew-symmetric contravariant second rank tensors, respectively. Formula **(c)** is trivial, and **(e)** follows from **(d)** and **(c)**. To prove **(d)**, suppose that A is diagonalizable, i.e., $SAS^{-1} = [\lambda_1, \lambda_2, \ldots, \lambda_n]$. Then **(d)** is evident, since it reduces to the identity $\displaystyle\sum_{i \le j} \lambda_i\lambda_j = \frac{1}{2}\left[\left(\sum_i \lambda_i\right)^2 + \sum_i \lambda_i^2\right]$.
However, any matrix A can be approximated arbitrarily closely by a diagonalizable matrix. (To see this, reduce A to canonical form and make slight modifications of the diagonal elements so that all the eigenvalues of A are distinct.) Thus, since both sides of **(d)** are continuous in the matrix A, the formula holds in general.

17. The result is obviously true if $\mathcal{G} = \mathcal{C}$, the only group of order 2. Suppose that the order of $\mathcal{G}$ is greater than 2, and let A and B be any two distinct elements of $\mathcal{G}$ other than the unit element I. Then by hypothesis, $A^2 = B^2 = I$, so that $A = A^{-1}$, $B = B^{-1}$. The element $AB \neq I$, since otherwise $A = B^{-1} = B$. Therefore, by hypothesis $(AB)^2 = A(BA)B = I$, i.e., $BA = A^{-1}B^{-1} = AB$, so that $\mathcal{G}$ is Abelian. Since $\mathcal{G}$ is finite, it is generated by a finite number of elements $A_1, A_2, \ldots, A_m$ of $\mathcal{G}$ (cf. Prob. 25 of Chap. 7). Thus, since $\mathcal{G}$ is Abelian, every element G of $\mathcal{G}$ can be written in the form

$$G = A_1^{\alpha_1}A_2^{\alpha_2}\cdots A_m^{\alpha_m},$$

where each α_i $(i = 1, 2, \ldots, m)$ is 0 or 1, since the square of every element of $\mathcal{G}$ equals I. The required result follows at once.

18. If $\mathcal{A}$ is a representation of $\mathcal{G}$, it obviously has dimension νn. To prove that $\mathcal{A}$ is a representation of $\mathcal{G}$, we first note that if $\mathcal{G}$ can be decomposed into the left cosets $G_i\mathcal{H}$, then $\mathcal{G}$ can also be decomposed into the right cosets $(G_i\mathcal{H})^{-1} = \mathcal{H}G_i^{-1}$ $(i = 1, 2, \ldots, \nu)$. Hence

$$\sum_{j,H} \sigma_{ij}(G, H) = \sum_{j,H} \delta_{G,G_iHG_j^{-1}} = \sum_{j,H} \delta_{G_i^{-1}G,HG_j^{-1}} = 1,$$

where $\delta_{A,B} = 1$ if $A = B$ and 0 otherwise, and we sum j from 1 to ν and H over all the elements of the subgroup $\mathcal{H}$. Thus, for fixed (i, G), there is exactly one pair (j_0, H_0) such that $\sigma_{ij_0}(G, H_0) \neq 0$. Now let

$$\tau_{ij}(G) = \sum_{H} \sigma_{ij}(G, H)B(H),$$

so that

$$A(G) = \|\tau_{ij}(G)\|.$$

Then

$$\sum_{j} \tau_{ij}(G)\tau_{jk}(G') = \sum_{j,H,H'} \sigma_{ij}(G, H)\sigma_{jk}(G', H')B(HH') = \sum_{H'} \sigma_{j_0k}(G', H')B(H_0H')$$

$$= \sum_{H'} \delta_{G',G_{j_0}H'G_k^{-1}}B(H_0H') = \sum_{H'} \delta_{G',G^{-1}G_iH_0H'G_k^{-1}}B(H_0H') = \sum_{H'} \delta_{GG',G_iH_0H'G_k^{-1}}B(H_0H')$$

$$= \sum_{H''} \delta_{GG',G_iH''G_k^{-1}}B(H'') = \sum_{H''} \sigma_{ik}(GG', H'')B(H'') = \tau_{ik}(GG'),$$

so that $A(G)A(G') = A(GG')$. (We have used the fact that $G_{j_0} = G^{-1}G_iH_0$.) Finally, we note that

$$\tau_{ij}(G_1) \equiv \tau_{ij}(I) = \sum_{H} \sigma_{ij}(I, H)B(H) = \sum_{H} \delta_{G_j,G_iH}B(H) = \delta_{ij}I,$$

so that $\mathcal{A}$ assigns the unit matrix to $G_1 = I$.

19. Every eigenvalue of a unitary matrix has absolute value 1.

20. $\mathcal{A}_1 \oplus \mathcal{A}_2$ has character $\chi_1 + \chi_2$, and $\mathcal{A}_1 \times \mathcal{A}_2$ has character $\chi_1\chi_2$.

21. The representation $\mathcal{A}$ can be chosen to be unitary. If $G \in \mathcal{Z}$, then the unitary matrix in $\mathcal{A}$ corresponding to G is a multiple of the unit matrix, since G commutes with every element of $\mathcal{G}$ (see Prob. 36 of Chap. 7). Hence

$$\sum_{G\in\mathcal{Z}} |\chi(G)|^2 = kd^2,$$

where χ is the character corresponding to $\mathfrak{A}$. But

$$\sum_{G \in \mathfrak{Z}} |\chi(G)|^2 \leq \sum_{G \in \mathfrak{G}} |\chi(G)|^2 = d,$$

so that $kd^2 \leq g$.

22. Let A be the matrix with elements

$$a_{ik} = \sqrt{\frac{g_k}{m}}\, \chi^{(i)}(C_k).$$

To prove (I), use (107) to show that A is unitary. To prove (II), write (112) as

$$d_i \sum_{l=1}^{r} a_{pql} g_l \chi^{(i)}(C_l) = g_p \chi^{(i)}(C_p) g_q \chi^{(i)}(C_q).$$

Divide by d_i, multiply by $\overline{\chi^{(i)}(C_k)}$, and sum over i. Then use (I).

24. Write (112) as

$$\sum_{k=1}^{r} a_{pqk} g_k \chi^{(i)}(C_k) = d_i^{-1} g_p \chi^{(i)}(C_p) g_q \chi^{(i)}(C_q).$$

Then multiply both sides by $\overline{\chi^{(j)}(C_p)\chi^{(k)}(C_q)\chi^{(i)}(C)}$ and sum over p, q, and i.

25. If $P(z) = z$, then by (106)

$$\frac{1}{g}\Phi = \frac{1}{g}\sum_{\alpha=1}^{g} \chi(G_\alpha)$$

is the number of times $\mathfrak{A}$ contains the trivial one-dimensional representation (which assigns the number 1 to every G_α). If $P(z) = z^n$, then

$$\frac{1}{g}\Phi = \frac{1}{g}\sum_{\alpha=1}^{g} \chi^n(G_\alpha)$$

is the number of times the trivial one-dimensional representation is contained in the direct product of $\mathfrak{A}$ with itself n times. The general result for $P(z) = \sum_{n=1}^{N} a_n z^n$, with integral a_n, follows at once.

26. Any representation of a finite group $\mathfrak{G}$ is equivalent to a unitary representation, and hence has the same character as a unitary representation. But if U is unitary, $\operatorname{tr} U^{-1} = \operatorname{tr} \bar{U}^* = \operatorname{tr} \bar{U}$.

28. By the solution of Prob. 25,

$$m = \frac{1}{g}\sum_{\alpha=1}^{g} \chi^2(G_\alpha) = \frac{1}{g}\sum_{\alpha=1}^{g} \chi(G_\alpha)\overline{\bar{\chi}(G_\alpha)},$$

which is 1 if $\bar{\chi}(G_\alpha) = \chi(G_\alpha)$ and 0 otherwise.

29. $\mathfrak{U}$ is equivalent to its own complex conjugate, as shown in Sec. 86. Therefore, $\mathfrak{U}$ is of the first or second kind. If $\mathfrak{U}$ were of the first kind, then by Prob. 6, it would be equivalent to a group of real orthogonal matrices. Since every matrix in $\mathfrak{U}$ has determinant 1, every matrix in $\mathfrak{A}$ must have determinant 1. Thus $\mathfrak{A}$

must be a group of rotation matrices, and hence $\mathfrak{A}$ must be Abelian. Since $\mathfrak{U}$ is not Abelian, such an equivalence is impossible. Therefore, $\mathfrak{U}$ is not of the first kind.

30. Let $\mathcal{G}$ be of order g, and let $\mathcal{G}$ have r classes of which α are ambivalent. Let k be the number of irreducible representations of $\mathcal{G}$ of the first and second kinds, and let $m(G)$ be the order of the class containing the element G of $\mathcal{G}$. Consider the quantity

$$s = \sum_G \sum_p [\chi^{(p)}(G)]^2,$$

where G is summed over all elements G of $\mathcal{G}$, and p is summed over all the irreducible representations of $\mathcal{G}$. Since

$$\sum_p [\chi^{(p)}(G)]^2 = \sum_p \chi^{(p)}(G)\overline{\chi^{(p)}(G^{-1})} = \frac{g}{m(G)}$$

if the class containing G is ambivalent, and 0 otherwise, it follows by formula (I) of Prob. 22 that $s = g\alpha$. On the other hand, if $\chi^{(p)}$ is real,

$$\sum_G [\chi^{(p)}(G)]^2 = \sum_G \chi^{(p)}(G)\overline{\chi^{(p)}(G)} = g,$$

and if not,

$$\sum_G [\chi^{(p)}(G)]^2 = \sum_G \chi^{(p)}(G)\overline{\bar{\chi}^{(p)}(G)} = 0,$$

by the orthogonality formulas (101) and (102) for characters. It follows that $s = kg$, and hence $\alpha = k$.

31. Since any Hermitian matrix H can be written as $H = UDU^{-1}$, where D is diagonal and U is unitary, we have $p_n(H) = Up_n(D)U^{-1}$. But obviously $p_n(D) \to D^{1/2}$; therefore, $p_n(H) \to H^{1/2}$. The fact that $H^{1/2}$ commutes with any matrix commuting with H follows from the fact that $p_n(H)$ commutes with any matrix commuting with H. If H is real, then so is $p_n(H)$, and hence so is $H^{1/2}$.

32. We have $B_\alpha = SA_\alpha S^{-1}$. Let $T = (S\tilde{S})^{-1/2}$, and let $U = TS$. It follows easily that U is unitary and that $S\tilde{S}$ commutes with B_α. Using the fact that T commutes with B_α (see Prob. 31), we obtain $U^{-1}B_\alpha U = S^{-1}T^{-1}B_\alpha TS = S^{-1}BS = A_\alpha$, i.e., $UA_\alpha U^{-1} = B_\alpha$.

33. By Prob. 4, we can find a real matrix S such that $SA_\alpha S^{-1} = B_\alpha$. Let $T = (SS^*)^{-1/2}$, and let $R = TS$. It follows easily that R is real and orthogonal and that SS^* commutes with B_α. Using the fact that T commutes with B_α (cf. Prob. 30), we obtain $R^{-1}B_\alpha R = S^{-1}T^{-1}B_\alpha TS = S^{-1}B_\alpha S = A_\alpha$, i.e., $RA_\alpha R^{-1} = B_\alpha$.

34. Let $\mathfrak{A}$ consist of the matrices A_α, so that $U^{-1}A_\alpha U = \bar{A}_\alpha$. Then $U\bar{U}A_\alpha = A_\alpha U\bar{U}$. Hence, by Schur's lemma (Prob. 7), $U\bar{U} = \lambda I$, where I is the unit matrix. Also $1 = |\det U|^2 = \det U \det \bar{U} = \det U\bar{U} = \lambda^d$, where d is the dimension of $\mathfrak{A}$. Hence $\lambda = e^{i\theta}$. Thus $U^*U^{-1} = \lambda^{-1}I = \bar{U}U$, or $(U^*)^2 = U^2$. But $U = \lambda U^*$, so $\lambda^2 = 1$ or $\lambda = \pm 1$. Hence $U^* = \pm U$.

35. By the preceding problem, we have only to show that $\mathfrak{A}$ is of the first kind if and only if U is symmetric. Thus, let $\mathfrak{A}$ be of the first kind, let $\mathfrak{A}$ consist of the matrices A_α, and let $B_\alpha = S^{-1}A_\alpha S$ be an equivalent set of real orthogonal matrices (cf. Prob. 6). Then it easily follows that $(SS^*)^{-1}A_\alpha SS^* = \bar{A}_\alpha$. Moreover, since $U^{-1}A_\alpha U = \bar{A}_\alpha$, it follows that $U(SS^*)^{-1}$ commutes with A_α. Hence, by Schur's lemma, $U = \lambda SS^*$, and thus $U^* = U$.

Conversely, if $U^* = U$, then $\bar{U} = U^{-1}$. Thus, if $U\mathbf{v} = \lambda\mathbf{v}$, we have $\bar{U}\mathbf{v} = U^{-1}\mathbf{v} = \bar{\lambda}\mathbf{v}$, so that $U\bar{\mathbf{v}} = \lambda\bar{\mathbf{v}}$, i.e., the complex conjugate of an eigenvector of U belonging to an eigenvalue λ also belongs to the eigenvalue λ. It follows that U can be diagonalized in an orthonormal basis consisting of real vectors. Hence $U = V^2$, where V is unitary and is diagonal in the same basis consisting of real vectors, so that $V^{-1} = \bar{V}$, or $V = V^*$. Now let $R_\alpha = V^{-1}A_\alpha V$. Then $\bar{R}_\alpha = \bar{V}^{-1}\bar{A}_\alpha\bar{V} = \bar{V}^{-1}U^{-1}A_\alpha U\bar{V} = VV^{-2}A_\alpha V^2V^{-1} = V^{-1}A_\alpha V = R_\alpha$, so that the representation $\mathfrak{A}$ is equivalent to a real representation.

36. Let $\mathfrak{A}$ consist of the matrices A_α, and let $\mathfrak{B}$ be another irreducible representation (consisting of the matrices B_α) which is not equivalent to $\mathfrak{A}$. Then, according to the orthogonality relation for unitary representations (see Sec. 82), we have

$$\sum_\alpha \{A_\alpha\}_{ij}\overline{\{B_\alpha\}_{lk}} = 0,$$

while

$$\sum_\alpha \{A_\alpha\}_{ij}\overline{\{A_\alpha\}_{lk}} = \frac{g}{d}\delta_{ik}\delta_{jl}.$$

If $\mathfrak{A}$ is of the third kind, $\bar{\mathfrak{A}}$ is not equivalent to $\mathfrak{A}$, and we have

$$\zeta = \frac{1}{g}\sum_\alpha \chi(G_\alpha^2) = \frac{1}{g}\sum_{\alpha,i,j}\{A_\alpha\}_{ij}\{A_\alpha\}_{ji} = \frac{1}{g}\sum_{\alpha,i,j}\{A_\alpha\}_{ij}\overline{\{\bar{A}_\alpha\}_{ji}} = 0.$$

If $\mathfrak{A}$ is of the first or second kind, then

$$\overline{\{A_\alpha\}_{lk}} = \sum_{i',j'}\{U^{-1}\}_{ki'}\{A_\alpha\}_{i'j'}\{U\}_{j'l},$$

and by the preceding problem $U^* = U$ if $\mathfrak{A}$ is of the first kind, while $U^* = -U$ if $\mathfrak{A}$ is of the second kind. Substituting into the orthogonality relation, we have

$$\sum_{\alpha,i',j'}\{U^{-1}\}_{ki'}\{U\}_{j'l}\{A_\alpha\}_{ij}\{A_\alpha\}_{i'j'} = \frac{g}{d}\delta_{ik}\delta_{jl}.$$

Multiply by $\{U^{-1}\}_{li}$ and sum over l, and then multiply by $\{U\}_{jk}$ and sum over k. The result is

$$\sum_{\alpha,i',j'}\delta_{i'j}\delta_{ij'}\{A_\alpha\}_{ij}\{A_\alpha\}_{i'j'} = \frac{g}{d}\{U^{-1}\}_{lk}\{U\}_{lk}$$

or

$$\sum_\alpha \{A_\alpha\}_{ij}\{A_\alpha\}_{ji} = \frac{g}{d}\{U^{-1}\}_{lk}\{U\}_{lk}.$$

Summing over i and j, we get

$$\zeta = \frac{1}{g}\sum_\alpha \chi(A_\alpha^2) = \pm 1,$$

where we have the plus sign if $U^* = U$ and the minus sign if $U^* = -U$.

37. By the preceding problem

$$\sum_\alpha \chi^{(p)}(G_\alpha^2) = \sum_\beta \rho(G_\beta) = gc_p.$$

Now multiply by $\overline{\chi^{(p)}(G_{\beta'})}$ and sum over p, using formula (I) of Prob. 22.

38. $A_\alpha^* F A_\alpha = F$ implies $\bar{A}_\alpha F = F A_\alpha$. If $\mathfrak{A}$ is of the first kind, then $U A_\alpha U^{-1} = \bar{A}_\alpha$, where U is unitary (Prob. 32), so that $\bar{A}_\alpha U = U A_\alpha$. Then $U^{-1}F$ commutes with A_α, and by Schur's lemma, $F = \lambda U$. Since $U^* = U$ (cf. Prob. 35), it follows that $F^* = F$. The argument for $\mathfrak{A}$ of the second kind is essentially the same. If $\mathfrak{A}$ is of the third kind, Schur's lemma leads to $F = 0$. Parts (**d**) and (**e**) follow from Parts (**a**), (**b**), and (**c**). To prove Part (**f**), note that if only the zero matrix satisfies $A_\alpha^* F A_\alpha = F$, then the only solution of $\bar{A}_\alpha F = F A_\alpha$ is $F = 0$, so that clearly $\mathfrak{A}$ is not equivalent to $\bar{\mathfrak{A}}$, and hence $\mathfrak{A}$ is of the third kind.

39. Parallel the proofs given in the text for the complex case.

40. (**a**) Let $\mathfrak{A}$ be an irreducible representation of dimension d consisting of matrices A_α, which is not of the first kind. If

$$S_\alpha = \begin{Vmatrix} A_\alpha & 0 \\ 0 & \bar{A}_\alpha \end{Vmatrix}, \quad U = \frac{1}{\sqrt{2}} \begin{Vmatrix} I & I \\ -iI & iI \end{Vmatrix},$$

where I is the unit matrix of order d, then clearly $R_\alpha = U S_\alpha U^{-1}$ belongs to a real representation $\mathfrak{R}$. If $\mathfrak{R}$ is not real-irreducible, then by Part (**b**) of the preceding problem, $\mathfrak{R}$ decomposes into the direct sum of at least two real-irreducible representations. However, since $\mathfrak{R} = \mathfrak{A} \oplus \bar{\mathfrak{A}}$ is the decomposition of $\mathfrak{R}$ into irreducible representations, it is clear that the decomposition of $\mathfrak{R}$ into real-irreducible representations can involve only two terms, and that each of these terms is irreducible; thus, $\mathfrak{R} = \mathfrak{A} \oplus \bar{\mathfrak{A}}$ is in fact the decomposition of $\mathfrak{R}$ into real-irreducible representations, i.e., $\mathfrak{A}$ and $\bar{\mathfrak{A}}$ are equivalent to real representations. Since $\mathfrak{A}$ is not of the first kind, this is impossible. Hence, $\mathfrak{R}$ is real-irreducible.

(**b**) If $\mathfrak{R}$ is a real-irreducible representation of $\mathfrak{G}$ which is not real-equivalent to a representation of the form $\mathfrak{A}_1$, $\mathfrak{A}_2 \oplus \mathfrak{A}_2$, or $\mathfrak{A}_3 \oplus \bar{\mathfrak{A}}_3$, then it follows by Part (**c**) of the preceding problem that

$$\sum_{G \in \mathfrak{G}} \chi_{\mathfrak{R}}(G)(\chi_{\mathfrak{A}}(G) + \overline{\chi_{\mathfrak{A}}(G)}) = 0$$

for every irreducible $\mathfrak{A}$, where $\chi_{\mathfrak{R}}$ denotes the character of $\mathfrak{R}$ and $\chi_{\mathfrak{A}}$ the character of $\mathfrak{A}$. Thus, $\mathfrak{R}$ contains no irreducible representation $\mathfrak{A}$, which is impossible.

CHAPTER 9

1. $\frac{1}{2}r^2(r-1)$.

2. We abbreviate the formula (1) to $\gamma_s = \varphi_s(\beta_i, \alpha_k)$ and introduce new quantities $\gamma_s' = \varphi_s(\tau_i, \beta_k)$, where the τ_i, like the α_i and β_i ($i = 1, 2, \ldots, r$), are independent variables. Then, by the associativity of the group operation, we have

$$\varphi_s(\tau_i, \gamma_k) = \varphi_s(\gamma_i', \alpha_k).$$

Differentiating this relation with respect to α_l, β_m, and τ_n in succession, and using the summation convention (cf. Prob. 22 of Chap. 3), we obtain

$$\frac{\partial^3 \varphi_s}{\partial \gamma_u\, \partial \gamma_v\, \partial \tau_n} \frac{\partial \gamma_u}{\partial \alpha_l} \frac{\partial \gamma_v}{\partial \beta_m} + \frac{\partial^2 \gamma_u}{\partial \alpha_l\, \partial \beta_m} \frac{\partial^2 \varphi_s}{\partial \gamma_u\, \partial \tau_n} = \frac{\partial^3 \varphi_s}{\partial \gamma_u'\, \partial \gamma_v'\, \partial \alpha_l} \frac{\partial \gamma_u'}{\partial \beta_m} \frac{\partial \gamma_v'}{\partial \tau_n} + \frac{\partial^2 \gamma_u'}{\partial \beta_m \partial \tau_n} \frac{\partial^2 \varphi_s}{\partial \gamma_u'\, \partial \alpha_l},$$

where in the left-hand side φ is a function of τ, γ and γ is a function of β, α, while in the right-hand side φ is a function of γ', α and γ' is a function of τ, β. (We omit the subscripts in statements like this.) We now set α, β, τ equal to zero, which implies that γ, γ' also vanish. If we use formulas (3) and (5) of Sec. 88, the above equation becomes

$$A_{uvn}^{(s)}\, \delta_{ul}\, \delta_{vm} + a_{lm}^{(u)} a_{un}^{(s)} = B_{luv}^{(s)}\, \delta_{um}\, \delta_{vn} + a_{lu}^{(s)} a_{mn}^{(u)},$$

where $A^{(s)}_{uvn}$, $B^{(s)}_{luv}$ are defined as in the statement of the problem. The desired result follows at once.

3. According to the definition (6), $C^{(t)}_{ij} = a^{(t)}_{ij} - a^{(t)}_{ji}$. Therefore, using the summation convention again, we have

$$\begin{aligned} C^{(t)}_{is}C^{(s)}_{jk} + C^{(t)}_{js}C^{(s)}_{ki} + C^{(t)}_{ks}C^{(s)}_{ij} &= (a^{(t)}_{is}a^{(s)}_{jk} - a^{(s)}_{ij}a^{(t)}_{sk}) + (a^{(t)}_{js}a^{(s)}_{ki} - a^{(s)}_{jk}a^{(t)}_{si}) \\ &+ (a^{(t)}_{ks}a^{(s)}_{ij} - a^{(s)}_{ki}a^{(t)}_{sj}) - (a^{(t)}_{js}a^{(s)}_{ik} - a^{(s)}_{ji}a^{(t)}_{sk}) - (a^{(t)}_{ks}a^{(s)}_{ji} - a^{(s)}_{kj}a^{(t)}_{si}) \\ &- (a^{(t)}_{is}a^{(s)}_{kj} - a^{(s)}_{ik}a^{(t)}_{sj}) = (A^{(t)}_{ijk} - B^{(t)}_{ijk}) + (A^{(t)}_{jki} - B^{(t)}_{jki}) + (A^{(t)}_{kij} - B^{(t)}_{kij}) \\ &- (A^{(t)}_{jik} - B^{(t)}_{jik}) - (A^{(t)}_{kji} - B^{(t)}_{kji}) - (A^{(t)}_{ikj} - B^{(t)}_{ikj}) = 0, \end{aligned}$$

since $A^{(t)}_{ijk}$ is symmetric in its first two subscripts and $B^{(t)}_{ijk}$ is symmetric in its last two subscripts.

6. For $\mathcal{G}_1$ we have

$$I_1 = \begin{Vmatrix} 0 & -1 & 0 \\ 1 & 0 & 0 \\ 0 & 0 & 0 \end{Vmatrix}, \qquad I_2 = \begin{Vmatrix} 0 & 0 & 1 \\ 0 & 0 & 0 \\ 0 & 0 & 0 \end{Vmatrix}, \qquad I_3 = \begin{Vmatrix} 0 & 0 & 0 \\ 0 & 0 & 1 \\ 0 & 0 & 0 \end{Vmatrix}$$

and $\qquad [I_1, I_2] = I_3, \qquad [I_2, I_3] = 0, \qquad [I_3, I_1] = I_2.$

For $\mathcal{G}_2$ we have

$$I_1 = \begin{Vmatrix} 0 & 1 & 0 \\ 1 & 0 & 0 \\ 0 & 0 & 0 \end{Vmatrix}, \qquad I_2 = \begin{Vmatrix} 0 & 0 & 1 \\ 0 & 0 & 0 \\ 0 & 0 & 0 \end{Vmatrix}, \qquad I_3 = \begin{Vmatrix} 0 & 0 & 0 \\ 0 & 0 & 1 \\ 0 & 0 & 0 \end{Vmatrix}$$

and $\qquad [I_1, I_2] = I_3, \qquad [I_2, I_3] = 0, \qquad [I_3, I_1] = -I_2.$

7. Since J commutes with I_1, I_2, and I_3, it follows from Schur's lemma that in an irreducible representation, J must be a scalar matrix, i.e., $J = \lambda E$, where E denotes the unit matrix. Let $2j + 1$ be the dimension of the irreducible representation. Then, taking traces, we obtain

$$\begin{aligned} \operatorname{tr} J = (2j + 1)\lambda = \operatorname{tr}(I_1^2 + I_2^2 + I_3^2) &= -\tfrac{1}{2} \operatorname{tr}(A_1A_2 + A_2A_1 + 2A_3^2) \\ &= -j(j + 1)(2j + 1), \end{aligned}$$

where the matrices A_1, A_2, A_3 are defined by (32) and satisfy the formulas (42). It follows that $J = -j(j + 1)E$ in a $(2j + 1)$-dimensional irreducible representation of $\mathcal{R}$.

8. For $\mathcal{G}_1$ we have

$$\begin{aligned} [I_1, J_1] &= [I_1, I_2]I_2 + I_2[I_1, I_2] + [I_1, I_3]I_3 + I_3[I_1, I_3] \\ &= I_3I_2 + I_2I_3 - I_2I_3 - I_3I_2 = 0. \end{aligned}$$

Moreover, it follows from Prob. 5 that

$$e^{-2\pi I_1}I_2e^{-2\pi I_1} = I_2 \cos n\pi - I_3 \sin n\pi = I_2$$

and

$$e^{-2\pi I_1}I_3e^{2\pi I_1} = e^{-2\pi I_1}[I_1, I_2]e^{2\pi I_1} = [e^{-2\pi I_1}I_1e^{2\pi I_1}, e^{-2\pi I_1}I_2e^{2\pi I_1}] = [I_1, I_2] = I_3.$$

For $\mathcal{G}_2$ we have

$$\begin{aligned} [I_1, J] &= [I_1, I_2]I_2 + I_2[I_1, I_2] - [I_1, I_3]I_3 - I_3[I_1, I_3] \\ &= I_3I_2 + I_2I_3 - I_2I_3 - I_3I_2 = 0. \end{aligned}$$

9. By Prob. 3 of Chap. 7, $R = e^A$, where A is a real skew-symmetric matrix. But any such A can be written as a real linear combination of I_1, I_2, I_3. The eigenvalues of $\boldsymbol{\lambda} \cdot \mathbf{I}$ are 0 and $\pm i\|\boldsymbol{\lambda}\|$; therefore $\operatorname{tr} R = 1 + 2 \cos \|\boldsymbol{\lambda}\|$. On the other hand (Sec. 42), $\operatorname{tr} R = 1 + 2 \cos \varphi$, so that $\varphi = \|\boldsymbol{\lambda}\| + 2n\pi$. Every vector lying along the axis of rotation is an invariant eigenvector of R and hence an eigenvector of $\boldsymbol{\lambda} \cdot \mathbf{I}$ belonging to the eigenvalue 0. But $(\boldsymbol{\lambda} \cdot \mathbf{I})\boldsymbol{\lambda} = 0$.

10. Solve equations (72) of Chap. 7 for x_1', x_2', x_3', x_0'.

11. As in Sec. 92, put $a = 1 + \alpha_1 + i\alpha_2$, $b = \alpha_3 + i\alpha_4$, $c = \alpha_5 + i\alpha_6$, $d = (1 + bc)/a$ in formula (I) of Prob. 10. To find I_1, set $\alpha_2 = \alpha_3 = \cdots = \alpha_6$, differentiate with respect to α_1, and then set $\alpha_1 = 0$. To find $I_2, I_3, \ldots, I_6$, proceed similarly.

12. Suppose that

$$A = \begin{Vmatrix} 0 & a_{12} & a_{13} & a_{14} \\ -a_{12} & 0 & a_{23} & a_{24} \\ -a_{13} & -a_{23} & 0 & a_{34} \\ -a_{14} & -a_{24} & -a_{34} & 0 \end{Vmatrix}.$$

Then

$$AG = \begin{Vmatrix} 0 & a_{12} & a_{13} & -a_{14} \\ -a_{12} & 0 & a_{23} & -a_{24} \\ -a_{13} & -a_{23} & 0 & -a_{34} \\ -a_{14} & -a_{24} & -a_{34} & 0 \end{Vmatrix} = -\tfrac{1}{2}a_{34}I_1 + \tfrac{1}{2}a_{12}I_2 - \tfrac{1}{2}a_{14}(I_3 + I_5) \\ -\tfrac{1}{2}a_{13}(I_3 - I_5) + \tfrac{1}{2}a_{23}(I_4 + I_6) \\ + \tfrac{1}{2}a_{24}(I_4 - I_6).$$

13. Let $\mathfrak{V}$ be the group of complex 2×2 matrices with determinant 1, and let $\mathfrak{L}_h$ be the group of all real, proper, orthochronous, 4×4 Lorentz matrices. Any $V_0 \in \mathfrak{V}$ can be reduced to canonical form by some $V \in \mathfrak{V}$, i.e., $V^{-1}V_0V = \Lambda$, where Λ is the canonical form of V_0 (cf. Sec. 87). Moreover, $\mathfrak{L}_h$ is homomorphic to $\mathfrak{V}$ (cf. Sec. 70). If under this homomorphism

$$V \to L, \quad V_0 \to L_0, \quad \Lambda \to \Lambda',$$

then $L^{-1}L_0L = \Lambda'$. As shown in Sec. 87, there are three possible forms of Λ:

$$\Lambda_1 = \begin{Vmatrix} \lambda & 0 \\ 0 & \lambda^{-1} \end{Vmatrix}, \qquad \Lambda_2 = \begin{Vmatrix} 1 & 0 \\ 1 & 1 \end{Vmatrix}, \qquad \Lambda_3 = \begin{Vmatrix} -1 & 0 \\ 1 & -1 \end{Vmatrix},$$

where λ is an arbitrary nonzero number. Using Prob. 10, we find that the corresponding matrices in $\mathfrak{L}_h$ are

$$\Lambda_1' = \begin{Vmatrix} \operatorname{Re}(\bar\lambda/\lambda) & -\operatorname{Im}(\bar\lambda/\lambda) & 0 & 0 \\ \operatorname{Im}(\bar\lambda/\lambda) & \operatorname{Re}(\bar\lambda/\lambda) & 0 & 0 \\ 0 & 0 & \tfrac{1}{2}(|\lambda|^2 + |\lambda|^{-2}) & \tfrac{1}{2}(|\lambda|^2 - |\lambda|^{-2}) \\ 0 & 0 & \tfrac{1}{2}(|\lambda|^2 - |\lambda|^{-2}) & \tfrac{1}{2}(|\lambda|^2 + |\lambda|^{-2}) \end{Vmatrix},$$

$$\Lambda_2' = \begin{Vmatrix} 1 & 0 & 1 & 1 \\ 0 & 1 & 0 & 0 \\ -1 & 0 & \tfrac{1}{2} & -\tfrac{1}{2} \\ 1 & 0 & \tfrac{1}{2} & \tfrac{3}{2} \end{Vmatrix}, \qquad \Lambda_3' = \begin{Vmatrix} 1 & 0 & -1 & -1 \\ 0 & 1 & 0 & 0 \\ 1 & 0 & \tfrac{1}{2} & -\tfrac{1}{2} \\ -1 & 0 & \tfrac{1}{2} & \tfrac{3}{2} \end{Vmatrix}.$$

The matrices $\Lambda_1', \Lambda_2', \Lambda_3'$ can be expressed in real exponential form. In fact, $\Lambda_1' = e^{S_1}$, where

$$S_1 = \begin{Vmatrix} 0 & -\theta & 0 & 0 \\ \theta & 0 & 0 & 0 \\ 0 & 0 & 0 & \varphi \\ 0 & 0 & \varphi & 0 \end{Vmatrix}$$

and $\bar\lambda/\lambda = e^{i\theta}$, $|\lambda|^2 = e^{\varphi}$, whereas $\Lambda_2' = e^{S_2}$, $\Lambda_3' = e^{S_3}$, where

$$S_2 = \begin{Vmatrix} 0 & 0 & 1 & 1 \\ 0 & 0 & 0 & 0 \\ -1 & 0 & 0 & 0 \\ 1 & 0 & 0 & 0 \end{Vmatrix}, \qquad S_3 = \begin{Vmatrix} 0 & 0 & -1 & -1 \\ 0 & 0 & 0 & 0 \\ 1 & 0 & 0 & 0 \\ -1 & 0 & 0 & 0 \end{Vmatrix}.$$

(For the theory of e^X, where X is a matrix, see Prob. 4 of the Appendix.) Thus if $\Lambda' = e^S$ (where we temporarily drop the subscripts), it follows from $L^{-1}L_0L = \Lambda'$ that

$$L_0 = Le^SL^{-1} = e^{LSL^{-1}} = e^{(LSGL^*)G} = e^{AG},$$

where we use the fact that $L^{-1} = GL^*G$ (cf. Prob. 11 of Chap. 7). It remains to show that $A = LSGL^*$ is skew-symmetric, i.e., that $A^* = -A$. This will be the case if SG is skew-symmetric. But this follows from the formulas

$$S_1G = \begin{Vmatrix} 0 & -\theta & 0 & 0 \\ \theta & 0 & 0 & 0 \\ 0 & 0 & 0 & -\varphi \\ 0 & 0 & \varphi & 0 \end{Vmatrix}, \qquad S_2G = \begin{Vmatrix} 0 & 0 & 1 & -1 \\ 0 & 0 & 0 & 0 \\ -1 & 0 & 0 & 0 \\ 1 & 0 & 0 & 0 \end{Vmatrix},$$

$$S_3G = \begin{Vmatrix} 0 & 0 & -1 & 1 \\ 0 & 0 & 0 & 0 \\ 1 & 0 & 0 & 0 \\ -1 & 0 & 0 & 0 \end{Vmatrix}.$$

14. Use Probs. 12 and 13.

15. Consider the Lorentz matrix

$$\begin{Vmatrix} \cos\theta & -\sin\theta & 0 & 0 \\ \sin\theta & \cos\theta & 0 & 0 \\ 0 & 0 & \cosh\varphi & \sinh\varphi \\ 0 & 0 & \sinh\varphi & \cosh\varphi \end{Vmatrix}.$$

16. Let $C = ABA^{-1}B^{-1}$ be the commutator of two elements A and B in $\mathfrak{H}_2$. Then since $\det C = \det A \det B \det A^{-1} \det B^{-1} = 1$, the commutator subgroup $\mathcal{K}$ of $\mathfrak{H}_2$ is a (normal) subgroup of the unimodular group $\mathcal{U}$ in two variables. Since $\mathcal{U}$ has no proper normal subgroups other than the group consisting of the two diagonal matrices $[1, 1]$ and $[-1, -1]$, $\mathcal{K}$ is identical with $\mathcal{U}$. In particular, $\mathfrak{H}_2$ is not perfect. (Cf. Sec. 87 and Prob. 44 of Chap. 7.)

17. Since a commutator subgroup is a normal subgroup and since $\mathcal{L}_h$ is simple, $\mathcal{L}_h$ is perfect. Hence the commutator subgroup $\mathcal{K}$ of $\mathcal{L}_i$ contains $\mathcal{L}_h$. Consider the element

$$H = \{L, \mathbf{a}\}\{L, 0\}\{L, \mathbf{a}\}^{-1}\{L, 0\}^{-1} = \{I, (I - L)\mathbf{a}\}$$

of $\mathcal{K}$, where I is the unit matrix of order 4, and let L be any Lorentz matrix which does not have an eigenvalue equal to 1 (see Prob. 15). Then $(I - L)^{-1}$ exists, so that we can write $\mathbf{a} = (I - L)^{-1}\mathbf{x}$ for any $\mathbf{x}$. With this choice of $\mathbf{a}$, H becomes the translation $\{I, \mathbf{x}\}$, i.e., $\mathcal{K}$ contains all translations. But then it follows from $\{L, \mathbf{a}\} = \{I, \mathbf{a}\}\{L, 0\}$ that every element of $\mathcal{L}_i$ is a product of elements of $\mathcal{K}$, i.e., $\mathcal{L}_i = \mathcal{K}$ and $\mathcal{L}_i$ is perfect.

18. It is shown in Sec. 95 that the left-invariant and right-invariant integrals of perfect groups are equal.

19. In each case $u'(G_\alpha) = u(G_\alpha) = 1$.

20. Use the lemma of Sec. 96 to calculate

$$\frac{\partial(c_0, c_1, c_2, c_3)}{\partial(a_0, a_1, a_2, a_3)}, \qquad \frac{\partial(c_0, c_1, c_2, c_3)}{\partial(b_0, b_1, b_2, b_3)},$$

where the c_i are given in terms of the a_i and b_i by the formulas (96) of Sec. 96. The result is $u'(G_\alpha) = u(G_\alpha) = (\det G_\alpha)^{-4}$ for G_α in $\mathfrak{H}_2$.

21. Use Prob. 2 of Chap. 8 to imbed the group of linear transformations of the form $z' = e^{\alpha_1}z + \alpha_2$ (Sec. 95) in $\mathfrak{H}_2$.

APPENDIX

1. (a) $x = (1 - 2t)e^{-2t}$, $y = (1 + 2t)e^{-2t}$; (c) $x = -e^{-t}$, $y = e^{-t}$, $z = 0$.

3. (b) $x = \frac{1}{2}t(e^t - e^{-t})$, $y = \frac{1}{2}(e^t - e^{-t}) + \frac{1}{2}t(e^t + e^{-t})$.

4. Let $X = \|x_{ik}\|$, and let ρ be such that $|x_{ik}| \leq \rho$ for $i, k = 1, 2, \ldots, n$. Then

$$|\{X^m\}_{ik}| \leq \sum_{i_1, i_2, \ldots, i_{m-1}} |x_{ii_1}|\,|x_{i_1 i_2}| \cdots |x_{i_{m-1}k}| \leq n^{m-1}\rho^m,$$

so that

$$|\{e^X\}_{ik}| = \delta_{ik} + \{X\}_{ik} + \frac{1}{2!}\{X^2\}_{ik} + \cdots + \frac{1}{m!}\{X^m\}_{ik} + \cdots +$$

$$\leq \delta_{ik} + \rho + \frac{1}{2!}n\rho^2 + \cdots + \frac{1}{m!}n^{m-1}\rho^m + \cdots,$$

where the series on the right converges for all ρ and n.

5. (b) $\begin{Vmatrix} 4e - 3 & 2 - 2e \\ 6e - 6 & 4 - 3e \end{Vmatrix}$. **6.** Cf. proof of Prob. 16 of Chap. 8.

Index

Mathematics-Bestsellers

HANDBOOK OF MATHEMATICAL FUNCTIONS: with Formulas, Graphs, and Mathematical Tables, Edited by Milton Abramowitz and Irene A. Stegun. A classic resource for working with special functions, standard trig, and exponential logarithmic definitions and extensions, it features 29 sets of tables, some to as high as 20 places. 1046pp. 8 x 10 1/2. 0-486-61272-4

ABSTRACT AND CONCRETE CATEGORIES: The Joy of Cats, Jiri Adamek, Horst Herrlich, and George E. Strecker. This up-to-date introductory treatment employs category theory to explore the theory of structures. Its unique approach stresses concrete categories and presents a systematic view of factorization structures. Numerous examples. 1990 edition, updated 2004. 528pp. 6 1/8 x 9 1/4. 0-486-46934-4

MATHEMATICS: Its Content, Methods and Meaning, A. D. Aleksandrov, A. N. Kolmogorov, and M. A. Lavrent'ev. Major survey offers comprehensive, coherent discussions of analytic geometry, algebra, differential equations, calculus of variations, functions of a complex variable, prime numbers, linear and non-Euclidean geometry, topology, functional analysis, more. 1963 edition. 1120pp. 5 3/8 x 8 1/2. 0-486-40916-3

INTRODUCTION TO VECTORS AND TENSORS: Second Edition--Two Volumes Bound as One, Ray M. Bowen and C.-C. Wang. Convenient single-volume compilation of two texts offers both introduction and in-depth survey. Geared toward engineering and science students rather than mathematicians, it focuses on physics and engineering applications. 1976 edition. 560pp. 6 1/2 x 9 1/4. 0-486-46914-X

AN INTRODUCTION TO ORTHOGONAL POLYNOMIALS, Theodore S. Chihara. Concise introduction covers general elementary theory, including the representation theorem and distribution functions, continued fractions and chain sequences, the recurrence formula, special functions, and some specific systems. 1978 edition. 272pp. 5 3/8 x 8 1/2. 0-486-47929-3

ADVANCED MATHEMATICS FOR ENGINEERS AND SCIENTISTS, Paul DuChateau. This primary text and supplemental reference focuses on linear algebra, calculus, and ordinary differential equations. Additional topics include partial differential equations and approximation methods. Includes solved problems. 1992 edition. 400pp. 7 1/2 x 9 1/4. 0-486-47930-7

PARTIAL DIFFERENTIAL EQUATIONS FOR SCIENTISTS AND ENGINEERS, Stanley J. Farlow. Practical text shows how to formulate and solve partial differential equations. Coverage of diffusion-type problems, hyperbolic-type problems, elliptic-type problems, numerical and approximate methods. Solution guide available upon request. 1982 edition. 414pp. 6 1/8 x 9 1/4. 0-486-67620-X

VARIATIONAL PRINCIPLES AND FREE-BOUNDARY PROBLEMS, Avner Friedman. Advanced graduate-level text examines variational methods in partial differential equations and illustrates their applications to free-boundary problems. Features detailed statements of standard theory of elliptic and parabolic operators. 1982 edition. 720pp. 6 1/8 x 9 1/4. 0-486-47853-X

LINEAR ANALYSIS AND REPRESENTATION THEORY, Steven A. Gaal. Unified treatment covers topics from the theory of operators and operator algebras on Hilbert spaces; integration and representation theory for topological groups; and the theory of Lie algebras, Lie groups, and transform groups. 1973 edition. 704pp. 6 1/8 x 9 1/4. 0-486-47851-3

Mathematics–Logic and Problem Solving

PERPLEXING PUZZLES AND TANTALIZING TEASERS, Martin Gardner. Ninety-three riddles, mazes, illusions, tricky questions, word and picture puzzles, and other challenges offer hours of entertainment for youngsters. Filled with rib-tickling drawings. Solutions. 224pp. 5 3/8 x 8 1/2. 0-486-25637-5

MY BEST MATHEMATICAL AND LOGIC PUZZLES, Martin Gardner. The noted expert selects 70 of his favorite "short" puzzles. Includes The Returning Explorer, The Mutilated Chessboard, Scrambled Box Tops, and dozens more. Complete solutions included. 96pp. 5 3/8 x 8 1/2. 0-486-28152-3

THE LADY OR THE TIGER?: and Other Logic Puzzles, Raymond M. Smullyan. Created by a renowned puzzle master, these whimsically themed challenges involve paradoxes about probability, time, and change; metapuzzles; and self-referentiality. Nineteen chapters advance in difficulty from relatively simple to highly complex. 1982 edition. 240pp. 5 3/8 x 8 1/2. 0-486-47027-X

SATAN, CANTOR AND INFINITY: Mind-Boggling Puzzles, Raymond M. Smullyan. A renowned mathematician tells stories of knights and knaves in an entertaining look at the logical precepts behind infinity, probability, time, and change. Requires a strong background in mathematics. Complete solutions. 288pp. 5 3/8 x 8 1/2. 0-486-47036-9

THE RED BOOK OF MATHEMATICAL PROBLEMS, Kenneth S. Williams and Kenneth Hardy. Handy compilation of 100 practice problems, hints and solutions indispensable for students preparing for the William Lowell Putnam and other mathematical competitions. Preface to the First Edition. Sources. 1988 edition. 192pp. 5 3/8 x 8 1/2. 0-486-69415-1

KING ARTHUR IN SEARCH OF HIS DOG AND OTHER CURIOUS PUZZLES, Raymond M. Smullyan. This fanciful, original collection for readers of all ages features arithmetic puzzles, logic problems related to crime detection, and logic and arithmetic puzzles involving King Arthur and his Dogs of the Round Table. 160pp. 5 3/8 x 8 1/2. 0-486-47435-6

UNDECIDABLE THEORIES: Studies in Logic and the Foundation of Mathematics, Alfred Tarski in collaboration with Andrzej Mostowski and Raphael M. Robinson. This well-known book by the famed logician consists of three treatises: "A General Method in Proofs of Undecidability," "Undecidability and Essential Undecidability in Mathematics," and "Undecidability of the Elementary Theory of Groups." 1953 edition. 112pp. 5 3/8 x 8 1/2. 0-486-47703-7

LOGIC FOR MATHEMATICIANS, J. Barkley Rosser. Examination of essential topics and theorems assumes no background in logic. "Undoubtedly a major addition to the literature of mathematical logic." – *Bulletin of the American Mathematical Society.* 1978 edition. 592pp. 6 1/8 x 9 1/4. 0-486-46898-4

INTRODUCTION TO PROOF IN ABSTRACT MATHEMATICS, Andrew Wohlgemuth. This undergraduate text teaches students what constitutes an acceptable proof, and it develops their ability to do proofs of routine problems as well as those requiring creative insights. 1990 edition. 384pp. 6 1/2 x 9 1/4. 0-486-47854-8

FIRST COURSE IN MATHEMATICAL LOGIC, Patrick Suppes and Shirley Hill. Rigorous introduction is simple enough in presentation and context for wide range of students. Symbolizing sentences; logical inference; truth and validity; truth tables; terms, predicates, universal quantifiers; universal specification and laws of identity; more. 288pp. 5 3/8 x 8 1/2. 0-486-42259-3

Mathematics–Algebra and Calculus

VECTOR CALCULUS, Peter Baxandall and Hans Liebeck. This introductory text offers a rigorous, comprehensive treatment. Classical theorems of vector calculus are amply illustrated with figures, worked examples, physical applications, and exercises with hints and answers. 1986 edition. 560pp. 5 3/8 x 8 1/2. 0-486-46620-5

ADVANCED CALCULUS: An Introduction to Classical Analysis, Louis Brand. A course in analysis that focuses on the functions of a real variable, this text introduces the basic concepts in their simplest setting and illustrates its teachings with numerous examples, theorems, and proofs. 1955 edition. 592pp. 5 3/8 x 8 1/2. 0-486-44548-8

ADVANCED CALCULUS, Avner Friedman. Intended for students who have already completed a one-year course in elementary calculus, this two-part treatment advances from functions of one variable to those of several variables. Solutions. 1971 edition. 432pp. 5 3/8 x 8 1/2. 0-486-45795-8

METHODS OF MATHEMATICS APPLIED TO CALCULUS, PROBABILITY, AND STATISTICS, Richard W. Hamming. This 4-part treatment begins with algebra and analytic geometry and proceeds to an exploration of the calculus of algebraic functions and transcendental functions and applications. 1985 edition. Includes 310 figures and 18 tables. 880pp. 6 1/2 x 9 1/4. 0-486-43945-3

BASIC ALGEBRA I: Second Edition, Nathan Jacobson. A classic text and standard reference for a generation, this volume covers all undergraduate algebra topics, including groups, rings, modules, Galois theory, polynomials, linear algebra, and associative algebra. 1985 edition. 528pp. 6 1/8 x 9 1/4. 0-486-47189-6

BASIC ALGEBRA II: Second Edition, Nathan Jacobson. This classic text and standard reference comprises all subjects of a first-year graduate-level course, including in-depth coverage of groups and polynomials and extensive use of categories and functors. 1989 edition. 704pp. 6 1/8 x 9 1/4. 0-486-47187-X

CALCULUS: An Intuitive and Physical Approach (Second Edition), Morris Kline. Application-oriented introduction relates the subject as closely as possible to science with explorations of the derivative; differentiation and integration of the powers of x; theorems on differentiation, antidifferentiation; the chain rule; trigonometric functions; more. Examples. 1967 edition. 960pp. 6 1/2 x 9 1/4. 0-486-40453-6

ABSTRACT ALGEBRA AND SOLUTION BY RADICALS, John E. Maxfield and Margaret W. Maxfield. Accessible advanced undergraduate-level text starts with groups, rings, fields, and polynomials and advances to Galois theory, radicals and roots of unity, and solution by radicals. Numerous examples, illustrations, exercises, appendixes. 1971 edition. 224pp. 6 1/8 x 9 1/4. 0-486-47723-1

AN INTRODUCTION TO THE THEORY OF LINEAR SPACES, Georgi E. Shilov. Translated by Richard A. Silverman. Introductory treatment offers a clear exposition of algebra, geometry, and analysis as parts of an integrated whole rather than separate subjects. Numerous examples illustrate many different fields, and problems include hints or answers. 1961 edition. 320pp. 5 3/8 x 8 1/2. 0-486-63070-6

LINEAR ALGEBRA, Georgi E. Shilov. Covers determinants, linear spaces, systems of linear equations, linear functions of a vector argument, coordinate transformations, the canonical form of the matrix of a linear operator, bilinear and quadratic forms, and more. 387pp. 5 3/8 x 8 1/2. 0-486-63518-X

Mathematics-Geometry and Topology

PROBLEMS AND SOLUTIONS IN EUCLIDEAN GEOMETRY, M. N. Aref and William Wernick. Based on classical principles, this book is intended for a second course in Euclidean geometry and can be used as a refresher. More than 200 problems include hints and solutions. 1968 edition. 272pp. 5 3/8 x 8 1/2. 0-486-47720-7

TOPOLOGY OF 3-MANIFOLDS AND RELATED TOPICS, Edited by M. K. Fort, Jr. With a New Introduction by Daniel Silver. Summaries and full reports from a 1961 conference discuss decompositions and subsets of 3-space; n-manifolds; knot theory; the Poincaré conjecture; and periodic maps and isotopies. Familiarity with algebraic topology required. 1962 edition. 272pp. 6 1/8 x 9 1/4. 0-486-47753-3

POINT SET TOPOLOGY, Steven A. Gaal. Suitable for a complete course in topology, this text also functions as a self-contained treatment for independent study. Additional enrichment materials make it equally valuable as a reference. 1964 edition. 336pp. 5 3/8 x 8 1/2. 0-486-47222-1

INVITATION TO GEOMETRY, Z. A. Melzak. Intended for students of many different backgrounds with only a modest knowledge of mathematics, this text features self-contained chapters that can be adapted to several types of geometry courses. 1983 edition. 240pp. 5 3/8 x 8 1/2. 0-486-46626-4

TOPOLOGY AND GEOMETRY FOR PHYSICISTS, Charles Nash and Siddhartha Sen. Written by physicists for physics students, this text assumes no detailed background in topology or geometry. Topics include differential forms, homotopy, homology, cohomology, fiber bundles, connection and covariant derivatives, and Morse theory. 1983 edition. 320pp. 5 3/8 x 8 1/2. 0-486-47852-1

BEYOND GEOMETRY: Classic Papers from Riemann to Einstein, Edited with an Introduction and Notes by Peter Pesic. This is the only English-language collection of these 8 accessible essays. They trace seminal ideas about the foundations of geometry that led to Einstein's general theory of relativity. 224pp. 6 1/8 x 9 1/4. 0-486-45350-2

GEOMETRY FROM EUCLID TO KNOTS, Saul Stahl. This text provides a historical perspective on plane geometry and covers non-neutral Euclidean geometry, circles and regular polygons, projective geometry, symmetries, inversions, informal topology, and more. Includes 1,000 practice problems. Solutions available. 2003 edition. 480pp. 6 1/8 x 9 1/4. 0-486-47459-3

TOPOLOGICAL VECTOR SPACES, DISTRIBUTIONS AND KERNELS, François Trèves. Extending beyond the boundaries of Hilbert and Banach space theory, this text focuses on key aspects of functional analysis, particularly in regard to solving partial differential equations. 1967 edition. 592pp. 5 3/8 x 8 1/2. 0-486-45352-9

INTRODUCTION TO PROJECTIVE GEOMETRY, C. R. Wylie, Jr. This introductory volume offers strong reinforcement for its teachings, with detailed examples and numerous theorems, proofs, and exercises, plus complete answers to all odd-numbered end-of-chapter problems. 1970 edition. 576pp. 6 1/8 x 9 1/4. 0-486-46895-X

FOUNDATIONS OF GEOMETRY, C. R. Wylie, Jr. Geared toward students preparing to teach high school mathematics, this text explores the principles of Euclidean and non-Euclidean geometry and covers both generalities and specifics of the axiomatic method. 1964 edition. 352pp. 6 x 9. 0-486-47214-0